Water Resources and Sustainable Development

Water Resources and Sustainable Development

Editors

Peiyue Li
Jianhua Wu

Basel • Beijing • Wuhan • Barcelona • Belgrade • Novi Sad • Cluj • Manchester

Editors

Peiyue Li
School of Water and
Environment
Chang'an University
Xi'an
China

Jianhua Wu
School of Water and
Environment
Chang'an University
Xi'an
China

Editorial Office
MDPI
St. Alban-Anlage 66
4052 Basel, Switzerland

This is a reprint of articles from the Special Issue published online in the open access journal *Water* (ISSN 2073-4441) (available at: www.mdpi.com/journal/water/special_issues/L263LQQ1OM).

For citation purposes, cite each article independently as indicated on the article page online and as indicated below:

Lastname, A.A.; Lastname, B.B. Article Title. *Journal Name* **Year**, *Volume Number*, Page Range.

ISBN 978-3-0365-9974-8 (Hbk)
ISBN 978-3-0365-9973-1 (PDF)
doi.org/10.3390/books978-3-0365-9973-1

Contents

About the Editors

Peiyue Li

Dr. Peiyue Li is a distinguished Full Professor specializing in hydrogeology and environmental sciences at Chang'an University in China. With his extensive expertise, he has made significant contributions to the field, particularly in groundwater quality assessment, hydrogeochemistry, and groundwater modeling. Dr. Li's influence extends beyond his research publications. He has edited and co-edited nine textbooks and academic monographs, providing valuable resources for students and researchers in the field. Additionally, he has published over 180 articles in reputable refereed journals, covering a wide range of topics, including groundwater quality assessment, groundwater hydrochemistry, groundwater pumping tests, and in situ tracer tests. Acknowledging his expertise and research impact, Dr. Li holds positions as an Associate Editor for esteemed journals such as *Exposure and Health, Mine Water and the Environment, Archive of Environmental Contamination and Toxicology, Environmental Monitoring and Assessment, Discover Water,* and *Human and Ecological Risk Assessment*. This involvement reflects his commitment to advancing the field through editorial contributions and peer review. Dr. Li's research excellence has garnered recognition from renowned organizations such as Clarivate and Elsevier, who have recognized him as one of the most highly cited researchers in the world. This recognition underscores the significance and influence of his research contributions within the scientific community. As a Full Professor, Dr. Peiyue Li's extensive experience, numerous publications, editorial roles, and international recognition affirm his status as a leading figure in the field of hydrogeology and environmental sciences. His work continues to shape the understanding of groundwater processes and contribute to the advancement of environmental science.

Jianhua Wu

Dr. Jianhua Wu is a Full Professor at Chang'an University in China, specializing in hydrology and water resources in arid areas, loess seepage and its disaster mechanisms, and groundwater environmental protection. With her extensive research experience, she has made notable contributions to these fields. Her research has been supported by numerous research projects funded by the National Natural Science Foundation of China, demonstrating her leadership and expertise in securing research funding. Her research output includes over 80 academic papers published in international journals, with 16 of them recognized as highly cited and popular papers by Essential Science Indicators. This recognition highlights the impact and significance of her research within the scientific community. In addition to her research contributions, Dr. Wu actively engages in the academic community as an Associate Editor for the international journal *Exposure and Health*. She also serves as a reviewer for various international journals, contributing to the peer review process and ensuring the quality of scientific publications. Dr. Wu's research excellence has garnered recognition from prestigious organizations such as Clarivate and Elsevier, who have selected her as one of the most highly cited researchers in the world. This recognition reflects the influence and impact of her research contributions on the global stage. As a Full Professor, Dr. Jianhua Wu's expertise, extensive publication record, editorial roles, and international recognition establish her as a prominent figure in the field of hydrology and water resources. Her research on arid areas, loess seepage, and groundwater environmental protection contributes to our understanding of these critical areas and informs strategies for water resource management and environmental protection.

Preface

As the lifeblood of our planet, water resources are integral to sustainable development. They underpin the survival of all life forms, the vitality of ecosystems, and the socio-economic prosperity of human societies. However, these invaluable resources are currently under siege from climate change, population growth, and unsustainable economic activities. The ensuing water-related challenges, encompassing pollution, scarcity, and subsequent socio-economic ramifications, have emerged as global issues demanding immediate and concerted action.

In response to this urgent call, we are proud to present this volume, titled "Water Resources and Sustainable Development". This Reprint is designed to serve as a platform for researchers, policymakers, and practitioners to disseminate their most recent findings, innovative ideas, and effective strategies to address the critical issue of water security. The volume encompasses sixteen chapters, organized into four principal thematic areas: Hydrochemical Characteristics and Water Quality, Impact of Human Activities on Water Resources, Water Management Strategies, and Technological Applications for Water Resource Management. By amalgamating a diverse range of studies across various geographical and contextual backgrounds, we aim to provide a sweeping overview of the current state of water resource management and its pivotal role in sustainable development.

The process of editing this Reprint has been a voyage of discovery and a testament to the power of collaborative knowledge and the significance of interdisciplinary research. We extend our profound gratitude to all the authors who have contributed their invaluable research to this volume. Their unwavering dedication and expertise have enabled this comprehensive exploration of water resources and sustainable development. We also wish to express our sincere appreciation to the reviewers, whose meticulous scrutiny and insightful feedback have greatly elevated the quality of this reprint. Their commitment to upholding high academic standards is deeply valued.

We cordially invite potential authors and researchers to engage with this Reprint. We hope that it will spark further research, stimulate dialogue, and incite action in the field of water resources and sustainable development. We firmly believe that through the sharing of knowledge and a united effort, we can surmount the water-related challenges of our era and forge a sustainable and water-secure future.

"Water Resources and Sustainable Development" is more than just a reprint. It is a clarion call to action, an appeal for the responsible stewardship of our most precious resource. We fervently hope that this volume will make a substantial contribution to the global conversation on water security and sustainability and that it will serve as a treasured resource for all those dedicated to this vital research topic.

Peiyue Li and Jianhua Wu
Editors

water

MDPI

Editorial

Water Resources and Sustainable Development

Peiyue Li [1,2,3,*] and Jianhua Wu [1,2,3]

1. School of Water and Environment, Chang'an University, No. 126 Yanta Road, Xi'an 710054, China; wjh2005xy@126.com
2. Key Laboratory of Subsurface Hydrology and Ecological Effects in Arid Region of the Ministry of Education, Chang'an University, No. 126 Yanta Road, Xi'an 710054, China
3. Key Laboratory of Eco-Hydrology and Water Security in Arid and Semi-Arid Regions of the Ministry of Water Resources, Chang'an University, No. 126 Yanta Road, Xi'an 710054, China
* Correspondence: lipy2@163.com

Abstract: This editorial introduces the Special Issue titled "Water Resources and Sustainable Development," underscoring the critical need for sustainable management of water resources in light of increasing demand, climate change impacts, and pollution. The issue delves into the intricate relationship between water availability, quality, utilization, and the socioeconomic determinants shaping these aspects, highlighting the necessity for novel, balanced strategies that cater to societal, economic, and environmental requirements. The research within this Special Issue is segmented into four key areas: understanding hydrochemical properties and water quality; evaluating anthropogenic effects on water resources; strategizing water resource management; and applying technological innovations in water resource management. Collectively, these studies broaden our comprehension of water resources and sustainable development, stressing the importance of continuous research in this sphere. As we look ahead, this editorial accentuates the importance of ongoing exploration and innovation in these pivotal areas, focusing on understanding climate change implications, mitigating human-induced impacts, refining water management strategies, and harnessing technological advancements. Its overarching aim is to propel worldwide initiatives towards achieving comprehensive water security and sustainability.

Keywords: water resources management; hydrochemical characteristics; water quality protection; hydrological processes; climate change; human activities; technological innovations

Citation: Li, P.; Wu, J. Water Resources and Sustainable Development. *Water* **2024**, *16*, 134. https://doi.org/10.3390/w16010134

Received: 25 December 2023
Accepted: 28 December 2023
Published: 29 December 2023

1. Introduction

Water, often described as the planet's lifeblood, faces mounting stress from escalating demand, climate change, and pollution [1]. Once perceived as limitless, this invaluable resource is now understood to be finite and increasingly scarce. The distribution of water resources across the globe is uneven, with some areas enjoying water abundance and others wrestling with severe scarcity. Countries endowed with extensive river systems, such as Canada and Brazil, stand in stark contrast to arid and semi-arid regions such as the Middle East and North Africa, where chronic water shortages persist [2,3]. The pressure on water resources intensifies with rapid urbanization and population growth, as seen in countries such as India and China, leading to the overuse of groundwater and stress on surface water bodies [4,5]. The challenges are further compounded by climate change, which introduces greater unpredictability in water availability, evidenced by an increase in both droughts and floods [6]. With the world's population on the rise and ever-progressing climate change, sustainable water resource management has become an urgent necessity [7].

Water resources, central to sustainable socioeconomic development, support a wide range of economic activities and are vital for human survival. The availability and management of these resources often dictate a nation's prosperity and societal wellbeing [8]. Agriculture, which consumes about 70% of the world's freshwater, epitomizes the economic dependence on water. It forms the economic mainstay of many developing countries, where a significant proportion of the population relies on farming. In regions such as Sub-Saharan Africa and parts of Asia, where water scarcity is a pressing issue, the absence of reliable water sources can drastically impede agricultural productivity, leading to food insecurity, poverty, and economic stagnation [9]. Conversely, effective water management can enhance agricultural output, improve food security, and stimulate economic growth. Beyond agriculture, water is integral to various industrial processes, emphasizing the economic consequences of water resource management [10]. A lack of sufficient or reliable water can disrupt industrial operations, dampen economic output, and result in job losses, whereas sustainable water management can encourage industrial growth and economic resilience [11].

Water resources significantly contribute to societal wellbeing by supporting a range of essential services [12]. One of the most vital is sanitation, where water is used to maintain cleanliness and prevent disease spread. Access to clean water and sanitation facilities is a fundamental human right and a crucial indicator of societal progress [13,14]. Yet, as of 2021, an estimated 2.2 billion people globally still lack access to safely managed drinking water, and 4.2 billion are without safely managed sanitation services [15]. This stark reality underscores the societal implications of water resource management and the urgent need for solutions that expand access to clean water and sanitation services. Additionally, water quality directly impacts human health, with contaminated water being a primary cause of diseases such as cholera, dysentery, and typhoid [16–19]. Water resources also enhance recreational activities and mental health, offering spaces for relaxation, exercise, and connection with nature. They play a pivotal role in maintaining environmental sustainability, supporting biodiversity, regulating the Earth's climate, and acting as a buffer against climate change [20–22].

Considering the pivotal role of water resources in socioeconomic development, it is crucial to examine their relationship with sustainable development. This involves understanding the complex interplay between water availability, quality, usage, and the socioeconomic factors that influence these dynamics. It also requires exploring innovative strategies for water resource management that balance societal, economic, and environmental needs [23,24]. This Special Issue on "Water Resources and Sustainable Development" in *Water* aims to address these critical areas. It brings together a collection of research that provides fresh insights into the challenges and opportunities in water resource management for sustainable development. The studies presented in this issue cover diverse geographies and contexts, offering a comprehensive view of the current state of knowledge in the field. With the publication of this Special Issue, we aim to stimulate further research and dialogue on this vital topic, contributing to global efforts towards achieving water security and sustainability.

2. Findings Reported in the Special Issue

This Special Issue includes 16 papers, among which 14 are research articles and 2 are review papers. These 16 papers can be clustered into four main topical groups: hydrochemical characteristics and water quality; impact of human activities on water resources; water management strategies; and technological applications for water resource management.

Hydrochemical Characteristics and Water Quality: This cluster of papers includes two papers and delves into the hydrochemical properties of notable water bodies. Pastukhov et al. (contribution 1) investigated the hydrochemical characteristics of the Irkutsk Reservoir, an important drinking water source in the Baikal region, while Lin et al. (contribution 2) examined the hydrochemical characteristics of surface water in the Danjiang River basin and assessed the risk to human health posed by polluted water bodies. They present a comprehensive examination of the distribution of key ions and trace elements, thereby enriching our understanding of water quality and its potential health implications. The studies underscore the necessity of consistent monitoring and evaluation of water quality, particularly in areas where these reservoirs are primary sources of potable water.

Impact of Anthropogenic Activities on Water Resources: This topical cluster includes three papers, focusing on the consequences of human activities such as antibiotic pollution, alterations in land use, and mining on water resources. The research in this area is critical in quantifying the scale of anthropogenic alterations to water resources and formulating strategies to alleviate negative impacts. For instance, the study by Jia et al. (contribution 3) provides a robust scientific basis for optimal water resource distribution and sustainable development in the Cele–Yutian Oasis in China. Chen et al. (contribution 4) studied the adsorption behavior of the mesoporous molecular sieve MCM-41 in relation to common antibiotics. Lyv et al. (contribution 5) quantified the impact of coal mining on the underground water resources in the New Shanghai No. 1 Coal Mine.

Water Management Strategies: This topical cluster discusses diverse strategies for water resource management. These papers highlight the need for sustainable water management practices and explore the challenges and opportunities associated with their implementation. For instance, Wen et al. (contribution 6) highlight the importance of collaborative governance in managing water resources, water conservancy facilities, and socioeconomic systems within a river basin. In addition, the comprehensive review of Jordan's water resources provides an in-depth understanding of the country's water management issues and potential remedies (contribution 7), while another paper (contribution 8) advocates for an integrated watershed management approach for sustainable development.

Technological Innovations in Water Resource Management: The remaining papers delve into the application of various technological solutions for managing water resources. These papers showcase the potential of digital tools and advancements in computational and monitoring techniques in enhancing water resource efficiency and sustainability. For example, the study by Bonilla et al. (contribution 9) offers a feasible methodology for small cities to digitize their water distribution systems, even with limited budgets. Similarly, the research by Lee et al. (contribution 10) developed an optimal water allocation model for reservoir system operation. Zhao et al. (contribution 11) proposed a method by which to calculate the base value range of the ecological compensation standard in transboundary river basins, providing a basis for negotiation on the determination of the ecological compensation standard. Al-Rashidi et al. (contribution 12) used a fuzzy logic technique to develop groundwater suitability maps for irrigation purposes in Kuwait. Xu et al. (contribution 13) incorporated the concept of virtual water into the water rights allocation model. Monitoring is important in hydrogeological studies and can provide vital information for groundwater resources management. For example, Lu et al. (contribution 14) found that groundwater depth significantly affects vegetation coverage in the Ulan Buh Desert. Du et al. (contribution 15) analyzed the trend of water resources in the groundwater overexploitation area of the North China Plain, and Makanda et al. (contribution 16) assessed the water quality status of the Blesbokspruit River Catchment using the total maximum daily loads (TMDLs) and chemical mass balance (CMB) techniques.

In conclusion, the collection of papers in this Special Issue collectively broadens our knowledge of water resources and sustainable development. They emphasize the critical role of ongoing research in understanding hydrochemical characteristics, assessing the impact of human activities, strategizing water management, and utilizing technology in water resource management. Each of these domains plays a pivotal role in safeguarding the sustainability of our water resources amidst escalating environmental and human pressures.

3. Looking Ahead: Future Directions

As we cast our gaze towards the future, the intersection of water resources and sustainable development looms large, underscoring the imperative for continued research and innovation in this domain. The wealth of research presented in this Special Issue illuminates several pivotal areas that warrant further exploration.

Continuing Surveillance of Hydrochemical Dynamics and Water Quality: The research on water quality and hydrochemical properties underscores the necessity for persistent and extensive monitoring of global water bodies. This calls for the creation of advanced, user-friendly monitoring tools and methodologies [25]. It also necessitates a stronger emphasis on international collaboration and data exchange [26,27]. Future investigations should strive to deepen our understanding of climate change's effects on water quality and availability and explore potential mitigation strategies.

Addressing Anthropogenic Influences on Water Resources: The escalating impact of human activities on water resources necessitates attention. Future research should aim to further quantify these effects, with a spotlight on emerging challenges such as novel forms of pollution and the repercussions of rapid urbanization and industrialization [28]. Concurrently, efforts should be channeled towards promoting sustainable practices across water-intensive sectors such as agriculture and industry.

Advancing Water Management Strategies: The studies focusing on water management strategies underscore the need for a more integrated and comprehensive approach to water resource management. Future endeavors should aim to refine these strategies, ensuring they are adaptable to local conditions and resilient in the face of climate change [29]. The role of policy-making and governance in enforcing sustainable practices is another crucial area that merits further investigation.

Leveraging Technological Innovations: The potential of technology in enhancing water resource efficiency and sustainability is a recurring theme in this Special Issue. Future research should continue to push the boundaries of these technologies, exploring ways to make them accessible and affordable to communities worldwide. The role of digital tools in revolutionizing water resource management is a particularly promising avenue that deserves additional exploration.

In summation, the future of research in water resources and sustainable development hinges on a multifaceted understanding of environmental, socioeconomic, and technological variables. As we forge ahead, we hope that the research encapsulated in this Special Issue will serve as a catalyst for further exploration in this critical field, driving global efforts towards achieving water security and sustainability for all.

Author Contributions: Conceptualization, P.L. and J.W.; investigation, P.L. and J.W.; writing—original draft preparation, P.L.; writing—review and editing, P.L. and J.W.; project administration, P.L. and J.W.; funding acquisition, P.L. and J.W. All authors have read and agreed to the published version of the manuscript.

Funding: Financial support has been received from the National Natural Science Foundation of China (42272302, 42072286, 41761144059, and 42090053), the National Key Research and Development Program of China (2023YFC3706901), and the Qinchuangyuan "Scientist + Engineer" Team Development Program of the Shaanxi Provincial Department of Science and Technology (2022KXJ-005).

Data Availability Statement: Not applicable.

Conflicts of Interest: The authors declare no conflicts of interest.

List of Contributions:

1. Pastukhov, M.V.; Poletaeva, V.I.; Hommatlyyev, G.B. Hydrochemical Characteristics and Water Quality Assessment of Irkutsk Reservoir (Baikal Region, Russia). *Water* **2023**, *15*, 4142. https://doi.org/10.3390/w15234142
2. Lin, L.; Zhang, Y.; Qian, X.; Wang, Y. Hydrochemical Characteristics and Human Health Risk Assessment of Surface Water in the Danjiang River Source Basin of the Middle Route of China's South-to-North Water Transfer Project. *Water* **2023**, *15*, 2203. https://doi.org/10.3390/w15122203
3. Jia, G.; Li, S.; Jie, F.; Ge, Y.; Liu, N.; Liang, F. Assessing Water Resource Carrying Capacity and Sustainability in the Cele–Yutian Oasis (China): A TOPSIS–Markov Model Analysis. *Water* **2023**, *15*, 3652. https://doi.org/10.3390/w15203652
4. Chen, J.; Yang, Y.; Yao, Y.; Huang, Z.; Xu, Q.; He, L.; Gong, B. Rapid Antibiotic Adsorption from Water Using MCM-41-Based Material. *Water* **2023**, *15*, 4027. https://doi.org/10.3390/w15224027
5. Lyv, Y.; Qiao, W.; Chen, W.; Cheng, X.; Liu, M.; Liu, Y. Quantifying the Impact of Coal Mining on Underground Water in Arid and Semi-Arid Area: A Case Study of the New Shanghai No. 1 Coal Mine, Ordos Basin, China. *Water* **2023**, *15*, 1765. https://doi.org/10.3390/w15091765

6. Wen, J.; Li, H.; Meseretchanie, A. Assessment and Prediction of the Collaborative Governance of the Water Resources, Water Conservancy Facilities, and Socio-Economic System in the Xiangjiang River Basin, China. *Water* **2023**, *15*, 3630. https://doi.org/10.3390/w15203630

7. Al-Addous, M.; Bdour, M.; Alnaief, M.; Rabaiah, S.; Schweimanns, N. Water Resources in Jordan: A Review of Current Challenges and Future Opportunities. *Water* **2023**, *15*, 3729. https://doi.org/10.3390/w15213729

8. Farid Ahmed, M.; Mokhtar, M.B.; Lim, C.K.; Suza, I.A.B.C.; Ayob, K.A.K.; Khirotdin, R.P.K.; Majid, N.A. Integrated River Basin Management for Sustainable Development: Time for Stronger Action. *Water* **2023**, *15*, 2497. https://doi.org/10.3390/w15132497

9. Bonilla, C.; Brentan, B.; Montalvo, I.; Ayala-Cabrera, D.; Izquierdo, J. Digitalization of Water Distribution Systems in Small Cities, a Tool for Verification and Hydraulic Analysis: A Case Study of Pamplona, Colombia. *Water* **2023**, *15*, 3824. https://doi.org/10.3390/w15213824

10. Lee, E.; Ji, J.; Lee, S.; Yoon, J.; Yi, S.; Yi, J. Development of an Optimal Water Allocation Model for Reservoir System Operation. *Water* **2023**, *15*, 3555. https://doi.org/10.3390/w15203555

11. Zhao, Y.; Li, F.; Chen, Y.; Chen, X.; Xu, X. The Base Value of the Ecological Compensation Standard in Transboundary River Basins: A Case Study of the Lancang–Mekong River Basin. *Water* **2023**, *15*, 2809. https://doi.org/10.3390/w15152809

12. Al-Rashidi, A.; Sabarathinam, C.; Samayamanthula, D.R.; Alsabti, B.; Rashid, T. Groundwater Management for Agricultural Purposes Using Fuzzy Logic Technique in an Arid Region. *Water* **2023**, *15*, 2674. https://doi.org/10.3390/w15142674

13. Xu, X.; Yuan, J.; Yu, Q. A Study on Water Rights Allocation in Transboundary Rivers Based on the Transfer and Inequality Index of Virtual Water. *Water* **2023**, *15*, 2379. https://doi.org/10.3390/w15132379

14. Lu, T.; Wu, J.; Lu, Y.; Zhou, W.; Lu, Y. Effects of Groundwater Depth on Vegetation Coverage in the Ulan Buh Desert in a Recent 20-Year Period. *Water* **2023**, *15*, 3000. https://doi.org/10.3390/w15163000

15. Du, L.; Wang, G.; Lei, B. Evolution Characteristics of Rainfall and Runoff in the Upper Reaches of Zhang River Basin. *Water* **2023**, *15*, 2521. https://doi.org/10.3390/w15142521

16. Makanda, K.; Nzama, S.; Kanyerere, T. Assessing Feasibility of Water Resource Protection Practice at Catchment Level: A Case of the Blesbokspruit River Catchment, South Africa. *Water* **2023**, *15*, 2394. https://doi.org/10.3390/w15132394

References

1. Vörösmarty, C.J.; Green, P.; Salisbury, J.; Lammers, R.B. Global Water Resources: Vulnerability from Climate Change and Population Growth. *Science* **2000**, *289*, 284–288. [CrossRef] [PubMed]

2. Papa, F.; Crétaux, J.F.; Grippa, M.; Robert, E.; Trigg, M.; Tshimanga, R.M.; Kitambo, B.; Paris, A.; Carr, A.; Fleischmann, A.S.; et al. Water Resources in Africa under Global Change: Monitoring Surface Waters from Space. *Surv. Geophys.* **2023**, *44*, 43–93. [CrossRef] [PubMed]

3. De Brito, P.L.C.; De Azevedo, J.P.S. Charging for Water Use in Brazil: State of the Art and Challenges. *Water Resour. Manag.* **2020**, *34*, 1213–1229. [CrossRef]

4. Han, Y.; Liu, Y.; Rong, X.; Wang, M.; Xue, Y.; Dai, H.; Jiang, H. Exposure, Distribution, and Ecological Risk of Four New Bisphenol Analogs in the Typical Lake Region of Taihu Lake. *Expo. Health*, 2023; *early access*. [CrossRef]

5. Zhao, H.; Li, P.; He, X.; Ning, J. Microplastics pollution and risk assessment in selected surface waters of the Wei River Plain, China. *Expo. Health* **2023**, *15*, 745–755. [CrossRef]

6. Wang, D.; Li, P.; He, X.; He, S. Exploring the response of shallow groundwater to precipitation in the northern piedmont of the Qinling Mountains, China. *Urban Clim.* **2023**, *47*, 101379. [CrossRef]

7. Hubbard, M.L. The risky business of water resources management: Assessment of the public's risk perception of Oregon's water resources. *Hum. Ecol. Risk Assess.* **2020**, *26*, 1970–1987. [CrossRef]

8. Zhang, Z.; Yin, Z.; Chen, Y.; Chen, J. Evaluation and prediction of water resources carrying capacity using a multiple linear regression model in Taizhou City, China. *Hum. Ecol. Risk Assess.* **2023**, *29*, 553–570. [CrossRef]

9. Schilling, J.; Hertig, E.; Tramblay, Y.; Scheffran, J. Climate change vulnerability, water resources and social implications in North Africa. *Reg. Environ. Chang.* **2020**, *20*, 15. [CrossRef]

10. Liu, K.-D.; Yang, G.-L.; Yang, D.-G. Investigating industrial water-use efficiency in mainland China: An improved SBM-DEA model. *J. Environ. Manag.* **2020**, *270*, 110859. [CrossRef]

11. López-Pacheco, I.Y.; Silva-Núñez, A.; Salinas-Salazar, C.; Arévalo-Gallegos, A.; Lizarazo-Holguin, L.A.; Barceló, D.; Iqbal, H.M.N.; Parra-Saldívar, R. Anthropogenic contaminants of high concern: Existence in water resources and their adverse effects. *Sci. Total Environ.* **2019**, *690*, 1068–1088. [CrossRef]

12. Errich, A.; El Hajjaji, S.; Fekhaoui, M.; Hammouti, B.; Azzaoui, K.; Lamhamdi, A.; Jodeh, S. Seawater Intrusion and Nitrate Contamination in the Fum Al Wad Coastal Plain, South Morocco. *J. Earth Sci.* **2023**, *34*, 1940–1950. [CrossRef]
13. Weststrate, J.; Dijkstra, G.; Eshuis, J.; Gianoli, A.; Rusca, M. The Sustainable Development Goal on Water and Sanitation: Learning from the Millennium Development Goals. *Soc. Indic. Res.* **2019**, *143*, 795–810. [CrossRef]
14. Bayu, T.; Kim, H.; Oki, T. Water governance contribution to water and sanitation access equality in developing countries. *Water Resour. Res.* **2020**, *56*, e2019WR025330. [CrossRef]
15. Sachs, J.D.; Kroll, C.; Lafortune, G.; Fuller, G.; Woelm, F. *Sustainable Development Report 2022*; Cambridge University Press: Cambridge, UK, 2022.
16. Feng, Y.; Na, L. Gray water footprint evaluation of arsenic in Central China: From the perspective of health risk theory. *Environ. Monit. Assess.* **2022**, *194*, 901. [CrossRef] [PubMed]
17. Mourão, A.O.; Santos, M.S.; Da Costa, A.S.V.; da Silva, H.T.; Maia, L.F.O.; Faria, M.C.D.S.; del Vale Rodriguez Rodriguez, M.; Rodrigues, J.L. Assessment of Health Risk and Presence of Metals in Water and Fish Samples from Doce River, Brazil, After Fundão Dam Collapse. *Arch. Environ. Contam. Toxicol.* **2023**, *84*, 377–388. [CrossRef] [PubMed]
18. Sun, H.; Sun, X.; Wei, X.; Huang, X.; Ke, G.; Wei, H. Geochemical Characteristics and Origin of Nuanquanzi Geothermal Water in Yudaokou, Chengde, Hebei, North China. *J. Earth Sci.* **2023**, *34*, 838–856. [CrossRef]
19. Sangaré, L.O.; Ba, S.; Diallo, O.; Sanogo, D.; Zheng, T. Assessment of potential health risks from heavy metal pollution of surface water for drinking in a multi-industry area in Mali using a multi-indices approach. *Environ. Monit. Assess.* **2023**, *195*, 700. [CrossRef]
20. Li, L.; Li, P.; He, S.; Duan, R.; Xu, F. Ecological security evaluation for Changtan Reservoir in Taizhou City, East China, based on the DPSIR model. *Hum. Ecol. Risk Assess.* **2023**, *29*, 1064–1090. [CrossRef]
21. Alnsour, M.; Moqbel, S. Enhancing environmental sustainability through a household pharmaceuticals take-back program in Jordan. *Environ. Monit. Assess.* **2023**, *195*, 1424. [CrossRef]
22. Zhang, L.; Zhang, Z.; Chen, Y.; Tao, F. Spatial pattern of surface water quality in China and its driving factors—Implication for the environment sustainability. *Hum. Ecol. Risk Assess.* **2019**, *25*, 1789–1801. [CrossRef]
23. Cosgrove, W.J.; Loucks, D.P. Water management: Current and future challenges and research directions. *Water Resour. Res.* **2015**, *51*, 4823–4839. [CrossRef]
24. Xiang, X.; Li, Q.; Khan, S.; Khalaf, O.I. Urban water resource management for sustainable environment planning using artificial intelligence techniques. *Environ. Impact Assess. Rev.* **2021**, *86*, 106515. [CrossRef]
25. Cucho-Padin, G.; Loayza, H.; Palacios, S.; Balcazar, M.; Carbajal, M.; Quiroz, R. Development of low-cost remote sensing tools and methods for supporting smallholder agriculture. *Appl. Geomat.* **2020**, *12*, 247–263. [CrossRef]
26. Mazer, K.E.; Erwin, A.; Popovici, R.; Bocardo-Delgado, E.; Bowling, L.C.; Ma, Z.; Prokopy, L.S.; Zeballos-Velarde, C. Creating a Collaboration Framework to Evaluate International University-led Water Research Partnerships. *J. Contemp. Water Res. Educ.* **2020**, *171*, 9–26. [CrossRef]
27. Gao, J.; Castelletti, A.; Burlado, P.; Wang, H.; Zhao, J. Soft-cooperation via data sharing eases transboundary conflicts in the Lancang-Mekong River Basin. *J. Hydrol.* **2022**, *606*, 127464. [CrossRef]
28. Snousy, M.G.; Helmy, H.M.; Wu, J.; Zawrah, M.F.; Abouelmagd, A. Nanotechnology for Water Treatment: Is It the Best Solution Now? *J. Earth Sci.* **2023**, *34*, 1616–1620. [CrossRef]
29. Xu, F.; Yan, X.; Wang, F.; Ma, X.; Yun, J.; Wang, H.; Xu, B.; Zhang, S.; Mao, D. Development Strategy and Countermeasures of China's CBM Industry under the Goal of "Carbon Peak and Neutrality". *J. Earth Sci.* **2023**, *34*, 975–984. [CrossRef]

Article

Hydrochemical Characteristics and Water Quality Assessment of Irkutsk Reservoir (Baikal Region, Russia)

Mikhail V. Pastukhov [ID], Vera I. Poletaeva *[ID] and Guvanchgeldi B. Hommatlyyev

Vinogradov Institute of Geochemistry SB RAS, 1A Favorsky Str., Irkutsk 664033, Russia; mpast@igc.irk.ru (M.V.P.); hom@igc.irk.ru (G.B.H.)
* Correspondence: alieva@igc.irk.ru

Abstract: The Irkutsk Reservoir, belonging to the largest unified freshwater Baikal–Angara system, is an important source of drinking water in the region. Therefore, studies of its hydrochemical characteristics are of prime importance in deciding on the role of anthropogenic activity in water quality. The water samples were collected across the reservoir in 2007, 2012, and 2021 and then were analyzed for major ions and trace elements. The data revealed that the distribution of HCO_3^-, SO_4^{2-}, Cl^-, Ca^{2+}, Mg^{2+}, Na^+ and K^+ is stable across the reservoir. Trace element concentrations varied from 1.13 to 15.39 $\mu g\ L^{-1}$ for Al, from <DL to 0.39 $\mu g\ L^{-1}$ for Cr, from 0.39 to 23.12 $\mu g\ L^{-1}$ for Mn, from 1.25 to 53.22 $\mu g\ L^{-1}$ for Fe, from 0.005 to 0.100 $\mu g\ L^{-1}$ for Co, from 0.20 to 1.98 $\mu g\ L^{-1}$ for Cu, from <DL to 13.40 $\mu g\ L^{-1}$ for Zn, from 0.25 to 0.48 $\mu g\ L^{-1}$ for As, from 0.004 to 0.127 $\mu g\ L^{-1}$ for Cd, from <DL to 0.195 $\mu g\ L^{-1}$ for Sn, from <DL to 0.0277 $\mu g\ L^{-1}$ for Cs, from <DL to 1.13 $\mu g\ L^{-1}$ for Pb, from <DL to 0.0202 $\mu g\ L^{-1}$ for Th, and from 0.27 to 0.75 $\mu g\ L^{-1}$ for U. The concentrations of all major ions and trace elements in water were below the drinking water standards. CF values showed considerable and high contamination of samples with Al, Mn, Fe, Co, Cu, Cd, Sn, Pb, and Th. PLI values classified the majority of water samples as water with baseline levels of pollutants, and part of the samples was classified as either polluted or highly polluted.

Keywords: Irkutsk Reservoir; major ions; trace elements; monitoring; water quality index

Citation: Pastukhov, M.V.; Poletaeva, V.I.; Hommatlyyev, G.B. Hydrochemical Characteristics and Water Quality Assessment of Irkutsk Reservoir (Baikal Region, Russia). *Water* **2023**, *15*, 4142. https:// doi.org/10.3390/w15234142

Academic Editor: Bruno Charrière

Received: 28 September 2023
Revised: 23 November 2023
Accepted: 27 November 2023
Published: 29 November 2023

1. Introduction

Surface water bodies, providing drinking water and water for economic use, are a vital resource and are of crucial importance for economic and social development [1,2]. Freshwater resources in these water bodies and their availability can significantly increase the industrial–agricultural and recreational potential of an area, leading to a higher rate of urbanization on the coast. The growing activity of the population, growth in agricultural areas, and industrial development in the reservoir basin led to the influx of pollutants, therefore leading to negative transformations in the quality of water resources [3–5]. At present, numerous studies are devoted to the ecological state of surface water bodies, including the identification of water pollution sources, spatial–temporal dynamics of water hydrochemical composition, and assessment of its quality [6–8]).

One of the main industries of human economic activity, which is developed at large surface water bodies is hydropower. Despite certain economic advantages, dams give rise to various environmental transformations: changes in the amount of regional precipitation, river discharge from different locations along the river, biodiversity, etc. [9–11]. The water hydrochemical composition of reservoirs, including the concentrations of major ions and trace elements, relies on natural (weathering processes, lithology of the basin, the composition of water in tributaries, etc.), and anthropogenic factors [12,13]. The anthropogenic enrichment of water chemical composition is determined both by the creation of the reservoir [14], and the entry of these elements with discharge water from different industries and agriculture, surface runoff, and atmospheric transport from urban areas. Irrespective of the origin, the accumulation of pollutants, in particular heavy metals, is a serious ecological threat [15]. First of all, it is a problem of deterioration in the quality of water used for drinking. Bioaccumulation and biomagnification of potentially toxic elements in hydrobionts of different trophic levels [16] and further consumption of biological resources from such a reservoir may negatively affect the health of the local population [17,18].

The water resources of the Baikal region, with Lake Baikal located in its center, are characterized by fresh and ultra-fresh surface and groundwater [19]. Lake Baikal itself is the largest freshwater reservoir on the planet. In terms of the concentrations of 58 trace elements, the lake is classified as the cleanest lake of the biosphere [20]. The only channel of the surface runoff of Lake Baikal is the Angara River, which carries about 61 km^3 annually. The anthropogenic transformation of the hydrological and hydrochemical regimes of the river is related to the construction of a series of water reservoirs, the so-called cascade of the Angara River reservoirs (Irkutsk, Bratsk, Ust-Ilimsk, and Boguchany). Water resources of the Irkutsk Reservoir, which is the first reservoir in the Angara cascade, are used for supplying drinking water to the population of large cities (Irkutsk and Shelekhov) and a number of smaller settlements of the Irkutsk district. The gathered resources of the Irkutsk Reservoir are also used for generating hydroelectric power and navigation, as well as for fisheries and a number of recreational activities.

The major ion composition and the components of the water trophic status were studied at different stages of the Irkutsk Reservoir operation [21,22]. However, the data on water trace element composition of the Irkutsk Reservoir are still scarce. The studies by [23] were conducted to assess average total Ni, Zn, Cu, Pb, V, Co, Mn, Al, and Cr concentrations and to compare them with the trace element contents in other reservoirs of the Angara cascade. The present study focuses on: (a) concentrations of major ions (HCO_3^-, SO_4^{2-}, Cl^-, Ca^{2+}, Mg^{2+}, Na^+ and K^+) and trace elements (Al, Cr, Mn, Fe, Co, Cu, Zn, As, Cd, Sn, Cs, Pb, Th, and U) in surface and near-bottom waters of the Irkutsk Reservoir; (b) spatial–temporal dynamics in the concentrations of major ions and trace elements; (c) main natural and anthropogenic factors affecting the hydrochemical composition; (d) assessment of water quality in the Irkutsk Reservoir in terms of trace element concentrations using the single factor pollution index (CF) and the pollution load index (PLI). The studies are of significant practical interest for assessing the water quality of the large unified Baikal–Angara freshwater system and therefore contribute to preserving it as the key source of clean drinking water.

2. Materials and Methods

2.1. Study Site

The creation of the Irkutsk Reservoir (Figure 1), formed by the dam of the Irkutsk Hydroelectric Power Plant in 1956, resulted in flooding the valley of the Angara River between the dam (Irkutsk city) and the river source (Listvyanka settlement). As a result of the filling of the Irkutsk Reservoir, the water level of Lake Baikal increased by approximately 1 m on average. The Irkutsk Reservoir has a water surface area of 154 km^2 at normal water level (457 m a.s.l.), a volume of 2.1 km^3, a length of 55 km, and a width ranging from 0.5 to 3.5 km. The maximum depth of the reservoir occurring close to the dam is 35 m. Water level fluctuations in the reservoir are determined, to a greater extent, by the annual water level changes in Lake Baikal and the regime of the Irkutsk Hydroelectric Power Plant. The amplitude of the Irkutsk Reservoir level variations can reach as high as 3 m; however, during most time of its operation, the reservoir is lowered by 0.14–1.4 m than the normal water level [24]. Since 2001, the water levels in the Irkutsk Reservoir have been limited to the range between 456 m a.s.l. and 457 m a.s.l. in order to minimize the impact of the retention management on Lake Baikal ecosystem [25]. As a result of filling the reservoir, 40 bays were formed. The largest among them are Ershi, Kurma, Kartakoi, Elovyi, and Uladova.

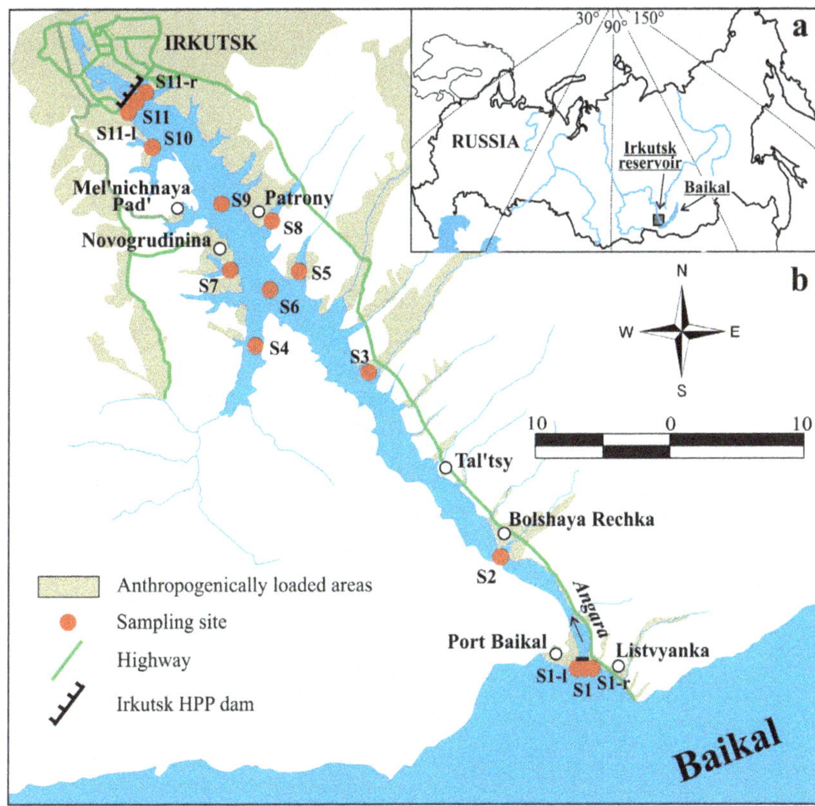

Figure 1. (**a**) Location of the Irkutsk Reservoir. (**b**) Schematic map with indicated sampling sites.

The northwestern part of the Irkutsk Reservoir lies within the Irkutsk–Cheremkhovo plain. At the southeast end of the basin, there are western spurs of the Primorsky Ridge. The northwestern part of the basin is characterized by a hilly, gently rolling landscape with incised valleys and vast flat watersheds. In the eastern part, there are groups of low hills and low ridges with intrefluves lying hypsometrically higher. The left shore is composed of bedrock, represented by the Jurassic sandstones, argillites, and siltstones. The right shore is made up of the Quaternary diluvial–alluvial sediments, containing loess-like loam and sandy loam as well as sand and pebbles, which were subject to abrasion along almost the entire shore [24].

The dam of the Irkutsk Hydroelectric Power Plant is located in Irkutsk city (Figure 1), the largest city in the Baikal region, whose leading industries are aircraft building, metalwork, and electric power generating. There are also metallurgy, forestry, food, pulp and paper, and woodworking industries. All major industrial enterprises are located below the dam of the Irkutsk Hydroelectric Power Plant. The left shore of the reservoir is very steep and therefore, less accessible and untapped. On the shores of such bays as Kurma, Kartakoi, Ershi, and Mel'nichnaya Pad', there are many settlements and tourist camps. However, from Kurma Bay to the source of the Angara River, settlements and tourist camps are less numerous. At the source of the Angara River, there is an abandoned shipyard that has been in operation since the 1890s. There is also an operating cargo and passenger port (Port Baikal) here. Along the right shore of the Irkutsk Reservoir, you can find a number of settlements, camps, and agricultural fields. The highway runs close to the reservoir connecting Irkutsk and Listvyanka settlements. The tourism industry is particularly developed in Listvyanka settlement, lying on the northeast shore of Lake Baikal, at the source of the Angara River. It should be noted that recreational attractiveness, in turn, entails an anthropogenic load on the environment. Around Listvyanka settlement and adjacent areas, the coastal zone is constantly developing from year to year. At the same time, the lack of a centralized wastewater treatment system in the settlement leads to the increasing anthropogenic load. The creation of the Irkutsk Reservoir provided good conditions for shipping from the dam to Lake Baikal. Currently, water transport for personal use is becoming more popular.

2.2. Sampling and Analytical Methods

There were five sampling campaigns to collect water samples from the Irkutsk Reservoir, conducted in July 2007, 2012, and in May (2021 (M)), July (2021 (J)), and September (2021 (S)) 2021. Sampling sites from S1 to S11 were distributed across the entire water area of the reservoir and were localized in the channel part and bays (Figure 1, Table S1). At the sampling sites with a water depth of over 2 m, the samples were collected from both the surface water layer (0.5 m) and the bottom one (in 1 m layer from the bottom). In 2021, the water samples were additionally collected along the right (S1-r, S11-r) and the left (S1-l, S11-l) shores of the reservoir, in the vicinity of Irkutsk city and Listvyanka settlement.

At each sampling site, three parallel water samples were collected. Replicate samples from the same site were taken within about 100 m of one another and then were well mixed. The mixed water samples were filtered through a 0.45 μm Millipore membrane filter and placed into polyethylene bottles which were prewashed with 3% nitric acid. The primary filtration part was discarded to clean the membrane. For the trace element analysis, the water samples were immediately acidified by addition of HNO_3 (ultrapure «Merk», Darmstandt, Germany). Prior to the analysis, the water samples were stored in the refrigerator.

The chemical analysis of water samples was accomplished at the Center for Collective Use «Isotope-Geochemical Research» located at the IGC SB RAS (Irkutsk, Russia). In the present study, SO_4^{2-}, Cl^-, Ca^{2+} and Mg^{2+}, Na^+, K^+ were analyzed by the method of capillary electrophoresis using devices of the "Drops" series (Lumex Ltd., St. Petersburg, Russia) and HCO_3^- was determined by the titrimetric method. The detection limits for elements were as follows: 6.1 mg L^{-1} for HCO_3^-; 0.5 mg L^{-1} for SO_4^{2-}; 0.5 mg L^{-1} for Cl^-; 0.5 mg L^{-1} for Ca^{2+}; 0.3 mg L^{-1} for Mg^{2+}; 0.5 mg L^{-1} for K^+; 0.5 mg L^{-1} for Na^+. Inductively coupled plasma mass spectrometry (ICP-MS) via a high-resolution double-focusing mass spectrometer ELEMENT-2 (Thermo Finnigan, Bremen, Germany) was used to analyze the total concentrations of trace elements. Multi-element standard samples such as ICP Multi-Element Standard Solution-Sol X CertiPUR for Surface Water Testing, Sol XII CertiPUR (MERCK, Darmstandt, Germany), and Combined Quality Control standard IQC-026 (NIST, North Kingstown, RI, USA) were applied to ensure the reliability of the analytical measurements. Water purified by the Millipore-ELIX-3 system (Millipore SA, Molsheim, France) was used to prepare washing, blank, calibration, and analyzed solutions. Three repetitions of the analysis yielded a relative standard deviation of <5%. Otherwise, the measurements were repeated until all the data reached the standard. The following metals were tested in water samples ^{27}Al, ^{52}Cr, ^{55}Mn, ^{56}Fe, ^{59}Co, ^{63}Cu, ^{66}Zn, ^{75}As, ^{111}Cd, ^{120}Sn, ^{133}Cs, ^{208}Pb, ^{232}Th and ^{238}U. The detection limits (DL) for the elements were as follows: 0.81 μg L^{-1} for Al; 0.05 μg L^{-1} for Cr; 0.37 μg L^{-1} for Mn; 0.75 μg L^{-1} for Fe, 0.003 μg L^{-1} for Co; 0.03 μg L^{-1} for Cu; 0.64 μg L^{-1} for Zn; 0.12 μg L^{-1} for As; 0.002 μg L^{-1} for Cd; 0.010 μg L^{-1} for Sn; 0.0003 μg L^{-1} for Cs, 0.013 μg L^{-1} for Pb, 0.0003 μg L^{-1} for Th; 0.001 μg L^{-1} for U.

The statistical data processing was performed using SPSS statistical software (IBM, Armonk, NY, USA, v. 20.0).

2.3. Pollution Indices

The following generally accepted pollution indices were used to calculate the water contamination in the Irkutsk Reservoir:

1. The single-factor pollution index (CF) was used to determine only one element in a sample [26]:

$$CF = \frac{C_i}{C_0},\tag{1}$$

where C_i is the concentration of the analyzed element and, C_0 is the concentration of metal in the control material. The CF value is divided into categories: CF < 1—low contamination, $1 \leq CF \leq 3$—moderate contamination, $3 \leq CF \leq 6$—considerable contamination, $6 \geq CF$—very high contamination.

2. The pollution load index (PLI) was used to calculate the total contamination in each sample [27]:

$$PLI = \sqrt[n]{CF_1 \cdot CF_2 \cdot \ldots \cdot CF_n},\tag{2}$$

where CF is an individual element pollution index. The PLI is divided into categories: PLI < 0—non-polluted, $0 < PLI \leq 1$—baseline levels of pollutants, $1 < PLI \leq 10$—polluted, $10 < PLI \leq 100$—highly polluted, PLI > 100—progressive deterioration of the environment.

3. Results and Discussion

3.1. Major Ion Composition

In samples collected in 2007, 2012, and 2021, the pH levels varied from 6.4 to 8.6, with an average value of 7.7. Table S2 presents the limits and mean concentrations of major ions and total mineralization (TDS) in 2007, 2012, May, June, and September 2021 in the Irkutsk Reservoir waters. Table 1 shows the data summarized for all sampling campaigns. The data available from monitoring studies of water in the Irkutsk Reservoir (Tables 1 and S2) showed insignificant spatial and temporal variations in major ion composition. The concentrations of the majority of main ions in the water samples collected during all sampling campaigns were found to lie within ±2 standard deviation (SD) of their mean. Only, the concentrations of SO_4^{2-} in three samples, Ca^{2+} in two samples, and Mg^{2+} in one sample, were found to fall within ±3 SD of the mean. In water samples taken from the Irkutsk Reservoir, the cations were dominated by Ca^{2+} (from 11.7 to 17.7 mg L^{-1}, with a mean value of 14.9 mg L^{-1}) and the anions were dominated by HCO_3^{-} (from 50.4 to 73.2 mg L^{-1}, 64.1 mg L^{-1} as a mean value). The water type was characterized as HCO_3-Ca.

Table 1. Comparison of major ion concentrations (mg L^{-1}) in water of the Irkutsk Reservoir and other ponds of the Baikal–Angara water system.

	HCO_3^{-}	Cl^{-}	SO_4^{2-}	Ca^{2+}	Mg^{2+}	Na^{+}	K^{+}	TDS	Reference
Lake Baikal	66.3 ± 1.6	0.4 ± 0.03	5.5 ± 0.1	16.4 ± 0.4	3.0 ± 0.1	3.3 ± 0.1	1.0 ± 0.1	~96	[28]
Source of the Angara River	66.2	0.6	5.7	15.4	3.3	4.2		95.6	[29]
Source of the Angara River	59.3–69.3 * 64.2 ± 3.7	<DL–1.21 0.76 ± 0.34	3.9–5.8 4.3 ± 0.5	14.4–15.6 14.0 ± 0.6	3.0–3.6 3.2 ± 0.2	2.9–3.4 3.2 ± 0.2	0.9–1.1 1.0 ± 0.1	86.1–100.1 92.4 ± 4.9	Present study
Irkutsk Reservoir	50.4–73.2 64.0 ± 3.9	<DL–1.27 0.85 ± 0.21	3.7–8.5 4.8 ± 0.8	11.7–17.7 14.9 ± 0.8	2.0–5.7 3.3 ± 0.4	2.7–3.6 3.2 ± 0.2	0.7–1.2 1.0 ± 0.1	50.4–73.2 64.0 ± 3.9	
Bratsk Reservoir	71.9	3.0	11.3	19.8	3.9	4.0	1.0	71.9	[30]
Ust-Ilimsk Reservoir	79.6	16.2	5.5	19.1	6.5	6.3	1.1	79.6	[31]

Notes: * Above the line—minimum–maximum value, below the line—mean value ± standard deviation.

In the study area, the water samples collected during all sampling campaigns demonstrated low TDS levels: from 78.5 to 104.0 mg L^{-1}. According to our determinations, the average TDS value in the Irkutsk Reservoir (92.5 mg L^{-1}) was less than the world river water median (127 mg L^{-1}) [32]. This value was close to the average TDS level in Lake Baikal (~96 mg L^{-1}) [28] and the source of the Angara River (95.6 mg L^{-1}) [29] (Table 1). The main natural sources of dissolved salts, which change the composition of water over a wide range, include cyclic salts and weathering of minerals [33–35]. Similar average concentrations of major ions and TDS values in water columns of Lake Baikal, the Angara River source (S1), and the Irkutsk HPP headwater (S11) (Tables 1 and S2) suggest that the main source of dissolved substances entering the reservoir is Lake Baikal runoff. Weathering processes and cyclic salts were found to insignificantly influence the major ion water chemistry in the reservoir. Higher concentrations of the major ions in the water column of the Bratsk and Ust-Ilimsk Reservoirs (Table 1), located downstream on the Angara River, suggest a lower contribution from the lake's runoff and a higher contribution from natural and anthropogenic sources.

As shown by monitoring studies conducted since the beginning of the 20th century, in the pelagic part of Lake Baikal major ion composition exhibits weak seasonal and interannual variability within the accuracy of methods [36,37]. In the Irkutsk Reservoir, in particular at the Angara River source, the levels of main ions, including HCO_3^-, Cl^-, and SO_4^{2-} were higher than those in Lake Baikal (Table 1). A similar distribution pattern of main ions was found in the Angara River source during the period of 1997–2003 [38]). Based on the results of long-term monthly investigations, it was shown that HCO_3^- concentration at the river's source could be affected by the fluctuations of water level, whose maxima correlate with minimum ion levels, while SO_4^{2-} concentrations were influenced by large seismic events. The long-term dynamics of the water level in the Irkutsk Reservoir are recorded as the alternation of maximum and minimum cycles. Thus, in the period from 2004 to 2018, the water level in the reservoir was close to the normal water level [39]. Larger water level variations were observed in 2021. Especially pronounced was the drop in the water level to 456.20 m in the first half of May resulting from both the increased discharges of the Irkutsk HPP, and contrast weather conditions of the spring of 2021, with frequent colder weather, extended snowmelt leading to losses of water by evaporation and, as a result, a slow increase in water inflow into the reservoir. In May, HCO_3^- exhibited the greatest variations. Moreover, its level was minimal throughout all sampling campaigns (Table S2). From June to September, there was a steady increase in the water level (up to 457.22 m) due to a decrease in discharge flows of the Irkutsk HPP and an increase in the water level of Lake Baikal.

In addition to natural factors, anthropogenic factors are also responsible for the water ionic composition in reservoirs. The anthropogenic impact may lead to increased concentrations of ions, mainly of Cl^- and SO_4^{2-} [5,40]. These ions were ascribed to the main pollutants in the southern part of Lake Baikal, whose runoff was found to significantly affect the chemical water composition at the source of the Angara River and in the Irkutsk Reservoir. For several decades, the sulfate concentrations in the water column of the South Baikal and in the Angara River source had been significantly affected by the wastewater from the Baikalsk Pulp and Paper Mill (BPPM) closed in 2013 [41]. In addition to SO_4^{2-}, large amounts of Cl^- enter (entered) the South Baikal from different sources: wastewater of BPPM, 7.3 ± 0.2 ton/year; wastewater of Ulan-Ude city—5274 ± 648 ton/year; polluted water of Selenga River (Lake Baikal largest tributary) 70.65 ± 9.91 tons/year [42]. A number of suburban settlements along the shores of the Irkutsk Reservoir were likely another source of Cl^- in the water of the reservoir. The potassium chloride used as a component of agricultural fertilizers is known to increase Cl^- ion concentration in the aquatic system [43]. With the data available, the contribution of fertilizers cannot be quantified. However, in the Irkutsk Reservoir, K^+ levels vary insignificantly over the entire monitoring period (0.91–0.96 mg L^{-1} as an average; Table S2); there is no correlation between the ions of chlorine and potassium. Therefore, this study revealed no contribution from agricultural fertilizers to the Cl^- concentration at all sampling localities within the Irkutsk Reservoir. A larger influence on Cl^- concentrations is likely to be recorded in the water of the bays as the areas adjacent to the bays are more developed.

3.2. Trace Elements Composition

Table S3 and Figure 2 present statistics of trace element concentrations in 2007, 2012, May, June, and September 2021 in the Irkutsk Reservoir waters. Table 2 shows the data summarized for all sampling campaigns.

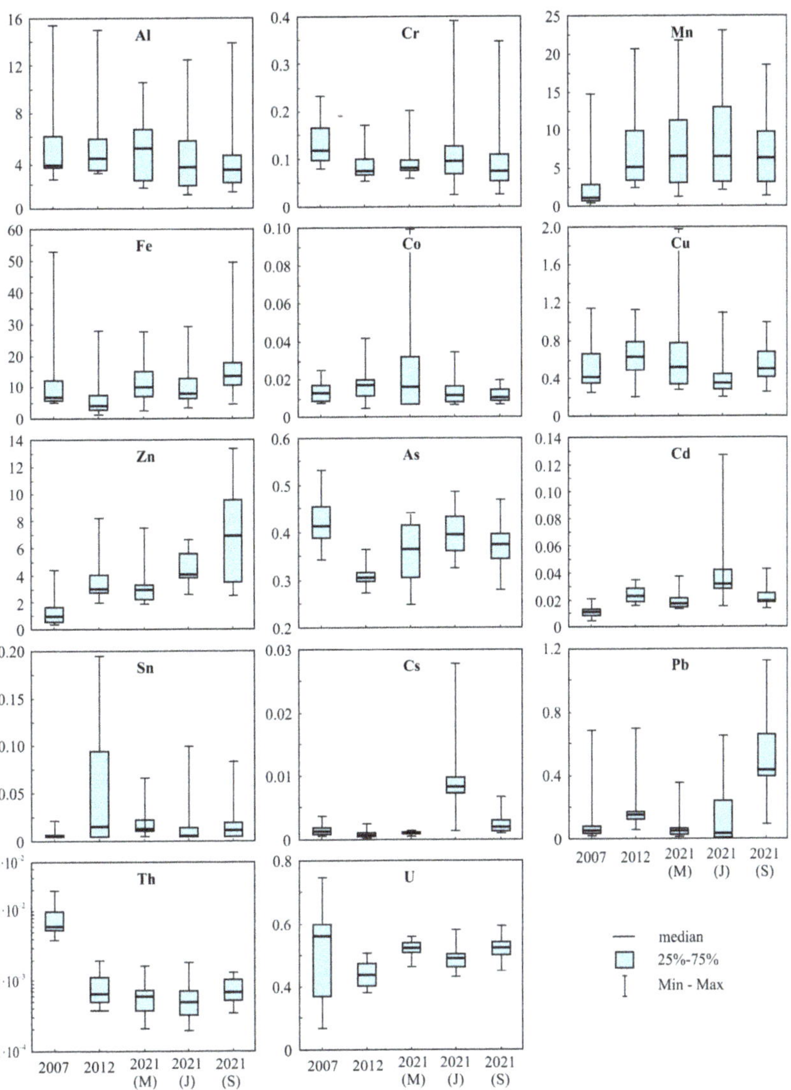

Figure 2. Concentrations of trace elements (μg L^{-1}) in water of the Irkutsk Reservoir in 2007, 2012, May 2021, July 2021, and September 2021.

A large collection of data on trace element geochemistry in surface waters of Africa, Europe, Asia, and North and South America are presented in [44]. The average concentrations of Al, Mn, Fe, Co, Cu, As, Cd, Cs, and Th in the Irkutsk Reservoir were markedly lower than the average world value; Cr levels were lower than their minimum values and the concentrations of Zn, Pb, and U exceeded the corresponding world average but were below the worldwide maximum values (Table 2).

14

Table 2. Comparison of trace elements concentrations ($\mu g\ L^{-1}$) in Irkutsk Reservoir and other ponds of the Baikal–Angara water system.

		Al	Cr	Mn	Fe	Co	Cu	Zn	References
Irkutsk Reservoir	Min–max	1.13–15.39	<DL–0.39	0.39–23.12	1.25–53.22	0.005–0.100	0.20–1.98	<DL–13.40	Present study
	Mean	4.60	0.10	7.02	12.09	0.017	0.55	4.21	
	Median	3.79	0.08	5.26	9.33	0.013	0.45	3.51	
	SD	2.93	0.05	5.42	9.51	0.016	0.30	2.71	
Source of the Angara River		1.13–6.09 * / 3.07	0.06–0.17 / 0.11	0.76–12.57 / 5.54	1.25–16.90 / 7.67	<DL–0.025 / 0.011	0.28–1.29 / 0.68	2.02–10.26 / 4.99	
Source of the Angara River		3.89	0.12	2.93	18.8	0.011	0.62	2.11	[45]
Lake Baikal		0.34–1.15	0.07	0.06–0.33	0.26–1.12	0.003	0.16–0.25	0.24–0.56	[46]
Lake Baikal		0.1–1.0 / 0.38	0.03–0.09 / 0.07	0.01–0.53 / 0.13	0.1–1.6 / 0.38	0.002–0.005 / 0.003	0.2–1.0 / 0.21	0.4–4.3 / 3.2	[20]
Bratsk Reservoir		12.9	0.19	22.9	22.4	–	0.84	3.80	[47]
World average values		0.5–480 / 32	0.5–11.5 / 0.7	0.41–113.52 / 34	1–525 / 66	0.006–0.260 / 0.148	0.23–3.53 / 1.48	0.04–27.0 / 0.6	[44]
MPC **		500	50	100	300	100	1000	5000	[48]

		As	Cd	Sn	Cs	Pb	Th	U	Reference
Irkutsk Reservoir	Min–max	0.25–0.53	0.004–0.127	<DL–0.195	<DL–0.0277	<DL–1.13	<DL–0.0202	0.27–0.75	Present study
	Mean	0.37	0.025	0.022	0.0034	0.23	0.0017	0.50	
	Median	0.37	0.021	0.011	0.0012	0.12	0.0007	0.51	
	SD	0.06	0.018	0.031	0.0050	0.25	0.0033	0.07	
Source of the Angara River		0.25–0.48 / 0.37	0.015–0.055 / 0.029	<DL–0.195 / 0.032	0.0006–0.0087 / 0.0043	<DL–1.13 / 0.33	<DL–0.0053 / 0.0009	0.40–0.55 / 0.50	
Source of the Angara River		0.46	0.011	0.028	0.0016	0.05	0.0014	0.58	[45]
Lake Baikal		0.40–0.41	0.008	<0.011	0.0017	0.010–0.036	0.0006	0.52	[46]
Lake Baikal		0.30–0.50 / 0.40	0.001–0.010 / 0.008	<0.01–0.04 / <0.01	0.002–0.008 / 0.0013	<0.02	0.002–0.020 / 0.004	0.4–0.7 / 0.55	[20]
Bratsk Reservoir		0.37	0.023	–	0.0022	0.111	–	0.52	[47]
World average values		0.11–2.71 / 0.62	0.0006–0.42 / 0.08	–	0.0006–0.016 / 0.011	0.006–3.8 / 0.079	0.001–4.3 / 0.0055	0.004–4.94 / 0.37	[44]
MPC		50	1	–	–	30	–	–	[48]

Notes: * Above the line—minimum–maximum value, below the line—mean value; ** Maximum permissible concentrations of trace elements.

When studying the trace element water chemistry of the Irkutsk Reservoir, it is reasonable to compare the trace element characteristics of the reservoir with those of Lake Baikal (Table 2). The data on hydrochemistry of Lake Baikal showed that like the ionic, the trace element composition was stable at all depths in the pelagic zone of the lake [46]. At the same time, its hydrochemical characteristics were influenced by natural (multi-component flows of more than three hundred rivers of its catchment area, hot and cold springs [49], and anthropogenic factors as well as by seismic activity [50], etc.). The effect of anthropogenic factors was demonstrated by the chemical composition of snow cover from the lake's water area near the settlements [51]. The results revealed that the snow chemical composition was characterized by higher concentrations of Mn, Al, Pb, Cu, Fe, and Zn in the southern part, increased Mn, Al, Pb, Cr, Fe, and Zn levels in the middle part, and higher Mn, Al, Pb, Cu, Cr, and Fe contents in the northern part of Lake Baikal. Long-term monitoring of concentrations of major ions [38] and mercury [50] in the Angara River source indicates that hydrochemical characteristics of water from the river's source characterize the average water composition of the entire Lake Baikal. Therefore, when studying trace element water composition in the Irkutsk Reservoir, the Angara River source is considered separately.

3.2.1. Angara River Source

In sampling campaigns of 2021, the element concentrations in water samples from the Angara River source demonstrated the following ranges: Al (1.13–6.09 μg L^{-1}); Cr (0.06–0.17 μg L^{-1}); Mn (1.22–12.57 μg L^{-1}); Fe (3.94–16.90 μg L^{-1}); Co (<DL–0.018 μg L^{-1}); Cu (0.28–1.29 μg L^{-1}); Zn (2.02–10.26 μg L^{-1}); As (0.25–0.48 μg L^{-1}); Cd (0.015–0.039 μg L^{-1}); Sn (<DL–0.036 μg L^{-1}); Cs (0.0012–0.0087 μg L^{-1}); Pb (<DL–1.13 μg L^{-1}); Th (<DL–0.0202 μg L^{-1}); U (0.27–0.61 μg L^{-1}) (Figure 3).

In different periods of studies [20,46,52], the concentrations of the vast majority of trace elements were in good agreement with each other, except for Zn and Th (Table 2). In the water of Lake Baikal, the Zn concentration measured by Sklyarova [46] (0.56 μg L^{-1}) is much lower than its abundance determined by Falkner et al. [52] (2.9 μg L^{-1}) and Vetrrov et al. [20] (4.3 μg L^{-1}). For thorium, this situation was quite different: Th levels given in [20] (0.004 μg L^{-1}) were an order of magnitude higher than its concentrations measured by Sklyarova [46] (0.0006 μg L^{-1}). In the water of the Angara River source, Zn concentrations were in line with its abundances in Lake Baikal obtained by Vetrov et al. [20], while Th levels were in good agreement with Th contents in Lake Baikal measured by Sklyarova [46]. The concentrations of As and U measured in the Angara source are well comparable with their levels in the water from the pelagic part of the lake (Table 2, Figure 3). The levels of Al, Cr, Mn, Fe, Co, Cu, Cd, Sn, Cs, and Pb in the water of the river source were higher than those in the lake's water. The results of this study are in line with the data on trace element compositions obtained from 3-year (2006–2008) monthly water monitoring of the Angara River source [45].

At the source of the Angara River in different sampling campaigns of 2021, higher Al, Mn, Fe, Co, Cu, As, Cd, Sn, Cs, Pb, Th concentrations were found along the left shore of the river as compared with its center, while increased levels of Cr, Mn, Fe, Co, Cu, Zn, As, Cd, Sn, and U were observed along the right shore (Figure 3). In June, the water of the middle part was characterized by higher Al, Cr, Fe, Cu, and U contents as compared with the shores. The anthropogenic impact on this part of the reservoir is primarily related to the activity of the cargo and passenger port, whose ships and ferries run between the left and right banks all year round. Antifouling paints, metals, and steel alloys, as well as petroleum products, are an important source of trace elements in the water bodies near the shipyards and yacht clubs [53–55]. The area close to the Listvyanka settlement with a well-developed tourist infrastructure may also be a main source of trace elements in this part of the Irkutsk Reservoir. The study of surface and groundwater in the vicinity of the settlement revealed that the anthropogenic impact resulted in the pollution of primarily groundwater, whose subaqual discharge led to higher Mn, Zn, and Pb concentrations in the near-bottom water of the lake's coastal zone [56]. The Baikal under-ice water taken close to the Listvyanka settlement was characterized by higher Mn, Cu, Co, Al, Fe, and Zn contents as compared with the deep-seated water [57]. The changes in the chemical composition of coastal waters are thought to be related to the dissolution of rocks and soils, mechanical transport of the detrital material from the coastline, and atmospheric transport.

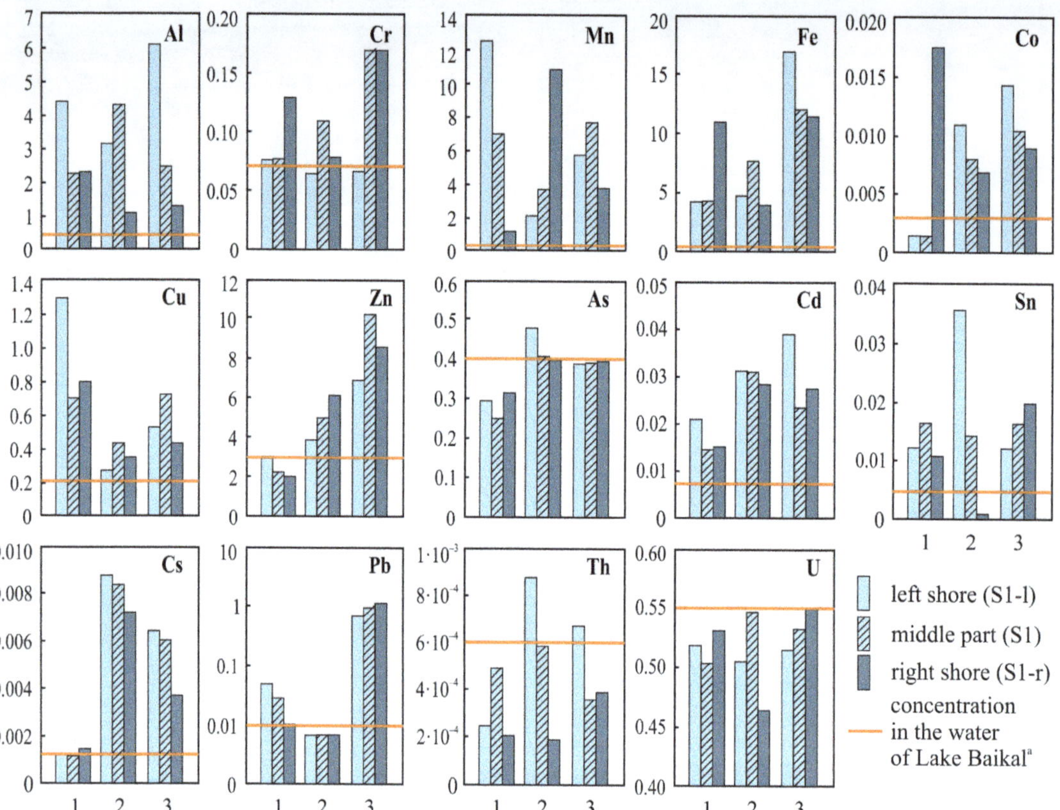

Figure 3. Trace element concentrations (μg L^{-1}) in the channel part of the Angara River source, 2021 (1—May, 2—June, 3—September). [a]—concentrations of trace elements, except for Th, in Lake Baikal water are given from [20], Th concentrations—are taken from [46].

3.2.2. Irkutsk Reservoir

As opposed to major ions, the trace element concentrations showed a more significant spatial–temporal variability (Tables 1 and 2). In the vast majority of water samples, the concentrations of trace elements were found to lie within mean $\pm$ 2SD. We calculated the median, which is unlike the mean and is not sensitive to the outlier measured values. Table 3 illustrates sampling sites, where the trace element concentrations were above 2SD of the median. In some samples, trace element concentrations were found to exceed the median + 2SD: Al—7 samples; Cr—6 samples; Mn—8 samples, Fe—8 samples, Co—6 samples, Cu—7 samples; Zn—12 samples; Cd—3 samples; Sn—12 samples, Cs—7 samples; Pb—11 samples; Th—4 samples.

Table 3. Sampling sites, where trace element concentrations in water exceed median + 2SD.

Sampling Site		Trace Element	Sampling Site		Trace Element
S1	s[1]	Al, Cu, Pb—2007[2], Sn—2012, Pb, Zn—2021 (S)	S1-r	s	Pb—2021 (S)
S1-1	s	Cu—2021 (M), Pb—2021 (S)	S2	s	Al, Cr, Th—2007, Sn—2012, Mn—2021 (J), Zn—2021 (S)
				b	Al, Cu—2021 (M), Zn—2021 (S)
S3	s	Cr, Fe, Co, Th, U—2007, Co—2021 (M), Cr, Zn, Pb—2021 (S)	S4	s	Cu—2007, Al, Mn, Co, Cu—2021 (M), Fe, Cs—2021 (J), Fe, Zn, Cr—2021 (S)
	b	Zn, Pb—2021 (S)		b	Sn—2021 (S), Al, Fe, Co, Zn—2021 (S)
S5	s	Al, Mn, Fe, Pb –2012, Mn, Fe—2021 (J), Mn—2021 (S)	S6	s	Sn—2012, Mn—2021 (M), Cr, Cs—2021 (J), Zn, Sn—2021 (S),
	b	Co—2021 (M), Cs—2021 (J), Fe, Zn—2021 (S)		b	Sn—2012, Sn—2021 (S)
S7	s	Zn—2021 (S)	S8	s	–
	b	Zn—2021 (S)		b	–
S9	s	Th—2007, Cu, Zn—2021 (M), Cr, Fe, Co, Sn, As—2021 (J), Sn—2021 (S)	S10	s	Fe, Th, U—2007, Al—2021 (J), Pb—2021 (S)
	b	–		b	Cu—2012, Mn, Cs—2021 (J)
S11	s	Sn—2012, Cd, Sn, Cs—2021 (J)	S11-r	s	Cd, Cs—2021 (J)
	b	Sn, Pb—2012, Cd, Cs, Pb—2021 (J), Pb—2021 (S)	S11-l	s	Sn—2012, Mn—2021 (S)

Notes: [1] s—surface water, b—bottom water; [2] sampling campaigns in 2007–2007, 2012–2012, May 2021–2021 (M), July 2021–2021 (J), and September 2021–2021 (S).

There are several features that characterize the influence of Lake Baikal runoff on the water trace element composition in the Irkutsk Reservoir. Firstly, during all sampling campaigns, median concentrations of Al, Cr, Mn, Co, As, Th, and U across the reservoir were close and the median Cu, Zn, Cd, Sn, Cs, and Pb contents were lower than those measured at the source of the Angara River. Secondly, the concentrations of Zn and U in the water of both the Irkutsk Reservoir and Lake Baikal exceeded the average value for the world's surface water (Table 2). Higher U concentrations (similar to maximum), as compared with the average world values, and increased Zn levels (similar to mean ones) in the water of Lake Baikal characterize its geochemical background in the area surrounding the lake [46]. Thirdly, the concentrations of trace elements in the water of the Irkutsk Reservoir decreased in the following order Fe > Mn > Al, Zn > Cu, U > As > Pb > Cr > Cd, Sn > Co > Cs > Th, which is similar to the pattern found for the water at the Angara River source.

When comparing the channel part and the bays of the reservoir (Table S3), the number of elements showing higher concentrations was larger in the surface water of the bays. It is, in particular, true for the Kurma Bay (S4), whose water area is widely used for fishery, Burduguz (S3), and Uladova (S5) Bays (Table 3). In the channel part, higher concentrations of trace elements are primarily found opposite the Mel'nichnaya Pad' settlement (S9). Water transport, which is rapidly developing at present, was another anthropogenic factor influencing the water hydrochemistry of both the Irkutsk Reservoir and the Angara River source. In addition, a number of suburban settlements, which are not connected to the urban sewer system or septic tanks for treating household waste, are located along the coastline of the bays in the immediate vicinity of the water. Untreated waste can significantly affect the water quality in waterways connected to the reservoir [58]. In the absence of wastewater treatment facilities, the entry of pollutants into the Irkutsk Reservoir with both surface (meltwater and rainwater flowing from coastal settlements) and underground runoffs can be significantly higher. The elements, whose concentrations were higher at the majority of sampling sites included Mn, Fe, Al, Zn, Sn, Pb, Cu, and Cs (Table 3). Giri and Singh [59] related the enhanced Mn, Pb, and Zn levels in the waters of the Subarnarekha River (India) to industrial wastes and transport pollution. In coastal waters of China, Pb and Zn abundances were found to be closely associated with motor transport [60]. The Pb and Zn levels in the Aibi Lake are mainly influenced by agricultural fertilizers, emissions from traffic, and/or urban construction in its basin [61].

The results from this study indicate that enhanced Al, Mn, Fe, Cu, Zn, As, and Cd levels were observed along the right and left banks of the Irkutsk Reservoir, close to Irkutsk city (Figure 4). In July and September, Pb concentrations significantly increased in relation to May. A variety of anthropogenic activities in urban areas led to the accumulation of a wide scope of trace elements, including Cd, Cu, Pb, and Zn [62] in the upper soil layer. From the area near Irkutsk, trace elements of anthropogenic origin enter the reservoir with the surface runoff. Therefore, the trace element concentrations can increase, primarily in the aquatic environment along the shores. In the vicinity of Irkutsk, the anthropogenic impact on the left shore is mainly due to the city beach, while on the right shore, this impact is related to a passenger port and parking for private water transport (about 200 small vessels and yachts). Herewith, near Listvyanka settlement, the navigation takes place all year round, but near the HPP dam, water transport activity and the growing number of tourists are observed in the warmer season (July–August).

Another source of trace elements in the reservoir, in particular major elements of the Earth's crust, is the abrasion of shores. The creation of the reservoir facilitated the processes of shore erosion, both in the reservoir itself and on the shores of Lake Baikal. Within the Irkutsk Reservoir, the shores, which are not exposed to erosion, lie close to the source of the Angara River, while the abrasive shores formed in the Jurassic sandstones and Quaternary sediments make up about 54% of the reservoir's shoreline [63]. The composition of the bottom sediments in the reservoir reflects the geochemical specifics of rocks, which compose the shores exposed to impacts from processes of erosion. Along the banks composed of the Jurassic sandstones, the bottom sediments showed the dominance of Mn, Cr, V, Zn, Cu, while on the banks composed of the Quaternary diluvial loess loams the bottom sediments contained higher abundances of Mn, Co [64]. It was found that, with the material from the coastal erosion, the reservoir receives 4700 tons of Fe, 154 tons of Mn, and about 220 tons of other trace elements each year [64]. The terrigenous material, entering the bottom of the reservoir due to erosion processes, is able to suspend in the water column as particles of different fractions for a long time leading to higher Al, Mn, Fe, Co, as well as Cr, Zn, Cu levels in the water of the Irkutsk Reservoir.

The retrospective impact of human activity on the reservoir can be assessed by studying the composition of bottom sediments, which are active accumulators of elements of anthropogenic origin [65,66]. In the Bratsk Reservoir, the decrease in anthropogenic impact led to better water quality characteristics [47]. However, the results of layer-by-layer sampling of its bottom sediments showed the accumulation of the vast majority of trace elements in the middle layers of the bottom sediments formed during the period of the greatest anthropogenic impact [67]. In this regard, the bottom sediments of the Irkutsk Reservoir have not been sufficiently studied. The most complete information on the chemical composition (30 elements) of its bottom sediments is given in Jagus et al. [68]. In four sectors of the reservoir (Tal'tsy, Patrony, Novogrudinina, and Mel'nichnaya Pad' settlements) we analyzed 10 samples of bottom sediments. Amongst the elements discussed in this study, Co, Pb, and Cr are worthwhile and noteworthy: Co and Pb demonstrate different spatial distribution patterns and Cr had higher concentrations as compared to its levels in other dammed reservoirs worldwide. At the same time, in Cartagena Bay, the bottom sediments taken at stations related to the repair and maintenance of ships were characterized by high concentrations of Cr, Cu, As, and Cd while sediments from stations receiving inputs from petroleum plants displayed high Pb levels [69]. Therefore, it can be suggested that higher concentrations of trace elements in near-bottom water layers of the Irkutsk Reservoir, particularly close to the dam (Table 3), may be a result of their release from bottom sediments.

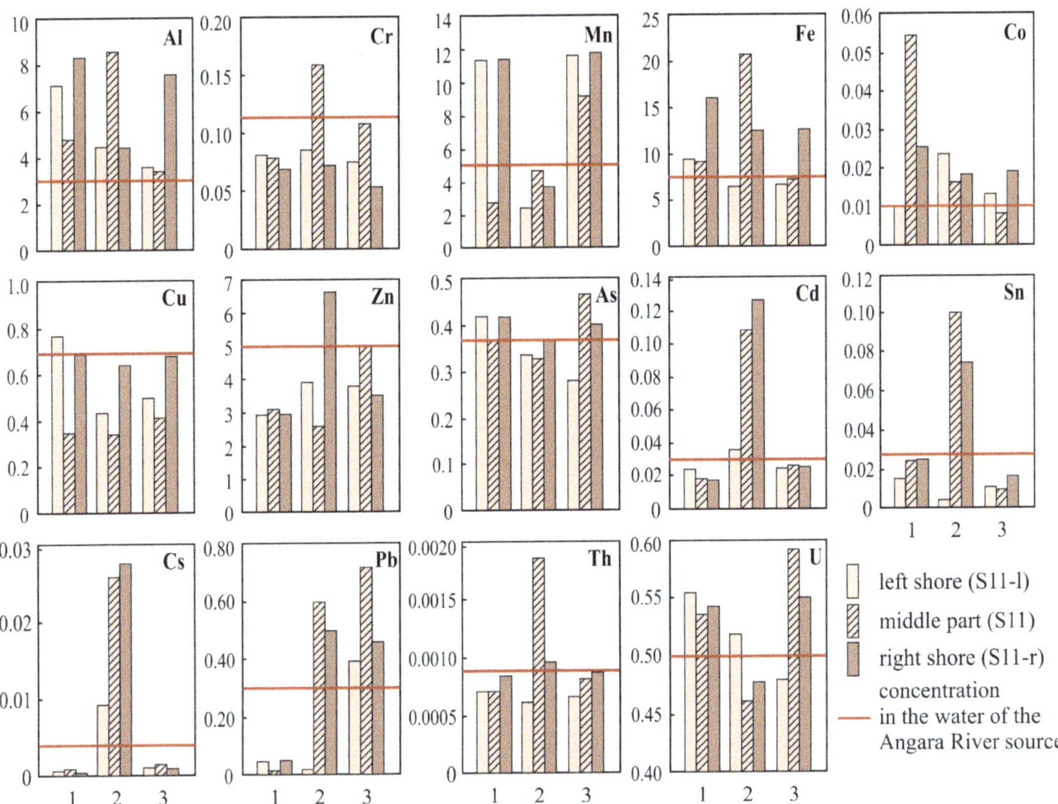

Figure 4. Concentrations of trace elements ($\mu g\ L^{-1}$) in the channel part (vicinity of Irkutsk city), 2021 (1—May, 2—June, 3—September).

3.3. Correlation Analysis

The Pearson correlation coefficient is widely used to determine general sources of elements' entry [70,71]. A number of significant positive and significant negative correlations were obtained using the information from the study of trace element composition of water in the Irkutsk Reservoir (Table 4). It was found that correlation relationships between elements were not stable across sampling campaigns in different years. At the same time, there were pairs of elements preserving significant correlation during 3 or 4 sampling campaigns. Amongst major ions, these pairs included HCO_3^- vs. Ca^{2+}, SO_4^{2-} vs. Mg^{2+}. The positive correlation between HCO_3^- vs. Ca^{2+} in the Irkutsk Reservoir, found also in the Boguchany Reservoir, being the Angara Cascade Reservoir as well [72]), is characteristic of freshwater calcium bicarbonate composition [73]. Of the trace elements, the pairs with a positive correlation included: Al–Mn, Al–Fe, Al–Co, Al–Th, Mn–Fe, Co–Fe, and Pb–Cr. Aluminum, which is the most abundant metal in the Earth's crust, is noteworthy [74]). The positive correlation of Al with Fe, Mn, Co, as well as of Fe with Mn, Co, Mg^{2+} over several sampling campaigns showed that these elements can enter the water environment of the

Irkutsk Reservoir as a result of weathering of the parent material, pedogenesis, and erosion activity. The data concerning Th concentrations in rocks, composing the drainage area of the Irkutsk Reservoir, are still lacking. At the same time, maximum Th levels in the bottom sediments of the Irkutsk Reservoir (12.9 ppm) exceeded the values of the geochemical background of sedimentary rocks in the region [68]. The significant positive correlation between Th and Al over three sampling campaigns might indicate that bedrock, composing the shores of the reservoir, is probably the main source of thorium entering the reservoir. The positive correlation between Pb and Cr, determining the relationships between these two elements, reflects that the combustion of petroleum or gasoline used for water transport is probably the main mechanism by which these trace elements enter the water reservoir.

3.4. Trace Element Pollution Status and Analysis of Water Quality

Water supply to the population in Irkutsk and smaller settlements of the Irkutsk district is provided from the water intake, located in the Ershi Bay. Residents of suburban settlements living on the shores of the reservoir, in the spring, summer, and autumn use submersible pumps for pumping water from nearby bays. Currently, in Russia, the requirements for the quality of drinking water are based on maximum permissible concentrations (MPC) of trace elements and are regulated by the Sanitary Rules and Norms for Drinking Water [48]. Concentrations of all the elements under consideration in the water of both the Irkutsk Reservoir and Lake Baikal are substantially below the standards for drinking water (Table 2).

An important tool that helps to assess surface water pollution is the choice of a criterion that should be used as a control material (geochemical background) [75]. At the same time, such a criterion is to be chosen on both local and global scales [71]. As the Irkutsk Reservoir is a part of the unified Baikal–Angara water system, the summarized hydrochemical characteristics of Lake Baikal which include the average concentrations of trace elements at the source of the Angara River, can be used as a control material.

The calculation of the single-factor pollution index (CF) is shown in Table 5. Water samples of the Irkutsk Reservoir were classified as low contaminated: in 2007 by the concentrations of Cd, Sn, Zn, and Cs; in 2012, May 2021—by the concentration of Cs, and in September 2021—by the concentration of Th. Water samples were classified as low and moderately contaminated: in 2007 by concentration of Cr, Mn, Co, Cu, As, Pb, and U, in 2012—by concentration of Cr, Cu, Zn, As, Cd, Pb, Th, and U, in May 2021—by concentration of Cr, Zn, As, Cd, Sn, Pb, Th, and U, in July 2021—by concentration of Cu, Zn, As, Pb, Th, and U, in September 2021—by concentration of Cr, Co, Cu, Zn, As, Cd, Sn, Cs, and U. Part of the water samples revealed considerable and very high contamination: in 2007 by concentrations of Al, Fe, Th, in 2012—by concentrations of Al, Mn, Fe, Co, and Sn, in May 2021—by concentrations of Al, Mn, Fe, Co, and Cu, in July 2021—by concentration of Al, Cr, Mn, Fe, Co, Cd, Sn, and Cs, in September 2021–by concentrations of Al, Mn, Fe, and Pb.

The PLI index in 2007 (0.000–0.407), 2012 (0.000–0.223), May 2021 (0.000–0.155), July 2021 (0.000–0.910), September 2021 (0.000–0.333) classified the vast majority of samples as waters with a baseline level of pollutants. The exceptions included samples of surface water from S5 site in 2012 (PLI = 7.2, polluted water), S11 in July 2021 (PLI = 39.6, highly polluted), S11-r July 2021 (PLI = 11.6, highly polluted), S5 July 2021 (PLI = 1.4, polluted), as well as surface (PLI = 1.9, polluted) and near-bottom (PLI = 5.3, polluted) waters S4 in September 2021.

Table 4. Statistically significant correlations between concentrations of major ions and trace elements in water of the Irkutsk Reservoir.

	2007	2012	2021 (M)	2021 (J)	2021 (S)
HCO_3^-	$(+**)$**Ca^{2+}**	$(+**)K^+$	$(+**)K^+$, **Ca^{2+}**, U; $(-**)SO_4^{2-}$, Mg^{2+}, Al, Cr, Fe, Th	$(+*)$**Ca^{2+}**; $(-*)$Al	$(+*)Mg^{2+}$, As; $(+**)$Fe
Cl^-	$(+**)SO_4^{2-}$, $(+*)Mg^{2+}$, $(-*) Ca^{2+}$		$(-**)$Zn	$(+**)SO_4^{2-}$, Cd, Cs; $(+*)Na^+$, Pb	$(+*)$Cu
SO_4^{2-}	$(+**)Cl^-$, **Mg^{2+}**	$(+*)$Co	$(+**)$**Mg^{2+}**, Al, Mn, Fe, Pb, Th; $(-**)HCO_3^-$, K^+, Ca^{2+}, U	$(+**)Cl^-$; $(+*)$ **Mg^{2+}**, Cr, Cu; $(-*)K^+$, U	$(+*)$**Mg^{2+}**, Al, Mn
K^+	$(+**)$Al, Co; $(+*)$Cd, Pb	$(+**)HCO_3^-$; $(+*)$U; $(-**)$Fe; $(-*)$Mn	$(+**)HCO_3^-$, Ca^{2+}, U; $(-**)SO_4^{2-}$, Al, Cr, Mn, Fe, Pb, Th	$(-*)SO_4^{2-}$, Zn	
Na^+			$(+**)$Cs; $(+*)$ Cd	$(+*)Cl^-$; $(-*)$ U	$(+*)$Fe: $(+*)Mg^{2+}$; $(-*)Ca^{2+}$
Ca^{2+}	$(+*)$**HCO_3^-**, Cr $(-*)Cl^-$	$(-**)Mg^{2+}$, Al, **Fe**, Zn	$(+**)$**HCO_3^-**, K^+, U; $(-**)SO_4^{2-}$, Mg^{2+}, Al, Cr, **Fe**, Pb, Th; $(-*)$Mn	$(+*)$**HCO_3^-**	$(+*)$Cd; $(-**)$**Fe**; $(-*)Na^+$
Mg^{2+}	$(+**)$**SO_4^{2-}**; $(+*)Cl^-$	$(+**)$**Fe**, Zn; $(+*)$ Al; $(-**)Ca^{2+}$	$(+**)$**SO_4^{2-}**, Ca^{2+}, Al, Mn, **Fe**, Th; $(+*)$ Pb, $(-**)HCO_3^-$, U	$(+**)$Th; $(+*)$**SO_4^{2-}**	$(+**)$**Fe**; $(+*)HCO_3^-$, **SO_4^{2-}**, Na^+
Al	$(+**)K^+$, **Mn, Fe, Co**	$(+**)$**Mn, Fe**, Zn; $(+*)Mg^{2+}$, **Co, Th**; $(-**)Ca^{2+}$	$(+**)SO_4^{2-}$, Mg^{2+}, Cr, **Mn, Fe**, Pb, **Th**; $(-**)HCO_3^-$, K^+, Ca^{2+}; $(-**)$U	$(+**)$Cu, **Th**	$(+**)$**Fe, Co**, Cu; $(+*)$ SO_4^{2-}
Cr	$(+*)Ca^{2+}$		$(+**)$Al, Fe, **Pb**, Th; $(-**)HCO_3^-$, K^+, Ca^{2+}, U	$(+**)$Fe, Co, Cu, As, Sn; $(+*)SO_4^{2-}$, **Pb**	$(+**)$Zn, **Pb**
Mn	$(+**)$**Al, Fe**, Co, Th	$(+**)$**Al**, $(+*)$**Fe**, Co, Zn $(-*)$ K^+,U,	$(+**)SO_4^{2-}$, Mg^{2+}, **Al, Fe**, Pb; $(+*)$Th; $(-**)K^+$; $(-*)Ca^{2+}$,U		$(+*)SO_4^{2-}$
Fe	$(+**)$**Al, Mn**, Co	$(+**)$**Mg^{2+}**, **Al, Co**; $(+*)$**Mn**, Zn; $(-**)K^+$, **Ca^{2+}**	$(+**)SO_4^{2-}$, **Mg^{2+}**, **Al**, Cr, **Mn**, Pb, Th; $(-**)HCO_3^-$, K^+, **Ca^{2+}**, U	$(+**)$ Cr, **Co**, Cu, As, Sn	$(+**)$ HCO_3^-, **Mg^{2+}**, **Al, Co**; $(+*)Na^+$, Zn; $(-**)$ **Ca^{2+}**
Co	$(+**)K^+$, **Al, Mn, Fe**	$(+**)$**Fe**, $(+*)SO_4^{2-}$, **Al**, Mn		$(+**)$Cr, **Fe**, Cu, Sn; $(+*)$As	$(+**)$**Al, Fe**; $(+*)$Cu
Cu	$(+**)$Cs		$(+**)$Sn, Pb	$(+**)$Al, Cr, Fe, Co, Sn; $(+*)SO_4^{2-}$, Zn	$(+**)$Al, $(+*)$Co, Cl^-
Zn	$(+**)$Cd, Pb	$(+**)Mg^{2+}$, Al, As; $(+*)$Mn, Zn, Cd; $(-*)Ca^{2+}$	$(-**)Cl^-$	$(+*)$Cu; $(-*)K^+$	$(+**)$Cr, Cs; $(+*)$Fe
As		$(+**)$Zn, Cd		$(+**)$Cr, Fe, Sn; $(+*)$Co	$(+*)HCO_3^-$, Pb
Cd	$(+**)$Zn, Sn, Pb; $(+*)K^+$	$(+**)$As; $(+*)$Zn	$(+*)Na^+$	$(+**)Cl^-$, Cs, Pb	$(+*)Ca^{2+}$
Sn	$(+**)$Cd, Pb	$(+**)$Th	$(+**)$Cu	$(+**)$Cr, Fe, Co, Cu, As	
Cs	$(+**)$Cu		$(+**)Na^+$	$(+**)Cl^-$, Cd, Pb	$(+**)$Zn, $(+*)$Pb
Pb	$(+**)$Zn, Cd, Sn; $(+*)K^+$		$(+**)SO_4^{2-}$, Al, **Cr**, Mn, Fe, Cu, Th $(+*)Mg^{2+}$; $(-**)K^+$, Ca^{2+}	$(+**)Cl^-$, Cd, Cs; $(+*)$**Cr**	$(+**)$**Cr**; $(+*)$As, Cs
Th	$(+**)$Mn	$(+**)$Sn; $(+*)$**Al**	$(+**)SO_4^{2-}$, Mg^{2+}, **Al**, Cr, Fe, Pb; $(+*)$Mn; $(-**)HCO_3^-$, K^+, Ca^{2+}, U	$(+**)$**Al**; $(+**)Mg^{2+}$	
U		$(+*)K^+$; $(-**)$Mn	$(+**)HCO_3^-$, K^+, Ca^{2+}, $(-**)SO_4^{2-}$, Mg^{2+}, Al, Cr, Fe, Th; $(-*)$Mn	$(-*)SO_4^{2-}$	

Notes: *—$p < 0.05$, **—$p < 0.01$. Elements that are correlated during 3–4 sampling campaigns are given in bold.

Table 5. Pollution indices (CF) in Irkutsk Reservoir water.

Trace Element	2007	2012	2021 (M)	2021 (J)	2021 (S)
Al	0.8–5.3	1.0–5.2	0.6–3.7	0.4–4.3	0.5–4.8
Cr	0.6–1.8	0.4–1.3	0.5–1.6	0.2–3.0	0.1–1.7
Mn	0.1–2.8	0.5–3.9	0.2–4.2	0.4–4.4	0.3–3.5
Fe	0.6–6.3	0.1–3.3	0.3–3.3	0.4–3.5	0.5–5.9
Co	0.8–2.5	0.5–4.2	0.1–10.0	0.6–3.5	0.6–1.9
Cu	0.4–1.7	0.3–1.7	0.4–3.0	0.3–1.7	0.4–1.5
Zn	0.1–0.9	0.4–1.7	0.4–1.6	0.5–1.4	0.5–2.8
As	0.9–1.4	0.7–1.0	0.7–1.2	0.9–1.3	0.7–1.2
Cd	0.2–0.8	0.6–1.3	0.5–1.4	0.6–4.7	0.5–1.6
Sn	0.2–0.7	0.2–6.3	0.2–2.1	0.2–3.2	0.2–2.7
Cs	0.1–0.9	0.0–0.6	0.1–0.4	0.3–6.9	0.2–1.7
Pb	0.1–2.2	0.2–2.3	0.0–1.1	0.0–2.1	0.3–3.6
Th	2.4–12.6	0.2–1.2	0.1–1.1	0.1–1.2	0.2–0.8
U	0.5–1.5	0.7–1.0	0.4–1.1	0.9–1.1	0.9–1.2

4. Conclusions

This study focused on the hydrochemical parameters of the Irkutsk Reservoir, which is a unique natural site due to it belonging to the unified Baikal–Angara freshwater system. The spatial–temporal dynamics in HCO_3^-, SO_4^{2-}, Cl^-, Ca^{2+}, Mg^{2+}, Na^+, and K^+ concentrations in the Irkutsk Reservoir showed that the major source of dissolved substances in the reservoir is Lake Baikal runoff. The comparison of Al, Cr, Mn, Fe, Co, Cu, Zn, As, Cd, Sn, Cs, Pb, Th, and U levels in the waters of Lake Baikal, Angara River source and Irkutsk Reservoir suggests that the water trace element composition of the reservoir also carries the main characteristics of Lake Baikal water. At the same time, spatial heterogeneity determined by local increases in the concentrations of trace elements in the reservoir water is highlighted. Transformation of Lake Baikal water is already evident in the vicinity of the Listvyanka settlement (Angara River source): the water samples taken here showed higher Al, Cr, Mn, Fe, Co, Cu, Cd, Sn, Cs, Pb, and Th concentrations. Downstream, from this settlement to the Irkutsk HPP dam, the frequently observable concentrations of Al, Cr, Mn, Co, As, Th, and U remained at a level close to their levels in the water of the Angara River source. The levels of Cu, Zn, Cd, Sn, Cs, and Pb were lower relative to the Angara River source, but they were still higher than the concentrations in the water of Lake Baikal. In the absence of significant industrial facilities along the shores of the Irkutsk Reservoir, the water trace element composition is mainly influenced by anthropogenically loaded areas and water transport. The concentrations of trace elements, mainly Al, Fe, Mn, Co, and Th, are greatly influenced by the shore abrasion processes, which became faster with the creation of the reservoir.

The concentrations of all major ions and trace elements in the surface and near-bottom waters of the Irkutsk Reservoir were substantially below the drinking water quality standards. Therefore, in terms of concentrations of components under study, the water is of high quality and may be used for drinking. At the same time, the pollution indices calculated relative to the Baikal water showed that the water samples from the Irkutsk Reservoir were contaminated with trace elements. By the CF values, considerable and very high contamination was found for Al, Fe, Mn, Fe, Co, Cu, Cr, Cd, Sn, Th, and Pb during different sampling campaigns. According to the PLI values, water samples from Uladova Bay (2012 and 2021, July) and Kurma Bay (2021, September) can be classified as polluted, while the samples from the middle and the right bank, in the vicinity of Irkutsk city, as highly polluted.

Our results show that dense constructions of houses along the reservoir coast, as well as the development of recreational infrastructure and navigation, have a significant impact on the water quality in the reservoir. Local higher concentrations of trace elements in the water of the Irkutsk Reservoir, which were not repeated in the interannual dynamics, suggest that the water of the Irkutsk Reservoir still copes with the anthropogenic load. The impact of anthropogenic activity could likely be more significant in the coastal areas of the reservoir. The lack of water treatment facilities makes the source of drinking water supply not protected from the anthropogenic impact. Human activity in the reservoir and on its coasts may affect the unique physical and chemical properties of the water, as well as the aquatic flora and fauna inhabiting the reservoir. All of the above underlines the need to take environmental protection measures aimed at preventing irreversible changes affecting the water quality and aquatic organisms, primarily endemic Lake Baikal species, poorly tolerant to habitat transformation.

Supplementary Materials: The following supporting information can be downloaded at: https://www.mdpi.com/article/10.3390/w15234142/s1, Table S1: Characterization of water sampling sites; Table S2: Spatial characterization of major ions and TDS in water of the Irkutsk Reservoir (unit in mg L^{-1}); Table S3: Comparisons of trace elements in water of channel part and bays of the Irkutsk Reservoir (unit in μg L^{-1}).

Author Contributions: Conceptualization and methodology, M.V.P. and V.I.P.; investigation, and data collection, M.V.P. and G.B.H.; writing—original draft preparation, M.V.P.; writing—review and editing, V.I.P., M.V.P. and G.B.H. All authors have read and agreed to the published version of the manuscript.

Funding: The work was supported by the Ministry of Science and Higher Education of the Russian Federation, grant No. 075-15-2020-787 (the project «Fundamentals, methods and technologies for digital monitoring and forecasting of the environmental situation on the Baikal natural territory»).

Data Availability Statement: The data presented in this study are available on request from the first author.

Acknowledgments: The author is grateful to Khomutova M. Yu. for editing the English version of the text.

Conflicts of Interest: The authors declare no conflict of interest.

References

1. Li, P.; Qian, H. Water resources research to support a sustainable China. *Int. J. Water Resour. Dev.* **2018**, *34*, 327–336. [CrossRef]
2. Lu, J.; Gu, J.; Han, J.; Xu, J.; Liu, Y.; Jiang, G.; Zhang, Y. Evaluation of Spatiotemporal Patterns and Water Quality Conditions Using Multivariate Statistical Analysis in the Yangtze River, China. *Water* **2023**, *15*, 3242. [CrossRef]
3. Carstens, D.; Amer, R. Spatio-temporal analysis of urban changes and surface water quality. *J. Hydrol.* **2019**, *569*, 720–734. [CrossRef]
4. Kumar, V.; Parihar, D.R.; Sharma, A.; Bakshi, P.; Preet Singh Sidhu, G.; Shreeya Bali, A.; Karaouzas, I.; Bhardwaj, R.; Thukral, A.K.; Gyasi-Agyei, Y.; et al. Global evaluation of heavy metal content in surface water bodies: A meta-analysis using heavy metal pollution indices and multivariate statistical analyses. *Chemosphere* **2019**, *236*, 124364. [CrossRef] [PubMed]
5. Kushwah, V.K.; Singh, K.R.; Gupta, N.; Berwal, P.; Alfaisal, F.M.; Khan, M.A.; Alam, S.; Qamar, O. Assessment of the Surface Water Quality of the Gomti River, India, Using Multivariate Statistical Methods. *Water* **2023**, *15*, 3575. [CrossRef]
6. Muhammad, S.; Ullah, I. Spatial and seasonal variation of water quality indices in Gomal Zam Dam and its tributaries of south Waziristan District, Pakistan. *Environ. Sci. Pollut. Res.* **2022**, *29*, 29141–29151. [CrossRef]
7. Cho, Y.-C.; Im, J.-K.; Han, J.; Kim, S.-H.; Kang, T.; Lee, S. Comprehensive Water Quality Assessment Using Korean Water Quality Indices and Multivariate Statistical Techniques for Sustainable Water Management of the Paldang Reservoir, South Korea. *Water* **2023**, *15*, 509. [CrossRef]
8. Kamani, H.; Hosseini, A.; Mohebi, S. Evaluation of water quality of Chahnimeh as natural reservoirs from Sistan region in southwestern Iran: A Monte Carlo simulation and Sobol sensitivity assessment. *Environ. Sci. Pollut. Res.* **2023**, *30*, 65618–65630. [CrossRef]
9. Dević, G. Environmental Impacts of Reservoirs. In *Environmental Indicators*; Armon, R., Hänninen, O., Eds.; Springer: Dordrecht, The Netherlands, 2015; pp. 561–575. [CrossRef]
10. Zhu, X.; Xu, Z.; Liu, Z.; Liu, M.; Yin, Z.; Yin, L.; Zheng, W. Impact of dam construction on precipitation: A regional perspective. *Mar. Freshw. Res.* **2022**, *74*, 877–890. [CrossRef]
11. Yin, L.; Wang, L.; Keim, B.D.; Konsoer, K.; Yin, Z.; Liu, M.; Zheng, W. Spatial and wavelet analysis of precipitation and river discharge during operation of the Three Gorges Dam, China. *Ecol. Indic.* **2023**, *154*, 110837. [CrossRef]

12. Zhao, Q.; Liu, S.; Deng, L.; Yang, Z.; Dong, S.; Wang, C.; Zhang, Z. Spatio-temporal variation of heavy metals in fresh water after dam construction: A case study of the Manwan Reservoir, Lancang River. *Environ. Monit. Assess.* **2012**, *184*, 4253–4266. [CrossRef] [PubMed]

13. Ochoa-Contreras, R.; Jara-Marini, M.E.; Sanchez-Cabeza, J.A.; Meza-Figueroa, D.M.; Pérez-Bernal, L.H.; Ruiz-Fernández, A.C. Anthropogenic and climate induced trace element contamination in a water reservoir in northwestern Mexico. *Environ. Sci. Pollut. Res.* **2021**, *28*, 16895–16912. [CrossRef] [PubMed]

14. Wiejaczka, Ł.; Prokop, P.; Kozłowski, R.; Sarkar, S. Reservoir's impact on the water chemistry of the Teesta River mountain course (Darjeeling Himalaya). *Ecol. Chem. Eng. S* **2018**, *25*, 73–88. [CrossRef]

15. Ali, H.; Khan, E.; Ilahi, I. Environmental chemistry and ecotoxicology of hazardous heavy metals: Environmental persistence, toxicity, and bioaccumulation. *J. Chem.* **2019**, *4*, 6730305. [CrossRef]

16. Griboff, J.; Wunderlin, D.A.; Horacek, M.; Monferrán, M.V. Seasonal variations on trace element bioaccumulation and trophic transfer along a freshwater food chain in Argentina. *Environ. Sci. Pollut. Res. Int* **2020**, *27*, 40664–40678. [CrossRef]

17. Okogwu, O.I.; Nwonumara, G.N.; Okoh, F.A. Evaluating Heavy Metals Pollution and Exposure Risk Through the Consumption of Four Commercially Important Fish Species and Water from Cross River Ecosystem, Nigeria. *Bull. Environ. Contam. Toxicol.* **2019**, *102*, 867–872. [CrossRef]

18. Garnero, P.L.; Bistoni, M.A.; Monferrán, M.V. Trace element concentrations in six fish species from freshwater lentic environments and evaluation of possible health risks according to international standards of consumption. *Environ. Sci. Pollut. Res.* **2020**, *27*, 27598–27608. [CrossRef]

19. Lomonosov, I.S.; Yanovsky, L.M.; Brukhanova, N. Major water quality indicators in Pribaikalye and their influence on man (Report 1). *Sib. Sci. Med. J.* **2009**, *3*, 110–113. (In Russian)

20. Vetrov, V.A.; Kuznetsova, A.I.; Sklyarova, O.A. Baseline Levels of Chemical Elements in the Water of Lake Baikal. *Geogr. Nat. Resour.* **2013**, *34*, 228–238. [CrossRef]

21. Nikolaeva, M.D. Hydrochemistry of the Angara River and the Irkutsk Reservoir. Ph.D. Thesis, State University of Irkutsk, Irkutsk, Russia, 1968. (In Russian).

22. Tarasova, E.N.; Mamontov, A.A.; Mamontova, E.A. Factors determining the the modern hydrochemical regime of Irkutsk reservoir. *Water Chem. Ecol.* **2015**, *7*, 10–17. (In Russian)

23. Karnaukhova, G.A. Hydrochemistry of the Angara and reservoirs of the Angara cascade. *Water Resour.* **2008**, *35*, 71–79. [CrossRef]

24. Ovchinnikov, G.I.; Pavlov, S.C.; Trzhtsinsky, U.B. *Change of Geological Environment in the Zones of Influence of the Angara-Yenisei Reservoirs*; Nauka: Novosibirsk, Russia, 1999; 254p. (In Russian)

25. Jaguś, A.; Rzetala, M.A.; Rzetala, M. Water storage possibilities in Lake Baikal and in reservoirs impounded by the dams of the Angara River cascade. *Environ. Earth Sci.* **2015**, *73*, 621–628. [CrossRef]

26. Hakanson, L. Ecological risk index for aquatic pollution control—A sedimentological approach. *Water Res.* **1980**, *14*, 975–1001. [CrossRef]

27. Tomlinson, D.L.; Wilson, J.G.; Harris, C.R.; Jeffrey, D.W. Problem in the assessment of heavy metals level in estuaries and the formation of a pollution index. *Helgol. Meeresunters* **1980**, *33*, 566–575. [CrossRef]

28. Khodzher, T.V.; Domysheva, V.M.; Sorokovikova, L.M.; Sakirko, M.V.; Tomberg, I.V. Current chemical composition of Lake Baikal water. *Inland Waters* **2017**, *7*, 250–258. [CrossRef]

29. Grebenshchikova, V.I.; Kuzmin, M.I.; Doroshkov, A.A.; Proydakova, O.A.; Tsydypova, S.B. The cyclicity in the changes in the chemical composition of the water source of the Angara River (Baikal Stock) in 2017–2018 in comparison with the last 20 years of data. *Environ. Monit. Assess.* **2019**, *191*, 728. [CrossRef]

30. Poletaeva, V.I.; Tirskikh, E.N.; Pastukhov, M.V. Hydrochemistry of sediment pore water in the Bratsk reservoir (Baikal region, Russia). *Sci. Rep.* **2021**, *11*, 11124. [CrossRef]

31. Poletaeva, V.I.; Dolgikh, P.G.; Pastukhov, M.V. Specifics of hydrochemical regime formation at the Ust-ilimsk water reservoir. *Water Chem. Ecol.* **2017**, *10*, 11–17. (In Russian)

32. Meybeck, M. Global occurrence of major elements in rivers. In *Treatise on Geochemistry*; Holland, H.D., Turekian, K.K., Eds.; Elsevier: Amsterdam, The Netherlands, 2003; Volume 5, pp. 207–223.

33. Berner, E.K.; Berner, R.A. *Global Environmental: Water, Air and Geochemical Cycles*; Prentice-Hall: Englewood Cliffs, NJ, USA, 1996; 376p.

34. Gaury, P.K.; Meena, N.K.; Mahajan, A.K. Hydrochemistry and water quality of Rewalsar Lake of Lesser Himalaya, Himachal Pradesh, India. *Environ. Monit. Assess.* **2018**, *190*, 84. [CrossRef]

35. Zhang, W.; Ma, L.; Abuduwaili, J.; Ge, Y.; Issanova, G.; Saparov, G. Hydrochemical characteristics and irrigation suitability of surface water in the Syr Darya River, Kazakhstan. *Environ. Monit. Assess.* **2019**, *191*, 572. [CrossRef]

36. Votintsev, K.K. *Hydrochemistry of Lake Baikal*; Publishing House of the Academy of Sciences of the USSR: Moscow, Russia, 1961; 311p. (In Russian)

37. Domysheva, V.M.; Sorokovikova, L.M.; Sinyukovich, V.N.; Onishchuk, N.A.; Sakirko, M.V.; Tomberg, I.V.; Zhuchenko, N.A.; Golobokova, L.P.; Khodzher, T.V. Ionic Composition of Water in Lake Baikal, Its Tributaries, and the Angara River Source during the Modern Period. *Russ. Meteorol. Hydrol.* **2019**, *44*, 687–694. [CrossRef]

38. Koval, P.V.; Udodov, Y.N.; Andrulaitis, L.D.; Gapon, A.; Sklyarova, O.E.; Chemigova, S.E. Hydrochemical characteristics of the surface runoff in Lake Baikal (1997–2003). *Dokl. Earth Sci.* **2005**, *401*, 452–455.

39. Karnaukhova, G.A. Mineralogical zoning of bottom sediments of the Irkutsk reservoir under conditions of unstable level regime. *Proc. Fersman Sci. Sess. GI KSC RAS* **2019**, *16*, 250–254. (In Russian)
40. Li, S.; Ye, C.; Zhang, Q. 11-Year change in water chemistry of large freshwater Reservoir Danjiangkou, China. *J. Hydrol.* **2017**, *551*, 508–517. [CrossRef]
41. Kuzmin, M.I.; Tarasova, E.N.; Mamontova, E.A.; Mamontov, A.A.; Kerber, E.V. Seasonal and interannual variations of water chemistry in the headwater streams of the Angara River (Baikal) from 1950 to 2010. *Geochem. Int.* **2014**, *52*, 523–532. [CrossRef]
42. Silow, E.A. Lake Baikal: Current Environmental Problems. In *Encyclopedia of Environmental Management*; Taylor and Francis: New York, NY, USA, 2014; pp. 1–9.
43. Ongley, E.D.; Zhang, X.; Yu, T. Current status of agricultural and rural non-point source Pollution assessment in China. *Environ. Pollut.* **2010**, *158*, 1159–1168. [CrossRef]
44. Gaillardet, J.; Viers, J.; Dupré, B. Trace Elements in River Waters. In *Treatise on Geochemistry, Surface and Groundwater, Weathering and Soils*; Drever, J.I., Ed.; Elsevier: Amsterdam, The Netherlands, 2003; pp. 225–272.
45. Alieva, V.I.; Grebenshchikova, V.I.; Zagorulko, N.A. Long-term monitoring and modern methods for studying the microelement composition of the waters of the Angara River. *Eng. Ecol.* **2011**, *3*, 24–34. (In Russian)
46. Sklyarova, O.A. Distribution of trace elements in the water column of middle Baikal. *Geogr. Nat. Resour.* **2011**, *32*, 34–39. [CrossRef]
47. Poletaeva, V.I.; Pastukhov, M.V.; Tirskikh, E.N. Dynamics of Trace Element Composition of Bratsk Reservoir Water in Different Periods of Anthropogenic Impact (Baikal Region, Russia). *Arch. Environ. Contam. Toxicol.* **2021**, *80*, 531–545. [CrossRef]
48. SanPiN 2.1.4.1074-01. Drinking Water Hygienic Requirements for Water Quality of Centralized Drinking Water Supply Systems. Quality Control. Hygienic Requirements for Provision of Safety of Hot Water Supply Systems. Available online: https://www.mast.is/static/files/library/Regluger%25C3%25B0ir/Russland/SanPin%25202_1_4_1074-01_ForWater.pdf (accessed on 2 November 2023).
49. Sklyarov, E.V.; Sklyarova, O.A.; Lavrenchuk, A.V.; Menshagin, Y. Natural pollutants of Northern Lake Baikal. *Environ. Earth Sci.* **2015**, *74*, 2143–2155. [CrossRef]
50. Koval, P.V.; Udodov, Y.N.; Andrulaitis, L.D.; San'kov, V.A.; Gapon, A.E. Mercury in the source of the Angara river: Fiver-year concentration trend and possible reasons of its variations. *Dokl. Earth Sci.* **2003**, *389*, 282–285.
51. Belozertseva, I.A.; Vorobyeva, I.B.; Vlasova, N.V.; Lopatina, D.N.; Yanchuk, M. Snow pollution in Lake Baikal water area in nearby land areas. *Water Resour.* **2017**, *44*, 471–484. [CrossRef]
52. Falkner, K.K.; Church, M.; Measures, C.I.; Lebaron, G.; Thouron, D.; Jeandel, C.; Stordal, M.C.; Gill, G.A.; Mortlock, R.; Froelich, P.; et al. Minor and Trace Element Chemistry of Lake Baikal, Its Tributaries and Surrounding Hot Springs. *Limnol. Oceanogr.* **1997**, *42*, 329–345. [CrossRef]
53. Pereira, T.L.; Wallner-kersanach, M.; Costa, L.D.; Costa, D.P.; Baisch, P.R. Nickel, vanadium, and lead as indicators of sediment contamination of marina, refinery, and shipyard areas. *Environ. Sci. Pollut. Res.* **2017**, *25*, 1719–1730. [CrossRef]
54. Valero, A.; Umbría-Salinas, K.; Wallner-Kersanach, M.; de Andrade, C.F.; Yabe, M.J.S.; Contreira-Pereira, L.; Wasserman, J.C.; Kuroshima, K.N.; Zhang, H. Potential availability of trace metals in sediments in southeastern and southern Brazilian shipyard areas using the DGT technique and chemical extraction methods. *Sci. Total. Environ.* **2020**, *710*, 136216. [CrossRef]
55. Umbría-Salinas, K.; Valero, A.; Wallner-Kersanach, M.; de Andrade, C.F.; Yabe, M.J.S.; Wasserman, J.C.; Kuroshima, K.N.; Zhang, H. Labile metal assessment in water by diffusive gradients in thin films in shipyards on the Brazilian subtropical coast. *Sci. Total Environ.* **2021**, *775*, 145184. [CrossRef]
56. Suturin, A.N.; Chebykin, E.P.; Malnik, V.V.; Khanaev, I.V.; Minaev, A.V.; Minaev, V.V. The role of anthropogenic factors in the development of ecological stress in lake Baikal littoral (the Listvyanka settlement lakescape). *Geogr. Nat. Resour.* **2016**, *6*, 43–54. [CrossRef]
57. Vorobyeva, I.B.; Naprasnikova, E.V.; Vlasova, N.V. Ecological and geochemical features of snow, ice and under ice water in the southern part of lake Baikal. *Geoecol. Eng. Geol. Hydrogeol. Geocryol.* **2009**, *1*, 54–60.
58. Gorme, J.B.; Maniquiz, M.C.; Song, P.; Kim, L.-H. The water quality of the Pasig River in the City of Manila, Philippines: Current status, management and future recovery. *Environ. Eng. Res.* **2010**, *15*, 173–179. [CrossRef]
59. Giri, S.; Singh, A.K. Assessment of Surface Water Quality Using Heavy Metal Pollution Index in Subarnarekha River, India. *Water Qual Expo. Health* **2014**, *5*, 173–182. [CrossRef]
60. Wu, J.; Lu, J.; Zhang, C.; Zhang, Y.; Lin, Y.; Xu, J. Pollution, sources, and risks of heavy metals in coastal waters of China. *Hum. Ecol. Risk Assess. Int. J.* **2019**, *26*, 2011–2026. [CrossRef]
61. Zhaoyong, Z.; Xiaodong, Y.; Shengtian, Y. Heavy metal pollution assessment, source identification, and health risk evaluation in Aibi Lake of northwest China. *Environ Monit Assess* **2018**, *190*, 69. [CrossRef] [PubMed]
62. Xia, X.; Chen, X.; Liu, R.; Liu, H. Heavy metals in urban soils with various types of land use in Beijing, China. *J. Hazard. Mater.* **2011**, *186*, 2043–2050. [CrossRef] [PubMed]
63. Solpina, N.G.; Cherkashina, A.A. Erosion processes on the banks of the Irkutsk reservoir and their consequences. *Bull. Irkutsk. State University. Ser. Earth Sci.* **2020**, *33*, 124–136. [CrossRef]
64. Karnaukhova, G.A.; Shtel'makh, S.I. Geochemical heterogeneities of lithosphere and hydrosphere in the Irkutsk Reservoir as the indicator of the geo-ecological state. *Limnol. Freshw. Biol.* **2020**, *4*, 849–850. [CrossRef]

65. Ammar, R.; Kazpard, V.; Wazne, M.; Samrani, A.G.; Nabil, A.; Saad, Z.; Chou, L. Reservoir sediments: A sink or source of chemicals at the surface water-groundwater interface. *Environ. Monit. Assess.* **2015**, *187*, 579. [CrossRef]
66. Pavoni, E.; Crosera, M.; Petranich, E.; Faganeli, J.; Klun, K.; Oliveri, P.; Covelli, S.; Adami, G. Distribution, Mobility and Fate of Trace Elements in an Estuarine System Under Anthropogenic Pressure: The Case of the Karstic Timavo River (Northern Adriatic Sea, Italy). *Estuaries Coasts* **2021**, *44*, 1831–1847. [CrossRef]
67. Pastukhov, M.V.; Poletaeva, V.I.; Tirskikh, E.N. Long-term dynamics of mercury pollution of the Bratsk reservoir bottom sediments, Baikal region, Russia. *IOP Conf. Ser. Earth Environ. Sci.* **2019**, *321*, 012041. [CrossRef]
68. Jaguś, A.; Khak, V.A.; Rzetala, M.A.; Mariusz, R. Trace Elements in the Bottom Sediments of the Irkutsk Reservoir. *Ecol. Chem. Eng. A* **2012**, *19*, 939–950. [CrossRef]
69. Caballero-Gallardo, K.; Alcala-Orozco, M.; Barraza-Quiroz, D.; De la Rosa, J.; Olivero-Verbel, J. Environmental risks associated with trace elements in sediments from Cartagena Bay, an industrialized site at the Caribbean. *Chemosphere* **2019**, *242*, 125173. [CrossRef]
70. Zhang, H.; Jiang, Y.; Wang, M.; Wang, P.; Shi, G.; Ding, M. Spatial characterization, risk assessment, and statistical source identification of the dissolved trace elements in the Ganjiang River-feeding tributary of the Poyang Lake, China. *Environ. Sci. Pollut. Res. Int.* **2017**, *24*, 2890–2903. [CrossRef] [PubMed]
71. Hussain, J.; Dubey, A.; Hussain, I.; Arif, M.; Shankar, A. Surface water quality assessment with reference to trace metals in River Mahanadi and its tributaries, India. *Appl. Water Sci.* **2020**, *10*, 193. [CrossRef]
72. Poletaeva, V.I. Angara river hydrochemical variability when biulding the Boguchany reservoir (Russia). *Bull. Tomsk. Polytech. Univ. Geo Assets Eng.* **2022**, *333*, 146–158. [CrossRef]
73. Wetzel, R.G. *Limnology: Lakes and River Ecosystems*; Academic Press: London, UK, 2001; 1006p.
74. Senze, M.; Kowalska-Góralska, M.; Czyż, K. Aluminum Bioaccumulation in Reed Canary Grass (*Phalaris arundinacea* L.) from Rivers in Southwestern Poland. *Int. J. Environ. Res. Public Health* **2022**, *19*, 2930. [CrossRef]
75. Gałuszka, A.; Migaszewski, Z.M. Geochemical background—An environmental perspective. *Mineralogia* **2011**, *42*, 7–17. [CrossRef]

Article

Hydrochemical Characteristics and Human Health Risk Assessment of Surface Water in the Danjiang River Source Basin of the Middle Route of China's South-to-North Water Transfer Project

Longjian Lin [1], Yafeng Zhang [2,*], Xinyu Qian [2] and Yingwei Wang [2]

1 School of Environmental Science and Engineering, Shaanxi University of Science & Technology, Xi'an 710021, China; longjl0130@163.com
2 Shaanxi Mineral Resources and Geological Survey, Xi'an 710068, China
* Correspondence: aimom84@163.com

Abstract: The Danjiang River basin is an important water source for the Middle Route of the South-to-North Water Diversion Project. With the shortage of water resources and the increase in pollution pressure, it is of great significance to study the hydrochemical characteristics of surface water in the Danjiang River basin and the risk to human health posed by polluted water bodies for the protection and utilization of water resources. In this paper, 40 surface water samples were collected and analyzed by innovatively adopting the sampling principle of "geological structure unit + landform unit + small watershed unit". Comprehensive mathematical statistical analysis, Piper trilinear diagrams, Gibbs diagrams, and ion ratio coefficients were used to analyze the hydrochemical composition, spatial distribution characteristics and influencing factors of surface water in the Danjiang River. The entropy weight comprehensive index method (EWQI) and the health risk assessment model recommended by the United States Environmental Protection Agency (HHRA model) were used to evaluate the water quality and potential non-carcinogenic risk of surface water in the Danjiang River source basin. The results showed that the pH of surface water in the study area was 7.02~8.77, with an average value of 8.26; and the TDS was ranged from 134 to 388 mg/L, with an average value of 252.75 mg/L. The main cations in the surface water were Ca^{2+} and Mg^{2+}, accounting for 71% and 20% of the total cations, respectively, while the main anions were HCO_3^- and SO_4^{2-}, accounting for 74% and 19% of the total anions, respectively. The hydrochemical type was HCO_3^- Ca·Mg. The hydrochemical genesis was mainly controlled by the weathering of carbonate rocks, while some ions were influenced by the weathering of silicate rocks, and human activities were also an important factor affecting the chemical characteristics of the water. The EWQI of surface water in the whole region was 8.95~25.69, and the health risk index (HI) of nitrate pollution ranged from 0.0122 to 0.2118, in which the HI ranges for children and adults were 0.0217~0.2118 and 0.0122~0.1333, respectively, indicating that the water quality of the entire study area met the Class I water standards, and the potential non-carcinogenic risk of nitrate was low. However, its impact on children was significantly higher than on adults, so it is recommended to monitor the water quality downstream of urban areas in the study area to reduce agricultural non-point source pollution and urban domestic sewage discharge and thereby reduce the potential health risks for young populations.

Keywords: hydrochemical characteristics; rock weathering; water quality evaluation; human health risk assessment; Danjiang River source

Citation: Lin, L.; Zhang, Y.; Qian, X.; Wang, Y. Hydrochemical Characteristics and Human Health Risk Assessment of Surface Water in the Danjiang River Source Basin of the Middle Route of China's South-to-North Water Transfer Project. *Water* **2023**, *15*, 2203. https://doi.org/10.3390/w15122203

Academic Editor: Dimitrios E. Alexakis

Received: 4 May 2023
Revised: 1 June 2023
Accepted: 4 June 2023
Published: 12 June 2023

1. Introduction

The hydrochemical composition of groundwater and river water is the result of the long-term interaction between water and the surrounding environment in the process of circulation [1–4]. The chemical composition is crucial in characterizing water environment quality, regional environmental chemical characteristics, water element distribution, and migration and transformation [5–7]. Analyzing the chemical composition and spatial variation in river water can provide vital information on the geochemical behavior of elements, rock weathering, and human activities in the basin, which is essential for the rational utilization and protection of surface water resources [8–10].

Scholars have utilized various analytical methods, including mathematical statistics, correlation analysis, simulation calculation, Piper three-line diagrams, Gibbs diagrams, ion proportional coefficient methods, and spatial variation characteristics analysis, to study the hydrochemical characteristics of major rivers worldwide and their relationship with climatic conditions and regional geological lithology [11,12]. Notable studies have been conducted on rivers such as the Amazon [13], Mackenzie [14], Lena [15], and Brahmani [16]. In China, the study of river water chemistry dates can be traced to the 1960s, when the focus was on the various types of river water chemistry in different regions of China. Since then, many scholars have studied the hydrochemical characteristics of major basins, lakes, reservoirs, wetlands, and springs in China, uncovering their formation processes and evolutionary patterns [17–23]. It has been demonstrated that chemical weathering of rocks is the primary natural control mechanism of the hydrochemical characteristics of rivers in China [24]. However, the relative contributions of various controlling factors of dissolved substances in river water, such as tectonic movement, lithology, landform, climate, hydrological characteristics, and vegetation types, vary under different climatic and lithological conditions [25]. Generally, the hydrochemical characteristics are primarily determined by the geological environmental factors of the basin. Furthermore, the study of river water hydrochemistry can provide insight into the types of rock that are weathering and the chemical weathering processes in the basin [19,26].

Excessive nitrate nitrogen in drinking water has been linked to gastric cancer and methemoglobinemia of infants in various studies [27,28]. Therefore, developing a Human Health Risk Assessment (HHRA) model based on water quality assessment is crucial to quantitatively evaluate the harm caused by nitrate nitrogen-polluted and gain a deeper understanding of its intrinsic link to human health [29]. Among the water quality evaluation models, the entropy weight comprehensive index (EWQI) is widely recognized as the most unbiased reflection of the real weight of each parameter [30]. Therefore, it is commonly employed for water quality assessment. The United States Environmental Protection Agency has released the Health Risk Assessment Model (HHRA model) for this purpose [31]. For instance, Wang et al. assessed the nitrate pollution of groundwater in Zhangjiakou City [32], Li et al. evaluated the health risk of nitrate in groundwater in the Songnen Plain [33], and Banajarani et al. [34] assessed the health risk of nitrate in aquifers in the foothills of the Western Ghats in southern India, and the corresponding water quality management recommendations and measures were proposed.

The Danjiang River, which originates in the southern foothills of the Qinling Mountains in the northwest of Shangluo City, Shaanxi Province, is the largest tributary of the Hanjiang River and a crucial water conservation area of the Middle Route of the South-to-North Water Diversion Project. Its water quantity and quality are critical to the sustainable development of the project and directly impact the water safety of Beijing, Tianjin, and Hebei. Previous studies have focused on the ecological environment, land use, and hydrology of the downstream in Danjiang River, specifically the Danjiangkou area [35–43]. However, in previous studies, the study of hydrochemical characteristics and water quality evaluation only consisted of macro-scale analysis based on the monitoring results of hydrological stations, without controlling for changes in water quality at the small watershed scale in the water source area. Moreover, the geological background and other factors were ignored in the study of the influencing factors of hydrochemical composition. At the same time, with the development of the economy and society, the pressure on the prevention and control of water resource pollution and water resource shortage in water source areas has increased. It is particularly important to obtain a comprehensive scientific understanding of the chemical composition and influencing factors of water in the area, objectively evaluate water quality, clarify water health risks and identify the sources of pollutants. Therefore, this study focuses on the surface water of the Danjiang small watershed sources. We examine the water chemistry and its influencing factors such as geology, geomorphology and hydrological process, evaluate the water quality, and quantitatively assess the risk of nitrate nitrogen pollution to human health using the HHRA model. The study aims to provide a theoretical foundation for the prevention and control of water resource pollution and risk control of drinking water health in the Danjiang source basin, thereby ensuring the safe operation of the Middle Route of the South-to-North Water Diversion Project.

2. Materials and Methods

2.1. Study Area

The study area is located in the northern region of Shangluo City, Shaanxi Province, at the source of the Danjiang River basin in the southern foothills of the Qinling Mountains. This region is an essential water source conservation area for the Middle Route Project of the South-to-North Water Diversion, with an elevation range of 699 to 1684 m and a topographical inclination towards the Danjiang River valley (Figure 1). It can be divided into three primary geomorphic units: the middle mountain area, low mountain and hilly area, and river valley and plateau area. The region comprises a well-developed river system, which is a typical mountain rainfed river system, with an average annual runoff of 482 million m^3 and an uneven distribution throughout the year. The runoff is mostly concentrated between July and October, with this period accounting for 73 to 82% of the annual total. The region experiences a continental humid climate in the southern margin of the warm temperate zone that is characterized by moderate heat and abundant rainfall. The area's average annual precipitation is 699.44 mm, and the average annual temperature is 12.8 °C (source: http://data.cma.cn/, accessed on 21 September 2022). The precipitation and temperature exhibit considerable vertical differences due to the topography. The mountainous area records an average annual rainfall of 798.8 mm and an average annual temperature of 10 to 11 °C, while the valley and plateau area records an average annual rainfall below 730 mm and an average annual temperature above 13 °C.

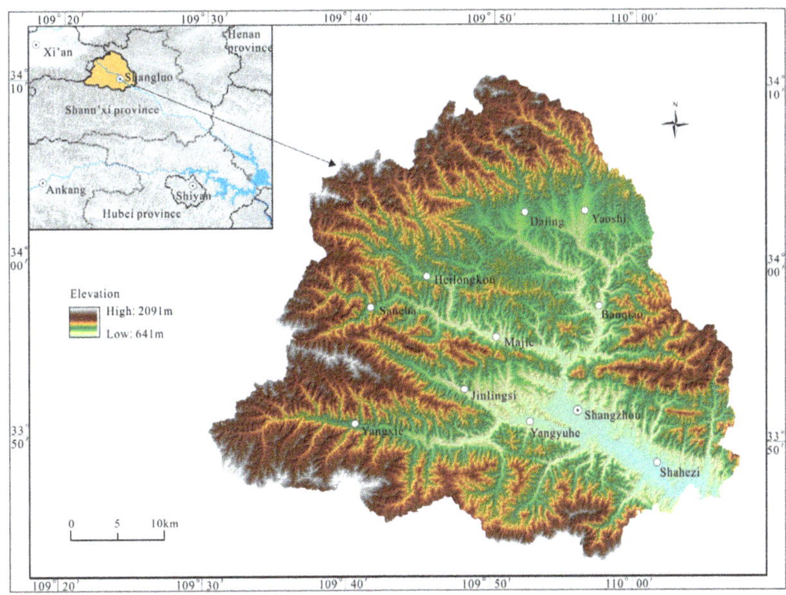

Figure 1. Location and digital elevation model (DEM) of the study area.

The study area is located at the intersection of the North China plate and the Yangtze plate and has undergone a lengthy and complex tectonic evolution [44]. As a result, the area comprises diverse rock assemblages, including Quaternary loose sediment, Cretaceous-Paleogene clastic rock, Paleozoic basic gabbro, Neoproterozoic-Triassic acid granite, Mesoproterozoic-Paleozoic basic volcanic rock, Meso-Neoproterozoic metamorphic clastic rock, Ordovician light metamorphic clastic rock, and Proterozoic-Ordovician carbonate rock (Figure 2). The resident population of the district was 472,978 in 2020. In terms of gender, the male population was 242,300 and the female population was 230,678. In terms of age distribution, the population aged 0–14 was 91,413, accounting for 19.33%, the population aged 15–59 was 290,497, accounting for 61.42%, and the population aged above 60 was 91,068, accounting for 19.25%, of whom 58,644 were aged above 65, accounting for 12.40% of the total population. The cultivated land in the area is concentrated mainly along the river valley terraces of the Danjiang River and its tributaries, with a total area of 28,742 hectares, accounting for approximately 10.76% of the total area.

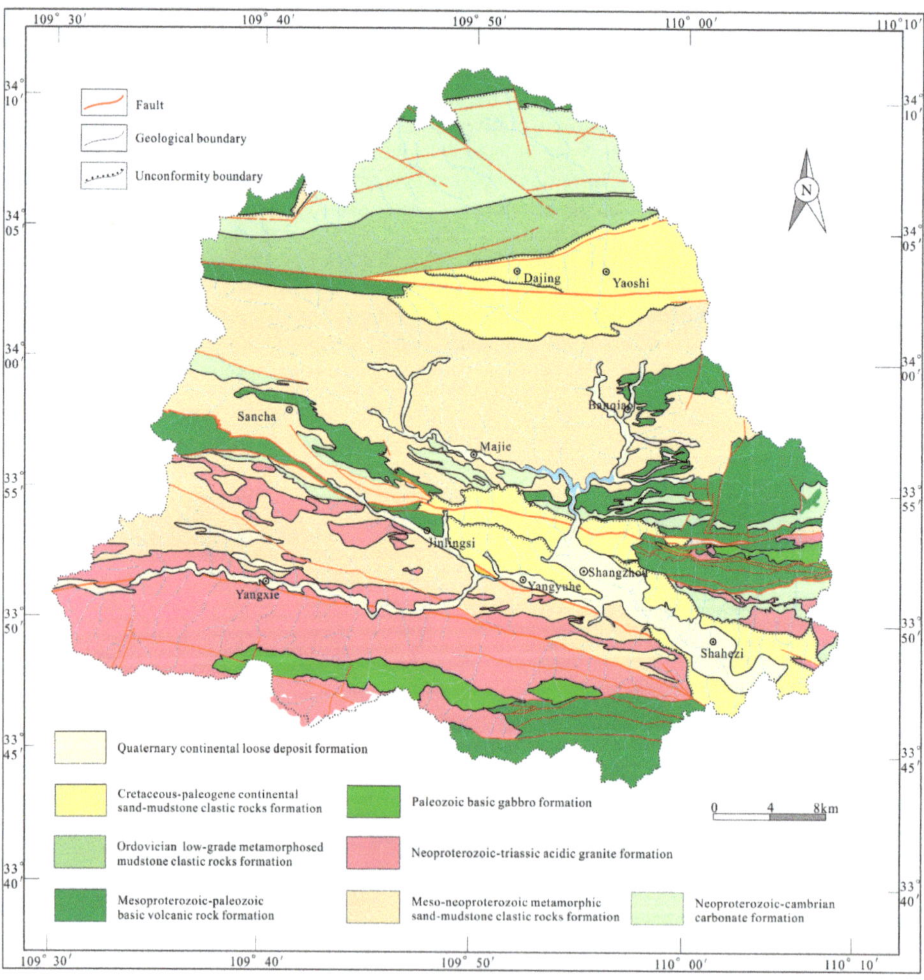

Figure 2. Map showing the geological formations and sampling sites of the Danjiang River source basin.

2.2. Sample Collection and Analysis

To comprehensively understand the hydrochemical composition, spatial variability, and water quality characteristics of the Danjiang River source basin, as well as to investigate the hydrochemical characteristics of different geological structures, landform units, and different sub-watersheds, the source region of the Danjiang river was divided into 37 small watersheds using the function of hydrological analysis module in ArcGIS. This division was based on the sampling principle of "geological structure unit + landform unit + small watershed unit". Forty surface water samples were collected in November 2021, with sampling points covering the geological structures, landform units, and small watershed units in the study area (Figure 3). After being rinsed three times with water, 5 L polyethylene plastic bottle was used to collect the samples, and the sampling depth was generally 10 cm below the water surface. At the time of sampling, the GPS location and surrounding environment of each sampling site were recorded. The on-site parameters such as pH and water temperature (T) were measured using a portable multi-parameter water quality analyzer, while Ca^{2+}, Mg^{2+}, Na^+, K^+ and other cations were measured using inductively coupled plasma atomic emission spectrometry (ICP-OES). HCO_3^- was measured using hydrochloric acid titration, and Cl^-, NO_3^-, SO_4^{2-} and other anions were measured using ion chromatography (ICS-1100). Total dissolved solids (TDS) were determined using the drying method. The detection limits for K^+, Na^+, Mg^{2+}, Cl^-, and SO_4^{2-} were 0.02 mg/L, 0.01 mg/L, 0.05 mg/L, 1.0 mg/L, and 0.05 mg/L, respectively. Standard samples, parallel samples and blank samples were regularly inserted in the sample analysis, and the accuracy was checked by anion and cation balance method to ensure the absolute error was within 5%. After inspection, the test results all met the quality requirements.

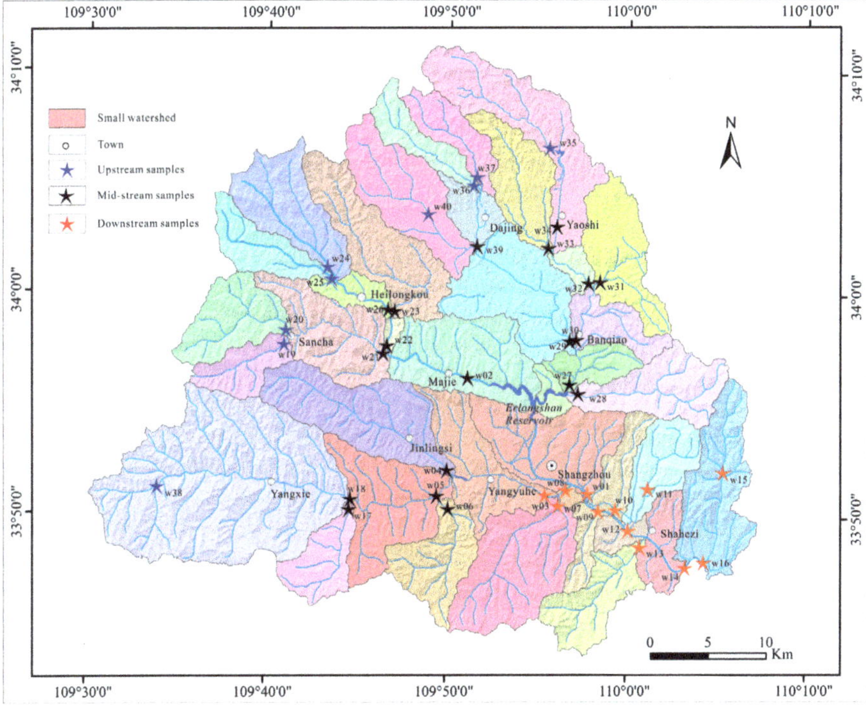

Figure 3. Map showing the drainage distribution and sampling locations in the Danjiang River source basin.

2.3. Research Methods

In the Danjiang River source zone, the hydrochemical parameters of surface water were measured and analyzed using statistical and ion correlation analysis methods in software packages such as SPSS v. 20 (IBM Corp., Armonk, NY, USA).Hydrochemical characteristics and controlling factors were examined using the Piper three-line diagram, Gibbs diagram, and ion ratio.

The entropy weight comprehensive index (EWQI) method was employed for water quality evaluation. Based on the statistical distribution characteristics of surface water chemical components and typical surface water pollution in the Danjiang River source basin, sodium, iron, manganese, sulfate, chloride, fluoride, total dissolved solids (TDS), chemical oxygen demand (COD), and nitrate nitrogen (NO_3-N) were selected as water quality analysis indicators. These indicators were used to establish an initial evaluation index matrix, determine the weight of each evaluation index, and calculate the comprehensive evaluation index (EWQI). The EWQI consolidates multiple indicators in the water quality evaluation into a single indicator that measures the overall quality of the environment. The EWQI value reflects the pollution degree of each sample. Further details on the evaluation process and water quality classification standard can be found in [45].

To quantitatively evaluate the potential health risk of nitrate nitrogen to surrounding water users through the oral intake of surface water, we used the health risk assessment model (HHRA) developed by the United States Environmental Protection Agency [31]. The population was divided into three groups based on age and gender characteristics: children (0–14 years old), adult males (>14 years old), and adult females (>14 years old). The evaluation model was:

$$HI = I_{CD}/D_{Rf} \tag{1}$$

where *HI* is the non-carcinogenic risk index of nitrate nitrogen (dimensionless), I_{CD} is the daily average exposure dose (mg·(kg·d)$^{-1}$), and D_{Rf} is the reference dose of nitrate nitrogen through drinking water intake, as given in Table 1. The non-carcinogenic risk threshold recommended by USEPA is 1. If *HI* < 1, the non-carcinogenic health risk is controllable; if 1 < *HI* < 4, there is a moderate risk of nitrate nitrogen; and if *HI* > 4, the non-carcinogenic health risk caused by pollution is uncontrollable. Moreover, as the *HI* value increases, the non-carcinogenic risk increases.

According to the definition of USEPA, there are three potential contact pathways for human health risk assessment, including direct drinking water intake, air inhalation and skin absorption. The target pollutant nitrate nitrogen in water mainly enters the human body through drinking water intake, while the dose of nitrate nitrogen inhaled through skin and air contact is less than one thousandth of that absorbed by drinking water [46]. Therefore, in this study, only the potential risk of nitrate nitrogen intake through drinking water to the human body was considered. The formula for calculating the daily average exposure dose (I_{CD}) is as follows:

$$I_{CD} = \frac{C \times IR \times EF \times ED}{BW \times AT} \tag{2}$$

$$AT = ED \times 365 \tag{3}$$

In the formula, *C* is the concentration of nitrate nitrogen in surface water (mg/L), *IR* is the daily drinking water intake, *EF* is the exposure frequency, *ED* is duration, *BW* is the average weight of residents, and *AT* is the average exposure time.

Table 1. Parameters employed for human health risk assessment [47].

Parameter	Meaning	Unit	Values for Different Age Groups		
			Children	Adult Males	Adult Females
D_{Rf}	Reference dose	mg/kg·d	1.6	1.6	1.6
IR	Ingestion rate	L/d	1.8	2.79	2.20
BW	Average body weight	kg	25.9	63.8	56.5
AT	Average exposure time	d	2190	10,950	10,950
ED	Exposure duration	a	6	30	30
EF	Exposure frequency	d/a	365	365	365

3. Results and Analysis

3.1. Hydrochemical Characteristics

The statistical analysis of 40 chemicals in surface waters in the Danjiang River source basin is shown in Table 2.

Table 2. Statistics of the main surface water quality indices in the Danjiang River source zone (unit: mg/L except pH).

Reaches	Statistic	pH	K^+	Na^+	Ca^{2+}	Mg^{2+}	Cl^-	SO_4^{2-}	HCO_3^-	TDS	NO_3-N	Total Alkalinity
Upper reaches	Min	8.00	0.87	1.47	36.64	4.94	1.74	9.76	95.87	134.00	0.50	78.63
	Max	8.77	2.32	3.89	61.07	28.39	6.95	58.55	287.62	272.00	2.64	235.89
	Mean	8.34	1.55	2.65	51.40	16.51	3.48	40.86	183.51	215.13	1.28	152.96
	SD	0.24	0.57	0.88	7.11	8.50	2.08	21.50	71.04	40.61	0.80	58.53
Middle reaches	Min	8.07	0.95	2.06	36.64	7.41	3.48	4.88	119.84	154.00	0.95	98.29
	Max	8.49	2.91	11.40	81.43	29.63	24.33	112.21	293.62	388.00	4.18	240.81
	Mean	8.28	1.64	5.19	60.21	17.54	8.51	48.79	207.52	257.95	2.51	171.75
	SD	0.13	0.48	2.84	10.84	7.31	4.63	24.12	56.43	53.16	0.99	44.49
Lower reaches	Min	7.02	0.91	4.24	38.68	3.70	3.48	14.64	77.90	150.00	1.25	63.89
	Max	8.57	2.51	10.56	73.28	24.69	15.64	121.97	245.68	350.00	4.88	211.32
	Mean	8.17	1.68	7.96	61.85	15.67	8.69	59.67	192.21	268.31	3.28	160.66
	SD	0.39	0.45	1.94	9.57	6.92	3.40	24.24	48.71	55.93	1.20	39.69

In the study area, the pH of surface water ranged from 7.02 to 8.77, with an average value of 8.26 and a variation coefficient of 3.19%. Although pH was relatively stable within the basin, it exhibited a slight decrease in the downstream direction. The average level of total dissolved solids (TDS) was 252.75 mg/L, ranging from 134 to 388 mg/L. This level was lower than the global average of 283 mg/L reported by [48], with a coefficient of variation of 21.48%. Notably, changes in TDS were more apparent than those in pH, and they showed an increasing trend from upstream to downstream and from tributaries to the mainstream. Therefore, the water body in the study area can be classified as a low-salinity alkaline environment.

The surface water was found to have a cation equivalent concentration mean value sequence of $Ca^{2+} > Mg^{2+} > Na^+ > K^+$. Ca^{2+} was identified as the dominant cation, with concentrations ranging from 36.64 to 81.43 mg/L and an average value of 58.98 mg/L, accounting for 65–86% of the total cation concentration, with an average value of 71%. The coefficient of variation was 17.52%, indicating a low degree of variation. Mg^{2+} was the second most abundant cation, with concentrations ranging from 3.70 to 29.63 mg/L and an average value of 16.73 mg/L. This accounted for 9–24% of the total cation concentration, with an average value of 20% and a variation coefficient of 43.55%. These findings suggest that the concentration of Ca^{2+} was more significant than that of Mg^{2+} in the basin.

Regarding anions, the average concentration order was $HCO_3^- > SO_4^{2-} > NO_3^- > Cl^-$, and the dominant anions were HCO_3^- and SO_4^{2-}. The HCO_3^- concentrations ranged from 77.90 to 293.62 mg/L with an average value of 197.74 mg/L. This represented 64% to 90% of the total anion concentration, with an average value of 74% and a variation coefficient of 28.62%. The SO_4^{2-} concentrations ranged from 4.88 to 121.97 mg/L, with an average of 50.74 mg/L. This accounted for 6% to 26% of the total anion concentration, with an average value of 19%, and a variation coefficient 47.50%.

The Piper diagram is a valuable tool that presents the main ion composition and hydrochemical features of water related to cations (Na^+, K^+, Ca^{2+}, Mg^{2+}) and anions (Cl^-, SO_4^{2-}, CO_3^{2-}, HCO_3^-) in terms of milligram equivalent percentages [49]. Figure 4 shows that the surface water in the Danjiang River source basin had a high concentration of Ca^{2+} in the lower left corner of the cation map closer to the magnesium type area, indicating that Ca^{2+} was the primary cation followed by Mg^{2+}. Additionally, the anion diagram shows that all the water sample points were biased towards the lower left corner axis, indicating that HCO_3^- ions were the dominant anion. Based on these findings, the HCO_3^- Ca·Mg type is the main hydrochemical category for the Danjiang River source basin.

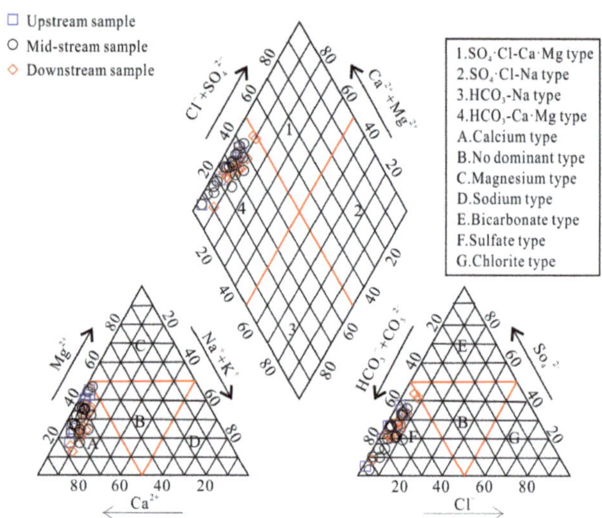

Figure 4. Piper diagram of surface water chemical composition in the Danjiang River source zone.

3.2. Spatial Variability Characteristics of Hydrochemistry

To investigate the spatial variability of hydrochemisty in the water bodies in the Danjiang River source zone, 10 surface water samples were collected and analyzed for their physicochemical properties. The samples were classified as upstream (W24 and W25), mid-stream (W26, W22, and W02), and downstream (W01, W10, W12, W14, and W16) (Figure 5). While pH and K^+ values remained relatively stable with low variation, SO_4^{2-} showed fluctuations. Moving from upstream to downstream, Na^+, Ca^{2+}, Mg^{2+}, Cl^-, NO_3-N, and HCO_3^- showed similar variation patterns with slight upward trends, indicating higher concentrations in the downstream samples. Notably, TDS exhibited a significant uptrend, which could be attributed to surface water runoff and mixing with other sources. Moreover, the concentrations of Ca^{2+} and Mg^{2+} remained relatively stable. It can be concluded that precipitation of calcium and magnesium minerals may have occurred due to enhanced evaporation and water-rock interactions in the middle and lower reaches, considering the significant upward trend of HCO_3^-. Loose rock mineral infiltration and discharge of industrial and agricultural water alter the ion concentration in the main stream of the Danjiang River source in Shangzhou District-Shahezi Town. Cl^- and SO_4^{2-} exhibited greater variation and larger ion coefficients of variation compared with the Heilongkou-Erlongshan Reservoir section, suggesting a certain impact of point source discharge (i.e., human activities) on the surface water of the Danjiang River source basin. Industrial wastewater discharge significantly affected the Cl^-, SO_4^{2-}, and TDS concentrations at the W10 sample point.

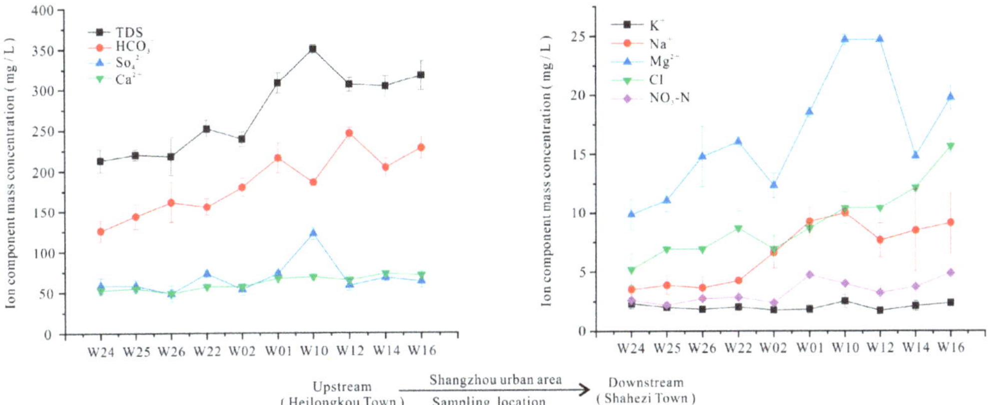

Figure 5. Spatial variations in major ion concentrations in surface water at the source of the Danjiang River.

3.3. Water Quality and Health Risk Assessment

The surface water quality of the Danjiang River source basin was classified and evaluated in this study using both the single factor index evaluation method and the entropy weight comprehensive index method. The single factor evaluation method was based on the water chemical index classification standard outlined in the 'Surface Water Environmental Quality Standard (GB3838-2002)'. The results show that all 40 samples met the Class I water standard. Similarly, using the entropy weight comprehensive index method, the calculated EWQI values of the samples ranged between 8.95 and 25.69 (Table 3), and belonged to the Class I water standard, which was consistent with the single factor evaluation method results.

Table 3. Assessment results according to computed WQI and HHRA in the Danjiang River source zone.

Sample Number	EWQI	Water Quality	Sample Number	EWQI	Water Quality	Sample Number	EWQI	Water Quality	Sample Number	EWQI	Water Quality
W01	22.68	I	W11	18.87	I	W21	16.72	I	W31	14.01	I
W02	15.72	I	W12	20.50	I	W22	17.77	I	W32	15.33	I
W03	18.12	I	W13	18.83	I	W23	16.33	I	W33	22.82	I
W04	21.99	I	W14	21.30	I	W24	15.26	I	W34	13.44	I
W05	14.33	I	W15	16.21	I	W25	15.13	I	W35	10.02	I
W06	18.51	I	W16	23.46	I	W26	16.01	I	W36	8.95	I
W07	15.74	I	W17	11.83	I	W27	17.90	I	W37	11.71	I
W08	21.92	I	W18	10.54	I	W28	14.85	I	W38	14.44	I
W09	14.63	I	W19	10.45	I	W29	15.55	I	W39	19.59	I
W10	25.69	I	W20	11.95	I	W30	20.66	I	W40	11.65	I

Furthermore, a health risk assessment was conducted using the HHRA model to evaluate the impact of nitrate nitrogen pollution in surface water. The health risk index (HI) was calculated for children as well as adult males and females based on the obtained samples. The results show that the HI for children ranged from 0.0217 to 0.2118, while for adult males and females, the HI ranged from 0.0137 to 0.1333 and 0.0122 to 0.1187, respectively (Table 4).

Table 4. Assessment results according to HHRA in the Danjiang River source basin.

Samples	Health Risk			Samples	Health Risk		
	Children	Adult Males	Adult Females		Children	Adult Males	Adult Females
W01	0.0217	0.0137	0.0122	W21	0.1061	0.0668	0.0595
W02	0.0226	0.0142	0.0127	W22	0.1065	0.067	0.0597
W03	0.0297	0.0187	0.0166	W23	0.1112	0.07	0.0623
W04	0.0401	0.0252	0.0225	W24	0.1146	0.0721	0.0642
W05	0.0411	0.0259	0.023	W25	0.1197	0.0753	0.0671
W06	0.0523	0.0329	0.0293	W26	0.1248	0.0786	0.0699
W07	0.0544	0.0342	0.0305	W27	0.1376	0.0866	0.0771
W08	0.0547	0.0344	0.0307	W28	0.1376	0.0866	0.0771
W09	0.0557	0.0351	0.0312	W29	0.1402	0.0882	0.0785
W10	0.063	0.0396	0.0353	W30	0.1525	0.0959	0.0854
W11	0.0632	0.0398	0.0354	W31	0.1541	0.097	0.0863
W12	0.0717	0.0451	0.0401	W32	0.1541	0.097	0.0863
W13	0.0755	0.0475	0.0423	W33	0.1624	0.1022	0.091
W14	0.0771	0.0485	0.0432	W34	0.1743	0.1097	0.0977
W15	0.0825	0.0519	0.0462	W35	0.18	0.1132	0.1008
W16	0.0946	0.0595	0.053	W36	0.18	0.1132	0.1008
W17	0.0967	0.0608	0.0542	W37	0.1816	0.1143	0.1018
W18	0.0984	0.0619	0.0551	W38	0.2042	0.1285	0.1144
W19	0.1042	0.0656	0.0584	W39	0.2057	0.1294	0.1153
W20	0.1046	0.0658	0.0586	W40	0.2118	0.1333	0.1187

4. Discussion

4.1. Influencing Factors of Main Ions

The use of Gibbs diagram, which relates TDS to the ratios of $(Na^+)/(Na^+ + Ca^{2+})$ and $(Cl^-)/(Cl^- + HCO_3^-)$, is a useful tool to determine the origin of ions in water and assess the effects of rock weathering, atmospheric precipitation, and evaporation-concentration on river water chemistry [50,51]. In the Danjiang River source basin, the TDS mass concentration in the surface water ranged from 134 to 388 mg/L, while the ratios of $(Na^+)/(Na^+ + Ca^{2+})$ and $(Cl^-)/(Cl^- + HCO_3^-)$ were in the ranges 0.03 to 0.16 and 0.01 to 0.09, respectively. The values for the majority of sampled locations were found in the middle and left regions of the Gibbs diagram (Figure 6), indicating that rock weathering largely influenced the hydrochemical components.

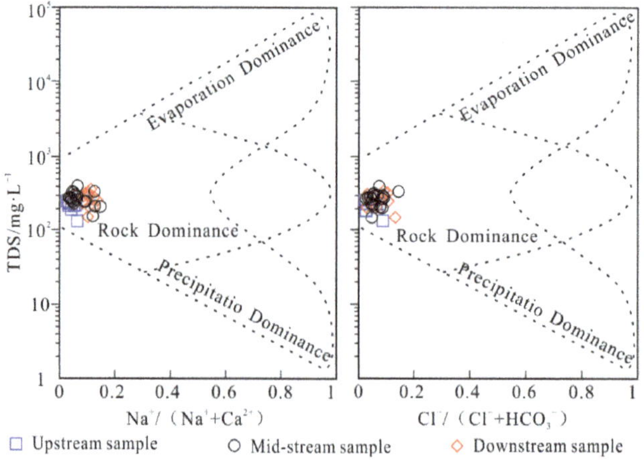

Figure 6. Gibbs diagram of surface water in the Danjiang River source zone.

The process of cation exchange adsorption plays a crucial role in shaping the composition of surface water chemistry [52]. The ratios of $(Na^+ - Cl^-)/(Ca^{2+} + Mg^{2+}) - (HCO_3^- + SO_4^{2-})$ and $(K^+ + Na^+ + Cl^-)/(HCO_3^- + SO_4^{2-})$ are used to indicate the presence of cation exchange in water [53]. In the case of ion exchange, the slope after fitting should be -1. However, the poor fitting of the aforementioned ratios in the surface water samples collected from the Danjiang River source basin suggests that the contribution of ion exchange to the surface water ions is insignificant.

4.2. Source of Main Ions

The origin of ion indices in surface water can be determined through their interdependence. Dong et al. [54] suggested that Pearson correlation analysis can provide initial insights into whether the chemical components of water have a shared source. Strong correlations among ion components indicate a common origin, while weak correlations suggest different sources. In Table 5, TDS (total dissolved solids) was found to be significantly correlated with Ca^{2+}, Mg^{2+}, HCO_3^-, and Cl^-, with correlation coefficients of 0.907, 0.733, and 0.681, respectively, suggesting that these ions predominantly contribute to TDS. Additionally, a significant correlation was observed among HCO_3^- and Ca^{2+} and Mg^{2+} ions (correlation coefficients of 0.685 and 0.851, respectively) and between Ca^{2+} and Mg^{2+}, indicating that these three ions likely originate from the weathering and dissolution of carbonate rocks. The correlation between Na^+ and K^+ was also evident, possibly derived from the dissolution of silicate rocks. In contrast, Cl^- and NO_3^- concentrations were low, which could be attributed to industrial and agricultural emissions or atmospheric precipitation. Based on the degree of correlation among ions, the sources are further discussed in groups.

Table 5. Correlation coefficients between major ions in the surface water of the Danjiang River source basin.

	pH	K^+	Na^+	Ca^{2+}	Mg^{2+}	Cl^-	SO_4^{2-}	HCO_3^-	NO_3^-	TDS
pH	1									
K^+	−0.130	1								
Na^+	−0.135	0.432 **	1							
Ca^{2+}	−0.196	0.203	0.337 *	1						
Mg^{2+}	−0.026	−0.013	−0.188	0.538 **	1					
Cl^-	−0.310	0.607 **	0.577 **	0.558 **	0.192	1				
SO_4^{2-}	−0.066	0.522 **	0.493 **	0.374 *	0.143	0.422 **	1			
HCO_3^-	−0.073	−0.188	−0.140	0.685 **	0.851 **	0.194	−0.182	1		
NO_3^-	−0.404 *	0.439 **	0.693 **	0.592 **	−0.009	0.598 **	0.488 **	0.074	1	
TDS	−0.155	0.310	0.372 *	0.907 **	0.733 **	0.598 **	0.564 **	0.681 **	0.544 **	1

Notes: ** significant correlation at the 0.01 level (two-tailed). * significant correlation at the 0.005 level (two-tailed).

4.2.1. Sources of Ca^{2+}, Mg^{2+}, and HCO_3^-

The primary control on surface water in the Danjiang River source basin is rock weathering, according to this study. According to the ratio of Ca^{2+}/Na^+ and HCO_3^-/Na^+ content in the water of 60 major rivers in the world, Gaillardet et al. [55] classified rock weathering into three types, including evaporate rock, silicate rock and carbonate rock. In the $(Ca^{2+}/Na^+)/(HCO_3^-/Na^+)$ diagram, the surface water samples from the Danjiang River source basin fall within the silicate rock to carbonate rock area range (Figure 7a), indicating that the chemical composition of surface water in the basin is primarily controlled by carbonate rock-dominated water-rock interaction. Natural water primarily obtains Ca^{2+} and Mg^{2+} from the dissolution of carbonate rocks, silicate rocks, and evaporites, whereas HCO_3^- mainly arises from the dissolution of carbonate rocks and silicates [52,56], with minimal influence from anthropogenic activities. The $(Ca^{2+} + Mg^{2+})/HCO_3^-$ ratio should be 1:1 when carbonate rocks and silicates are the primary sources of these ions. If the ratio is less than 1, silicate rock dissolution is the primary source of these ions, whereas a ratio greater than 1 indicates that carbonate rocks are the primary source. The molar concentration ratios of $(Ca^{2+} + Mg^{2+})/HCO_3^-$ in surface water samples from the Danjiang River source basin ranged from 1.08 to 1.80 (Figure 7b), indicating that Ca^{2+}, Mg^{2+}, and HCO_3^- are primarily derived from the dissolution of carbonate rocks. The (Ca^{2+}/Na^+) and (Mg^{2+}/Na^+) ratios are often used to identify the source of major ions in water [57] and are not influenced by flow rate, dilution, or evaporation [58]. Surface water samples from the Danjiang River source basin primarily fall in the carbonate rock area (Figure 7c), supporting the conclusion that carbonate rock weathering controls the chemical composition of surface water in the Danjiang River source basin. This finding is consistent with previous analyses of the hydrochemical characteristics of the Danjiangkou Reservoir and its upstream region [35,36,41]. It is noteworthy that the upstream samples are closer to the carbonate rock weathering area, whereas the middle and downstream samples are closer to the silicate rock weathering area. This phenomenon is related to the geological formations in the study area, with the northern part comprising Proterozoic-Ordovician carbonate formations, while the central and southern surfaces are predominantly covered by (shallow metamorphic) clastic rocks, magmatic rocks, and their weathering products. This observation is consistent with the conclusion that geological formations significantly influence hydrochemical genesis in the area.

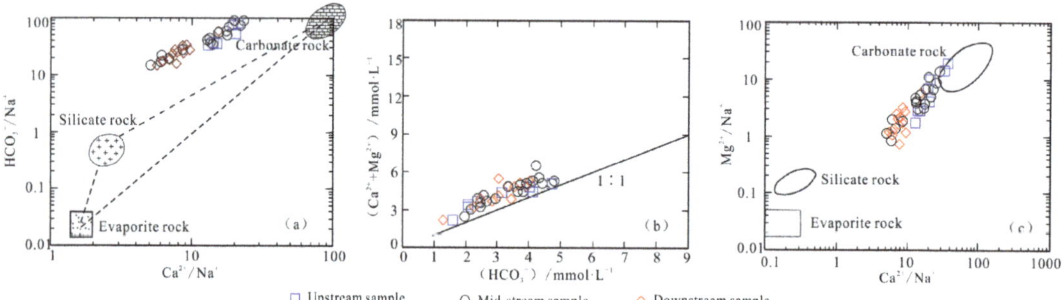

Figure 7. Relationship diagrams of (**a**) Ca^{2+}/Na^+ vs. HCO_3^-, (**b**) HCO_3^- vs. Ca^{2+}/Mg^+ and (**c**) Ca^{2+}/Na^+ vs. Mg^{2+}/Na^+ in the Danjiang River source basin.

4.2.2. Sources of Na^+, K^+, Cl^-, and NO_3^-

The study conducted by Guo et al. [59] indicated that the origin of Na^+ and K^+ in natural water is primarily attributed to the weathering products of rock salt or silicate, while Cl^- and SO_4^{2-} mostly come from evaporite dissolution. When basin evaporites control surface water cations, the $(K^+ + Na^+)/Cl^-$ equivalent ratio tends to be close to 1. However, in the Danjiang River source basin, the $(K^+ + Na^+)/Cl^-$ equivalent ratio varies between 0.59 and 3.68, with an average value of 1.56 (Figure 8a). Most samples are located above the $(K^+ + Na^+)/Cl^-$ ratio 1:1 contour line, indicating that Cl^- is insufficient to balance the presence of K^+ and Na^+. This suggests that K^+ and Na^+ may originate from the dissolution of silicate minerals such as plagioclase and potassium feldspar [60,61]. It is worth noting that sampling point 9 has a significantly higher Na^+/Cl^- value than the other sampling points, possibly due to human activities such as point source emissions from factories in the lower reaches of the Shangzhou urban area.

Cl^- is relatively stable in surface water and is almost unaffected by physical, chemical and biological processes. Its concentration only changes when water bodies with different Cl^- concentrations are mixed [62]. Therefore, the relationship between n(NO_3^-), n(NO_3^-)/n(Cl^-) and n (Cl^-) is widely used to determine the mixing or biological process [63] of nitrate in the watershed. Specifically, higher n(NO_3^-)/n(Cl^-) and low Cl^- concentrations indicate that the nitrate in the water mainly comes from agricultural activities, while lower n(NO_3^-)/n(Cl^-) and high Cl^- concentrations indicate that the nitrate in the water mainly originates from domestic sewage and manure. Lower concentrations of n(NO_3^-) and n(Cl^-) indicate that nitrate mainly comes from soil nitrogen [64]. As shown in Figure 8b,c, the concentration of surface water samples collected from the study area of the Danjiang River source basin mainly falls within high $n(NO_3^-)$, high $n(NO_3^-)/n(Cl^-)$ and low $n(Cl^-)$ categories, indicating that the NO_3^- in the region primarily comes from agricultural activities, such as the application of agricultural fertilizer, the discharge of rural domestic sewage and soil organic nitrogen. However, compared with the upstream, some sample points in the mid-stream and downstream showed a correlation trend for decreasing $n(NO_3^-)/n(Cl^-)$ and increasing $n(Cl^-)$, indicating that the surface water qualities in the mid-stream and downstream were affected by the mixed effects of agricultural activities and urban domestic sewage. Overall, the nitrate concentration in the surface water shows a gradually increasing trend from upstream to downstream. This is mainly due to the fact that most of the towns and farmland in the study area are distributed along rivers, and rainfall leaching, runoff erosion of riverbank soil, sewage and feces discharged by human activities all converge into the river, resulting in the gradual accumulation of nitrogen load and increasing nitrate concentration in the river. In addition, the Cretaceous Paleogene Shanyang clastic rock formation is widespread around Yaoshi Town and Shahezi Town in the study area. The degree of rock consolidation is low, and it is more prone to soil erosion under the influence of human activities, resulting in nitrogen migration to the river and accelerating nitrogen enrichment [65].

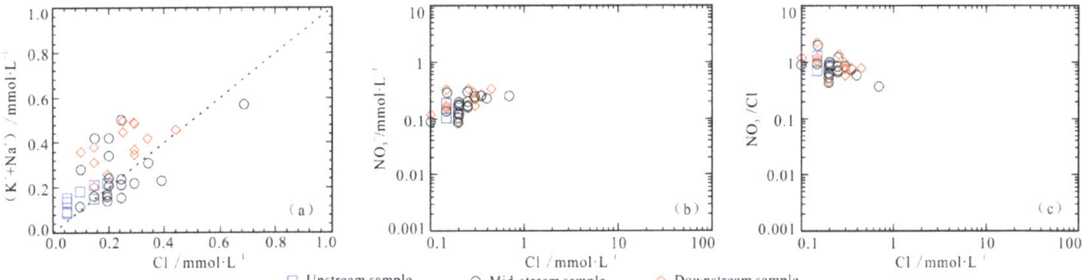

Figure 8. Relationship diagrams of (**a**) Cl^- vs. $K^+ + Na^+$, (**b**) Cl^- vs. NO_3^- and (**c**) Cl^- vs. NO_3^-/Cl^- in the Danjiang River source basin.

4.3. Water Environment Quality

4.3.1. Influence of Human Activities on Water Environmental Quality

Human activities, particularly industrial wastewater and farmland drainage, significantly influence the chemical composition and evolution of surface water, altering the contents of SO_4^{2-}, Cl^-, NO_3^-, and Na^+ ions [66]. Under natural conditions, the ratio of total hardness to alkalinity during water chemistry formation is approximately 1:1, with carbonic acid resulting from atmospheric CO_2 dissolution dominating rock weathering in the basin. However, acidic substance discharges from urban activities accelerate rock erosion, resulting in a total hardness/alkalinity ratio in water bodies greater than 1 [67]. In the Danjiang River source basin, the surface water's total hardness/alkalinity ratio ranges from 1.07 to 1.80 (Figure 9a), with an average value of 1.35, indicating the impact of acidic substances. The high values are mainly concentrated in sections with intensive human industrial and agricultural activities, such as Shangluo City, Shahezi Town, and Yaoshi Town, indicating that acidic substances emitted by human activities also have an impact on the Danjiang River's surface water in addition to natural weathering. According to He et al. [68], the undeveloped evaporite in the Danjiang river source basin and the $(Ca^{2+} + Mg^{2+})/(HCO_3^- + SO_4^{2-})$ ratio of surface water samples tend to be 1:1 (Figure 9b), indicating the involvement of sulfuric acid in carbonate dissolution. Moreover, the average $(Ca^{2+} + Mg^{2+})/HCO_3^-$ and SO_4^{2-}/HCO_3^- equivalent ratios of the samples were 1.37 and 0.36, respectively, indicating that both sulfuric acid and carbonic acid are involved in the chemical weathering process of carbonate rock salt in the basin (Figure 9c). Sulfuric acid is primarily derived from sulfur oxides produced from human activities, and diffuses into the atmosphere and dissolves in water [67].

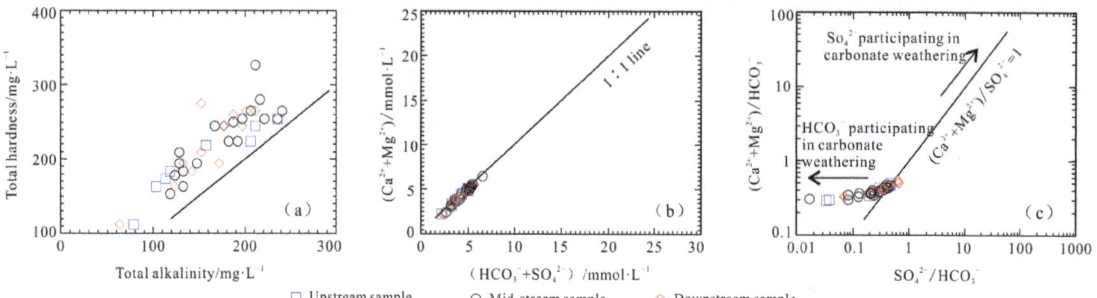

Figure 9. Relationship diagrams of (**a**) total hardness vs. Alkalinity, (**b**) Ca^{2+} + Mg^{2+} vs. HCO_3^- + SO_4^{2-} and (**c**) (Ca^{2+} + Mg^{2+})/HCO_3^- vs. SO_4^{2-}/HCO_3^- in the Danjiang River source zone.

Numerous previous studies have investigated the water quality of the Danjiang River source basin. For instance, Wang et al. [69] conducted a study on the water quality of 14 sections of the Hanjiang River and the main branch of the Danjiang River in Shaanxi Province from 2008 to 2012. They found that the water quality of the Danjiang section was primarily Class II and Class III, with notable spatial differences and regional distribution characteristics of comprehensive pollution. Pollution was more significant in the urban areas than in the suburbs, and the downstream pollution was higher than the upstream pollution in each administrative section. In another study, Zhang Jin [70] used the fuzzy comprehensive index method to comprehensively evaluate the water quality of the Majie, Zhangcun, Danfeng, and Yueri sections in Shangluo City in the upper reaches of the Danjiang River from 2018 to 2021. The study concluded that the water quality of the Yueri section was lower than that of the Majie section in the upper reaches.

This study used the entropy weight comprehensive index evaluation method to assess the water quality of 40 surface water samples in the Danjiang River basin. The obtained EWQI values ranged from 8.95 to 25.69, indicating that the water in the study area met the Class I water quality standard and was of excellent quality. This was attributed to the implementation of robust water resource protection measures by the local government in recent years. When sorting the EWQI values of the water quality samples, it was observed that the samples with EWQI values above 20 were mainly distributed in Shangzhou City and Shahezi Town along the main stream of the Danjiang River. These areas had dense residential populations, significant discharge of industrial and domestic wastewater, and slightly higher EWQI values than the tributaries in mountainous areas with less impact from human activities. Therefore, to safeguard water quality, water quality protection and monitoring efforts should be enhanced in areas with higher levels of human economic and social activities.

4.3.2. Health Risk Assessment

In this study, nitrate nitrogen was included in the human health risk assessment. The risk index calculation model showed that the health risk index (HI) due to nitrate nitrogen in surface water consumed as drinking water in the Danjiang River source area ranged from 0.0217 to 0.2118 for children, from 0.0137 to 0.1333 for adult males, and from 0.0122 to 0.1187 for adult females. It is thus clear that the risk of nitrate pollution in surface water in the Danjiang River source basin was at a relatively low level, and was significantly lower than the acceptable risk level for non-carcinogenic chronic toxic effects, indicating minimal risk to human health. However, the study found that children were more vulnerable to polluted water sources than adults, with a significantly higher health risk index scores, indicating that the younger population are more sensitive with relatively weak resistance to polluted water, which warrants further attention. The limit concentration of nitrate nitrogen in drinking water stipulated by the World Health Organization (WHO) and China is 10 mg/L, and the concentration of nitrate nitrogen in the study area ranged from 0.52 to 4.88 mg/L, with an average value of 2.56 mg/L, which was lower than the WHO limit. However, the value of nitrate nitrogen showed an increasing trend in downstream urban areas, and the human health risk index of nitrate nitrogen was also higher than that in the upstream areas. According to the statistics of the Shangzhou Center for Disease Control and Prevention in 2020, there are 36 cases of goiter patients in the district, mainly distributed in urban areas. As mentioned earlier, this may be related to the unreasonable discharge of agricultural waste and urban sewage. After the nitrate in enters the mouth, it is converted into nitrite by the nitrate reducing bacteria in saliva, which increases the risk of diseases such as methemoglobinemia, diabetes, spontaneous abortion, thyroid disease and gastric cancer [71]. Therefore, it is recommended to focus on protecting and monitoring the water quality in the downstream area and to pay more attention to the health risks of younger age groups with poor resistance.

4.4. Uncertainty Analysis

Water environmental health risk assessment mainly includes the harm to human health caused by toxic pollutants directly ingested through contact routes such as food in the water, drinking water, and breathing. In this paper, the evaluation of the human health risk is only focused on the impact of drinking water intake, without considering other toxic substances and exposure routes. Thus, the actual health risk level of water nitrate pollution may be greater than the risk value reported in this study. In addition, the health risk assessment of nitrate pollution lacks data for the wet season and long periods with surface water, and the model parameters used are mainly determined based on previous research results in Shaanxi Province. However, in fact, there are differences in human characteristics and lifestyles among southern, northern and central Shaanxi regions, so there may be some errors in the evaluation parameters of health risks. Therefore, the research results of this paper are preliminary and need to be further improved in future work to provide better support for scientific management and decision-making related to water resources protection and utilization.

5. Conclusions

1. The surface water in the Danjiang River source basin is characterized by a low mineralization and slightly alkaline environment. The average concentration of cation equivalents is in the order of $Ca^{2+} > Mg^{2+} > Na^+ > K^+$, with Ca^{2+} and Mg^{2+} being the dominant cations. The average concentration of anions is in the order of HCO_3^- > SO_4^{2-} > NO_3^- > Cl^-, with HCO_3^- and SO_4^{2-} being the dominant anions. The hydrochemisty of the water is characterized as HCO_3^- Ca·Mg type.
2. Our comprehensive study based on Gibbs diagrams, ion ratio coefficients, and factor analysis shows that the solutes in the water of the Danjiang River are mainly controlled by rock weathering, while agricultural activities and urban domestic sewage discharge also have certain impacts on the chemical components of the river water.
3. The results of the single-factor evaluation method and the entropy weight comprehensive index method show that the surface water quality in the Danjiang River source basin meets the Class I water quality standards. Overall, the water quality characteristics of the tributaries are better than the mainstream sections, and the upstream sections are better than those of the downstream areas. Additionally, the non-carcinogenic potential risks of nitrate nitrogen were low and within controllable ranges, but the impact on children was significantly higher than that on adults. Therefore, it is recommended to strengthen water quality management downstream of urban areas within the region, rationally plan the structure of industrial and agricultural production, and reduce the concentration of nitrate nitrogen in water bodies to mitigate health risks for vulnerable young populations with weaker resistance.

Author Contributions: L.L., conceptualization, writing—original draft preparation; Y.Z., methodology, writing—review and editing; X.Q. and Y.W., investigation. All authors have read and agreed to the published version of the manuscript.

Funding: This study was supported by the special project of Geological Survey Finance of Shaanxi Province (No. 202002), the Natural Science Basic Research Program of Shaanxi (No. 2022JQ-292), and the Young Talent Fund of Association for Science and Technology in Shaanxi, China (No. NYHB202208).

Data Availability Statement: The data used for this research may be provided upon reasonable request.

Conflicts of Interest: The authors declare no conflict of interest.

References

1. Li, P.; Zhang, Y.; Yang, N.; Jing, L.; Yu, P. Major ion chemistry and quality assessment of groundwater in and around a mountainous tourist town of China. *Expo. Health* **2016**, *8*, 239–252. [CrossRef]
2. Li, W.; Wu, J.; Zhou, C.; Nsabimana, A. Groundwater pollution source identification and apportionment using PMF and PCA-APCS-MLR receptor models in Tongchuan City, China. *Ach. Environ. Contam. Toxicol.* **2021**, *81*, 397–413. [CrossRef] [PubMed]
3. Xu, D.; Li, P.; Chen, X.; Yang, S.; Zhang, P.; Guo, F. Major ion hydrogeochemistry and health risk of groundwater nitrate in selected rural areas of the Guanzhong Basin, China. *Hum. Ecol. Risk Assess* **2023**, *29*, 701–727. [CrossRef]
4. Wang, D.; Li, P.; He, X.; He, S. Exploring the response of shallow groundwater to precipitation in the northern piedmont of the Qinling Mountains, China. *Urban Clim.* **2023**, *47*, 101379. [CrossRef]
5. Xia, X.Q.; Yang, Z.F.; Wang, Y.P.; Ji, J.F.; Li, W.M.; Yuan, X.Y. Major ion chemistry in the Yangtze River. *Earth Sci. Front.* **2008**, *15*, 194–202. (In Chinese)
6. Zhang, B.; Song, X.F.; Zhang, Y.H.; Ma, Y.; Tang, C.Y.; Yang, L.H.; Wang, Z.L. The interaction between surface water and groundwater and its effect on water quality in the Second Songhua River basin, northeast China. *J. Earth Syst. Sci.* **2016**, *125*, 1495–1507. [CrossRef]
7. Li, P.Y.; Wu, J.H.; Tian, R.; He, S.; He, X.D.; Xue, C.Y.; Zhang, K. Geochemistry, Hydraulic Connectivity and Quality Appraisal of Multilayered Groundwater in the Hongdunzi Coal Mine, Northwest China. *Mine Water Environ.* **2018**, *37*, 222–237. [CrossRef]
8. Markich, S.J.; Brown, P.L. Relative importance of natural and anthropogenic influences on the fresh surface water chemistry of the Hawkesbury-Nepean River south-eastern Australia. *Sci. Total Environ.* **1999**, *217*, 201–230. [CrossRef]
9. Hindshaw, R.S.; Tipper, E.T.; Reynolds, B.C.; Lemarchand, E.; Wiederhold, J.G.; Magnusson, J.; Bernasconi, S.M.; Kretzschmar, R.; Bourdon, B. Hydrologicalcontrol of stream water chemistry in a glacial catchment (DammaGlacier, Switzerland). *Chem. Geol.* **2011**, *285*, 215–230. [CrossRef]
10. He, X.D.; Li, P.Y. Surface water pollution in the middle Chinese Loess Plateau with special focus on hexavalent chromium (Cr^{6+}): Occurrence, sources and health risks. *Expo. Health* **2020**, *12*, 385–401. [CrossRef]
11. Li, P.Y.; Tian, R.; Liu, R. Solute geochemistry and multivariate analysis of water quality in the GuohuaPhosphorite Mine, Guizhou Province, China. *Expo. Health* **2019**, *11*, 81–94. [CrossRef]
12. Wu, J.H.; Li, P.Y.; Wang, D.; Ren, X.F.; Wei, M.J. Statistical and multivariate statistical techniques to trace the sources and affecting factors of groundwater pollution in a rapidly growing city on the Chinese Loess Plateau. *Hum. Ecol. Risk Assess* **2020**, *26*, 1603–1621. [CrossRef]
13. Gibbs, R.J. Water chemistry of the Amazon River. *Geochim. Cosmochim. Acta* **1972**, *36*, 1061–1066. [CrossRef]
14. Millot, R.; Gaillardet, J.; Dupré, B.; Allègre, C.J. Northern latitudechemical weathering rates: Clues from the Mackenzie River basin. *Geochim. Cosmochim. Acta* **2003**, *67*, 1305–1329. [CrossRef]
15. Huh, Y.; Tsoi, M.Y.; Zaitsev, A.; Edmond, J.M. The fluvial geochemistry ofthe rivers of the eastern Siberia: Tributaries of the Lena River draining the sedimentary platform of the Siberian Craton. *Geochim. Cosmochim. Acta* **1998**, *62*, 1657–1676. [CrossRef]
16. Sundaray, S.K. Application of multivariate statistical techniques in hydrogeochemical studies—A case study: Brahmani-Koel River (India). *Environ. Monit. Assess* **2010**, *164*, 297–310. [CrossRef]
17. Hu, M.H.; Stallard, R.F.; Edmond, J.M. Major ion chemistry of some large Chinese rivers. *Nature* **1982**, *298*, 550–553. [CrossRef]
18. Chen, J.S.; Wang, F.Y.; He, D.W. Geochemistry of water quality of the Yellow River basin. *Earth Sci. Front.* **2006**, *13*, 58–73. (In Chinese)
19. Wang, Y.P.; Wang, L.; Xu, C.X.; Yang, Z.F.; Ji, J.F.; Xia, X.Q.; An, Z.Y.; Yuan, J. Hydro-geochemistry and genesis of major ions in the Yangtze River, China. *Geol. Bull. China* **2010**, *29*, 446–456. (In Chinese)
20. Guo, Y.W.; Tian, F.Q.; Hu, H.C.; Liu, Y.P.; Zhao, S.H. Characteristics and Significance of Stable Isotopes and Hydrochemistry in Surface Water and Groundwater in Nanxiaohegou Basin. *Environ. Sci.* **2020**, *41*, 682–690. (In Chinese)
21. Wang, W.X.; Li, W.P.; Cai, Y.M.; An, Y.H.; Shao, X.M.; Wu, X.; Yin, D.C. The hydrogeochemical evolution of groundwater in the middle reaches of the Heihe River Basin. *Earth Sci. Front.* **2021**, *28*, 184–193. (In Chinese)

22. Zhang, J.T.; Shi, Z.M.; Wang, G.C.; Jiang, J.; Yang, B.C. Hydrochemical characteristics and evolution of groundwater in the Dachaidan area, Qaidam Basin. *Earth Sci. Front.* **2021**, *28*, 194–205. (In Chinese)

23. Liu, C.Y.; Huang, G.X.; Jing, J.H.; Liu, J.T.; Zhang, Y.; Guo, J.X. Characteristics and driving mechanisms of evolution of groundwater chemistry in Huang-Huai-Hai Plain and suggestions for its exploitation and utilization. *Geol. China* **2022**. (In Chinese)

24. Ma, B.J.; Zhang, Q.F.; Li, S.Y. Hydrochemical characteristics and controlling factors of trans-boundary rivers in China. *Quat. Sci.* **2023**, *43*, 425–438. (In Chinese)

25. Hagedorn, B.; Cartwright, I. Climatic and lithologic controls on thetemporal and spatial variability of CO_2 consumption via chemicalweathering: An example from the Australian Victorian Alps. *Chem. Geol.* **2009**, *260*, 234–253. [CrossRef]

26. Qin, X.Q.; Jiang, Z.C.; Huang, Q.B.; Zhang, L.K.; Liu, P.Y.; Liang, Y.P. The influenceofsulfideacidon rock weathering and carbon cycle in catchment scale: Acase study in Sanchuan River basin of Huanghe River tributary. *Quat. Sci.* **2020**, *40*, 1070–1082. (In Chinese)

27. Chen, J.Y.; Wang, Y.; Zhang, H.B.; Zhao, X.F. Overview on the studies of nitrate pollution in Groundwater. *Prog. Geogr.* **2006**, *25*, 34–44. (In Chinese)

28. Davidson, E.A.; David, M.B.; Galloway, J.N.; Goodale, C.L.; Haeuber, R.; Harrison, J.A.; Howarth, R.W.; Jaynes, D.B.; Lowrance, R.R.; Nolan, B.T.; et al. Excess Nitrogen in the U.S. Environment: Trends, Risks, and Solutions. *Issues Ecol.* **2012**, *15*, 1–16.

29. Zhang, G.L.; Liu, H.Y.; Guo, H.M.; Sun, Z.X.; Wang, Z.; Wu, T.H. Occurrences and health risks of high nitrate groundwater in the typical piedmont areas of the North China Plain. *Earth Sci. Front.* **2023**; *early access.* (In Chinese). [CrossRef]

30. Ukah, B.U.; Ameh, P.D.; Egbueri, J.C.; Unigwe, C.O. Impact of effluent-derived heavy metals on the groundwater quality in Ajao industrial area, Nigeria: An assessment using entropy water quality index (EWQI). *Int. J. Energy Water Resour.* **2020**, *4*, 231–244. [CrossRef]

31. USEPA. *Risk Assessment Guidance for Superfund, Volume I: Human Health Evaluation Manual (Part A)*; US Environmental Protection Agency: Washington, DC, USA, 1989.

32. Wang, H.; Lu, K.; Shen, C.; Song, X.; Hu, B.; Liu, G. Human health risk assessment of groundwater nitrate at a two geomorphic units transition zone in northern China. *J. Environ. Sci.* **2021**, *110*, 38–47. [CrossRef]

33. Li, D.F.; Zhai, Y.Z.; Lei, Y.; Li, J.; Teng, Y.G.; Lu, H.; Xia, X.L.; Yue, W.F.; Yang, J. Spatiotemporal evolution of groundwater nitrate nitrogen levels and potential human health risks in the Songnen Plain, Northeast China. *Ecotox. Environ. Saf.* **2021**, *208*, 111524. [CrossRef]

34. Panda, B.; Chidambaram, S.; Snow, D.; Malakar, A.; Singh, D.K.; Ramanathan, A.L. Source apportionment and health risk assessment of nitrate in foothill aquifers of Western Ghats, South India. *Ecotox. Environ. Saf.* **2022**, *229*, 113075. [CrossRef] [PubMed]

35. Li, S.Y.; Cheng, X.L.; Gu, S.; Li, J.; Zhang, Q.F. Hydro-chemical Characteristics in the Danjiangkou Reservoir (Water Source Area of the Middle Route of the South to North Water Transfer Project) China. *Environ. Sci.* **2008**, *29*, 2111–2116. (In Chinese)

36. Xu, Z.F.; Tang, Y. Hydrogeochemistry of the water source area of the Middle Route of China's South to North Water Transfer Project. *Bull. Mineral. Petrol. Geochem.* **2011**, *30*, 26–30. (In Chinese)

37. Meng, Q.P.; Zhang, J.; Zhang, Z.Y.; Wu, T.R. Geochemistry of dissolved trace elements and heavy metals in the Dan River Drainage (China): Distribution, sources, and water quality assessment. *Environ. Sci. Pollut. R.* **2016**, *23*, 8091–8103. [CrossRef] [PubMed]

38. Xu, G.C.; Li, P.; Lu, K.X.; Zhan, T.T.; Zhang, J.X.; Ren, Z.P.; Wang, X.K.; Yu, K.X.; Shi, P.; Cheng, Y.T. Seasonal changes in water quality and its main influencing factors in the Dan River basin. *Catena* **2019**, *173*, 131–140. [CrossRef]

39. Li, X.G.; Huang, C.C.; Zhang, Y.Z.; Pang, J.L.; Ma, Y.G. Hydrological reconstruction of extreme palaeoflood events 9000–8500a BP in the Danjiang River Valley, tributary of the Danjiangkou Reservoir, China. *Arab. J. Geosci.* **2020**, *13*, 137. [CrossRef]

40. Zhao, K.P.; Yang, W.J.; Sha, J.; Shang, Y.T.; Li, X. Spatial Characteristics of Nutrient Status in Danjiangkou Reservoir. *Environ. Sci. Technol.* **2020**, *43*, 51–58. (In Chinese)

41. Zhang, Q.Z.; Deng, H.J.; Lu, Y.; Zhou, H.M.; Gao, Q.; Zhou, Y.H. Status Quo of Hydrochemical Characteristics in Danjiangkou Reservoir. *J. Yangtze River Sci. Res. Inst.* **2020**, *37*, 49. (In Chinese)

42. Guo, X.M.; Zhang, Q.M.; Zhao, T.Q.; Chao, J. Fluxes, characteristics and influence on the aquatic environment of inorganic nitrogen deposition in the Danjiangkou reservoir. *Ecotox. Environ. Saf.* **2022**, *241*, 113814. [CrossRef]

43. Wang, Y.R.; Ding, W.F.; Zhang, G.H. Study on the impact of land use change on runoff in Danjiang watershed based on Swat Model. *Res. Soil Water Conserv.* **2022**, *29*, 62–74. (In Chinese)

44. Zhang, G.W.; Guo, A.L.; Dong, Y.P.; Yao, A.P. Rethinking of the Qinling Orogen. *J. Geomech.* **2019**, *25*, 746–768. (In Chinese)

45. Li, P.Y.; Qian, H.; Wu, J.H. Ground water quality assessment based on improved water quality index in Pengyang County, Ningxia, Northwest China. *E-J. Chem.* **2010**, *7*, S209–S216. [CrossRef]

46. Wu, J.J.; Bian, J.M.; Wan, H.L.; Wei, N.; Ma, Y.X. Health risk assessment of groundwater nitrogen pollution in Songnen Plain. *China Environ. Sci.* **2019**, *39*, 3493–3500. (In Chinese) [CrossRef] [PubMed]

47. Zhai, Y.Z.; Lei, Y.; Wu, J.; Teng, Y.G.; Wang, J.S.; Zhao, X.B.; Pan, X.D. Does the groundwater nitrate pollution in China pose a risk to human health? A critical review of published data. *Environ. Sci. Pollut. Res.* **2017**, *24*, 3640–3653. [CrossRef]

48. Han, G.L.; Liu, C.Q. Water geochemistry controlled by carbonate dissolution: A study of the river waters draining karst-dominated terrain, Guizhou Province, China. *Chem. Geol.* **2004**, *204*, 1–21. [CrossRef]

49. Piper, D.Z. Seawater as the source of minor elements in black shales, phosphorites, and other sedimentary rocks. *Chem. Geol.* **1994**, *114*, 95–114. [CrossRef]

50. Gibbs, R.J. Mechanisms controlling world water chemistry. *Science* **1970**, *170*, 1088–1090. [CrossRef]

51. Kilham, P. Mechanisms controlling the chemical composition of lakes and rivers: Data from Africa. *Limnol. Oceanogr.* **1990**, *35*, 80–83. [CrossRef]

52. Zhang, T.; Wang, M.G.; Zhang, Z.Y.; Liu, T.; He, J. Hydrochemical characteristics and possible controls of the surface water in Ranwu Lake basin. *Environ. Sci.* **2020**, *41*, 4003–4010. (In Chinese)

53. Yan, Z.X.; Feng, M.Q. Hydrochemical characteristics and driving factors of surface water in the mining area of Changhe River Basin. *Environ. Chem.* **2022**, *41*, 632–642. (In Chinese)

54. Dong, W.H.; Meng, Y.; Wang, Y.S.; Wu, X.C.; Lü, Y.; Zhao, H. Hydrochemical characteristics and formation of the shallow groundwater in Fujin, Sanjiang plain. *J. Jilin Univ. (Earth Sci. Ed.)* **2017**, *47*, 542–553. (In Chinese)

55. Gaillardet, J.; Dupré, B.; Louvat, P.; Allègre, C.J. Global silicate weathering and CO_2 consumption rates deduced from the chemistry of large rivers. *Chem. Geol.* **1999**, *159*, 3–30. [CrossRef]

56. Zhou, J.X.; Ding, Y.J.; Zeng, G.X.; Wu, J.K.; Qin, J. Majorion chemistry of surface water in the upper reach of Shule river basin and the possible controls. *Environ. Sci.* **2014**, *35*, 3315–3324. (In Chinese)

57. Chen, L.; Zhong, J.; Li, C.; Wang, W.F.; Xu, S.; Yan, Z.L.; Li, S.L. The chemical weathering characteristics of different litholoyic mixed small watersheds in Southwest China. *Chin. J. Ecol.* **2020**, *39*, 1288–1299. (In Chinese)

58. Yuan, J.F.; Xu, F.; Liu, H.Z.; Deng, G.S. Application of hydrochemical and isotopic analysis to research a typical karst ground water system: Acase study at Xianrendong, Xichang city. *Sci. Technol. Eng.* **2019**, *19*, 76–83. (In Chinese)

59. Guo, Y.; Gan, F.P.; Yan, B.K.; Wang, F.; Bai, J. Hydrochemical-isotopic characteristics of surface water in southwest Tibetan Plateau and controlling factors analysis. *J. North China Univ. Water Resour. Electr. Power (Nat. Sci. Ed.)* **2021**, *43*, 96–107. (In Chinese)

60. Feng, Q.C.; Yang, W.H.; Wen, S.M.; Wang, H.; Zhao, W.J.; Han, G. Flotation of copper oxide minerals: A review. *Int. J. Min. Sci. Technol.* **2022**, *32*, 1351–1364. [CrossRef]

61. Feng, Q.C.; Wang, M.L.; Zhang, G.; Zhao, W.J.; Han, G. Enhanced adsorption of sulfide and xanthate on smithsonite surfaces by lead activation and implications for flotation intensification. *Sep. Purif. Technol.* **2023**, *307*, 122772. [CrossRef]

62. Matiatios, I. Nitrate source identification in groundwater of multiple land-use area by combining isotopes and multivariate statistical analysis: A case stydy of Asopos Basin (Central Greece). *Sci. Total. Environ.* **2016**, *541*, 802–814. [CrossRef]

63. Yue, F.-J.; Li, S.-L.; Liu, C.-Q.; Zhao, Z.-Q.; Hu, J. Using dual isotopes to evaluate sources and transformation of nitrogen in the Liao River, northeast China. *Appl. Geochem.* **2013**, *36*, 1–9. [CrossRef]

64. Xu, Q.F.; Xia, Y.; Li, S.J.; Wang, W.Z.; Li, Z. Temporal and Spatial Distribution Characteristics and Source Analysis of Nitrate in Surface Water of Wuding River Basin. *Environ. Sci.* **2022**; *early access*. (In Chinese). [CrossRef]

65. Tian, H.Q.; Lu, C.Q.; Melillo, J.; Ren, W.; Huang, Y.; Xu, X.F.; Liu, M.L.; Zhang, C.; Chen, G.S.; Pan, S.F.; et al. Food benefit and climate warming potential of nitrogen fertilizer uses in China. *Environ. Res. Lett.* **2012**, *7*, 044020. [CrossRef]

66. Tu, C.L.; Yin, L.H.; He, C.Z.; Cun, D.X.; Ma, Y.Q.; Linghu, C.W. Hydrochemical composition characteristics and control factors of Xiaohuangni River basin in the upper pearl river. *Environ. Sci.* **2022**, *43*, 1885–1897. (In Chinese)

67. Xu, Q.J.; Lai, C.Y.; Ding, Y.; Wang, Z.L.; Cheng, Z.H.; Yu, T. Natural water chemical change in the surface water of Chengdu and impact factors. *Environ. Sci.* **2021**, *42*, 5364–5374. (In Chinese)

68. He, J.Y.; Zhang, D.; Zhao, Z.Q. Spatial and temporal variations in hydrochemical composition of river water in Yellow River basin, China. *Chin. J. Ecol.* **2017**, *36*, 1390–1401. (In Chinese)

69. Wang, L.; Guan, J.L.; Yao, Z.P.; Ding, Q.; Luo, Y.N.; Cheng, J.X. The variation characteristics of water quality in the Han River and Dan River of Shaanxi region. *Environ. Monit. China* **2015**, *31*, 73–77. (In Chinese)

70. Zhang, J. Evaluation of water quality in the upper reaches of Danjiang River based on fuzzy comprehensive index method. *Shaanxi Water Resour.* **2022**, *7*, 98–100. (In Chinese)

71. Hord, N.G. Dietary nitrates, nitrites, and cardiovascular disease. *Curr. Atheroscler. Rep.* **2011**, *13*, 484–492. [CrossRef]

Article

Assessing Water Resource Carrying Capacity and Sustainability in the Cele–Yutian Oasis (China): A TOPSIS–Markov Model Analysis

Guangwei Jia, Sheng Li *, Feilong Jie, Yanyan Ge, Na Liu and Fuli Liang

School of Geology and Mining Engineering, Xinjiang University, Urumqi 830017, China; 18393834486@163.com (G.J.); jie@xju.edu.cn (F.J.); geyanyan0511@xju.edu.cn (Y.G.); liuna110615@163.com (N.L.); fuli8060@163.com (F.L.)
* Correspondence: lisheng2997@163.com

Abstract: This study employs the *Driving Force–Pressure–State–Response* (*DPSR*) framework to establish an evaluation index system for the water resource carrying capacity (*WRCC*) in the Cele–Yutian Oasis (China). Utilizing the TOPSIS and obstacle degree models, we analyze the trends in the WRCC and its main hindrance factors in the Cele–Yutian Oasis from 2005 to 2020. Additionally, we employ the Markov model to investigate the dynamic changes in the land use types. The findings reveal that the most unfavorable WRCC status occurred in 2007, with a Grade IV rating (a mild overload). By 2020, the WRCC improved to a Grade III rating (critical), indicating a positive trajectory. However, persistent challenges for water resources remain, with a prolonged critical state. Over the past 15 years, the grassland area has decreased by 15.18%, and the forest area has decreased by 50%. The dynamic degree of grassland, forests, and water bodies is negative, signifying shifts to other land types, with water bodies undergoing the most significant change at −10.16%. Based on the outcomes of these two models, we propose regionally tailored measures to support sustainable development. These research results provide a scientific foundation for optimal water resource allocation and sustainable development in the Cele–Yutian Oasis Economic Belt.

Keywords: water resource carrying capacity; TOPSIS model; barrier degree; Markov model; Celle–Yutian Oasis

Citation: Jia, G.; Li, S.; Jie, F.; Ge, Y.; Liu, N.; Liang, F. Assessing Water Resource Carrying Capacity and Sustainability in the Cele–Yutian Oasis (China): A TOPSIS–Markov Model Analysis. *Water* **2023**, *15*, 3652. https://doi.org/10.3390/w15203652

Academic Editors: Peiyue Li and Jianhua Wu

Received: 22 September 2023
Revised: 13 October 2023
Accepted: 16 October 2023
Published: 18 October 2023

1. Introduction

Water resources constitute essential natural assets pivotal for both economic and social advancement [1–3]. Nevertheless, the relentless surge in economic and social progress and expedited industrialization and urbanization coupled with persistent population expansion have precipitated an incessant upswing in the demand for water resources [4,5]. At the same time, water pollution and scarcity issues are becoming increasingly serious, which is to the detriment of sustainable development initiatives and effective water resource utilization [6]. Especially in the desert regions of the northwest, access to water resources is key to driving economic, social, and ecological development [7]. Hence, it is necessary to conduct a rigorous scientific assessment of the water resource carrying capacity to foster a robust growth in desert oasis economies, to safeguard ecological environments, and to promote the sustainable utilization of water resources [8].

In assessing the water resource carrying capacity, the initial step entails the selection of the appropriate indicators to be evaluation factors. Different scholars have different perspectives on the study of the water resource carrying capacity, resulting in a variety of indicator systems [9]. For example, Gao et al. [10] developed a regional water-resource-carrying-capacity evaluation indicator system encompassing four dimensions: the water resource quantity, virtual water, ecological environment, and socioeconomic factors. Liu et al. [11] examined the selection of the water-resource-carrying-capacity indicators in Tibet based on the three functions of water resources, namely, production, living, and ecology, and the PSR model, and they explored the coupling and coordination relationship between the three subsystems. Dang et al. [9] used the *DPSR (Driving Force–Pressure–State–Response)* framework to comprehensively evaluate the water resource carrying capacity of Longnan City. The *DPSR* model has several advantages, including comprehensiveness, systematicity, clear logic, and strong practicality, making it a favored choice among scholars. The water resource carrying capacity refers to the maximum capacity of the industrial, agricultural, and urban scales as well as the population that a region can bear without damaging its society and ecosystem, which is a comprehensive indicator that is adaptable to social, economic, and scientific and technological development [12–14]. The methods to evaluate the water resource carrying capacity include the TOPSIS method, fuzzy comprehensive evaluation method, principal component analysis method, system dynamics method, and multiobjective linear programming method [15–19]. For example, Li et al. [20] used an improved TOPSIS model to comprehensively evaluate the water resource carrying capacity in Jiangsu Province, and they used the obstacle degree model to diagnose the obstacles to the water resource carrying capacity. Ren et al. integrated a fuzzy comprehensive evaluation with the analytic hierarchy process to assess the water resource carrying capacity of Datong City [21]. Li et al. [22] proposed a water resource allocation method based on the characteristics of the crop and livestock water demand combined with the agricultural water resource carrying capacity. Xu et al. [23] predicted the dynamic changes in the water resource carrying capacity based on a system dynamics model with different development modes. Although these methods have been widely used in the evaluation of the water resource carrying capacity, there are still some problems. For example, the analytic hierarchy process and fuzzy comprehensive evaluation can be susceptible to subjective preferences when assigning indicator weights, while system dynamics modeling demands a multitude of variables and parameters, resulting in intricate model development. In comparison, the TOPSIS method is simple and intuitive, can consider multiple criteria, and can handle fuzzy and uncertain problems.

Researchers from around the world have made progress in studying the water resource carrying capacity, but there are still some problems and deficiencies that need to be addressed. Some of these are that (1) water-resource-carrying-capacity research primarily concentrates on individual projects and that comprehensive studies encompassing the socioeconomic aspects, ecology, and environment are limited. Particularly, there is a dearth of research that explores the influence of human activities on the water resource carrying capacity [24]. (2) The studies mostly pertain to developed provinces, river basins, or prefecture-level cities, with less focus on individual oases [25]. (3) Previous studies predominantly centered on uncovering the year-to-year fluctuations in water resources within the water resource carrying capacity, with comparatively limited exploration into the profound effects of dynamic alterations in various evaluation factors on the water resource carrying capacity [26]. (4) Fewer studies have been conducted to evaluate regional sustainable development using both the water resource carrying capacity and land use change trends. Spatiotemporal variations are high in land use and water-cycle processes. Moreover, inadequate vegetation restoration and excessive human intervention can lead to dramatic shifts in the present land use. The impact of land use on the water resource carrying capacity is further enhanced by human activities. Therefore, the scientific quantification of the dynamic relationship between the water resource carrying capacity and land use types is a prerequisite for protecting oasis ecologies and ensuring healthy sustainable development.

In light of the aforementioned challenges and requirements, this study primarily concentrates on the Cele–Yutian Oasis (China) Economic Belt, which is characterized by somewhat delayed economic development. It is located at the northern foot of Kunlun Mountain and the southern edge of Taklimakan Desert, with an extremely arid continental desert climate. The spatial and temporal distribution of water resources in the oasis is uneven, mainly in agriculture, the low utilization rate of irrigation water, the prominent shortage of water resources, and the weak potential of sustainable development. This study aims to establish a water-resource-carrying-capacity evaluation index system grounded in the *DPSR* framework. The Technique for Order Preference by Similarity to an Ideal Solution (TOPSIS) model is used to calculate the water resource carrying capacity of the Cele–Yutian Oasis from 2005 to 2020, and the impact of the major obstacle factors is evaluated based on the obstacle degree model system. In addition, the Markov model is introduced to calculate the dynamics among the different land use types. Through a comprehensive analysis of the water resource carrying capacity and land use transitions, this study assesses the state of sustainable development in the Cele–Yutian Oasis and puts forward constructive recommendations. The results of this study provide a solid theoretical foundation for the efficient allocation and judicious utilization of water resources in desert oases.

2. Research Area and Research Methods

2.1. Research Area Overview

Cele–Yutian Oasis (80°50′ E–82° E longitude, 36°50′ N–37°20′ N latitude) is located on the southern edge of the Tarim Basin, with an area of approximately 8257 square kilometers. It ranks as one of the world's most remote regions from the ocean (Figure 1). The mean annual temperature in this region stands at 11.8 degrees Celsius, with a precipitation of 35 mm and an evaporation of 2480 mm. The evaporation is more than 70 times the precipitation, resulting in an extremely harsh environment. The oasis is home to rivers such as the Keriya River, Cele River, Pisgah River, and Nur River. The water in these rivers mainly comes from the melting of mountain snow and ice, with over 70% of the annual runoff occurring during the hot season from June to August. Agriculture is significant to the oasis, contributing over 30% to the GDP. However, due to uneven seasonal distribution of water resources, inadequate agricultural irrigation facilities, low water resource utilization efficiency, and unreasonable water resource development and utilization patterns, the oasis has severe water shortages and unsustainable use patterns.

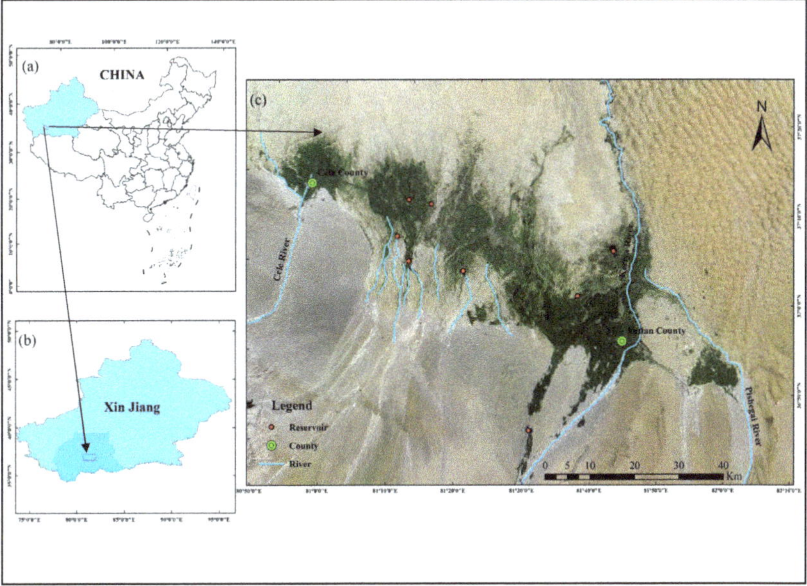

Figure 1. Location of study area. (**a**,**b**) http://datav.aliyun.com/portal/school/atlas/area_selector (accessed on 1 July 2022). (**c**) https://www.ovital.com/ (accessed on 1 July 2022).

2.2. Data Source

The data used in this study includes meteorological data, population data, economic data, and hydrological data. The meteorological data came from the historical meteorological data of two stations, Cele and Yutian. The population, economic, and hydrological data came from the "China County Statistical Yearbook", "Xinjiang Water Resources Bulletin", and "Hotan Statistical Yearbook" (2005–2020). The land use data came from Geo-Remote Sensing Ecology Website (gisrs.cn), and the land use data at three time points, 2005, 2010, and 2020, were downloaded.

2.3. Research Methods

2.3.1. Construction of Water-Resource-Carrying-Capacity Index System

In the process of water-resource-carrying-capacity evaluation, it is very important to establish a suitable evaluation index system, which needs to consider all relevant factors and their mutual influence. The selection of evaluation index system in this paper followed the principles of scientificalness, representativeness, and operability, and it referred to the evaluation index system of sustainable use of water resources. Based on the *DPSR* model, this study constructed an evaluation index system for the water resource carrying capacity of the Cele–Tarim Oasis. The *DPSR* model is based on the PSR model, which was proposed by Canadian statisticians David J. Rapport and Tony Friend in 1979 [27,28]. The *DPSR* model is based on causal relationships and comprehensively reflects the interaction and relationship between human beings and the natural environment [29–31]. In the *DPSR* model, *D* (*driving force*) represents the influence of external factors and internal dynamics on the water resource system, including population growth, economic development, climate change, and water resource management policies. The driving force reflects the dynamics and trends of water resource system development. *P* (*pressure*) represents the pressure and impacts faced by the water resource system, including water resource demand, water pollution, and water ecological destruction. Pressure reflects the pressures and challenges borne by the water resource system. *S* (*state*) represents the current situation and characteristics of the water resource system, including the availability of water resources, water quality, and water ecological health. *State* reflects the actual situation and status of the water resource system. *R* (*response*) represents the measures and response strategies taken in response to the pressures and challenges faced by the water resource system.

Based on existing research [32–34] and combined with the actual situation of Cele–Yutian Oasis, this article selected 16 indicators from 4 levels to evaluate the water resource carrying capacity of the oasis. The specific indicator system and weights were calculated using the entropy weight method, as shown in Table 1.

Table 1. Evaluation indicator system and weights of water resource carrying capacity in Cele–Yutian Oasis.

Target Layer	Ruler Layer	Indicator Layer	Code	Number	Unit	Indicator Properties	Weights
Water resource carrying capacity	Driving	Total population	×1	10^4	person	−	0.086
		Natural population growth rate	×2	/	‰	−	0.041
		Total GDP	×3	/	CNY	+	0.081
		Per capita GDP	×4	/	CNY	+	0.039
		Urbanization rate	×5	/	%	−	0.058
	Pressure	Water resources per capita	×6	/	$m^3/person$	+	0.051
		Domestic water quotas	×7	/	m^3/d	+	0.061
		Sewage Discharge	×8	10^4	m^3	−	0.051
	State	Total water supply	×9	10^8	m^3	+	0.114
		Water use in agriculture	×10	10^8	m^3	−	0.059
		Industrial water consumption	×11	10^8	m^3	−	0.026
		Ecological water use	×12	10^8	m^3	+	0.052
		Total groundwater	×13	10^8	m^3	+	0.064

Table 1. *Cont.*

Target Layer	Ruler Layer	Indicator Layer	Code	Number	Unit	Indicator Properties	Weights
Water resource carrying capacity	*Response*	Sewage treatment capacity	×14	10^4	m^3	+	0.115
		Length of water pipeline	×15	/	km	+	0.056
		Water supply capacity of built storage projects	×16	10^4	m^3	+	0.046

2.3.2. Entropy Weight Method

The entropy weight method calculates the information entropy of indicators considering the information and differences between indicators. The information entropy quantifies the contribution of indicators to decision making, making the allocation of weights more reflective of the importance and differences of indicators. The main calculation process is as follows [35,36]:

(1) Standardization of data

Firstly, a raw data matrix is created:

$$A = \begin{bmatrix} x_{11} & x_{12} & \cdots & x_{1j} \\ x_{21} & x_{22} & \cdots & x_{2j} \\ \vdots & \vdots & \vdots & \vdots \\ x_{i1} & x_{i2} & \cdots & x_{ij} \end{bmatrix} \tag{1}$$

where x_{ij} represents the j-th indicator of the i-th year.

(2) Because of the different dimensions of the original evaluation data, this paper adopted the range normalization method to standardize the original evaluation data matrix.

For the positive indicators, bigger value represents better indicator:

$$X_{ij}(+) = \frac{x_{ij} - min(x_{ij})}{max(x_{ij}) - min(x_{ij})} \tag{2}$$

For the reverse indicators, smaller value represents better indicator:

$$X_{ij}(-) = \frac{max(x_{ij}) - x_{ij}}{max(x_{ij}) - min(x_{ij})} \tag{3}$$

The decision matrix is obtained after data standardization:

$$M = \begin{bmatrix} X_{11} & X_{12} & \cdots & X_{1j} \\ X_{21} & X_{22} & \cdots & X_{2j} \\ \vdots & \vdots & \vdots & \vdots \\ X_{i1} & X_{i2} & \cdots & X_{ij} \end{bmatrix} \tag{4}$$

(3) Calculate the proportion of the j-th indicator p_{ij}:

$$p_{ij} = \frac{X_{ij}}{\sum_{i=1}^{n} X_{ij}} (i = 1, 2, ..., m) \tag{5}$$

(4) Calculate the entropy of the j-th indicator e_j:

$$e_j = -\frac{1}{\ln n} \sum_{i=1}^{n} p_{ij} \ln(p_{ij}) \tag{6}$$

(5)   Calculate information entropy redundancy d_j:

$$d_j = 1 - e_j \tag{7}$$

(6)   Determine the weights of indicators W_j:

$$W_j = \frac{1 - e_j}{\sum_{j=1}^{m} d_j} \tag{8}$$

where p_{ij} is the probability of the j-th factor in year i, e_j is the entropy value of the j-th indicator, and W_j is the normalized weight of the j-th indicator.

2.3.3. TOPSIS Model

The TOPSIS method, also known as the "Technique for Order Preference by Similarity to an Ideal Solution", is an evaluation method that determines the relative superiority or inferiority of multiple evaluation objects by their relative closeness to the ideal solution. It has a flexible and convenient calculation process that yields precise and rational evaluation outcomes, aligning well with the focus of this paper's research. The specific calculation steps of this method are as follows [37]:

$$Z_{ij} = \begin{bmatrix} Z_{11} & Z_{12} & \cdots & Z_{1j} \\ Z_{21} & Z_{22} & \cdots & Z_{2j} \\ \vdots & \vdots & \vdots & \vdots \\ Z_{i1} & Z_{i2} & \cdots & Z_{ij} \end{bmatrix} = \begin{bmatrix} W_1 X_{11} & W_2 X_{12} & \cdots & W_j X_{1j} \\ W_1 X_{21} & W_2 X_{22} & \cdots & W_j X_{2j} \\ \vdots & \vdots & \vdots & \vdots \\ W_1 X_{i1} & W_2 X_{i2} & \cdots & W_j X_{ij} \end{bmatrix} \tag{9}$$

Determine the positive and negative ideal solutions. If the value of Z_{ij} has a larger value in the decision matrix Z_{ij}, it represents a better solution.

Positive indicator are as follows:

$$\begin{cases} Z^+ = (Z_1{}^+, Z_2{}^+, \ldots Z_j{}^+) = \{ maxZ_{ij} | j = 1, 2, \ldots 16 \} \\ Z^- = (Z_1{}^-, Z_2{}^-, \ldots Z_j{}^-) = \{ minZ_{ij} | j = 1, 2, \ldots 16 \} \end{cases} \tag{10}$$

Negative indicators are as follows:

$$\begin{cases} Z^+ = (Z_1{}^+, Z_2{}^+, \ldots Z_j{}^+) = \{ minZ_{ij} | j = 1, 2, \ldots 16 \} \\ Z^- = (Z_1{}^-, Z_2{}^-, \ldots Z_j{}^-) = \{ maxZ_{ij} | j = 1, 2, \ldots 16 \} \end{cases} \tag{11}$$

Calculate the distance to the positive ideal solution $(D_i{}^+)$ and the distance to the negative ideal solution $(D_i{}^-)$ for each indicator:

$$D_i{}^+ = \sqrt{\sum_{j=1}^{16} (Z_{ij} - Z_j{}^-)^2}\ i = 1, 2, \ldots, n \tag{12}$$

$$D_i{}^- = \sqrt{\sum_{j=1}^{16} (Z_{ij} - Z_j{}^-)^2} i = 1, 2, \ldots, n \tag{13}$$

Calculate the closeness C_i:

$$C_i = \frac{D_i{}^-}{D_i{}^- + D_i{}^+} \tag{14}$$

The closer C_i is to 1, the better the assessment is.

In this paper, the closeness of each year could be used to judge the level of water resource carrying capacity and thus the carrying capacity status. Based on the results of previous studies [11,38] combined with the socioeconomic and natural ecological conditions of the Celle–Yutian Oasis, the five-level evaluation classification criteria of water resource carrying capacity were established as follows.

2.3.4. Obstacle Degree Model

Using the obstacle degree model, we conducted an obstacle degree index evaluation for various indicators of water resource carrying capacity in the Cele–Yutian Oasis and identified the obstructive factors [39]. With respect to this, factor contribution (F_{ij}), deviation degree (I_{ij}), and obstacle degree (p_{ij}) were introduced, and the calculation formula is as follows:

$$F_{ij} = W_j \, I_j \tag{15}$$

$$I_{ij} = 1 - X_{ij} \tag{16}$$

$$p_{ij} = \frac{F_{ij} I_{ij}}{\sum_1^n \left(F_{ij} I_{ij} \right)} \times 100\% \tag{17}$$

where W_j is the weight of each indicator and X_{ij} is the value of a single indicator after standardization.

2.3.5. Land Use Transfer Matrix

Compared to the well-established research on traditional land carrying capacity, the study of water resource carrying capacity is relatively new and continuously evolving in terms of theoretical methods. Despite this, significant progress has been made. However, there is still a need for further research on the evolving patterns of spatial and temporal transfer of regional water resource carrying capacity.

To address these knowledge gaps, it is imperative to consider the coupled relationship between water resource carrying capacity and land resource carrying capacity in oasis regions. By integrating the management and sustainable use of both water and land resources, a more holistic approach can be taken to ensure the preservation and sustainable supply of resources within these limited ecological regions.

Oasis regions are characterized by the close interconnection and interaction between their water resources and land resources. By recognizing and exploring the interdependence of these factors, it becomes possible to implement better strategies for the management and protection of oasis ecosystems. This integrated approach helps to achieve sustainable utilization of water and land resources, ensuring their long-term availability.

(1) Markov model

The mutual conversion of land use types is mainly achieved through the use of land use transfer matrix [40]. The land use transfer matrix is an application of the Markov model in analyzing land use changes. It provides a comprehensive depiction of the direction and quantity of land use type transitions within the region. This method finds extensive application in land use change research and effectively illustrates the spatiotemporal evolution of land use patterns [41,42].

$$S_{ij} = \begin{bmatrix} S_{11} & \cdots & S_{1n} \\ \vdots & \vdots & \vdots \\ S_{n1} & \cdots & S_{nn} \end{bmatrix} \tag{18}$$

where S_{ij} is the $n \times n$ matrix; S is the area; n is the number of land types; and i and j are the land types at the beginning and end of the study period, respectively.

(2) Land use dynamic index

The annual change rate of land use types can be obtained by calculating the dynamic index of a single land use type using this expression [43]:

$$K = \frac{U_b - U_a}{U_b} \frac{1}{T} \times 100\%$$

(19)

where K is the dynamic index of land use type movement; U_a and U_b are the area of a land type at the beginning and at the end of the study period, respectively; T is the length of the study; and when the time period of T is set to be years, K is the annual rate of change of a land use type during the study period.

3. Results and Discussion

3.1. Evaluation of Water Resource Carrying Capacity in Cele–Yutian Oasis

3.1.1. Time-Variation Model for Water Resource Carrying Capacity in the Cele–Yutian Oasis

Based on the comprehensive evaluation findings of the water resource carrying capacity in the Cele–Yutian Oasis spanning from 2005 to 2020 (refer to Figure 2 and Table 2), it is evident that, despite a general upward trajectory, the region still resides in a critical carrying state. The overall situation regarding the water resource carrying capacity remains far from optimistic. This is consistent with the research results of Wei [44] and Zhao et al. [45]. The main reason for this is that the Cele–Yutian Oasis has a fragile ecological environment, an uneven spatial and temporal distribution of water resources, and limited resources [46]. Amidst the rapid national economic growth, a local infrastructure lag and the overexploitation of water resources caused a sharp drop in the water resource carrying capacity from 2005 to 2007. Since 2010, intensive national support for Southern Xinjiang's economic and social development has led to enhanced infrastructure, industrial upgrades, resource recycling, and positive shifts in the water resource carrying capacity. Nonetheless, 2018 marked a nadir due to accelerated urbanization. A significant portion of arable land was transformed into construction zones, diminishing the agricultural land, while desert grasslands were converted to agricultural land, substantially heightening the water resource demand in agriculture. By observing the changes in the trend of the positive ideal solution and negative ideal solution, we found that the positive ideal solution gradually decreased from 0.24 in 2005 to 0.18 in 2020, indicating that the water resource carrying capacity is gradually approaching the positive ideal solution. On the contrary, the value of the negative ideal solution first decreased and then increased, reaching a minimum of 0.14 in 2008, and it has been increasing since then, indicating that the water resource carrying capacity is gradually moving away from the negative ideal solution. Overall, the water resource carrying capacity of the Cele–Yutian Oasis is trending positively, yet it remains in a critical state with a relatively limited water resource capacity. Therefore, the water-resource-carrying-capacity situation in this oasis is still severe, and further measures need to be taken to improve the water resource utilization efficiency and to protect water resources.

3.1.2. Evaluation of the Water-Resource-Carrying-Capacity Subsystem in the Cele–Yutian Oasis

(1) *Drive force subsystem*

According to Figure 3, the value of the *drive force subsystem* shows a steady increase from 0.17 in 2005 to 0.92 in 2020, and the composition of the *drive force subsystem* consists of demographic and economic indicators, with a significant growth in the total population and per capita GDP. The acceleration of the urbanization process, the significant contribution of agriculture to the oasis' GDP, and the rapid growth of the secondary industry all indicate an increasing influence of the *drive force subsystem* on the water resource carrying capacity.

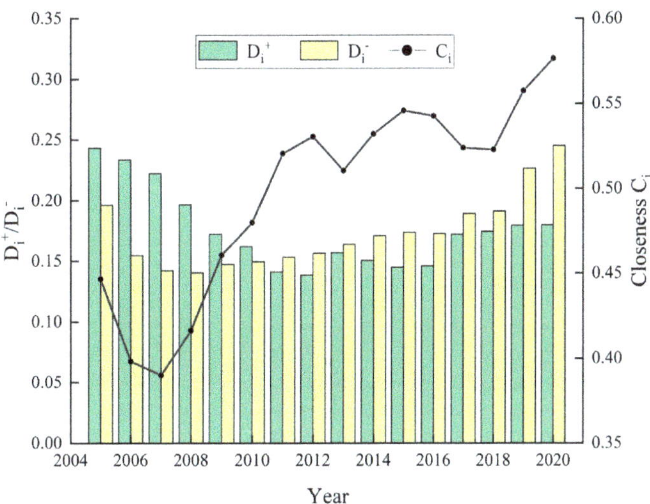

Figure 2. Comprehensive evaluation of water resource carrying capacity in Cele–Yutian Oasis from 2005 to 2020.

Table 2. Classification standards for water resource carrying capacity in Cele–Yutian Oasis.

Level	V	IV	III	II	I
Level Description C_i	Severe overload (0–0.2)	Mild overload (0.2–0.4)	Critical (0.4–0.6)	Good (0.6–0.8)	Excellent (0.8–1.0)

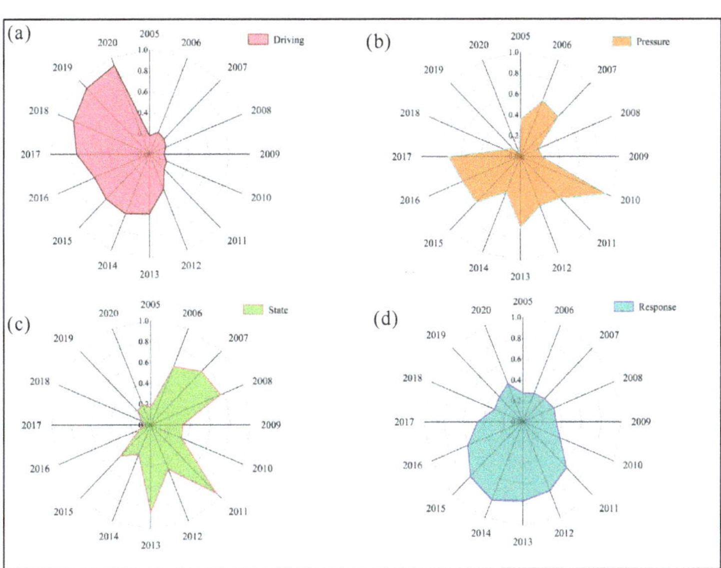

Figure 3. Changes in water-resource-carrying-capacity subsystem from 2005 to 2020. (**a**) *Driving subsystem* (**b**) *Pressure subsystem* (**c**) *State subsystem* (**d**) *Response subsystem.*

(2)   *Pressure subsystem*

The *pressure subsystem* shows a strong fluctuation trend, reaching its peak value of 0.94 in 2010. This is mainly due to a 25% decrease in the water volume of the oasis' river channels compared to previous years. Within the *pressure subsystem*, the reduction in the total water resource quantity resulted in a corresponding decline in per capita water resources. Furthermore, as residents' living standards have improved, the volume of domestic sewage discharge has also increased, significantly impacting the *pressure subsystem* of the water resource carrying capacity in the Cele–Yutian Oasis.

(3)   *State subsystem*

During the period of 2005 to 2020, the calculated values of the *state subsystem* fluctuated, with the minimum value occurring in 2018 at 0.09, while the maximum value occurred in 2011. Since 2011, climatic changes have increased the water inflow of the river, leading to a significant increase in the ecological water consumption and total water supply, thereby reflecting an improved condition within the *state subsystem*. Overall, the *state subsystem* shows significant fluctuations, which are closely related to the fragile ecological environment of the oasis.

(4)   *Response subsystem*

The *response subsystem* demonstrates stability with minimal fluctuations in its impact on the water resource carrying capacity. As the oasis' economy and society rapidly develop, improvements in sewage treatment, the water supply network, and water storage facilities address water scarcity from extreme weather, ensuring an overall improved water resource carrying capacity. In conclusion, compared to the evaluation index system established by the *DPSR* framework and Liu et al. [15], Zhang et al. [47] established an evaluation index system for the water resource carrying capacity that encompasses water resources, society, the economy, and the ecological environment. This comprehensive system takes into account multiple factors in water resource management, including human activity drivers, water resource pressures, water body conditions, and government and societal response measures. This enables the model to provide a comprehensive analysis of water resource management. In addition, the *DPSR* model establishes a logical and clear framework that can help researchers and managers understand and analyze the key factors and relationships in water resource systems. Through a model analysis, it is possible to better understand the impact of different policies and management interventions on water resource systems and to promote sustainable water resource planning and management.

3.2. Determination of the Obstacles to the Carrying Capacity of Water Resources in the Cele–Yutian Oasis

3.2.1. Index Obstacle Degree Analysis

After using the obstacle degree model to calculate the obstacle factors of the water resource carrying capacity of the Cele–Yutian Oasis, ranking the calculation results, and screening the top five obstacle factors, the results are shown in Table 3.

According to the index obstacle degree calculations, between 2005 and 2014, the primary factors hindering the water carrying capacity of the Cele–Yutian Oasis were the sewage treatment volume ×14, per capita GDP ×3, and total water supply ×9. These factors exhibited obstacle degrees exceeding 12%. The main reason for this is that the economic development of the Cele–Yutian Oasis is relatively slow, and although limited water resources are used extensively for agricultural irrigation, the utilization rate of agricultural water resources is low. In addition, the lack of effective domestic sewage and industrial wastewater purification facilities has led to the direct discharge of some sewage into the ecological environment, making it the main factor that affects water security. From 2015 to 2020, the main obstacle factors changed to the total population ×1 and industrial water supply ×11. This is mainly due to the accelerated urbanization process and the transformation of the agricultural sector. The rise in living standards among the population has resulted in a strained supply of water resources. Additionally, the continued growth of the industry has led to a significant consumption of agricultural water resources, with the industrial water supply emerging as the primary obstacle factor.

Table 3. Top five obstacles in the Celle–Yutian Oasis from 2005 to 2020.

	Ranking of Indicators									
Annum	1	Obstacle Degree	2	Obstacle Degree	3	Obstacle Degree	4	Obstacle Degree	5	Obstacle Degree
2005	×14	19.00%	×3	13.46%	×9	12.78%	×13	10.70%	×7	9.41%
2006	×14	21.01%	×3	13.46%	×9	12.40%	×7	11.25%	×15	9.62%
2007	×14	21.43%	×3	14.73%	×9	10.79%	×7	10.28%	×15	9.45%
2008	×14	19.73%	×9	13.75%	×3	13.10%	×13	11.09%	×7	9.06%
2009	×14	19.73%	×9	12.23%	×3	11.68%	×13	8.25%	×7	7.98%
2010	×14	17.35%	×9	14.24%	×3	11.38%	×13	9.42%	×7	7.73%
2011	×14	24.20%	×3	15.03%	×9	11.20%	×12	10.69%	×1	8.93%
2012	×14	18.77%	×9	16.23%	×9	16.23%	×5	8.48%	×13	8.31%
2013	×14	15.43%	×5	14.31%	×3	11.98%	×1	11.85%	×8	8.77%
2014	×14	13.80%	×1	12.3%	×9	10.16%	×13	9.95%	×5	9.76%
2015	×1	15.53%	×5	11.29%	×9	11.19%	×12	9.24%	×14	8.08%
2016	×1	14.93%	×9	10.52%	×11	10.37%	×12	10.33%	×8	8.06%
2017	×1	18.00%	×11	12.92%	×12	9.70%	×8	7.75%	×16	7.0%
2018	×1	16.78%	×11	10.11%	×13	10.05%	×8	9.13%	×12	8.64%
2019	×1	17.57%	×13	11.24%	×6	10.78%	×11	10.49%	×8	9.53%
2020	×1	19.83%	×11	13.09%	×8	11.90%	×13	11.65%	×6	10.25%

3.2.2. Obstacle Analysis of Subsystems

The analysis of the obstacle degrees within the subsystems reveals varying trends in the challenges to the water resource carrying capacity of the Cele–Yutian Oasis across different years, as shown in Table 4. Before 2011, the *state subsystem* was the main obstacle to the water resource carrying capacity, whereas since 2011, the *drive force subsystem* has become the main obstacle. This shift primarily stems from the implementation of poverty alleviation policies in the developed coastal areas since 2011. This has triggered the expansion of industrial enterprises within the oasis, resulting in heightened industrial water consumption and significant water resource depletion. Simultaneously, advancements in healthcare, population growth, rapid urbanization, and the influx of rural labor into towns have contributed to a continuous rise in the total water usage of the residents.

Table 4. Obstacle degree of the Cele–Yutian Oasis subsystem from 2005 to 2020.

Annum	*Drive Force*	*Pressure*	*State*	*Response*
2005	0.198	0.234	0.327	0.194
2006	0.235	0.225	0.336	0.189
2007	0.254	0.248	0.347	0.203
2008	0.252	0.288	0.373	0.247
2009	0.259	0.276	0.348	0.241
2010	0.259	0.276	0.342	0.242
2011	0.376	0.293	0.270	0.161
2012	0.308	0.282	0.273	0.212
2013	0.433	0.315	0.304	0.271
2014	0.358	0.308	0.292	0.250
2015	0.416	0.300	0.274	0.244
2016	0.382	0.249	0.205	0.213
2017	0.440	0.268	0.216	0.229
2018	0.366	0.283	0.218	0.264
2019	0.346	0.279	0.204	0.288
2020	0.386	0.291	0.197	0.314

3.3. Analysis of Land Use Transfer Mode

Based on the land use distribution maps obtained from remote sensing interpretation in 2005, 2010, and 2020 (Figure 4) as well as the proportion of land use types (Table 5) and the land use transfer matrix (Table 6), the main land use types are grassland, cropland, woodland, water bodies, construction land (urban and rural residential areas), and unused land. Excluding unused land, the Cele–Yutian Oasis consists mainly of grassland, cropland, and woodland. Between 2005 and 2020, both grassland and woodland exhibited a declining trend in their area. The grassland area saw the most significant decrease, plummeting from 35.68% in 2005 to 20.50% in 2020, marking a substantial decline of 15.18%. In comparison, the woodland area decreased by more than half, dropping from 3.19% in 2005 to 1.43% in 2020. Conversely, the cropland area experienced an upward trajectory, increasing from 8.31% in 2005 to 10.39% in 2020. Notably, a substantial portion of grassland has been converted into cropland, aligning with the findings of the research of Gao et al. [48], especially in areas surrounding urban areas. This is mainly due to the increase in population-accelerated urbanization, making flat grassland more suited for construction. At the same time, grassland and woodland far from urban areas have been used extensively for cultivation due to easier access to water resources, expanding along rivers and reservoirs. Furthermore, in the middle reaches of the Keriya River, a substantial portion of grassland on the eastern bank has undergone a transformation into desert and saline–alkali land. This is mainly due to the population growth and the unreasonable development and utilization of water resources. This has led to a decrease in ecological flow in the river, an increase in groundwater depth in the middle and lower reaches, and a reduction in vegetation.

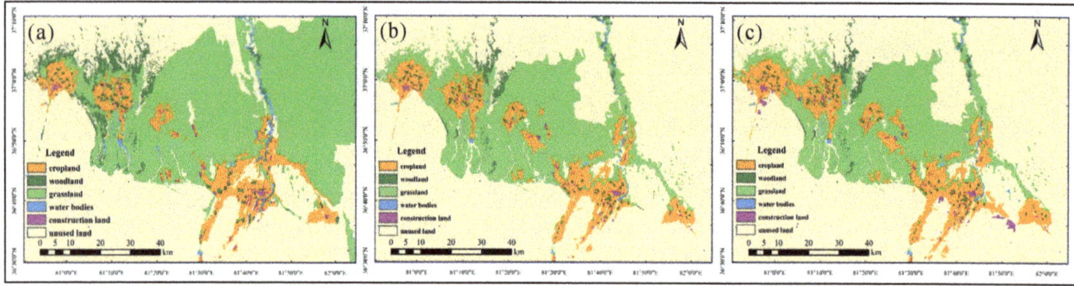

Figure 4. Land use types in 2005, 2010, and 2020. (**a**) Land use types in 2005 (**b**) Land use types in 2010 (**c**) Land use types in 2020.

Table 5. Area and share of each land use type in the Cele–Yutian Oasis in 2005, 2010, and 2020.

Land Use Type	2005		2010		2020	
	Area (km²)	Percentage (%)	Area (km²)	Percentage (%)	Area (km²)	Percentage (%)
grassland	2947.42	35.70%	1789.68	21.67%	1693.22	20.51%
cropland	686.49	8.31%	715.28	8.66%	857.95	10.39%
woodland	262.93	3.18%	118.98	1.44%	118.44	1.43%
water bodies	52.50	0.64%	19.77	0.24%	21.27	0.26%
construction land	23.74	0.29%	25.10	0.30%	45.46	0.55%
unused land	4283.92	51.88%	5588.19	67.68%	5520.66	66.86%

Table 6. Land use transfer matrix of Cele County in Hetian Oasis from 2005 to 2020 (km^2).

Land Use Types 2005	Land Use Types 2020					
	Grassland	Cropland	Woodland	Water Bodies	Construction Land	Unused Land
grassland	1350.24	154.27	3.25	10.82	1.76	1426.14
cropland	75.69	530.71	9.52	28.33	3.24	38.81
woodland	0.77	46.92	14.39	5.49	0.08	0.47
water bodies	151.09	20.37	0.16	65.23	0.39	25.69
construction land	13.56	3.83	0.00	2.93	13.70	18.48
unused land	101.87	101.81	18.14	5.61	2.03	4011.39

Over the past 15 years (Table 7), grassland, forests, water bodies, and unused land have all seen negative changes, indicating various conversions to other land use types. Water bodies, showing the most significant dynamics at −10.16%, are particularly susceptible to transformation, which is linked to the fragile oasis ecosystem. Conversely, cultivated and construction land areas have increased, mainly due to the cropland conversion of grassland and forest land. The growth in construction land is concentrated around towns, taking advantage of the flat terrain of cultivated land, especially during urbanization. In summary, the land use patterns reflect a decrease in grassland, forests, and water bodies along with an increase in cultivated and construction land areas. These shifts primarily resulted from population growth, rapid urbanization, and unsustainable water resource utilization.

Table 7. Land use dynamic index in Cele–Yutian Oasis from 2005 to 2020.

2005–2020	Land Use Types					
	Grassland	Cropland	Woodland	Water Bodies	Construction Land	Unused Land
Area of change (km^2)	−1254.2	171.46	−144.49	−31.23	21.72	1236.74
Dynamic index (%)	−4.94%	1.33%	−8.13%	−9.79%	3.19%	1.49%

3.4. Sustainability Analysis

The evaluation results of the TOPSIS model and Markov model of the land use type transfer show that the overall level of the water resource carrying capacity of the Cele–Yutian Oasis is level III (a critical state); the worst water resource carrying capacity appears in 2007, with a close degree value of 0.39, and the best water resource carrying capacity appears in 2020. The closeness value is 0.57. On the whole, the carrying capacity of water resources shows a benign development trend, but the situation of the water resource carrying capacity in the future is still severe. As can be seen from the weight results of Table 1, the sewage treatment capacity ×14, total water supply ×9, total population ×1, and total GDP ×3 have a great impact on the water resource carrying capacity, which are 0.115, 0.114, 0.086, and 0.081, respectively. This is precisely in line with the results of the index barrier calculation (the sewage treatment capacity ×14, gross domestic product per capita ×3, and total water supply ×9); the three barrier factors have a barrier of more than 12%. According to the analysis results of the land use transfer matrix from 2005 to 2020, grassland, forests, and water bodies are rapidly shrinking, with a dynamic attitude of −4.93%, −8.19%, and −10.16%, respectively, and the reduced part is basically completely used as cultivated land and building land. This is in line with the obstacles to the carrying capacity of water resources. In recent years, the acceleration of urbanization, industrialization, and the rapid growth in the population have brought serious challenges to the sustainable development of the oasis. Due to the fragile ecological environment of the oasis, in future development planning, the government of Cele and Yutian should take a series of countermeasures to promote the sustainable development of the oasis:

(1) Establish a scientific and rational water resource management system, formulate and implement policies for the rational allocation of water resources, and strengthen water resource monitoring and forecasting.

(2) Vigorously promote water-saving equipment, plan population distribution reasonably, enhance the awareness of water conservation, introduce advanced water-saving irrigation technology, reduce the total consumption of water resources, and thereby improve the efficiency of water resource utilization.

(3) Strengthen environment-friendly construction, protect and restore water sources and wetland systems, strengthen the construction of sand control projects, and protect the surrounding ecological areas and residential areas from harm.

(4) Give full play to the role of water conservation projects; regulate the uneven distribution of water resources through the construction of reservoirs, various canal systems, and other engineering projects; and by prioritizing drinking water and ecological water use, allocate industrial and agricultural water use to achieve the rational allocation of water resources.

(5) Strengthen the prevention and control measures for water pollution; control the discharge of industrial wastewater, domestic sewage, pesticides, and fertilizers; increase the intensity of sewage treatment; and improve the reuse rate of wastewater.

Coupled with the dynamic change in the water resource carrying capacity and land interest type, it is first necessary to design a scientific and reasonable research program; select indicators to follow the principles of science, representativeness, operability, and regionalism; establish an appropriate evaluation index system of the water resource carrying capacity; and then combine the Markov model to assist in evaluating regional sustainable development. This evaluation model is especially suitable for the long-term analysis of desert oases, river basins, etc. This evaluation research method will open up new ideas and methods for the evaluation of the regional water resource carrying capacity and sustainable development analysis, and it is of great significance for promoting the efficient and sustainable utilization of water resources and land and for coordinating regional development.

4. Conclusions

This study selected the Cele–Yutian Oasis, located in the arid desert area, as the research area. By introducing the *DPSR* model, TOPSIS model, and obstacle degree model, the water resource carrying capacity of the oasis from 2005 to 2020 was evaluated, and the main obstacle factors were calculated. At the same time, the Markov model of the land use type transfer matrix was used to compare and analyze the distribution of and changes in the surface coverage types in the oasis. The following conclusions were drawn from the comprehensive research results:

(1) From 2005 to 2020, the water resource carrying capacity of the Cele–Yutian Oasis generally improved. The lowest point occurred in 2007, with a proximity value of 0.39, categorizing it as level IV (a mild overload). By 2020, the highest proximity value reached 0.58, but the carrying capacity still remained at level III (critical), indicating a persistently severe water resource situation.

(2) The analysis results of the carrying capacity of the subsystems show that, from 2005 to 2020, the influence of the *drive force subsystem* decreased, the influence of the *pressure subsystem* and the *state subsystem* fluctuated, and the influence of the *response subsystem* increased and then decreased, reaching a maximum value of 0.98 in 2014.

(3) The results of the analysis of the obstacle degree subsystem show that the *drive force subsystem* gradually replaced the *state subsystem* as the main obstacle subsystem affecting the Cele–Yutian Oasis between 2005 and 2020; the main obstacle factors affecting the water resource carrying capacity included the amount of sewage treatment, the gross domestic product per capita, and the total water supply.

(4) During the period of 2005 to 2020, the results of the land use transfer matrix show that the grassland area decreased by 15.18% and that the forest area decreased by half. The dynamics of grassland, forests, and water bodies are all negative, indicating that these three types of land have been transferred to other types of land in different ways, with the largest change occurring in water bodies, with a dynamic value of -10.16%.

This study comprehensively considers factors such as water resources, society, the economy, and ecology and analyzes the changing patterns of the water resource carrying capacity and land use types. To further improve the water resource carrying capacity, it is advisable for local governments to strongly embrace the key concept of "determining land, people, and production based on water." This requires the implementation of strategies aimed at optimizing the allocation of water resources, actively promoting the development of water-saving industries, and ultimately achieving the efficient utilization of water resources.

However, this study only focuses on the current annual water resource carrying capacity and dynamic attitude of the land use change in the study area. If the water resource carrying capacity and the evolution trend of the land use type can be simulated and predicted in the future, it will be beneficial to the optimal allocation and sustainable development of regional water resources, which will be the focus of our follow-up research.

Author Contributions: Methodology, G.J.; Software, G.J. and N.L.; Validation, Y.G.; Investigation, F.J., N.L. and F.L.; Resources, Y.G.; Data curation, G.J., N.L. and F.L.; Writing—original draft, G.J.; Visualization, F.J.; Supervision, S.L.; Funding acquisition, S.L. All authors have read and agreed to the published version of the manuscript.

Funding: This research was funded by the National Natural Science Foundation of China (grant number U1603243).

Data Availability Statement: The general data are included in this article. Additional data are available on request.

Acknowledgments: We acknowledge the National Natural Science Foundation of China for funding this project.

Conflicts of Interest: The authors declare no conflict of interest.

References

1. Guo, Q.; Wang, J.; Zhang, B. Comprehensive evaluation of the water resource carrying capacity based on DPSIRM. *J. Nat. Res.* **2017**, *32*, 484–493.
2. Xia, J.; Zuo, Q. China's decade summary and prospect of water resources academic exchange. *J. Nat. Resour.* **2013**, *28*, 1488–1497.
3. Yan, L.; Qu, Z.; Fang, S. Discussion on the operation of ecological capitalization of water resources. *China Popul. Resour. Environ.* **2011**, *21*, 81–84.
4. Jing, J.; Song, H.L.; Jiang, G. Dynamic assessment of sustainable development of water ecological economic system based on emergy analysis in Ordos, China. *Water Sav. Irrig.* **2015**, *1*, 48–51.
5. Du, X.; Li, Y.; Zhang, X. Research on water resources carrying capacity of Zhengzhou city based on TOPSIS model. *People's Yellow River* **2022**, *44*, 84–88.
6. Jia, Y.; Wang, H. Study on water resource carrying capacity of Zhengzhou city based on DPSIR model. *Int. J. Environ. Res. Public Health* **2023**, *20*, 1394.
7. Zhang, Z.; Chen, F.; Long, A.; He, X.; He, Z. Evaluation of water resources security in arid zones based on topable cloud modeling—A case study of Shiheze Reclamation Area. *Arid Zone Res.* **2020**, *37*, 847–856.
8. Jia, Y.; Shen, J.; Wang, H.; Dong, G.; Sun, F. Evaluation of the spatiotemporal variation of sustainable utilization of water resources: Case Study from Henan Province (China). *Water* **2018**, *10*, 554.
9. Dang, X.; Zhao, X.; Kang, Y.; Liu, X.; Song, J.; Zhang, Y. Analysis of Carrying Capacity and Obstacle Factors of Water Resources in Longnan City, China, Based on Driving–Pressure–State–Response and Technique for Order Preference by Similarity to an Ideal Solution Models. *Water* **2023**, *15*, 2517.
10. Gao, Y.; Li, Y.; Gao, J.; Du, C.; Zhang, M.; Wu, J. Evaluation of regional water resources carrying capacity under the influence of virtual water. *Prog. Water Resour. Hydropower Sci. Technol.* **2022**, *42*, 22–27+77.
11. Liu, Q.; Wang, X.; Zhu, Q.; Song, J.; Yan, Q.; Zhao, Y. Coupling relationship of water resources carrying capacity system in Tibet Autonomous Region based on the function of "three lives". *J. Nat. Resour.* **2023**, *38*, 1618–1631.
12. Li, W.; Han, P.; Zhao, X. Research on sustainable development of tourism in ecologically fragile areas based on energy-value analysis--taking Gannan Tibetan Autonomous Prefecture as an example. *J. Ecol.* **2018**, *38*, 5894–5903.
13. Ni, T. Study on the Carrying Capacity and Sustainable Utilization of Water Resources in Jining City. Master's Thesis, Shandong Normal University, Jinan, China, 2011.
14. Li, Z.; Jin, J.; Cui, Y.; Zhou, R.; Ning, S.; Zhou, Y.; Zhou, L. Evaluation method of regional water resources carrying capacity based on semi-partial linkage number and dynamic subtractive set pair potential. *Lake Sci.* **2022**, *34*, 1656–1669.

15. Liu, H. Comprehensive evaluation of water resources carrying capacity of Yangtze River Economic Zone based on entropy weight TOPSIS-coupling coordination degree-gray correlation. *J. Wuhan Univ. (Eng. Ed.)* **2023**, *56*, 532–541.

16. Chen, L.; Zhou, H. Evaluation of water resources carrying capacity of karst basin based on fuzzy comprehensive evaluation and principal component analysis. *Saf. Environ. Eng.* **2021**, *28*, 159–173.

17. Ren, J.; Bing, J.; Zhang, X. Evaluation of water resources carrying capacity and adaptation analysis in Tianmen City. *J. Chang. Acad. Sci.* **2018**, *35*, 27–31.

18. Wang, Y.; Li, G.; Yan, B. Simulation study of water resources carrying capacity of a river basin based on SD and AHP model. *Syst. Eng.* **2022**, *40*, 24–32.

19. Liu, S.; Chen, J. Calculation of Water Resources Carrying Capacity in Urban Planning—Taking Huizhou City as an Example. *Water Resour. Prot.* **2006**, *22*, 47–50.

20. Li, S.; Zhao, H.; Wang, F.; Yang, D. Evaluation of water resources carrying capacity of Jiangsu Province based on AHP-TOPSIS model. *Water Resour. Prot.* **2021**, *37*, 20–25.

21. Ren, B.; Zhang, Q.; Ren, J.; Ye, S.; Yan, F. A Novel Hybrid Approach for Water Resources Carrying Capacity Assessment by Integrating Fuzzy Comprehensive Evaluation and Analytical Hierarchy Process Methods with the Cloud Model. *Water* **2020**, *12*, 3241.

22. Li, H.; Du, Y.; Wu, S.; Zhang, Z. Evaluation of the agricultural water resource carrying capacity and optimization of a planting-raising structure. *Agric. Water Manag.* **2021**, *243*, 106456.

23. Xu, H.; He, Y.; Elias, T.; Si, H.; Lu, X. Simulation and scenario simulation of water resources carrying capacity in Wusu city based on SD model. *J. Chang. Acad. Sci.* **2023**, 1–8.

24. Li, F.; Liu, W.; Dong, Z.; Jia, Z.; Li, Q. Comprehensive evaluation of ecological footprint and sustainable utilization of water resources in Sichuan Province. *Environ. Pollut. Prev.* **2023**, *45*, 245–249+256.

25. Fu, J.; Cao, R.; Zeng, W.; Zhuo, Y.; Wang, L. Bayesian network-based risk assessment of overloading of water environmental carrying capacity in a watershed--a case study of the North Canal Basin. *J. Environ. Sci.* **2023**, *43*, 516–528.

26. Ma, T.; Han, S.; Li, F.; Li, W.; Zhao, M.; Li, H. Water resources carrying capacity and coordinated development in Yulin City, Shaanxi Province. *Soil Water Conserv. Bull.* **2023**, *43*, 248–255.

27. Adriaanse, A. *Environmental Policy Performance Indicators*; S.D.U. Uitgevery: The Hague, The Netherlands, 1993.

28. Tong, C. Review on environmental indicator research. *Res. Environ. Sci.* **2000**, *13*, 53–55.

29. Zare, F.; Elsawah, S.; Bagheri, A. Improved integrated water resource modelling by combining DPSIR and system dynamics conceptual modelling techniques. *J. Environ. Manag.* **2019**, *246*, 27–41.

30. Wang, J.; Jiang, D.; Xiao, W.; Chen, Y.; Hu, P. Study on theoretical analysis of water resources carrying capacity: Definition and scientific topics. *J. Hydraul. Eng.* **2017**, *48*, 1399–1409.

31. Liu, Y.; Luo, Y.; Zhang, W.; Wu, Y.; Wang, Q. Research on the construction of urban water resources carrying capacity evaluation index system based on pressure-state-response model. *Environ. Pollut. Prev.* **2016**, *38*, 100–104. [CrossRef]

32. Xu, Y.; Chen, J.; Xia, H.; Chu, L.; Zhang, X. Evaluation of water resources carrying capacity in Huai'an city based on DPSR-improved TOPSIS model. *J. Water Resour. Water Eng.* **2019**, *30*, 47–52+62.

33. Hu, Z.; Pang, Y.; Xu, R.; Yu, H.; Niu, Y.; Wu, X.; Liu, Y. Systematic Evaluation and Influencing Factors Analysis of Water Environmental Carrying Capacity in Taihu Basin, China. *Water* **2023**, *15*, 1213.

34. Sun, Y.; Liu, W.; Sheng, Y. Analysis of Spatial and Temporal Differences in Economic and Ecological Resilience of Water Resources and Influencing Factors in Xinjiang Based on PSR Model. *Arid Zone Geogr.* 2023. Available online: http://kns.cnki.net/kcms/detail/65.1103.X.20230719.1853.001.html (accessed on 18 September 2023).

35. Liu, H.; Liu, Y.; Meng, L.; Jiao, K.; Zhu, M.; Chen, Y.; Zhang, P. Research progress of entropy weight method in water resources and water environment evaluation. *Glacial Permafr.* **2022**, *44*, 299–306.

36. Zhou, C.; Wang, Y.; Tian, S.; Xue, Z.; Huang, R. Evaluation of Water Environment and Diagnosis of Obstacle Factors in Baoji City Based on DPSR Model. *J. Xi'an Univ. Technol.* 2023. Available online: http://kns.cnki.net/kcms/detail/61.1294.N.20230717.1357.002.html (accessed on 18 September 2023).

37. Zuo, Q.; Zhang, Z.; Wu, B. Evaluation of water resources carrying capacity of nine provinces and districts in the Yellow River Basin based on the combined weights TOPSIS model. *Water Resour. Prot.* **2020**, *36*, 1–7.

38. Yang, Z.; Song, J.; Cheng, D.; Xia, J.; Li, Q. Muhammad Irfan Ahamad. Comprehensive evaluation and scenario simulation for the water resources carrying capacity in Xi'an city, China. *J. Environ. Manag.* **2019**, *230*, 221–233.

39. Gu, J.; Hu, W.; Tian, S. Evaluation of ecological carrying capacity and diagnosis of obstacle factors in Jiangsu Province based on DPSIR-TOPSIS model. *Soil Water Conserv. Bull.* **2019**, *39*, 246–252.

40. Liu, R.; Zhu, D. Exploration of land use change information mining method based on transfer matrix. *Resour. Sci.* **2010**, *32*, 1544–1550.

41. Tao, C.; Xu, G.; Le, H. Simulation study of land use change in Sino-Vietnamese border area based on Markov chain model. *South. Nat. Resour.* **2022**, 44–49.

42. Chen, Q.; Chen, A.; Ye, Y.; Min, J.; Zhang, D. Impact of land use change on groundwater quality in the Dianchi basin. *China Environ. Sci.* **2023**, *43*, 301–310. [CrossRef]

43. Zhang, J.; Li, J.; Yin, B.; Gao, Y.; Liu, X. Analysis of land use change in Junggar Banner based on transfer matrix. *Soil Water Conserv. Bull.* **2018**, *38*, 131–134. [CrossRef]

44. Wei, X. Research on Urbanisation and Water Resources Carrying Capacity in Southern Xinjiang Region. Master's Thesis, Xinjiang University, Ürümqi, China, 2020. [CrossRef]
45. Zhao, Y.; Yang, J. Spatial and temporal patterns of water resources carrying capacity and coupled coordination of subsystems in South Xinjiang region. *Arid Zone Res.* **2023**, *40*, 213–223. [CrossRef]
46. Han, X. Exploration of water resources status and countermeasures in Celle County, Xinjiang. *Groundwater* **2017**, *39*, 184–185+194.
47. Zhang, X.; Cao, Y.; Bao, T.; Wang, Y.; Shi, F. Study on water resources carrying capacity of yellow diversion receiving area in Henan Province based on combined weights TOPSIS model. *People's Yellow River* **2023**, *45*, 73–78.
48. Gao, Y.; Yang, Y.; Sun, L.; Fan, L.; Yu, R. Study on spatial and temporal changes of groundwater and surface cover in Celle Oasis. *Res. Agric. Arid Reg.* **2020**, *38*, 200–208.

Article

Rapid Antibiotic Adsorption from Water Using MCM-41-Based Material

Jie Chen [1,†], Yao Yang [2,†], Yuanyuan Yao [1], Zhujian Huang [2], Qiaoling Xu [2], Liping He [1] and Beini Gong [2,*]

[1] CCCC Fourth Harbor Engineering Institute Co., Ltd., Guangzhou 510230, China; cjie12@cccc4.com (J.C.); yyuanyuan@cccc4.com (Y.Y.); e001@cccc4.com (L.H.)

[2] Guangdong Provincial Key Laboratory of Agricultural & Rural Pollution Abatement and Environmental Safety, College of Natural Resources and Environment, South China Agricultural University, Guangzhou 510642, China; yangyao@stu.scau.edu.cn (Y.Y.); zjhuang@scau.edu.cn (Z.H.); amy.198510@163.com (Q.X.)

* Correspondence: bngong@scau.edu.cn; Tel.: +86-13632245773

† These authors contributed equally to this work and should be considered co-first authors.

Abstract: The contamination of antibiotics in the environment has raised serious concerns, impacting both human life and ecosystems. This has led to a growing focus on the development of cost-effective and environmentally friendly adsorbent materials. Mesoporous molecular sieve MCM-41, known for its strong adsorption capacity, low cost, and efficient regenerative properties, holds significant promise for addressing this issue. In this study, we investigated the adsorption behavior of demolded MCM-41 materials in relation to tetracycline, doxycycline, and levofloxacin at different temperatures and pH levels. Our experiments encompassed the adsorption of these three common antibiotics, revealing that a neutral or weakly acidic pH environment promoted adsorption, whereas alkaline conditions hindered it. Utilizing the equilibrium isotherm model, we determined the theoretical maximum adsorption capacities for tetracycline (TC), doxycycline (DOX), and levofloxacin (LFX) as 73.41, 144.83, and 33.67 mg g^{-1}, respectively. These findings underscore the significant potential of MCM-41 in mitigating antibiotic wastewater contamination.

Keywords: MCM-41; antibiotics; rapidly adsorption; wastewater

Citation: Chen, J.; Yang, Y.; Yao, Y.; Huang, Z.; Xu, Q.; He, L.; Gong, B. Rapid Antibiotic Adsorption from Water Using MCM-41-Based Material. *Water* **2023**, *15*, 4027. https://doi.org/10.3390/w15224027

Academic Editors: Peiyue Li and Jianhua Wu

Received: 21 September 2023
Revised: 2 November 2023
Accepted: 6 November 2023
Published: 20 November 2023

1. Introduction

As urbanization progresses, the transformation of people's lifestyles and production methods is introducing new challenges. Consequently, the prevalence of emerging contaminants like antibiotics and dyes in water sources has surged. This surge in pollutants directly contributes to the degradation of water quality, impeding its recovery and efficient utilization. Simultaneously, antibiotic pollution exerts a profoundly detrimental impact on ecosystems, including animals, plants, and human populations. Notably, antibiotics like doxycycline (DOX), tetracycline (TC), and levofloxacin (LFX) have emerged as environmental pollutants, posing threats to diverse life forms [1]. Furthermore, the unregulated discharge of pollutants such as pharmaceutical waste, dyes, and heavy metals directly into water bodies, devoid of proper classification and pretreatment, represents a significant global environmental challenge within the context of sewage treatment [2].

Antibiotics are widely used to prevent and treat diseases and supplement animal feed. Presently, there are more than 20 types of tetracyclines. However, tetracycline (TC), chlortetracycline, oxytetracycline, and doxycycline (DOX) are the most commonly used antibiotics in the poultry industry [3]. Tetracyclines are used for disease treatment and prevention and as growth promoters for entire populations. Therefore, they deserve special attention due to their crucial role in health and the environment. Tetracycline (TC) is an antibiotic that humans and veterinarians use against various harmful bacteria. Maintaining activity after poor intestinal absorption will lead to residues in edible products. These residues cause problems with the emergence and spread of antibiotic-resistant bacteria. DOX is also an antibiotic belonging to the tetracycline family, widely used in human and animal health. It has effectively treated infections in the human gut, kidneys, lungs, respiratory, and reproductive organs [4]. It is also used as an animal veterinary antibiotic [5]. Commonly used medical antibiotics include doxycycline, tetracycline, and levofloxacin [6]. They enter the environment through animal diets. In the body, they are usually not completely absorbed and metabolized. Between 30% and 90% of them can be discharged and released into the environment through animal excretion [7,8]. In addition to biological sources, antibiotics also come from wastewater in chemical manufacturing, mining, pharmaceutical, textile, and other industries. As a result, it contains high levels of toxic substances, organic pollutants, and many other complex compounds that destroy the integrity of the surface and groundwater [9–11]. These pollutants often are antibiotics such as levofloxacin and dyes such as Congo red, which may accumulate in the environment [12]. The wastewater treatment plant cannot completely remove antibiotics because they are non-biodegradable, thereby possessing a stable chemical structure. Then, the polluted wastewater flows into surface water, seeps into groundwater, and eventually into drinking water, which endangers the environment [13,14]. There are no standardized criteria for safe levels of antibiotic residues. However, in general, for some antibiotics that are difficult to degrade (e.g., tetracyclines), the residue level should not exceed 100 ng/L, and the total residue level of all types of antibiotics should not exceed 500 mg/L [15]. Current methods for removing organic compounds such as antibiotics include electrocoagulation [16], photocatalysis, chemical degradation, advanced oxidation, adsorption, solidification, membrane coagulation, ion exchange, reverse osmosis, and bioremediation [17–21]. Adsorption has been proven to be a feasible and economical method to remove water pollutants such as antibiotics and dyes. This prevalent research method has several advantages, including simpleness, fast kinetics, low cost, and high efficiency [22]. Various materials like clay minerals, activated carbon, zeolites, silica gels, polymeric resins, bioadsorbents, etc. can be used as adsorbents. The mechanism involves the adsorbents removing the target pollutants from the aqueous solution. Studies have shown adsorption can effectively reduce the emissions of antibiotics and organic contaminants in wastewater. The adsorption method has been widely used in wastewater treatment [12].

A molecular sieve is a new type of selective adsorbent medium with a high adsorption rate, which can selectively adsorb unsaturated, polar, and polarizable molecules according to molecular size and configuration. Common molecular sieve materials include zeolites, aluminophosphates, metal–organic frameworks, etc. The open framework structure and large internal and external specific surface area of mesoporous molecular sieve materials make them demonstrate unique adsorption functions [23]. At the same time, the molecular sieve has an excellent regeneration function because of its good physical and chemical stability. Mesoporous molecular sieve MCM-41 is a new synthetic molecular sieve with a regular hexagonal arrangement and uniform pore size. Studies have shown that MCM-41 has rich pore channels and good adsorption capacity. Also, MCM-41 has the advantages of low cost, strong adsorbent regeneration, and high application value. Presently, the research on mesoporous material MCM-41 mostly stays in the adsorption of antibiotic pollutants after modification. In contrast, the adsorption research on MCM-41 without modification mostly focuses on dye organics. Up to now, the adsorption of antibiotic pollutants has not been reported.

Previous studies have demonstrated the advantages of using MCM-41 as an adsorbent for antibiotic pollutants. In this study, we explore the adsorption performance of these demolded MCM-41 materials concerning three different pollutants across varying temperatures. Factors such as adsorption duration, MCM-41 surface charge, and the underlying physical properties of the adsorption mechanism are meticulously examined and confirmed. Utilizing adsorption kinetics and isotherm models, we assess the adsorption data for these three distinct pollutants, providing valuable insights into the adsorption mechanism's intricacies. The zeta potential characterization of MCM-41 materials serves as a pivotal tool in enhancing our understanding of the adsorption process and its associated mechanisms.

2. Materials and Methods

2.1. Materials

The analytical reagents employed in this study consisted of MCM-41, tetracycline (C22H24N2O8•HCl), sodium hydroxide (NaOH), levofloxacin, doxycycline, and hydrochloric acid. Initially, the MCM-41 molecular sieve was procured from Tianjin Yuanli Chemical Co., Ltd. (Tianjin, China). Subsequently, tetracycline, levofloxacin, and doxycycline antibiotics were obtained from Shanghai Yuanye Biotechnology Co., Ltd. (Shanghai, China). Hydrochloric acid (HCL) and sodium hydroxide (NaOH) were purchased from Sinopharm Chemical Reagent Co. (Shanghai, China). Moreover, ultrapure water served as the primary solvent throughout the experimental procedures.

2.2. Mesoporous MCM-41 Demolding Treatment

The acquired full-silica MCM-41 molecular sieve was loaded into a ceramic crucible weighing 4.0 g. The crucible, containing the molecular sieve, was then placed inside a box furnace. in the demolded samples were calcined at the temperature of 550 °C, 600 °C, 700 °C, 800 °C and 900 °C, respectively. Following this temperature ramp-up sequence, a constant temperature was maintained for a duration of six hours. Subsequently, the contents were allowed to cool, and the materials were packed into bags.

2.3. Adsorption Experiments on Three Kinds of Pollutants

Approximately 0.06 g of tetracycline was dissolved in a beaker and subsequently transferred to a 2-L volumetric flask, resulting in a tetracycline solution with a concentration of 30 mg/L. Simultaneously, a 10 mL centrifuge tube was prepared to hold the sample. In a separate 800 mL beaker equipped with a 250 mL measuring tube, approximately 200 mL of the tetracycline solution was added. The solution was then subjected to constant temperature magnetic stirring. The stirring parameters were initially set to 200 r/min with synchronous operations. Furthermore, six distinct beakers, each labeled as 550, 600, 700, 800, 900, and a blank control, were loaded with 0.1 g of MCM-41 calcined at temperatures of 550 °C, 600 °C, 700 °C, 800 °C, and 900 °C, respectively. These mixtures were stirred with 8 mL of sampling liquid in a 10 mL centrifuge tube for varying durations, specifically 10 min, 20 min, 30 min, 40 min, 70 min, 100 min, and 130 min, using a disposable syringe for uniform distribution. Subsequently, the absorbance values were measured at wavelengths of 375 nm, 351 nm, and 287 nm using a UV spectrophotometer to determine the concentrations of tetracycline, doxycycline, and levofloxacin, respectively.

The adsorption isotherms for three antibiotics adsorption by MCM-41 mesoporous materials were modelled using the Freundlich, Langmuir, and Temkin equations.

$$\text{Freundlich isotherm:} \qquad Q_e = K_F C_e{}^n \tag{1}$$

where Q_e equals the amount of adsorption, and K_F is the Freundlich adsorption coefficient, relating to the capacity and the intensity of adsorption. C_e is the equilibrium concentration of three antibiotics in the solution. n is Freundlich adsorption index.

$$\text{Langmuir isotherm}: \qquad Q_e = \frac{Q_m K_L C_e}{1 + K_L C_e} \tag{2}$$

where Q_m is the limited monolayer adsorption capacity. K_L is the Langmuir adsorption coefficient representing the affinity of the adsorbent for the adsorbate.

$$\text{Temkin isotherm}: \quad Q_e = \left(\frac{RT}{b_T}\right)\ln A_T + \left(\frac{RT}{b_T}\right)\ln C_e \tag{3}$$

where b_T is Temkin heat of adsorption, and A_T is the Temkin adsorption potential.

A series of adsorption experiments using solutions containing tetracycline, doxycycline, and levofloxacin at varying initial concentrations were conducted: 0 mg·L^{-1}, 10 mg·L^{-1}, 15 mg·L^{-1}, 20 mg·L^{-1}, 25 mg·L^{-1}, 30 mg·L^{-1}, 35 mg·L^{-1}, 40 mg·L^{-1}, and 50 mg·L^{-1}. These experiments were designed to establish a correlation between the initial concentration of the solutions and the adsorption capacity. Consequently, we aimed to assess how the initial concentration influences the adsorption capacity of the mesoporous material MCM-41.

About 0.05 g of tetracycline, doxycycline, and levofloxacin were dissolved in 1 L volumetric flask. The concentration of pollutants was 50 mg·L^{-1}. Different solutions of 0 mL, 4 mL, 6 mL, 8 mL, 10 mL, 12 mL, 14 mL, 16 mL, and 20 mL were put into nine 50 mL beakers using a 25 mL measuring tube. Then, a certain amount of distilled water was added to the 20 mL solution. Approximately 0.1 g MCM-41 adsorbent calcined at 800° was added to the solution and filtered using a 0.45 μm filter. Subsequently, the absorbance value was determined.

At room temperature, adsorption experiments were conducted using 50 mg·L^{-1} solutions of tetracycline, doxycycline, and levofloxacin. The aim was to assess the adsorption kinetics at various time intervals and ascertain the point of adsorption equilibrium. Each group consisted of three parallel experiments. To initiate the experiments, 200 mL of the pollutant solution was poured into a 500 mL beaker. Subsequently, 0.1 g of mesoporous material MCM-41 was introduced into the beaker, and the solution was placed on a magnetic stirrer. The stirring commenced at a predetermined time, maintaining a constant rotational speed of 250 r/min. Stirring intervals included 0 min, 1 min, 2 min, 3 min, 4 min, 5 min, 7 min, 9 min, 20 min, 30 min, 40 min, 50 min, 60 min, and 70 min. Following the designated stirring time, the solution underwent filtration using a 0.45 μm filter membrane. Subsequently, it was transferred into a centrifuge tube, and the pollutant concentrations in the solution were quantified using a spectrophotometer. The concentrations of pollutants remained stable after reaching adsorption equilibrium.

3. Results

3.1. Physicochemical Properties of MCM-41

As shown in Figure 1, different calcination temperatures had little effect on the adsorption performance of MCM-41 material, but the adsorption performance was significantly highest at 800 °C. As shown in Figure 2, the peak tends toward an increasing angle and then to a decreasing angle with an increase in the calcination demolding temperature. This shows that the empty surface structure of the material becomes smaller and larger. When the temperature is too high, many pores are melted, and the peak disappears. The X-ray diffraction (XRD) images for MCM-41 calcined within the temperature range of 550–800 °C closely resemble those of the original MCM-41, indicating the preservation of its ordered structure. However, a significant alteration in the image becomes apparent at the 900 °C calcination temperature. While the pure adsorption efficacy is optimal at 900 °C calcination, this temperature jeopardizes the integrity of the ordered mesoporous material structure. It is worth noting that the calcination process demands substantial energy consumption and prolonged durations, rendering it impractical for certain applications.MCM-41 has a strong peak at $2\theta = 5.5°$ corresponding to the (100) of hexagonal mesoporous structure, of which the spacing distance is 1.589 nm. After 900 °C of calcination, the reflection (100) collapsed.

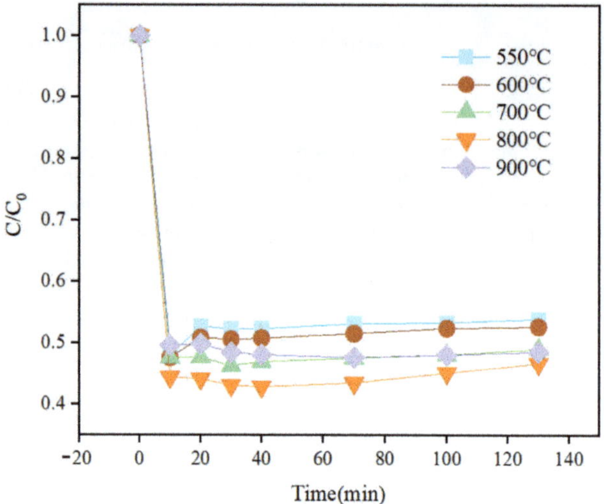

Figure 1. Comparison of adsorption performance on MCM-41 materials at different calcination temperatures (demold temperature: 550 °C, 600 °C, 700 °C, 800 °C, 900 °C; TC concentration: 30 mg/L; MCM-41 concentration: 0.5 g/L).

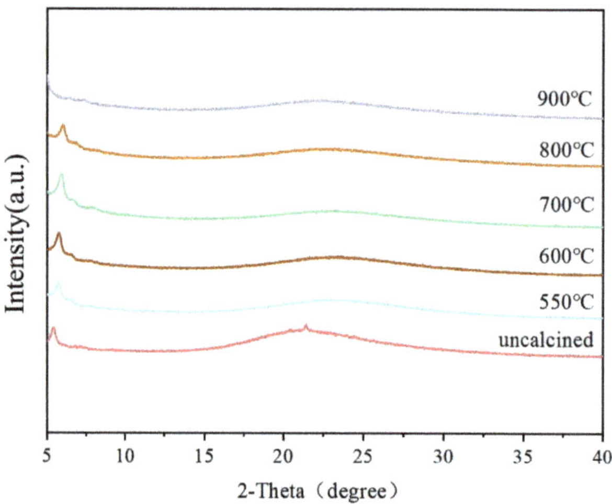

Figure 2. XRD patterns of MCM-41 materials at different calcination temperatures.

3.2. Adsorption of Antibiotics

Figure 3 illustrates the temporal evolution of pollutant adsorption by MCM-41, revealing a gradual increase in adsorption with time, culminating in a plateau at a specific time point. This plateau signifies that the available pollutants are retained in the solution, establishing a dynamic equilibrium between adsorbed and unadsorbed pollutants within the adsorbent. The duration required to attain this equilibrium state is referred to as the equilibrium time, and the quantity of pollutants adsorbed at this juncture represents the maximum adsorption capacity under these unique conditions. Remarkably, the equilibrium state is reached in less than 10 min, underscoring the rapid adsorption observed during the initial 10 min. This phenomenon can be attributed to the abundant availability of active sites on the surface of MCM-41. As these sites gradually become occupied, the efficiency of adsorption diminishes, and the rate of adsorption slows as it approaches equilibrium.

Prior research [24] has demonstrated that MCM-41 possesses a notably high saturated adsorption capacity. The generous pore size on the surface of MCM-41 allows pollutants to swiftly enter and undergo adsorption, leading to a mechanism characterized as continuous adsorption–diffusion–adsorption [25], where adsorption progresses from the exterior to the interior of the material.

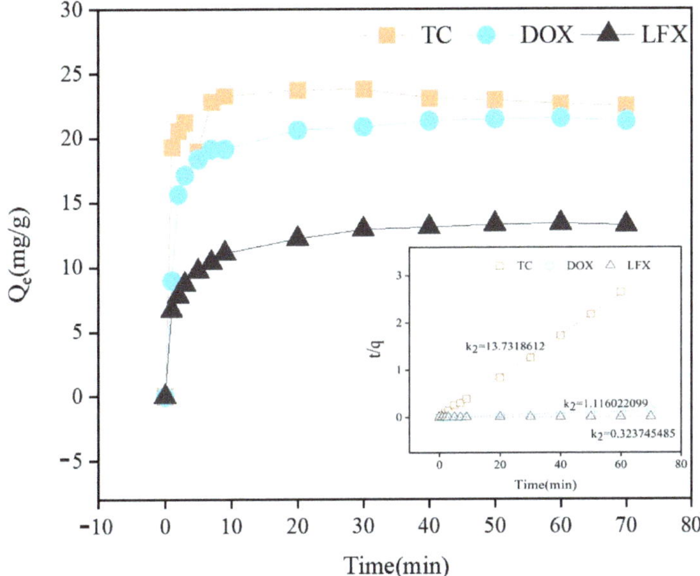

Figure 3. Adsorption effect of MCM-41 (800 °C) on different pollutants and kinetic equation of pseudo-second-order adsorption rate kinetic model diagram (initial TC, DOX, LFX concentration: 50 mg/L; adsorbent dose: 0.5 g/L; pH: 7.0; temperature: 25 °C).

The kinetic study of the adsorption process mainly describes the adsorbent's absorption rate, which controls the retention time of the adsorbent at solid–liquid interface. The adsorption kinetics model can be described using Lagrange's first-order rate equation:

$$\frac{dq}{dt} = k_1(q_e - q) \tag{4}$$

The Ho and McKay equation describes the pseudo-second-order model of adsorption kinetics. It is based on the rate control step. In addition, it is a second-order kinetic equation for chemical adsorption via chemical reactions involving the sharing, gaining, or losing of electrons. The expression formula of the second-order dynamic equation is as follows:

$$\frac{dq}{dt} = k_2(q_e - q)^2 \tag{5}$$

where mg/g is the equilibrium adsorption quantity, mg/g is the adsorption quantity for a certain time, min is the first-order adsorption kinetic rate constant, and g/(mg·min) is the second-order adsorption kinetic rate constant. For Formulas (4) and (5), from $t = 0$ to $t > 0$ (=0 to >0), the linear form is as follows:

$$\ln(q_e - q) = lnq_e - k_1t \tag{6}$$

$$\frac{t}{q} = \frac{1}{k_2q_e} + \frac{t}{q_e} \tag{7}$$

The adsorption experiments for tetracycline (TC), doxycycline (DOX), and levofloxacin (LFX) at a concentration of 50 mg/L were conducted under identical conditions, including stirring speed, temperature (25 °C), and MCM-41 concentration. Subsequently, the adsorption kinetics data obtained for MCM-41 were fitted to kinetic models, specifically the first-order and second-order kinetic adsorption models, as illustrated in Figure 3. The dynamic parameters resulting from the correlation calculations are presented in Table 1. The fitting outcomes reveal that the adsorption of TC, DOX, and LFX by MCM-41 conforms to the second-order adsorption kinetics model, as indicated by R2 values of 0.999. This observation suggests that the adsorption of these three pollutants by MCM-41 is characterized as chemical adsorption. Therefore, the adsorption of the three antibiotics onto MCM-41 demonstrated high efficiency, and this mode of adsorption can be attributed to chemical adsorption.

Table 1. Kinetic parameters for the adsorption of TC, DOX, and LFX onto MCM-41.

Pollutants	Adsorption Capacity Determined by Experiment qe, exp (mg/g)	Kinetic Equation of Pseudo-First-Order Adsorption Rate			Kinetic Equation of Pseudo-Second-Order Adsorption Rate		
		k_1	q_e	R^2	k_2	q_e	R^2
TC	23.77	0.02	2.30	0.126	13.73	22.97	0.999
DOX	21.54	0.06	6.39	0.843	1.12	495.04	0.998
LFX	13.35	0.06	0.93	0.941	0.32	6794.07	0.997

k_1: min, k_2: g/(mg·min), and q_e: mg/g.

The equilibrium relationship existing between MCM-41 and pollutants was described by the adsorption isotherms. The adsorption isotherms of mesoporous material MCM-41 for the three types of antibiotics are shown in Figure 4. The values of adsorption constants (K_F, n, K_L, Q_m, b_T, A_T, R^2) are shown in Table 2. The correction coefficient R^2 shows that the Langmuir model (0.96–0.99) can better describe the adsorption process than the Temkin model (0.95–0.98). This result showed that antibiotic adsorption on MCM-41 could be a multilayer. Therefore, it is concluded that the adsorption force of MCM-41 for these three types of pollutants is similar to the surface bond force forming a certain chemical bond. The obtained results are consistent with the second-order kinetic fitting results, which are related to the characteristics of MCM-41 and the adsorbed substances. MCM-41 provides a specific uniform site for the adsorption of pollutants [26,27]. According to the values of Q_m, K_L, and K_F, the maximum adsorption capacity and affinity of MCM-41 for these three antibiotics were in the order of TC > DOX > LFX. Therefore, it is speculated that MCM-41 has more adsorption capacity for tetracycline and doxycycline, composed of hydrocarbon elements. In contrast, it has a different adsorption capacity for levofloxacin not composed of hydrocarbon elements. It also shows that a single layer can control the whole adsorption process. These findings are consistent with the kinetic experiments.

Table 2. Equilibrium isotherm model parameters for TC, DOX, and LFX adsorption onto MCM-41 composites.

Pollutants	Freundlich			Langmuir			Temkin		
	K_F	n	R^2	K_L	Q_m	R^2	b_T	A_T	R^2
TC	12.89	0.33	0.95	0.204	73.410	0.96	373.01	8.90	0.96
DOX	13.39	0.27	0.99	0.100	144.835	0.99	809.60	532.80	0.95
LFX	18.93	0.13	0.81	1.47	33.676	0.98	456.52	12.89	0.98

K_F: L/g, K_L: L/µmol, b_T: kJ/mol, and A_T: L/mmol s.

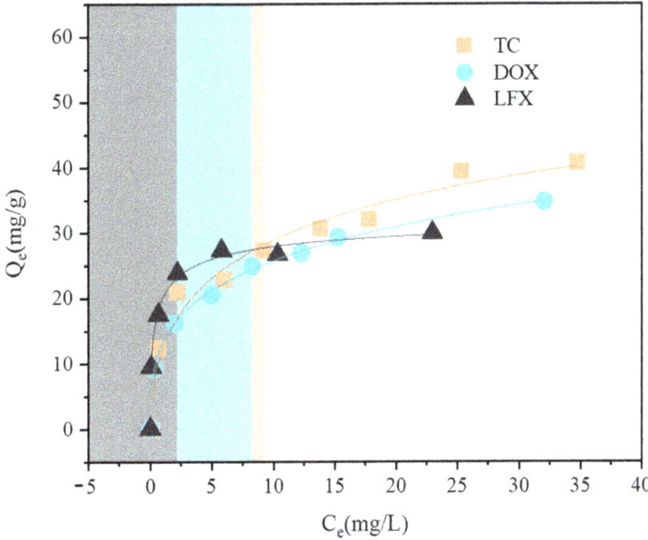

Figure 4. Langmuir isotherms of demolded MCM-41 at 800 °C for TC, DOX, and LFX. (pH: 7.0; adsorbent dosage: 0.5 g/L; temperature: 25 °C; contact time: 70 min).

The pH of a solution plays a pivotal role in influencing the adsorption capacity of mesoporous materials. However, it also has a bearing on the characteristics of the adsorbents and the charge state and presence of ions. In our study, we manipulated the pH value by employing hydrochloric acid and sodium hydroxide solutions. The objective was to investigate the adsorption behavior of the four different pollutants on mesoporous material MCM-41 under varying pH conditions. The pH values ranged from approximately 3 to 11, in conjunction with the zeta potential diagram of MCM-41 (Figure 5).

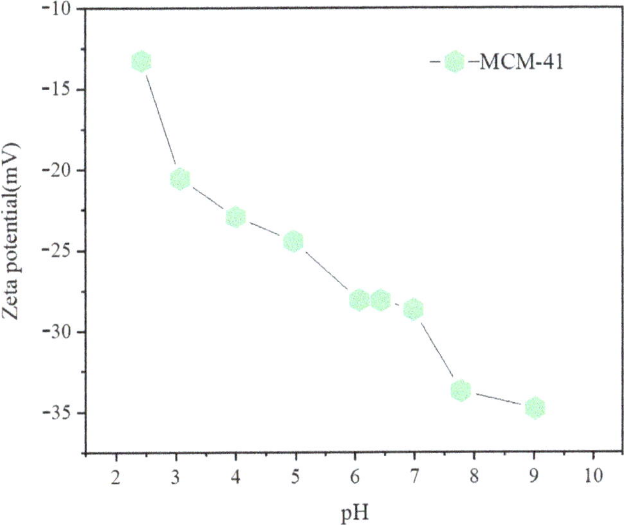

Figure 5. 800 °C MCM-41 Zeta potential diagram (pH: 7.0; adsorbent dosage: 0.5 g/L; temperature: 25 °C).

Regarding tetracycline (TC) adsorption (as shown in Figure 6), an impressive removal rate of 97.88% was achieved at a pH of 4. However, as the pH increased, the removal rate of TC exhibited a decline. The adsorption process typically involves several mechanisms, including electrostatic interactions, hydrogen bond formation, electron donor–acceptor interactions, and π-π dispersion interactions [28]. These phenomena are strongly influenced by pH values, consequently impacting the adsorption process. The underlying rationale for this behavior may be linked to the molecular structure of TC and the functional groups present on the surface of MCM-41. In aqueous solutions, TC comprises three distinct functional groups. Depending on the pH, these groups can undergo protonation or deprotonation reactions, rendering the TC molecule positively charged (under acidic conditions), neutrally charged (within the pH range of 4–8), or negatively charged (under alkaline conditions).

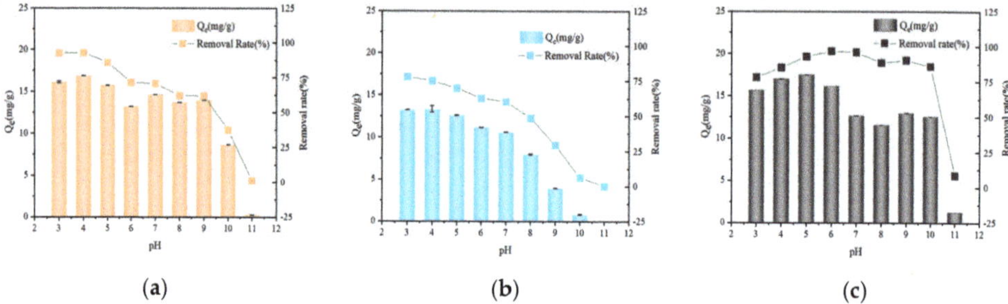

(a)   (b)   (c)

Figure 6. Effect of initial pH on adsorption of (**a**) TC, (**b**) DOX, and (**c**) LFX on MCM-41 (initial concentration: 10 mg/L; adsorbent dose: 0.5 g/L; temperature: 25 °C; contact time: 70 min).

When the pH is below the pKa value (7.78) of tetracycline (TC), TC molecules carry a positive charge, specifically as TCH3+ and TCH2± species. Conversely, at low pH values, MCM-41 also exhibits a positive charge [29]. As the pH increases, TC molecules acquire negative charges. Considering the properties of TC and the zeta potential diagram of MCM-41, it becomes evident that the zeta potential of the material decreases as the pH increases. This phenomenon indicates that the material becomes electronegative in an aqueous solution, resulting in a diminished electrostatic attraction between the surface charge of MCM-41 and TC, thereby reducing adsorption capacity. Consequently, both the adsorption rate and capacity decrease with increasing pH. Overall, the electrostatic repulsion between TC and MCM-41 is more pronounced under lower pH conditions, whereas higher pH levels intensify electrostatic negative repulsion. Furthermore, the pH-dependent adsorption behavior highlights that the MCM-41 adsorbent is capable of maintaining significant adsorption capacity over a broad range of low pH conditions.

For doxycycline (DOX) adsorption (as illustrated in Figure 6), the removal rate exhibited a decline as the pH increased. The highest removal rate, 78.1%, was achieved at pH 3, and it substantially decreased at pH 7. This suggests that DOX exhibits a pronounced affinity for MCM-41 at lower pH levels. The zeta potential and pH data further confirm that under the experimental conditions, the MCM-41 surface carried a negative charge, primarily due to pH-dependent variations in surface hydroxyl sites [30]. Consequently, the adsorption capacity of DOX decreased with rising pH. This phenomenon can be attributed to two key factors: (a) the increase in the number of negatively charged sites on the surface of MCM-41 due to the loss of H+ ions; (b) research has shown that DOX molecular species include DCH3+ (above pH 3.5, pKa1), DCH2± (above pH 7.7, pKa2), and DCH− (above pH 9.5, pKa3), with their charge states varying with pH.

As the pH level rises, doxycycline (DOX) undergoes deprotonation, resulting in a heightened net negative charge. These dual effects combine to increase the electrostatic repulsion between DOX and the adsorbent. It is important to note that aside from electrostatic interactions, the adsorption process also encompasses non-electrostatic interactions between organic compounds and solid surfaces. These non-electrostatic interactions encompass phenomena like hydrogen bonding, surface complexation, and van der Waals forces, which can occur with cations (DCH3+) as well as DCH2± and DCH−, all of which can contribute to the adsorption process [31]. However, the significance of these contributions diminishes notably with increasing pH.

For LFX, the maximum removal rate reached 96.73% at pH = 6, close to pKa1 of LFX. Also, the removal rate decreased significantly at pH = 10. LFX molecule is the cation below pH = 6 (protonated piperazine, H2L+), anion above pH = 8.5 (deprotonated carboxyl, L−), and amphoteric ion at pH between pKa1 and pKa2 [25]. The morphology of LFX molecules at different pH levels greatly influences adsorption. In the range of pH 6.0–8.0, levofloxacin can undergo various acid-base equilibrium, forming anions, cations, and dipoles. LFX changes with pH. The formation of the protonated piperazine group and dissociation equilibrium of the carboxylic acid group occurred at a pH value close to pKa1 (pH = 6.02). However, the partial deprotonation of the pyrazinyl group occurred at pH near pKa2 (pH = 8.15).

When the pH falls below 6, levofloxacin (LFX) exhibits an increasing positive charge as pH decreases. In this pH range, LFX carries a positive charge, whereas MCM-41 surfaces bear a negative charge. Consequently, the adsorption rates between these two entities are notably high. The enhanced adsorption of LFX is a consequence of the intensified attraction between the protonated amine groups, whether they are adsorbed or free, present on the LFX molecule. This attraction is also influenced by the negatively charged surface of MCM-41. As the pH increases, the deprotonation of the pyrazine group in LFX diminishes the positive charge of LFX. This, in turn, hinders the adsorption of LFX onto MCM-41 and heightens the electrostatic repulsion between them, particularly at higher pH levels. As a result, the maximum adsorption of LFX occurs at pH = 6 (as depicted in Figure 6).

In the aqueous solution utilized in this study, the three distinct molecules underwent protonation and deprotonation reactions in response to changes in pH values. Consequently, they formed varying chemical species [32]. As the pH increased, the deprotonation process (pH > pKa) followed a particular order, typically involving the initial deprotonation, followed by the second deprotonation, and finally, the third deprotonation step. These deprotonation steps emanate from the surface charge characteristics of the pollutants and exert a discernible influence on the adsorption rate of the adsorbent, in accordance with the pKa values associated with each individual pollutant.

4. Conclusions

The adsorbent medium was prepared by demolding mesoporous material MCM-41 at elevated temperatures. This adsorbent medium proves to be a remarkably efficient and effective system for the removal of antibiotics in water, encompassing tetracycline, doxycycline, and levofloxacin. Consequently, it facilitates the purification of wastewater. The experimental findings unequivocally demonstrate that the adsorption efficiency for all three organic pollutants consistently exceeds 95%, and the maximum adsorption capacities for tetracycline (TC), doxycycline (DOX), and levofloxacin (LFX) were 73.41, 144.83, and 33.67 mg g^{-1}, respectively. The adsorption of these three organic pollutants by MCM-41 predominantly falls under the category of chemical adsorption. This multifaceted process involves chemical ion exchange, intraparticle diffusion, hydrogen bonding, and electrostatic interactions, it accounts for the excellent adsorption performance of MCM-41 and provides insights into designing highly efficient adsorbent materials. Furthermore, the results underscore the favorable impact of a neutral or weakly acidic pH on adsorption, while the presence of alkaline conditions inhibits adsorption, which can be directly applied in engineering design and represents the first quantitative result in this field. In summary, MCM-41 emerges as a highly efficient adsorbent characterized by a rapid adsorption rate and short adsorption time, rendering it suitable for use in a fluidized bed setup. Simultaneously, this material offers the advantages of being effective, environmentally friendly, non-toxic, easily recoverable, and capable of selectively removing antibiotics and dye pollution from water.

Author Contributions: J.C. (first author): conceptualization, experiment, formal analysis, writing—original draft, software, and visualization. Y.Y. (Yao Yang) (co-first author): conceptualization, visualization, investigation, and writing—review and editing. Y.Y. (Yuanyuan Yao): investigation and writing—review and editing. Z.H.: conceptualization, funding acquisition, resources, supervision, and writing—review and editing. Y.Y. (Yuanyuan Yao): investigation and writing—review and editing. Q.X.: resources and supervision. L.H.: conceptualization, methodology, and supervision. B.G. (corresponding author): conceptualization, methodology, and supervision. All authors have read and agreed to the published version of the manuscript.

Funding: The study was supported by the Key-Area Research and Development Program of Guangdong Province (2020B0202080001), the Natural Science Foundation of Guangdong Province (2019A1515011659), and the National Natural Science Foundation of China (No. 51509093). The characterization results were supported by Beijing Zhongkebaice Technology Service Co., Ltd.

Data Availability Statement: All data employed in support of the outcomes of the study are included in this article.

Conflicts of Interest: Author Jie Chen, Yuanyuan Yao and Liping He was employed by the company CCCC Fourth Harbor Engineering Institute Co., Ltd. The remaining authors declare that the research was conducted in the absence of any commercial or financial relationships that could be construed as a potential conflict of interest.

References

1. Wang, J.; Xu, S.; Zhao, K.; Song, G.; Zhao, S.; Liu, R. Risk control of antibiotics, antibiotic resistance genes (ARGs) and antibiotic resistant bacteria (ARB) during sewage sludge treatment and disposal: A review. *Sci. Total Environ.* **2023**, *877*, 162772. [CrossRef]
2. Apreja, M.; Sharma, A.; Balda, S.; Kataria, K.; Capalash, N.; Sharma, P. Antibiotic residues in environment: Antimicrobial resistance development, ecological risks, and bioremediation. *Environ. Sci. Pollut. Res.* **2021**, *29*, 3355–3371. [CrossRef]
3. Kalli, M.; Noutsopoulos, C.; Mamais, D. The Fate and Occurrence of Antibiotic-Resistant Bacteria and Antibiotic Resistance Genes during Advanced Wastewater Treatment and Disinfection: A Review. *Water* **2023**, *15*, 2084. [CrossRef]
4. McClain, J.B.L.; Ballou, W.R.; Harrison, S.M.; Steinweg, D.L. Doxycycline therapy for leptospirosis. *Ann. Intern. Med.* **1984**, *100*, 696–698. [CrossRef]
5. Bhattacharyya, P.; Parmar, P.R.; Basak, S.; Dubey, K.K.; Sutradhar, S.; Bandyopadhyay, D.; Chakrabarti, S. Metal organic framework-derived recyclable magnetic coral Co@Co$_3$O$_4$/C for adsorptive removal of antibiotics from wastewater. *Environ. Sci. Pollut. Res.* **2023**, *30*, 50520–50536. [CrossRef] [PubMed]
6. Mohy-U-Din, N.; Farhan, M.; Wahid, A.; Ciric, L.; Sharif, F. Human health risk estimation of antibiotics transferred from wastewater and soil to crops. *Environ. Sci. Pollut. Res.* **2023**, *30*, 20601–20614. [CrossRef]
7. Zhou, C.-S.; Cao, G.-L.; Wu, X.-K.; Liu, B.-F.; Qi, Q.-Y.; Ma, W.-L. Removal of antibiotic resistant bacteria and genes by nanoscale zero-valent iron activated persulfate: Implication for the contribution of pH decrease. *J. Hazard. Mater.* **2023**, *452*, 131343. [CrossRef] [PubMed]
8. Ajduković, M.; Stevanović, G.; Marinović, S.; Mojović, Z.; Banković, P.; Radulović, K.; Jović-Jovičić, N. Ciprofloxacin Adsorption onto a Smectite-Chitosan-Derived Nanocomposite Obtained by Hydrothermal Synthesis. *Water* **2023**, *15*, 2608. [CrossRef]
9. Raper, E.; Stephenson, T.; Anderson, D.R.; Fisher, R.; Soares, A. Industrial wastewater treatment through bioaugmentation. *Process Saf. Environ. Prot.* **2018**, *118*, 178–187. [CrossRef]
10. Hasan, R.; Chong, C.; Setiabudi, H.; Jusoh, R.; Jalil, A. Process optimization of methylene blue adsorption onto eggshell-treated palm oil fuel ash. *Environ. Technol. Innov.* **2019**, *13*, 62–73. [CrossRef]
11. Gurgenidze, D.; Romanovski, V. The Pharmaceutical Pollution of Water Resources Using the Example of the Kura River (Tbilisi, Georgia). *Water* **2023**, *15*, 2574. [CrossRef]
12. Ighalo, J.O.; Adeniyi, A.G. Mitigation of Diclofenac Pollution in Aqueous Media by Adsorption. *ChemBioEng Rev.* **2020**, *7*, 50–64. [CrossRef]
13. Wang, X.; Yin, R.; Zeng, L.; Zhu, M. A review of graphene-based nanomaterials for removal of antibiotics from aqueous environments. *Environ. Pollut.* **2019**, *253*, 100–110. [CrossRef]
14. Lim, S.; Shi, J.L.; von Gunten, U.; McCurry, D.L. Ozonation of organic compounds in water and wastewater: A critical review. *Water Res.* **2022**, *213*, 118053. [CrossRef]
15. Madan, S.; Shaw, R.; Tiwari, S.; Tiwari, S.K. Adsorption dynamics of Congo red dye removal using ZnO functionalized high silica zeolitic particles. *Appl. Surf. Sci.* **2019**, *487*, 907–917. [CrossRef]
16. Zaidi, S.; Chaabane, T.; Sivasankar, V.; Darchen, A.; Maachi, R.; Msagati, T. Electro-coagulation coupled electro-flotation process: Feasible choice in doxycycline removal from pharmaceutical effluents. *Arab. J. Chem.* **2019**, *12*, 2798–2809. [CrossRef]

17. Khan, M.A.; Alothman, Z.A.; Naushad, M.; Khan, M.R.; Luqman, M. Adsorption of methylene blue on strongly basic anion exchange resin (Zerolit DMF): Kinetic, isotherm, and thermodynamic studies. *Desalination Water Treat.* **2015**, *53*, 515–523. [CrossRef]
18. Ganiyu, S.O.; dos Santos, E.V.; Costa, E.C.T.d.A.; Martínez-Huitle, C.A. Electrochemical advanced oxidation processes (EAOPs) as alternative treatment techniques for carwash wastewater reclamation. *Chemosphere* **2018**, *211*, 998–1006. [CrossRef]
19. Tagliavini, M.; Schäfer, A.I. Removal of steroid micropollutants by polymer-based spherical activated carbon (PBSAC) assisted membrane filtration. *J. Hazard. Mater.* **2018**, *353*, 514–521. [CrossRef] [PubMed]
20. Olusegun, S.J.; Freitas, E.T.F.; Lara, L.R.S.; Stumpf, H.O.; Mohallem, N.D.S. Effect of drying process and calcination on the structural and magnetic properties of cobalt ferrite. *Ceram. Int.* **2019**, *45*, 8734–8743. [CrossRef]
21. Ramlow, H.; Machado, R.A.F.; Bierhalz, A.C.K.; Marangoni, C. Dye synthetic solution treatment by direct contact membrane distillation using commercial membranes. *Environ. Technol.* **2020**, *41*, 2253–2265. [CrossRef]
22. Okoli, C.P.; Ofomaja, A.E. Development of sustainable magnetic polyurethane polymer nanocomposite for abatement of tetracycline antibiotics aqueous pollution: Response surface methodology and adsorption dynamics. *J. Clean. Prod.* **2019**, *217*, 42–55. [CrossRef]
23. Li, X.-D.; Zhai, Q.-Z. Use of nanometer mesoporous MCM-41 for the removal of Pb(II) from aqueous solution. *Appl. Water Sci.* **2020**, *10*, 1–10. [CrossRef]
24. Qin, X.; Liu, F.; Wang, G.; Weng, L.; Li, L. Adsorption of levofloxacin onto goethite: Effects of pH, calcium and phosphate. *Colloids Surf. B Biointerfaces* **2014**, *116*, 591–596. [CrossRef]
25. Baran, W.; Adamek, E.; Jajko, M.; Sobczak, A. Removal of veterinary antibiotics from wastewater by electrocoagulation. *Chemosphere* **2018**, *194*, 381–389. [CrossRef]
26. Sun, J.; Cui, L.; Gao, Y.; He, Y.; Liu, H.; Huang, Z. Environmental application of magnetic cellulose derived from Pennisetum sinese Roxb for efficient tetracycline removal. *Carbohydr. Polym.* **2021**, *251*, 117004. [CrossRef]
27. Zhou, J.; Wu, P.; Dang, Z.; Zhu, N.; Li, P.; Wu, J.; Wang, X. Polymeric Fe/Zr pillared montmorillonite for the removal of Cr(VI) from aqueous solutions. *Chem. Eng. J.* **2010**, *162*, 1035–1044. [CrossRef]
28. Vargas, A.M.M.; Cazetta, A.L.; Kunita, M.H.; Silva, T.L.; Almeida, V.C. Adsorption of methylene blue on activated carbon produced from flamboyant pods (Delonix regia): Study of adsorption isotherms and kinetic models. *Chem. Eng. J.* **2011**, *168*, 722–730. [CrossRef]
29. Zhang, D.; Yin, J.; Zhao, J.; Zhu, H.; Wang, C. Adsorption and removal of tetracycline from water by petroleum coke-derived highly porous activated carbon. *J. Environ. Chem. Eng.* **2015**, *3*, 1504–1512. [CrossRef]
30. Brigante, M.; Avena, M. Biotemplated synthesis of mesoporous silica for doxycycline removal. Effect of pH, temperature, ionic strength and Ca^{2+} concentration on the adsorption behaviour. *Microporous Mesoporous Mater.* **2016**, *225*, 534–542. [CrossRef]
31. Lagaly, G.; Ogawa, M.; Dékány, I. Chapter 7.3 Clay Mineral Organic Interactions. *Dev. Clay Sci.* **2006**, *1*, 309–377.
32. Zhao, Y.; Cao, B.; Lin, Z.; Su, X. Synthesis of $CoFe_2O_4$/C nano-catalyst with excellent performance by molten salt method and its application in 4-nitrophenol reduction. *Environ. Pollut.* **2019**, *254*, 112961. [CrossRef] [PubMed]

Article

Quantifying the Impact of Coal Mining on Underground Water in Arid and Semi-Arid Area: A Case Study of the New Shanghai No. 1 Coal Mine, Ordos Basin, China

Yuguang Lyv [1,2], Wei Qiao [1,*], Weichi Chen [3,*], Xianggang Cheng [1], Mengnan Liu [1] and Yingjie Liu [1]

[1] Institute of Mine Water Hazards Prevention and Controlling Technology, School of Resources and Geosciences, China University of Mining and Technology, Xuzhou 221116, China; lvyg691208@126.com (Yuguang Lyv)

[2] Shandong Energy Xinwen Mining Inner Mongolia Energy Group Co., Ltd., Ordos 750336, China

[3] School of Transportation Engineering, Jiangsu Vocational Institute of Architectural Technology, Xuzhou 221116, China

* Correspondence: qiaowei@cumt.edu.cn (W.Q.); weichichen159@163.com (W.C.)

Abstract: The new Shanghai No. 1 Coal Mine is located in arid and semiarid area of northwest China, which is characterized by scarce rainfall, intense evaporation, and limited water resources. High-intensity coal mining has caused severe damage to groundwater resources. The Baotashan sandstone aquifer of the Jurassic system has abundant water resources, and they are stored in the floor strata of mining coal seams. This poses the risk of high-pressure build-up and water inrush hazards during the mining of coal. To avoid these, the Baotashan sandstone aquifer needs to be drained and depressurized, which can result in a huge waste of water resources. Thus, taking the New Shanghai No. 1 Coal Mine as the basis for the case study, the impact of coal mining on the underground water resources was quantified. Large-scale water release tests were performed under the shaft to determine the hydrogeological properties of the Baotashan sandstone aquifer and a three-dimensional numerical model of the groundwater system was established. The dynamic phenomenon of water drainage was simulated and the drained water discharge was predicted under the condition of safe mining.

Keywords: arid and semiarid area; quantitative assessment; groundwater; floor water inrush hazard; northwest China

Citation: Lyv, Y.; Qiao, W.; Chen, W.; Cheng, X.; Liu, M.; Liu, Y. Quantifying the Impact of Coal Mining on Underground Water in Arid and Semi-Arid Area: A Case Study of the New Shanghai No. 1 Coal Mine, Ordos Basin, China. *Water* **2023**, *15*, 1765. https://doi.org/10.3390/w15091765

Academic Editors: Peiyue Li and Jianhua Wu

Received: 5 April 2023
Revised: 29 April 2023
Accepted: 1 May 2023
Published: 4 May 2023

1. Introduction

Water resource plays a vital role in industrial and agricultural development, ecological environmental protection, and vegetation growth [1–3], especially in arid and semiarid area of northwest China, where average annual precipitation is less than 400 mm and evaporation is greater than 2000 mm [4,5]. Underground water resources are extremely valuable in northwest China due to the scarcity of rivers and surface water bodies. On the contrary, this vast region is rich in oil, natural gas, and coal resources [6]. The coal reserves of northwest China accounted for approximately 73% of the national coal reserves [7], while coal output was estimated to be 1.83 billion tons in 2022, accounting for 58% of the total output of the country [8]. The resource strategy focused on the western regions and the energy structure of poor oil resources, limited gas resources, and abundant coal resources has driven up the mining intensity of northwest China [9,10].

Numerous studies have ascertained that coal mining has a negative impact on both surface and underground water resources, especially in arid and semiarid area [11–13]. Subsidence and surface cracks caused by high-intensity coal mining have changed the runoff generation and confluence within the catchment, resulting in decreased streamflow in the mining area [14–17]. The largest desert freshwater lake in China, Hongjiannao Lake, has seen a dramatic decline in its reservoir capacity and surface area due to the impact of coal mining [18]. Guo et al. (2019) quantified the effects of climate change, coal mining, and soil and water conservation on streamflow in the Kuye River Basin located in the arid and semiarid area of the Loess Plateau, China. They detected that the streamflow reduction induced by coal mining reached 29.88 mm, accounting for 54.24% of the total reduction [19]. Water-conducting fractures were found to have penetrated the aquifuges between the surface and the goaf causing massive water leakage, death of rivers, and eventually vegetation deterioration and ecological hazards [20,21]. Furthermore, the stress redistribution of overburden caused by underground coal mining led to the formation of caved zones and fractured zones in the roof strata, resulting in a dramatic increase in the permeability of the strata [22,23]. Thus, a complete groundwater drainage zone was formed in the roof aquifers [24]. Variations in the hydraulic characteristics of the strata such as the increase in the porosity, the opening of joints, and the separation of bedding planes resulted in the reduction of groundwater resources [25,26]. Furthermore, pre-drainage is adopted for the deep mining of the coal seams overlying the high-pressure confined aquifers to decrease the hydraulic pressure and prevent floor water inrush hazards [27–29]. However, the discharge of draining mine water from the roof and floor aquifers will cause the wastage of precious water resources and the contamination of the surface water systems which threatens human health [30–35].

Floor water inrush accidents have occurred frequently in the Permo-Carboniferous coal mining in north China, which caused massive casualties and economic loss [36]. Two main factors are responsible for these water hazards: high-pressure and ultrahigh-pressure of Ordovician limestone aquifers and the limited thickness of aquifuges underneath the mining coal seams [37]. Correspondingly, drainage and depressurization for the Ordovician limestone aquifers and grouting reinforcement for the limited-thickness aquifuges were proposed to prevent the floor water inrush [38]. Floor water inrush was a rare incidence in northwest China since the Ordovician limestone aquifers and Jurassic coal seams are separated by a considerable distance. However, on 25 November 2015, a severe floor water inrush hazard from the Jurassic Baotashan aquifer occurred in the New Shanghai No. 1 Coal Mine of Yinchuan city, Ningxia, northwest China [39]. The water inrush from the aquifer occurred from the floor strata during the advancement of deep roadway in the coal mine, wherein the initial water discharge was about 1500 m^3/h. The discharge of water inrush gradually increased with time. On 26 November 2015, the water discharge reached 3600 m^3/h, with the instantaneous discharge reaching 10,000 m^3/h, submerging the mine. To ensure safety during the mining activities, the Baotashen aquifer is planned to be drained and depressurized. This necessitates the quantification of the impact of coal mining on the Baotashan water resource, which is a valuable underground water resource in arid and semiarid areas.

2. Study Area

The New Shanghai No. 1 Coal Mine is located on the northwestern edge of the Mu Us Desert in the Ordos Basin (Figure 1), which has a semi-arid and semi-desert continental climate with annual precipitation of 150 mm and evaporation of 2770 mm. Due to the scarcity of surface water resources, industrial and agricultural water use is mainly from underground water sources. In recent years, the Ordos Basin has become the largest coal-producing region in China. It has coal reserves of 658.99 billion tons, which accounts for 38.8% of the total national reserves. Several national-level large coal development bases have been established, which have intensified the conflict over water resources.

The mining area of New Shanghai No. 1 Coal Mine is 26.6 km^2, with a length of 12.5 km and a width ranging from 2.0 to 3.5 km, approximately. The coal mine has 3.45 million tons of recoverable reserves and two mining levels are designed. The first level is at an altitude of +880 m, from which the coal seams of No. 2, No. 5, No. 8, No. 15, and No. 16 are mined. The second level is at an altitude of +700 m, and the coal seams of No. 18, No. 19, No. 20, and No. 21 are mined at this level. Combined mechanized mining and top-coal caving methods are adopted to excavate the coal seams. The coal seams are located in the strata of the Jurassic Yan'an formation. The lithology mainly consists of gray sandstone, gray black and black siltstone, and mudstone. The thickness of the Yan'an Formation layer ranges from 159 m to 345 m, with an average thickness of 288 m. The overlying strata of coal are in the order of Jurassic Zhiluo Formation, Cretaceous system, Paleogene system, and Quaternary system. The strata, lithology, and hydrogeological characteristics of overlying strata are described in detail in Figure 2.

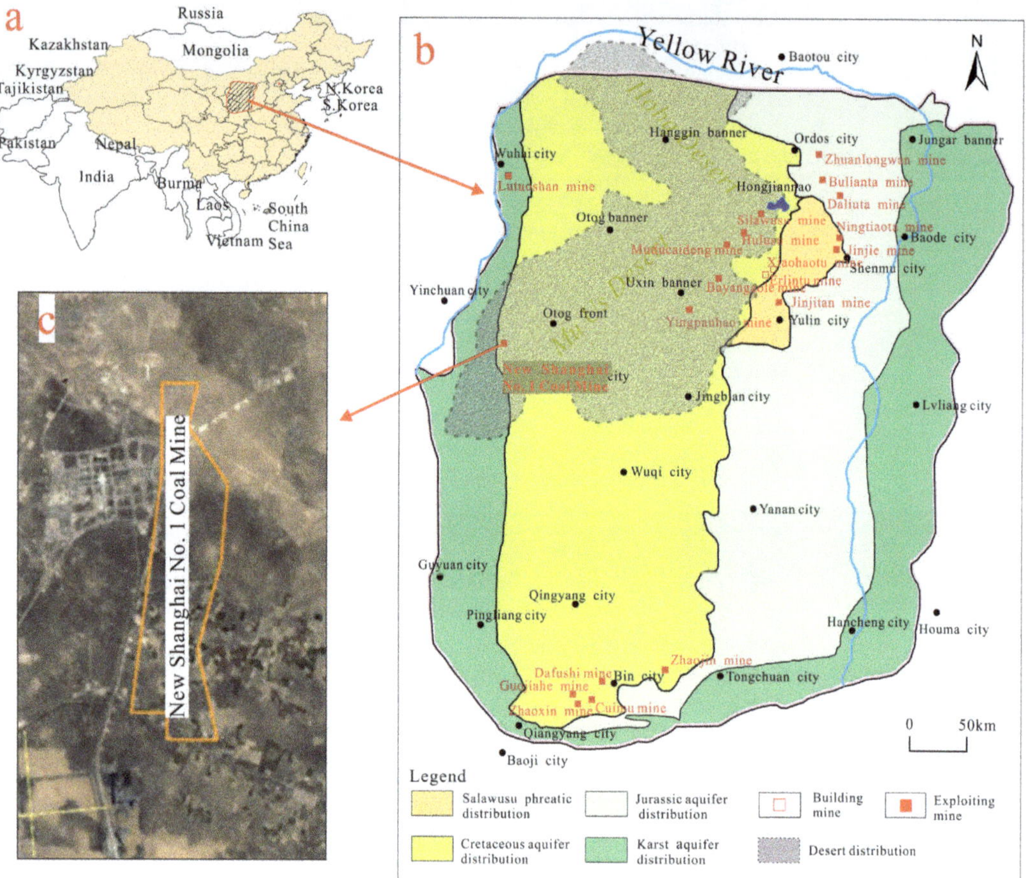

Figure 1. (**a**) Location of the study area in China; (**b**) water resources distribution of the Ordos Basin; (**c**) aerial photo of New Shanghai No. 1 Coal Mine.

At present, the No. 5, No. 8, and No. 15 coal seams, which are located in the upper part of the coal measure strata, have been mined. Penetrated by the water-conducted fractures, the groundwater stored in the sandstone aquifer of the Yan'an Formation and Zhiluo Formation is drained. Due to the low water abundance of the overlying strata, coal mining has little effect on groundwater resources. The No. 18 coal seam is planned to be mined in the next three years. The Jurassic Baotashan sandstone serving as the floor aquifer makes the coal mining highly probable of water inrush hazards.

Stratigraphic unit		Thickness (m)	Columnar legend	Lithologic characteristics	Hydrogeological characteristics	Legend
System	Formation					
Quaternary (Q)	Salawusu Formation (Q_{3s})	$\frac{9\sim72}{36}$		Composed of aeolian sand, siltstone, fine sandstone and silty clay with grayish-yellow, greyish-green, bluish-yellow, gray color.	The depth of phreatic water table ranges from 10 to 17 m and elevation ranges from 1294.8 to 1310m. The permeability ranges from 2.12 to 5.93 m/d.	Aeolian sand / Siltstone
Cretaceous (K)	Zhidan Formation (K_{1zh})	$\frac{122\sim300}{182}$		The bottom part mainly consists of sandy conglomerate with greyish-green, pale red, brownish red color, while the upper part comprised of siltstone, fine sandstone, medium sandstone and coarse sandstone with dark red color.	The elevation of hydraulic head ranges from 1179 to 1291 m. The permeability ranges from 0.0054 to 0.28 m/d. The specific yield derived from a single-bore pumping test ranges from 0.006 to 0.05 L/s·m which is assessed as low water-bearing.	Silty clay / Fine sandstone
Jurassic (J)	Zhiluo Formation (J_{2z})	$\frac{0\sim270}{119}$		Composed of sandy mudstone, siltstone, and fine sandstone with greyish-green, blue grey, purplish-gray color which was mingled with lilac medium sandstone.	The elevation of confined-aquifer hydraulic head ranges from 1171 to 1255 m. The permeability ranges from 0.02 to 0.28m/d. The specific yield derived from a single-bore pumping test ranges from 0.008 to 0.11 L/s·m.	Medium sandstone / Coarse sandstone
	Yanan Formation (J_{1-2y})	$\frac{159\sim345}{288}$		The strate belong to the coal measure strata which are comporised of sandy mudstone, siltstone, and fine sandstone with ashen, gray color. The minable coal seams include the 2, 5, 8, 15, 16, 18, 19, 20, and 21 coal seams.	The elevation of confined-aquifer hydraulic head ranges from 1061 to 1235 m. The permeability ranges from 0.0008 to 0.75 m/d. The specific yield derived from a single-bore pumping test ranges from 0.00054 to 0.0097 L/s·m, which is assessed as low water-bearing.	Mudstone / Sandy mudstone / Cole seam
Triassic (T)	Yanchang Formation (T_{3y})	>500		Composed of mudstone, sandy mudstone, siltstone, and fine sandstone with greyish-green, light gray, pale red, brown color.	The elevation of confined-aquifer hydraulic head was 1222 m. The permeability was 0.36 m/d. The specific yield derived from a single-bore pumping test was 0.11 L/s·m.	

Figure 2. Stratigraphic column of the study area. The values in the thickness column refer to the thickness range and the average thickness.

3. Methodology

3.1. Water Release Test of Multiple Holes

Large-scale water release tests were conducted to investigate the hydrogeological property of the Baotashan sandstone aquifer and evaluate the impact of coal mining on groundwater resources. Four underground draining boreholes were drilled in the No. 8 coal mining roadway and were named as F1, F2, F3, and F4, respectively (Figure 3a). The spacing between adjacent draining holes was about 24 m. The diameter of the first-level drilling hole was 190 mm and the drilling depth was 16 m, where a sealing casing of 168 mm hole diameter and 6 mm wall thickness was installed. The second-level borehole was made by drilling a hole of 133 mm diameter to a depth of 1 m below the floor of the No. 21 coal seam in the Yan'an formation, where the sealing casing with a hole diameter of 108 mm and the wall thickness of 6 mm was installed. Finally, a borehole with a diameter of

85 mm was drilled to 5 m below the floor of the Baotashan sandstone aquifer. The structure of borehole is shown in Figure 3b.

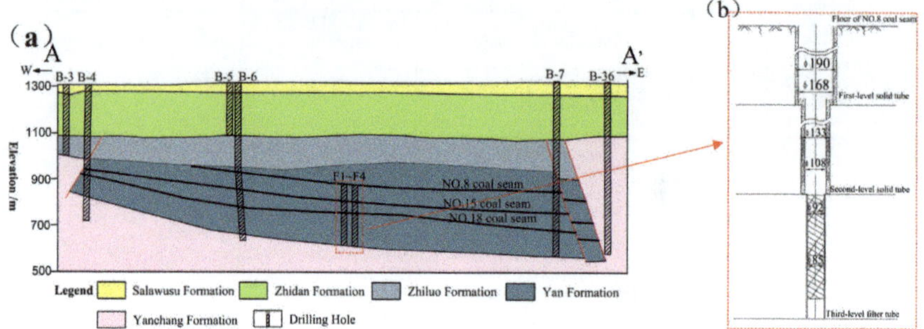

Figure 3. (**a**) cross-sectional location of water release holes from the Figure 4; (**b**) structure of drilling holes.

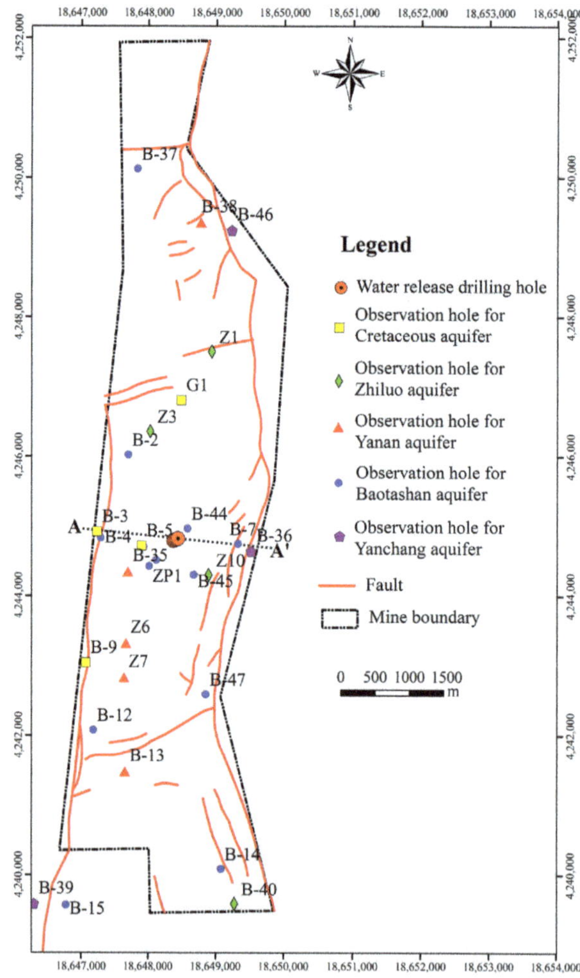

Figure 4. Layout of the water release test.

The water release test was divided into two parts: single-hole draining test and multi-hole draining test. In the single-hole draining test, the valve of the F2 hole was opened. The test started at 12 a.m. on 23 August 2019, and ended at 12 a.m. on 18 September 2019, lasting 624 h. During the test, the water discharge ranged from 180 m^3/h to 307 m^3/h, with an average of 237 m^3/h. In the multi-hole draining test, the valves of F1, F2, F3, and F4 holes were opened successively. On 8 October, the valve of the F2 hole was opened, which was referred to as the first draining stage. Then, the valve of F3 hole was opened on 12 October, which was named the second draining stage. Finally, the valves of F1 and F4 holes were opened on 16 October making the third draining stage. The multi-hole draining test ended on 6 November. The draining test lasted for 696 h and the hydraulic head recovery was 600 h. The average draining amounts of the first stage, second stage, and third stage were 206 m^3/h, 332 m^3/h, and 444 m^3/h, respectively.

Observation holes of hydraulic head to the Cretaceous aquifer, Zhiluo aquifer, Yan'an aquifer, Baotashan aquifer, and Triassic aquifer were used during the water release test of the Baotashan sandstone aquifer. There were four observation holes for the Cretaceous aquifer, namely G1, B-3, B-5, and B-9, four observation holes for the Zhiluo aquifer, namely holes Z1, Z3, Z10, and B-40, and six observation holes for the Yan'an Formation aquifer, namely holes Z6, Z7, B-13, B-24, B-35, and B-38. For the Baotashan sandstone aquifer, there were 11 observation holes, namely holes B-2, B-4, B-6, B-7, B-8, B-12, B-14, B-37, B-44, B-45, and B-47. The layout of the draining holes and observation holes are shown in Figure 4.

3.2. Numerical Simulation

To quantify the impact of coal mining on the groundwater resources of Baotashan sandstone, FEFLOW software was used to establish a three-dimensional (3D) numerical model of the groundwater system, as shown in Figure 5. Computational mesh densification was carried out in the area near the draining hole and the main faults in the well field. The model was divided on the plane into 18,420 calculation units and 9471 nodes. The model was divided vertically into seven layered structures, each representing a different aquifer.

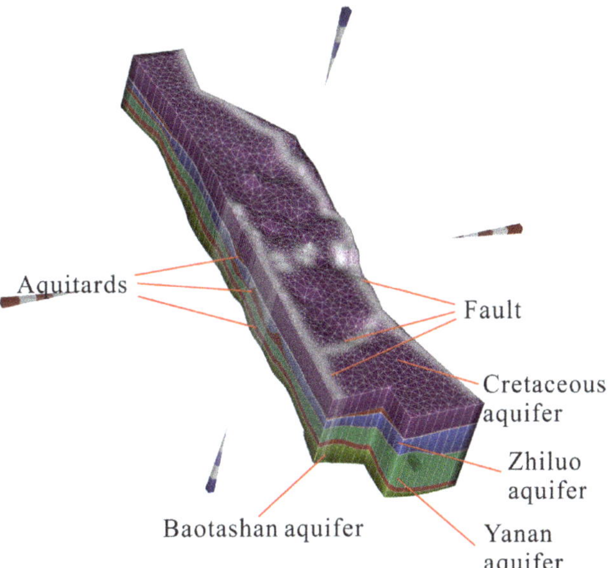

Figure 5. Three-dimensional (3D) model.

The first layer of the 3D numerical model was composed of the sandy mudstone and mudstone at the bottom of the Paleogene strata and the upper part of the Cretaceous strata, which blocks the hydraulic recharge of shallow water to the groundwater system. Thus, zero-flow boundary condition was assigned to the top boundary of the model. Similarly, zero-flow boundary condition was applied on the bottom boundary as well, since the seventh layer of the model was composed of the Baotashan sandstone aquifer, the floor of which consisted of mudstone of the Triassic strata.

Horizontally, the center fault segment of the western boundary was designated as the permeable boundary, while the southern and northern segments of the fault were generalized as impermeable boundaries. As illustrated in Figure 6a, the north fault segment of the eastern boundary was generalized as the permeable boundary, while the south segment was assigned as the impermeable boundary. The hydrogeological zoning of the Baotashan sandstone aquifer was divided into five zones according to its permeability and water-bearing capacity, as shown in Figure 6b. The parameters of zonings were inversely calculated during the fitting and verification of the numerical model.

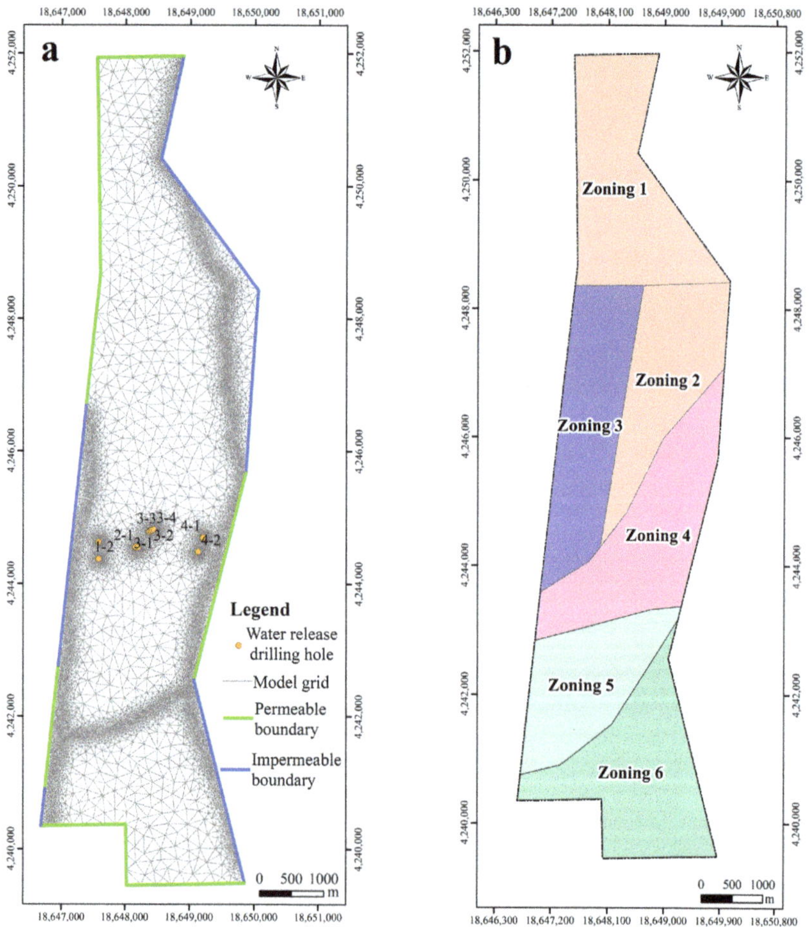

Figure 6. Numerical model generalization: (**a**) boundary generalization; (**b**) zoning of hydrogeological parameters.

4. Results

4.1. Hydrogeological Characteristics of the Baotashan Aquifer

The flow field of the natural, the single-hole draining test, and the multi-hole draining test are shown in Figure 7. The natural hydraulic head of the Baotashan sandstone aquifer ranges from 1184 m to 1228 m, with the lowest head occurring at the southwestern boundary of the well field. This indicates that the aquifer flows from the south and north of the well field to the southwest boundary. Figure 7b,c shows that the head drops for the single-hole draining test and the multi-hole draining test were more than 50 m and 100 m, respectively. This caused the depression-cone to develop in the middle of the draining holes. Additionally, the head contour distribution between observation holes B-12, B-47, and draining holes forms an extremely sparse area. It is analyzed that the predominant seepage zone is formed due to the existence of faults, which makes the area near observation holes B-12 quickly recharge to the draining area.

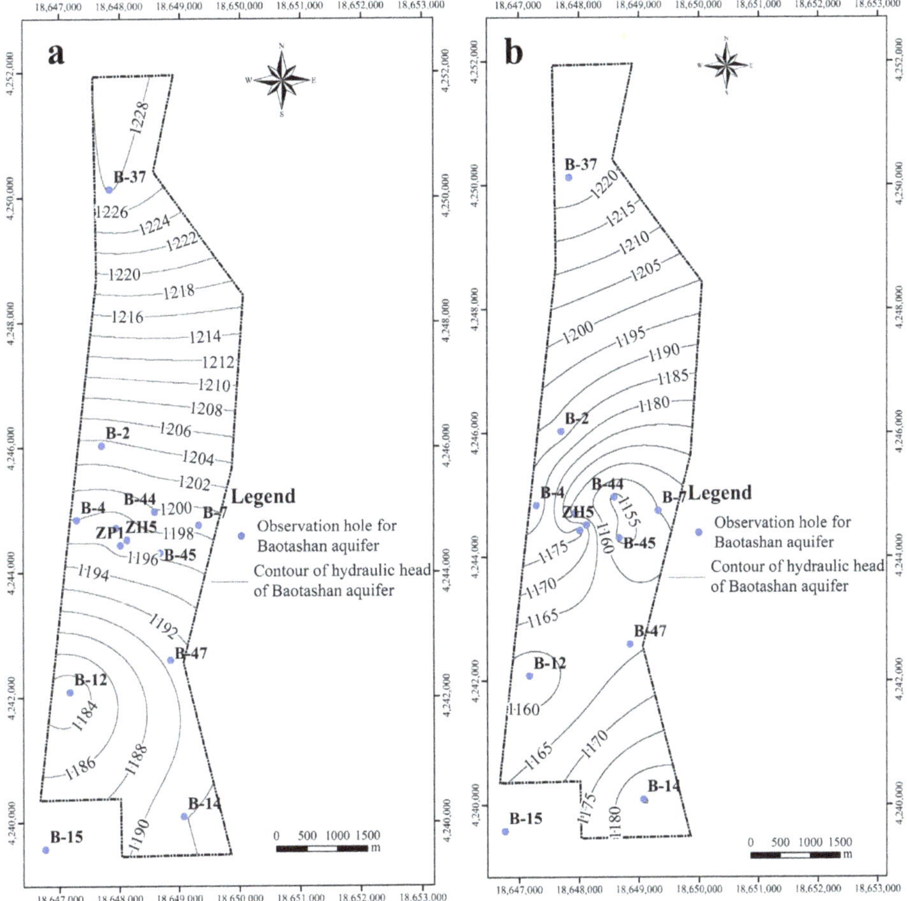

Figure 7. *Cont.*

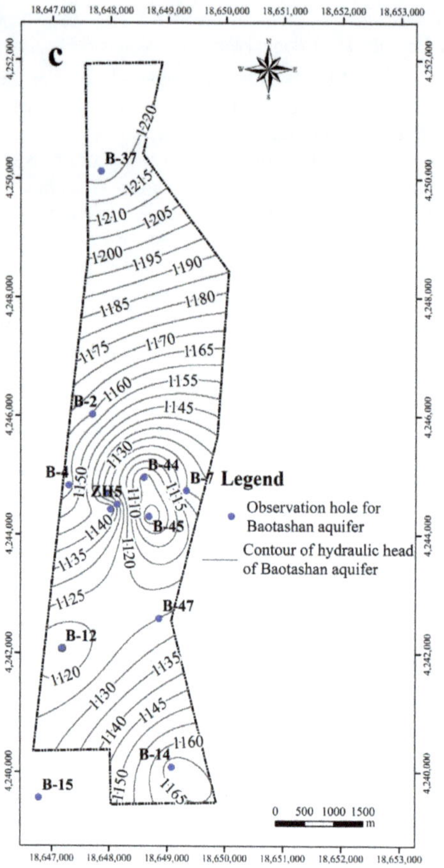

Figure 7. Variation of flow field during the water release tests: (**a**) before the water release tests; (**b**) single-hole water release tests; (**c**) multi-hole water release tests.

The permeability coefficient of the Baotashan sandstone aquifer is calculated based on the variation of water heads, as shown in Table 1. To accurately evaluate the water abundance of the aquifer, the single-hole unit inflow is calculated based on the previous pumping tests, which refers to the water inflow amount when the head is decreased a meter in the borehole during pumping tests. It is an important index for evaluating the degree of the water hazard in coal mines in China.

Table 1. Hydrogeological parameters of the Baotashan aquifer.

Borehole	Aquifer Thickness (m)	Water Level (m)	Permeability (m/d)	Unit Inflow (L/(s·m))
B-2	62.10	1200.03	0.3299	0.2048
B-4	127.10	1195.83	0.1057	0.1233
B-6	79.70	1183.862	0.4839	0.4669
B-7	53.50	1180.87	2.0247	0.9483
B-8	53.55	1187.241	1.7726	0.6783
B-12	56.85	1185.63	1.9688	1.0026
B-14	14.21	1184.609	0.288	0.0372
B37	56.35	1233.792	0.2541	0.1803
B44	81.00	1198.322	1.4290	1.5380
B45	58.98	1193.871	2.0603	0.9978
B47	42.18	1171.521	1.0955	0.4553

The permeability and unit inflow cloud maps of the Baotashan sandstone aquifer are shown in Figure 8. The permeability coefficient ranges from 0.1 m/d to 2.0 m/d and the high-permeability area is mainly caused by the tensile normal faults, which act as channels for water flow exchange.

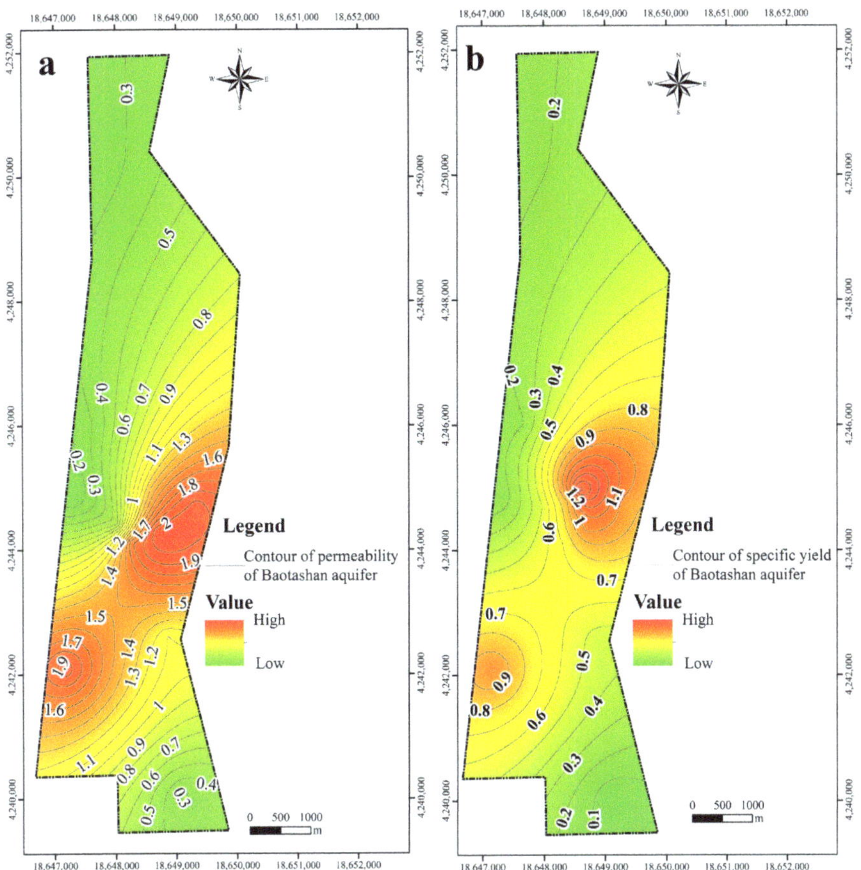

Figure 8. Hydrogeological parameters distribution of Baotashan aquifer: (**a**) permeability coefficient; (**b**) single-hole specific yield.

According to the classification standards of unit inflow in China, the majority of the areas near the Baotashan sandstone aquifer have medium water abundance, some local areas to the west and south have weak water abundance, and the area near the fault has strong water abundance, as shown in Figure 8b. Analyzing the data based on the permeability distribution, it is found that the area with high permeability also has good water abundance, suggesting that the water inrush amount of the Baotashan sandstone aquifer would be huge when the coal mining is carried out.

4.2. Influence on Water Resources of Baotashan Aquifer

4.2.1. Simulation Model Fitting

The comparison between the simulated and observed flow fields of the Baotashan aquifer is shown in Figure 9. It is difficult to effectively predict flow change using zoning parameters in faulted regions because the prevalent seepage channels resulted in the fast decrease of hydraulic head in local area. For example, in the area near the FD5 fault, the fitting error is about 2.46 m, indicating that the simulated head is higher than the observed head. Except for the area near the FD5 fault, the fitting of all the hydraulic heads is accurate.

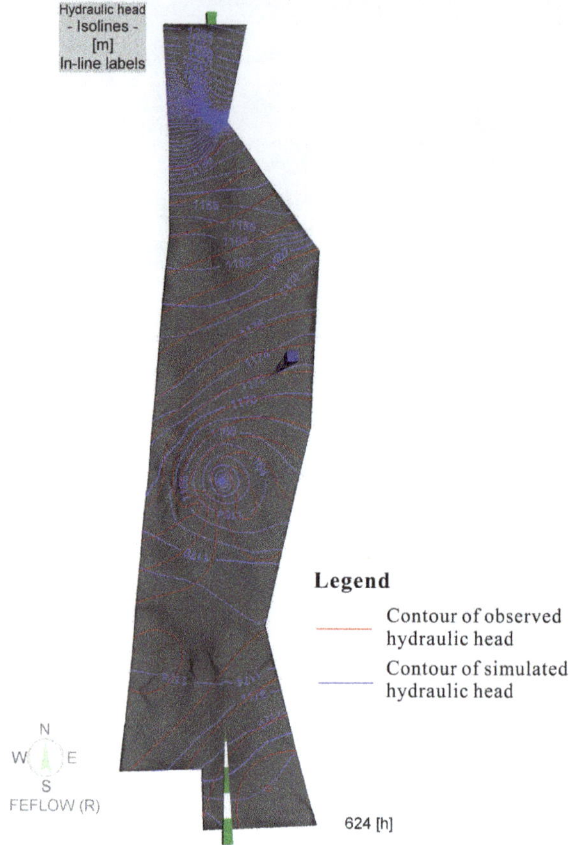

Figure 9. Fitting flow field of Baotashan aquifer between the observed and simulated hydraulic head.

Taking the hydrogeological parameters in Table 1 as initial parameters, the zoning parameters of the aquifer were inversely calculated, as shown in Table 2. The horizontal permeability coefficient of the Baotashan aquifer ranges from 0.67 m/d to 3.2 m/d, and the vertical permeability coefficient is found to be one-tenth of the horizontal permeability coefficient. The water-specific storage ranges from 1.5×10^{-7} m^{-1} to 1.3×10^{-6} m^{-1}.

Table 2. Parameter zoning of Baotashan sandstone aquifer.

Zone	Kx (m/d)	Ky (m/d)	Kz (m/d)	Ss (1/m)
Zone 1	1.25	1.25	0.12	2.1×10^{-7}
Zone 2	2.3	2.3	0.23	3.7×10^{-7}
Zone 3	0.67	0.67	0.06	1.5×10^{-7}
Zone 4	3.4	3.4	0.34	1.3×10^{-6}
Zone 5	2.9	2.9	0.29	7.8×10^{-6}
Zone 6	1.05	1.05	0.1	2.7×10^{-7}

4.2.2. Safe Mining Condition

The coefficient of water inrush has been widely used in coal mine production and plays an important role in evaluating the floor water inrush. The calculation formula for the water inrush coefficient is as follows:

$$T = \frac{p}{M} \tag{1}$$

where T is the water inrush coefficient (MPa/m), p is the water pressure (MPa), and M represents the thickness of the aquifuge floor (m).

According to the water inrush coefficient stipulated in China, the area with a water inrush coefficient below 0.06 MPa/m belongs to the safe mining zone when the impermeable floor is complete, whereas it is designated as a dangerous mining area when the water inrush coefficient is greater than 0.1 MPa/m. Under the latter condition, water inrush may occur during coal mining.

The hydraulic pressure distribution of the Baotashan aquifer was calculated, which ranges from 2.63 MPa to 7.06 MPa, with an average value of 4.72 MPa. The water inrush coefficient distribution of the Baotashan sandstone aquifer was obtained using the effective thickness of the impermeable floor and the hydraulic pressure, as shown in Figure 10. The water inrush coefficient of the Baotashan aquifer in most areas is found to be greater than 0.1 MPa/m. No specific area is found with a water inrush coefficient below 0.06 MPa/m. The water inrush coefficient of the Baotashan aquifer within the planned mining panel ranges from 0.08 Mpa/m to 0.51 MPa/m, indicating that there is a water inrush risk during coal mining. Thus, it is necessary to discharge water from the Baotashan aquifer to ensure safe mining.

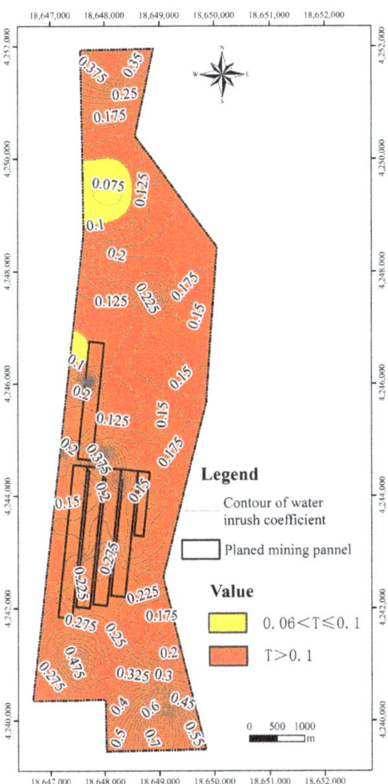

Figure 10. Distribution of water inrush coefficient of Baotashan aquifer.

4.2.3. Scheme of Water Drainage

Three additional draining fields with eight draining holes were arranged. The elevation of draining fields was 910, 880, and 750 m, respectively. Adding the previous four draining holes, a total of ten boreholes were arranged to release the water from the Baotashan aquifer. Since the new draining holes were spaced at a large distance from each other, the estimated drainage water amount of each hole was 3000 m^3/d. The estimated amount of the previous four draining holes was 2000 m^3/d. Thus, the total drainage water amount was 26,000 m^3/d. The numerical simulation was performed to simulate the dewatering scheme of the Baotashan aquifer.

Due to the varying elevation of each draining field, the draining holes would stop dewatering when the hydraulic head fell below the filed elevation. Therefore, initially, ten draining holes in four fields were functioning, and the draining water amount was 2.6×10^4 m^3/d. This was named the first draining stage. When the highest elevation surpassed the head, two holes were withdrawn. At this stage, eight boreholes in the remaining three draining fields were used to dewater the Baotashan aquifer, with an estimated water amount of 2×10^4 m^3/d. This stage was named the second draining stage. Finally, the remaining two draining fields and four holes were used to dewater the Baotashan aquifer, with an estimated amount of 1.2×10^4 m^3/d. The last stage was named the third draining stage.

As shown in Figure 11, the simulation results of the three draining stages reveal that the hydraulic pressure of the Baotashan aquifer satisfies the requirements of safe mining conditions after 241 days of large-scale drainage. The water inrush coefficient is less than 0.06 MPa/m in most areas, and the water inrush coefficient in most working faces of the No. 18 coal seams is less than 0.06 MPa/m, except for the local eastern areas.

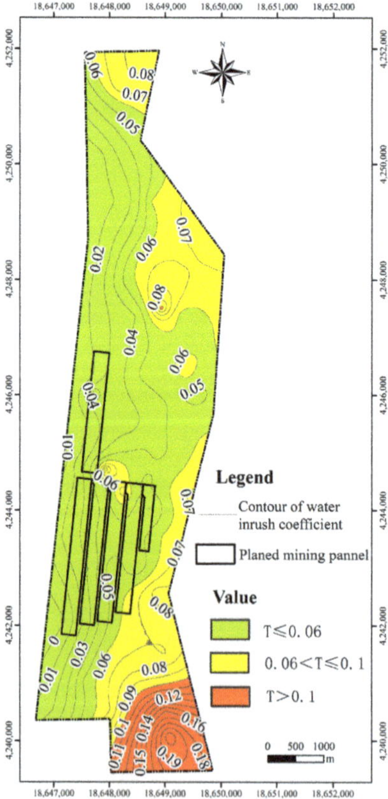

Figure 11. Distribution of water inrush coefficient of Baotashan aquifer after the draining depressurization.

The draining water amount of the Baotashan aquifer was calculated, as shown in Table 3. An amount of 4.314 million m^3 was released in 241 days, of which the elastic hydrostatic reserve of Baotashan sandstone was 367,600 m^3 and the boundary recharge was 3.9461 million m^3 accounting for 91.47% of the total water amount. In the first draining stage, the elastic water release capacity of Baotashan sandstone was 2991 m^3/d, accounting for 11.5% of the total amount of 26,000 m^3/d. Whereas, the water recharge amount was 23,009 m^3/d, accounting for 88.5% of the total amount. As the draining water advanced, the proportion of model static reserves in the total released water reduced progressively due to the expansion of the falling funnel.

Table 3. Simulated water balance statistics of Baotashan drainage water.

Stage	Draining Amount (m^3/d)	Time (d)	Static Storage (m^3/d)	Boundary Recharge (m^3/d)
First stage	26,000	73	2991 (11.5%)	23,009 (88.5%)
Second stage	20,000	50	1822 (9.12%)	18,177 (90.88%)
Third stage	12,000	118	495 (4.12)	11,505 (95.88%)
Total	431.4 × 10^4 m^3	241	36.76 × 10^4 m^3	394.61 × 10^4 m^3

5. Discussion

The Baotashan aquifer was drained to extract the No. 18 coal seam, wherein the water amount reached 4.31 million m^3, causing significant damage to groundwater in arid and semiarid areas. The fault created a major seepage channel to the west and south of the mining area. To reduce the impact of coal mining on the groundwater system, grouting was proposed to block the hydraulic recharge to the Baotashan aquifer. Distinguishing the grouting effects, the permeability coefficient of the fault was designed to be 2 × 10^{-2} m/d, 2 × 10^{-4} m/d, 2 × 10^{-6} m/d, and 2 × 10^{-8} m/d.

As shown in Table 4, when the permeability of the fault was designed as 2 × 10^{-2} m/d, the first, second, and third stages of the drainage took 70 days, 44 days, and 110 days, respectively, for completion, and the drainage water amount was 4.02 million m^3. Compared to the original scheme, the draining time was reduced by 19 days and the amount was reduced by 294,000 m^3. The period of water drainage was gradually shortened and the volume of water was lowered as the grouting effect was progressively improved. When the permeability of fault after grouting was designed as 2 × 10^{-6} m/d, the drainage time was 195 days and the drainage water was 3.502 million m^3. Compared to the original scheme, the drainage time was reduced by 46 d and the drainage amount was reduced by 812,000 m^3. The grouting of the fault prevented the hydraulic recharge of the southern region of the well field of the Baotashan aquifer. This led to the gradual reduction of the boundary recharge with the enhancement of the grouting effect.

Table 4. Optimization for the Baotashan aquifer drainage simulation.

Grouting Effect of Fault (m/d)	Stage 1 (d)	Stage 2 (d)	Stage 3 (d)	Discharge Water (×10^4 m^3)	Static Reserves of Model (×10^4 m^3)	Boundary Supply (×10^4 m^3)
No grouting	73	50	118	431.4	36.76	394.61
2 × 10^{-2}	70	44	110	402	38.53	363.47
2 × 10^{-4}	68	41	106	386	41.32	344.68
2 × 10^{-6}	63	35	97	350.2	45.19	305.01
2 × 10^{-8}	63	35	95	347.8	45.57	302.23

6. Conclusions

High-intensity coal mining has a profound impact on groundwater resources, especially in arid and semi-arid parts of western China, with there is a scarcity of rainfall and evaporation is intense. Thus, to evaluate the influence of No. 18 coal seam mining on the groundwater water resources in New Shanghai No. 1 Coal Mine, a series of large-scale drainage tests were carried out on the Baotashan aquifer to obtain its hydrogeological parameters. Based on the field data, a 3D numerical model was established. On the premise of safe mining, the influence of coal mining on groundwater resources was quantitatively evaluated. The following conclusions were derived:

According to the variation of the flow field of the Baotashan aquifer, the southwest boundary of the well field acted as the drainage boundary, and the north and south of the well field were recharge boundaries. The permeability coefficient of the Baotashan sandstone aquifer ranged from 0.1 to 2.0 m/d, and the permeability coefficient in the central region of the well field was found to be greater than that in the south and north regions. The working faces planned for mining were located in an area with a high permeability coefficient. The water abundance of most areas of the Baotashan aquifer was found to be in the medium-rich water abundance category.

The fitting error between the calculated water level and the water level obtained from the observation hole B-12 was relatively large, about 2.46 m. The water inrush coefficient of the Baotashan aquifer was further calculated. The water inrush coefficient of the No. 18 coal seam floor was between 0.08 Mpa/m and 0.51 MPa/m, indicating that there was a risk of floor water inrush during the mining. Therefore, water from the Baotashan aquifer had to be released before coal mining. The simulation results showed that the water pressure of the Baotashan aquifer meets the safe mining requirements after 241 days of large-scale drainage. The total amount of water discharged was about 4.314 million m^3. Subsequently, grouting of the fault was suggested to reduce the influence of groundwater resources. The drainage water steadily decreased as the grouting effect was improved. When the permeability of the fault was designed as 2×10^{-6} m/d, the drainage time and water amount were 195 days and 3.502 million m^3, respectively. Furthermore, the drainage time was 46 days less than the original scheme, and the amount was reduced by 812,000 m^3.

Author Contributions: Methodology, Y.L. (Yuguang Lyv) and W.Q.; simulation, W.C.; validation, W.C. and W.Q.; formal analysis, W.C.; investigation and resources, Y.L. (Yuguang Lyv); writing—original draft preparation, Y.L. (Yingjie Liu); writing—review and editing, X.C. and M.L.; supervision, W.Q.; project administration, Y.L. (Yuguang Lyv); funding acquisition, W.Q. All authors have read and agreed to the published version of the manuscript.

Funding: This research was funded by [National Natural Science Foundation] grant number [41772302] and funded by [National Natural Science Foundation] grant number [42007240]. The APC was funded by [National Natural Science Foundation] grant number [41772302].

Institutional Review Board Statement: Not applicable.

Informed Consent Statement: Not applicable.

Data Availability Statement: Data used in this study will be available upon request from the first author.

Conflicts of Interest: The authors declare no conflict of interest.

References

1. Afzal, M.; Battilani, A.; Solimando, D.; Ragab, R. Improving water resources management using different irrigation strategies and water qualities: Field and modelling study. *Agr. Water Manag.* **2016**, *176*, 40–54. [CrossRef]
2. Islam, M.B.; Firoz, A.B.M.; Foglia, L.; Marandi, A.; Khan, A.R.; Schüth, C.; Ribbe, L. A regional groundwater-flow model for sustainable groundwater-resource management in the south Asian megacity of Dhaka, Bangladesh. *Hydrogeol. J.* **2017**, *25*, 617–637. [CrossRef]
3. Kotir, J.H.; Smith, C.; Brown, G.; Marshall, N.; Johnstone, R. A system dynamics simulation model for sustainable water resources management and agricultural development in the Volta River Basin, Ghana. *Sci. Total Environ.* **2016**, *573*, 444–457. [CrossRef] [PubMed]
4. Liu, S.L.; Li, W.P.; Qiao, W.; Wang, Q.Q.; Hu, Y.B.; Wang, Z.K. Effect of natural conditions and mining activities on vegetation variations in arid and semiarid mining regions. *Ecol. Indic.* **2019**, *103*, 331–345. [CrossRef]
5. Zhang, D.S.; Li, W.P.; Lai, X.P.; Fan, G.W.; Liu, W.Q. Development on basic theory of water protection during coal mining in northwest of China. *J. China Coal Soc.* **2016**, *42*, 36–43. (In Chinese)
6. Wang, S.M. *Coal Accumulating and Coal Resource Evaluation of Ordos Basin*; China Coal Industry Publication House: Beijing, China, 1996. (In Chinese)

7. Qiao, W.; Li, W.P.; Li, T.; Chang, J.Y.; Wang, Q.Q. Effects of coal mining on shallow water resources in semiarid regions: A case study in the Shennan mining area, Shaanxi, China. *Mine Water Environ.* **2017**, *36*, 104–113. [CrossRef]
8. China National Energy Administration. Available online: http://www.nea.gov.cn (accessed on 6 March 2023).
9. Wang, J.H.; Kang, H.P.; Liu, J.Z.; Chen, P.P.; Fan, Z.Z.; Yuan, W.M.; Liu, Y.P. Layout strategic research of green coal resource development in China. *J. China Univ. Min. Technol.* **2018**, *47*, 15–20. (In Chinese)
10. Yuan, L.; Zhang, N.; Kan, J.G.; Wang, Y. The concept, model and reserve forecast of green coal resources in China. *J. China Univ. Min. Technol.* **2018**, *47*, 1–8. (In Chinese)
11. Booth, C.J.; Spande, E.D.; Pattee, C.T.; Miller, J.D.; Bertsch, L.P. Positive and negative impacts of longwall mine subsidence on a sandstone aquifer. *Environ. Geol.* **1997**, *34*, 223–233. [CrossRef]
12. Bruns, D.A. Macroinvertebrate response to land cover, habitat, and water chemistry in a mining-impacted river ecosystem: A GIS watershed analysis. *Aquati. Sci.* **2005**, *67*, 403–423. [CrossRef]
13. Candeias, C.; Ávila, P.F.; Ferreira, S.E.; Paulo, J.T. Integrated approach to assess the environmental impact of mining activities: Estimation of the spatial distribution of soil contamination (Panasqueira mining area, central Portugal). *Environ. Monit. Assess.* **2015**, *187*, 135–158. [CrossRef] [PubMed]
14. Chiew, F.H.S.; Fu, G.B.; Post, D.A.; Zhang, Y.Q.; Wang, B.; Viney, N.R. Impact of coal resource development on streamflow characteristics: Influence of climate variability and climate change. *Water* **2018**, *10*, 1161. [CrossRef]
15. Koch, H.; Kaltofen, M.; Grünewald, U. Scenarios of water resources management in the Lower Lusatian mining district, Germany. *Ecol. Eng.* **2005**, *24*, 49–57. [CrossRef]
16. Raghavendra, N.S.; Deka, P.C. Sustainable development and management of ground water resources in mining affected area: A review. *Procedia Earth Planet. Sci.* **2015**, *11*, 598–604. [CrossRef]
17. Sosa1, M.; Zwarteveen, M. The institutional regulation of the sustainability of water resources within mining contexts: Account-ability and plurality. *Curr. Opin. Environ. Sustain.* **2014**, *11*, 19–25. [CrossRef]
18. Yue, H.; Liu, Y. Water balance and influence mechanism analysis: A case study of Hongjiannao Lake, China. *Environ. Monit. Assess.* **2021**, *193*, 219–228. [CrossRef] [PubMed]
19. Guo, Q.L.; Han, Y.Y.; Yang, Y.S.; Fu, G.B.; Li, J.L. Quantifying the Impacts of Climate Change, Coal Mining and Soil and Water Conservation on Streamflow in a Coal Mining Concentrated Watershed on the Loess Plateau, China. *Water* **2019**, *11*, 1054. [CrossRef]
20. Cooper, D.J.; Sanderson, J.S.; Stannard, D.I.; Groeneveld, D.P. Effects of long-term water table drawdown on evapotranspiration and vegetation in an arid region phreatophyte community. *J. Hydrol.* **2006**, *325*, 21–34. [CrossRef]
21. Newman, C.; Agioutantis, Z.; Leon, G. Assessment of potential impacts to surface and subsurface water bodies due to longwall mining. *Int. J. Min. Sci. Technol.* **2017**, *27*, 57–64. [CrossRef]
22. Adhikary, D.P.; Guo, H. Modelling of longwall mining-induced strata permeability change. *Rock Mech. Rock Eng.* **2015**, *48*, 345–359. [CrossRef]
23. Tammetta, P. Estimation of the change in hydraulic conductivity above mined longwall panels. *Ground Water* **2015**, *53*, 122–129. [CrossRef] [PubMed]
24. Booth, C.J. The effects of longwall coal mining on overlying aquifers. *Geol. Soc. Spec. Publ.* **2002**, *198*, 17–45. [CrossRef]
25. Chen, W.C.; Li, W.P.; He, J.H.; Qiao, W.; Wang, Q.Q.; Yang, R.Y. Impact of mining-induced bed separation space on Cretaceous aquifer: A case study of Yingpanhao coal mine, Ordos Basin, China. *Hydrogeo. J.* **2022**, *30*, 691–706. [CrossRef]
26. Fan, K.F.; He, J.H.; Li, W.P.; Chen, W.C. Dynamic evolution and identification of bed separation in overburden during coal mining. *Rock Mech. Rock Eng.* **2022**, *55*, 4015–4030. [CrossRef]
27. Hu, Y.B.; Li, W.P.; Wang, Q.Q.; Liu, S.L.; Wang, Z.K. Evaluation of water inrush risk from coal seam floors with an AHP-EWM algorithm and GIS. *Environ. Earth Sci.* **2019**, *78*, 290–305. [CrossRef]
28. Li, W.P.; Liu, Y.; Qiao, W.; Zhao, C.X.; Yang, D.D.; Guo, Q.C. An improved vulnerability assessment model for floor water bursting from a confined aquifer based on the water inrush coefficient method. *Mine Water Environ.* **2018**, *37*, 196–204. [CrossRef]
29. Wu, Q.; Zhao, D.K.; Wang, Y.; Shen, J.J.; Liu, H.L. Method for assessing coal-floor water-inrush risk based on the variable weight model and unascertained measure theory. *Hydrogeol. J.* **2017**, *25*, 2089–2104. [CrossRef]
30. Singh, R.; Venkatesh, A.S.; Syed, T.H.; Reddy, A.G.S.; Kumar, M.; Kurakalva, R.M. Assessment of potentially toxic trace elements contamination in groundwater resources of the coal mining area of the Korba coalfield, central India. *Environ. Earth Sci.* **2017**, *76*, 566–573. [CrossRef]
31. Wessman, H.; Salmi, O.; Kohl, J.; Kinnunen, P.; Saarivuori, E.; Mroueh, U. Water and society: Mutual challenges for eco-efficient and socially acceptable mining in Finland. *J. Clean Prod.* **2014**, *84*, 289–298. [CrossRef]
32. Zipper, C.E.; Donovan, P.F.; Jones, J.W.; Li, J.; Price, J.E.; Stewart, R.E. Spatial and temporal relationships among watershed mining, water quality, and freshwater mussel status in an eastern USA river. *Sci. Total. Environ.* **2016**, *541*, 603–615. [CrossRef]
33. Howladar, M.F.; Deb, P.K.; Muzemder, A.S.H. Monitoring the underground roadway water quantity and quality for irrigation use around the Barapukuria Coal Mining Industry, Dinajpur, Bangladesh. *Groundw. Sustain Dev.* **2017**, *4*, 23–34. [CrossRef]
34. Guo, W.; Li, P.; Du, Q.; Zhou, Y.; Xu, D.; Zhang, Z. Hydrogeochemical Processes Regulating the Groundwater Geochemistry and Human Health Risk of Groundwater in the Rural Areas of the Wei River Basin, China. *Expo. Health* **2023**, *4*, 1–16. [CrossRef]

35. Yang, Y.; Li, P.; Elumalai, V.; Ning, J.; Xu, F.; Mu, D. Groundwater Quality Assessment Using EWQI With Updated Water Quality Classification Criteria: A Case Study in and Around Zhouzhi County, Guanzhong Basin (China). *Expo. Health* **2022**, *8*, 1–16. [CrossRef]
36. Sun, W.; Zhou, W.; Jiao, J. Hydrogeological classification and water inrush accidents in China's coal mines. *Mine Water Environ.* **2016**, *35*, 214–220. [CrossRef]
37. Yin, S.X.; Zhang, J.; Liu, D. A study of mine water inrushes by measurements of in situ stress and rock failures. *Nat. Hazards.* **2015**, *79*, 1961–1979. [CrossRef]
38. Chen, W.C.; Li, W.P.; Qiao, W.; Li, L.F. Beneficial Use of Deep Ordovician Limestone Water from Mine Safety Dewatering at the Xinglongzhuang Coal Mine, North China. *Mine Water Environ.* **2020**, *39*, 42–56. [CrossRef]
39. Zhao, B.F.; Lv, Y.G. Research on characteristics and prevention and control technology of sandstone water disaster in coal seam floor in Jurassic coalfield. *China Coal.* **2021**, *47*, 56–63. (In Chinese)

Article

Assessment and Prediction of the Collaborative Governance of the Water Resources, Water Conservancy Facilities, and Socio-Economic System in the Xiangjiang River Basin, China

Jie Wen [1], Hongmei Li [2] and Abate Meseretchanie [3,*]

[1] Physical Education Institute, Hunan University of Finance and Economics, Changsha 410205, China; wenjie142857@163.com
[2] School of Economics, Hunan Agricultural University, Changsha 410127, China
[3] Research Institution of Rural Revitalization, Hunan University of Science and Engineering, Yongzhou 425101, China
* Correspondence: abatemeseretchanie3407@huse.edu.cn; Tel.: +86-15773158953

Abstract: The collaborative governance of subsystems within a river basin can play a critical role in addressing challenges, such as water scarcity, soil erosion, flooding, sedimentation, and water pollution, to achieve sustainable utilization of water resources. However, the current literature only focuses on isolated observations of these subsystems, leading to uncertainty and water resource destruction. This paper examines the evolution of the collaborative governance of water resources, water conservancy facilities, and socio-economic systems through self-organization theory in the Xiangjiang River Basin, China. The coupling theory and gray Grey Model (1,1) model were utilized with panel data from 2000 to 2019 to assess and predict the governance synergies of five subsystems: natural water, water conservancy facilities, water resource development and utilization, ecological environment, and socio-economic systems. There are 22 indicators contributing to these subsystems that were selected. The results indicate an S-shaped trend in collaborative governance for water resources, water conservancy facilities, and socio-economic systems. The elements of each subsystem exhibit both synergistic and competitive relationships. The unpredictable precipitation triggers a butterfly effect, changing systemic governance coordination, which closely relates to developing the natural water subsystem. Effective water conservation and regulation of water conservancy facilities are the keys to improving water-use efficiency and safeguarding water ecology. This study provides insights into the collaborative governance among subsystems and the evolution of the water resources, water conservancy facilities, and socio-economic systems in the Xiangjiang River Basin to promote sustainable water resource utilization.

Keywords: collaborative governance; water resources; water conservancy facilities; socio-economic development

Citation: Wen, J.; Li, H.; Meseretchanie, A. Assessment and Prediction of the Collaborative Governance of the Water Resources, Water Conservancy Facilities, and Socio-Economic System in the Xiangjiang River Basin, China. *Water* **2023**, *15*, 3630. https://doi.org/10.3390/w15203630

Academic Editors: Peiyue Li, Andrea G. Capodaglio and Jianhua Wu

Received: 14 June 2023
Revised: 19 July 2023
Accepted: 31 July 2023
Published: 17 October 2023

1. Introduction

In the face of climate change, population growth, and socio-economic development, the sustainable utilization of water resources while preserving nature's self-restoration ability is critical for sustainable development [1,2]. The United Nations World Water Resources Report emphasizes the need to replicate nature's self-organizing development process, focus on constructing water conservancy and water quality facilities, and implement a collaborative governance approach for nature and society. China's 2021 Central Document No.1 similarly highlights the importance of promoting water system connectivity and comprehensive improvement, as well as strengthening the construction of water conservancy facilities and water source protection. However, previous investigations into rebuilding self-organization systems within water systems have often ignored collaboration, resulting in separate and non-synergistic observations, increasing uncertainty, and significant water resource destruction. Therefore, it is crucial to take a collaborative approach to examine the interaction between subsystems, such as water resources, water conservancy facilities, and socio-economic activities.

Water resources, along with water conservancy facilities, are crucial to achieving sustainable socio-economic development [3]. To effectively develop and utilize water resources, a systematic and integrative approach is necessary for water resources, water conservancy facilities, and socio-economic activities. Achieving sustainable utilization of water resources for socio-economic development requires optimizing infrastructure and respecting the carrying capacity of the environment. As such, it is critical to promote natural resource protection and socio-economic activities in a coordinated manner to achieve harmonious coexistence between humans and nature [4,5]. However, the combined effects of climate change [6] and economies of scale [7] have led to increasing water scarcity [8], aging and wear of water conservancy facilities, and degradation of water ecology [9]. These are fundamental issues that restrict sustainable socio-economic development and require urgent coordination and governance in the water sector.

River basin water management systems need to be integrative, inclusive, and consider both natural and man-made factors to effectively manage water resources [10,11]. Stakeholders may have different demands when it comes to water resource utilization, which is why previous studies have explored different aspects such as socio-economic [12], environmental [13], and water conservation facilities [3]. However, few studies have examined the interactions between water, ecology, and socio-economy [14]. There are also gaps in the existing literature, particularly in observing river-basins as a system and their constituents as a subsystem to enhance the efficiency of water use. To ensure sustainable water resource management, it is essential to take a holistic approach that considers all factors that impact water availability and utilization. Only by understanding the complex interactions between natural and human factors can we develop effective strategies for managing water resources and ensuring equitable distribution among all stakeholders.

Effective water resource governance in a river basin is crucial for organizing and coordinating multiple stakeholders towards sustainable water utilization [15]. This involves identifying the various stakeholders in the river basin, improving their ability to share knowledge [16], and enhancing water resource distribution efficiency [17]. Collaborative governance in a river basin has been found to play a significant role in achieving positive ecological outcomes [18]. Transforming from other forms of governance to collaborative governance can improve integrated planning and decision-making processes [19]. This article specifically examines the role of collaborative governance in water resource management, water conservancy facilities, and socio-economic systems based on an evolutionary logic mechanism. By adopting a collaborative approach, stakeholders can work together to achieve common goals and address challenges more effectively. Through collaboration, stakeholders can also share knowledge, build trust, and foster mutual understanding, which are all critical for achieving sustainable water resource management.

Over the last three decades, there has been significant research on multi-user allocation and multi-purpose water resource development and utilization projects by water conservancy engineers and economists. More recent studies have focused on joint scheduling and the optimization of regional water resources, taking into account factors such as water conservation project design and supply-and-demand balance [20]. For example, Arfanuzzaman Md and Atiq Rahman (2017) [21] analyzed various factors, including water resource supply-and-demand, system losses, groundwater levels, and per capita water consumption, to propose a series of social–ecological resilience-building measures, such as the comprehensive regulation of surface water and groundwater, water conservation, water footprint reduction, and a continuous-water-demand governance plan. By optimizing river basin water conservancy, Girard Corentin et al. (2016) establish a minimum-cost and robust investment portfolio model for the development of water resources at scale [20]. These types of studies provide valuable insights into how we can enhance the efficiency of water resource allocation and improve the overall management of water resources.

Several studies have focused on improving the assessment of water resource utilization and developing innovative solutions to enhance the efficiency of water utilization. For example, Cunha Henrique et al. (2019) [22] improved the existing assessment method of single-water-resource utilization, developed a calculation method for the comprehensive utilization of water resources, and created an intelligent collective irrigation system.

Tiwari Ashwani Kumar et al. (2021) [23] utilized geographic information systems to collect data on factors such as slope, elevation, landform, and drainage to assess water pollution levels and provide policy recommendations for the comprehensive management of urban and rural water resources. However, few studies have placed significant emphasis on assessing the safety and carrying capacity of water resources, measuring the sustainable development quantity and ecological quality of water resources in different regions, evaluating the level of modern water conservancy development, or analyzing the interaction between regional water conservancy facilities and socio-economic development. Li Xiaoyun (2017) [24] analyzed the concept, characteristics, and composition of water resource carrying capacity and provided a more comprehensive summary of water resource carrying capacity evaluation methods. Such studies are essential in providing a better understanding of water resource management and can help in developing policies and strategies that promote sustainable water resource utilization.

Several studies have used different models and evaluation methods to assess the relationship between water conservancy facilities and socio-economic development. For example, Mao H H et al. (2011) [25] used a coordinated development model to quantitatively evaluate the level of the water conservancy facilities and economic society in the Haihe River Basin. Yi X B et al. (2013) [26] used the Lorenz curve and Gini coefficient to calculate the coordinated development of water conservancy and socio-economic development in various cities in Guangdong and concluded that there is a positive correlation between water conservancy facilities and socio-economic development. In a different study [27], Huang X F et al. (2017) applied a qualitative and quantitative cloud evaluation model with both fuzzy and random factors to analyze the development level of water conservancy modernization. Finally, Li Y L et al. (2019) [28] constructed a system with input-response-output dimensions and analyzed the spatial heterogeneity and main controlling factors of the green development of water. They concluded that attention should be paid to the water conservancy-informational and production ecology that affects the economic benefits. These studies provide valuable insights into the complex relationship between water resource management and socio-economic development and can help inform policy and decision-making processes aimed at promoting sustainable water resource utilization.

The literature on the use of collaborative governance to manage water resources in river basins is relatively limited, with few studies addressing ecological, socio-economic, and air pollution perspectives [18,29,30]. However, there are some shortcomings in previous research. First, existing studies have not fully considered spatial differences within the same geographic unit and whether synergy is spatially balanced in the region. Second, previous research has not integrated water resources, water conservancy facilities, and socio-economic systems as a whole to study their coordination and change trends. Third, there is a lack of theoretical analysis of the logical relationship between the system and its internal interactions under the coordinated governance of the WCS system (hereafter, the WCS system). To address these gaps, this article uses the self-organization theory to examine the evolutionary logic mechanism of the collaborative governance of water resources, water conservancy facilities, and socio-economic systems, as well as the interrelationships and interweaving effects of the various subsystems. The authors also use coupling theory and the gray GM (1,1) model to build a system governance synergy evaluation and prediction model. Finally, they evaluate and interpret the coordination degree and change law of the WCS system in the Xiangjiang River Basin. This study provides valuable insights into the complex relationship between water resource management, water conservancy facilities, and socio-economic development. It highlights the importance of considering spatial differences and the need for an integrated approach to studying the coordination and change trends of these subsystems. The authors' use of self-organization theory and the gray GM (1,1) model provides a useful framework for analyzing and predicting the synergies of these subsystems.

This article makes a significant contribution to the literature in three ways. First, it systematically examines how to create coordinated governance relationships between the subsystems of the WCS system, which has not been extensively studied previously. The authors draw connections between the changes in the WCS system and its subsystems in the Xiangjiang River Basin from 2000 to 2019. Second, the article provides insights into the dynamic evolution logic of coordinated governance of the WCS system from a systematic perspective. Finally, the study explores differences in governance synergies between water resources, water conservancy facilities, and socio-economic development within the same geographic unit at different times and spaces. These findings have important implications for decision-making and provide a reference point for balancing regional water conservancy facility construction, water resource environmental protection, and socio-economic development. The article highlights the significance of considering the interrelationships and interactions between these subsystems, as well as spatial differences within the same region, to create effective and sustainable governance mechanisms for water resources. Overall, this study provides valuable insights into promoting collaborative governance and sustainable management of water resources in river basins.

2. Theoretical Framework

2.1. The Composition of the WCS System

In this article, the water resources, water conservancy facilities, and socio-economic activities of the Xiangjiang River Basin are studied holistically under the umbrella term "WCS system". This system is made up of a number of interconnected components, including surface water and groundwater, precipitation, water environment, water conservancy engineering facilities, socio-economic activities, and other natural environments [31]. Various humanistic activities such as resource utilization, socio-economic development, and ecological protection create interactive relationships between the different elements within the WCS system. Natural water resources such as precipitation, surface water, and groundwater serve as the fundamental building blocks of this system. The efficient and sustainable utilization of these resources for supporting socio-economic development through the construction of water conservancy facilities and safeguarding the water ecology serves as the central aspect of the WCS system [32,33].

The application of water resources in the WCS system of the Xiangjiang River Basin is fraught with two significant challenges.

1. The uneven distribution of the natural water volume and productivity over time and space is a significant issue that worsens regional socio-economic development. The construction of transitional water-conservancy connectors is inadequate in addressing this problem, leading to insufficient regulation in controlling regional water resources. This, in turn, causes a shortfall in meeting domestic water demands [18];
2. The discharge of pollutants resulting from the utilization of water resources for socio-economic development casts a shadow on the ecological health of the water envi-ronment.

The gradual progress of socio-economic development offers important financial and material resources to support the development and utilization of water resources and to restore the ecological balance of water systems. Therefore, this article examines and integrates the interdependence of water resources, water conservancy facilities, and socio-economic activities under a unified framework referred to as the WCS system. The WCS system is comprised of five subsystems, namely natural water, water conservancy facilities, water resources development and utilization, ecological environment, and socio-economic activities [34]. These subsystems are intricately connected and dependent, working together to govern the entire WCS system. Figure 1 illustrates the composition of the WCS system and the interactions between its subsystems.

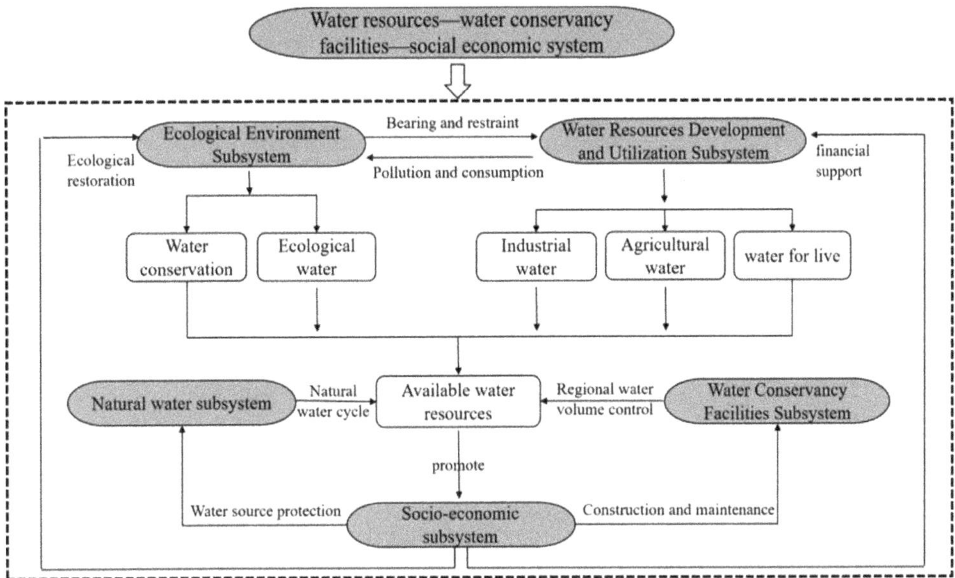

Figure 1. The composition of the WCS system and the interaction between the subsystems.

2.1.1. The Evolutionary Logic Analysis of the Governance Synergy of the WCS System

According to the theory of self-organization, complex systems comprising many elements tend to achieve a steady state of balance in response to external fluctuations without the need for external intervention [35]. The components within the system are capable of self-structuring and repairing without external instructions. The theory of self-organization is grounded on four main aspects. Firstly, Prigogine's dissipation theory, proposed in 1960, suggests that orderly and stable systems can emerge from more disordered ones [36]. Secondly, Hermann Haken developed the concept of synergistic effects in 1971, which revealed that the interaction of an overall or collaborative effect generates numerous open systems [37]. Thirdly, Rene Thom established the catastrophe theory in the 1960s, which states that slight changes in specific parameters of a nonlinear system can cause equilibrium to appear or disappear, leading to significant and abrupt changes in the system's behavior. Finally, Egan created the system evolution theory during the 1960s, which posits that interdependence within a system arises from determinate relationships among its parts or variables, as opposed to random variability [38].

In this study, we derive a self-organization theory that states that the open system is in a dissipative state, each subsystem and the system as a whole will exchange with the external environment in multiple ways through matter, energy, and information over time. The exchange process will produce entropy (dS), expressed as $dS = d_iS + d_eS$, where d_iS is the positive entropy generated inside the system, and d_eS is the entropy produced by the exchange between the system and the outside world, which can be positive or negative [39]. When the $dS < 0$ and $d_iS < |d_eS|$ performance develops over time, the negative entropy obtained by the system will be greater than the internal positive entropy and the total entropy will be reduced. When the total entropy reaches a certain threshold, the system will undergo a sudden change, the transformation from disorder to order will be realized, and the degree of coordination will become higher. Conversely, the system can evolve in the direction of disorder and the degree of coordination will decrease.

The self-organization theory suggests that the WCS system, due to its openness, will exchange various forms of material, energy, and information with the external environment. This is manifested in the fact that natural rainfall, water resources utilization, water environment carrying capacity, and water conservancy facilities construction, among other factors (Figure 2), affect the synergy of the WCS system's governance. During the period of ecological bearing, these factors promote each other, resulting in obvious positive effects on each subsystem, which continuously enhances synergy, leading to an upward trend in the system's governance. However, as social and economic development continues, the water conservancy facilities become worn and aged, water resources become scarce, and the water ecology falls into deficit. At this point, negative effects begin to appear in each subsystem, although they have yet to reach the critical stage and exhibit only small fluctuations. The WCS system can assimilate and absorb these minor fluctuations through self-organization mechanisms, thus retaining its original structure and governance synergy unchanged. Therefore, during this stage, the WCS system's governance synergy is altered by small fluctuations but remains predominantly stable and positive overall.

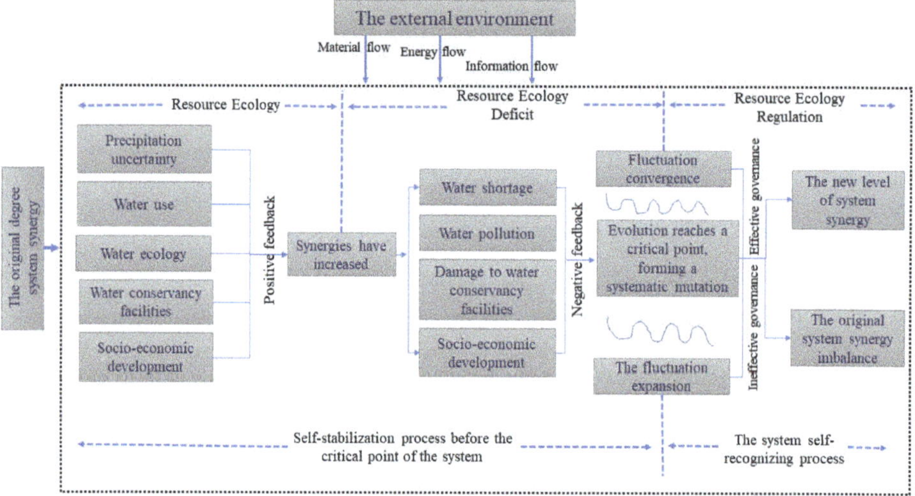

Figure 2. The evolution mechanism of the WCS system's governance synergy.

When the water conservancy facilities reach a state of aging, the availability of water resources becomes severely limited, and the water ecological environment is significantly damaged, the negative effects on each subsystem become pronounced. These factors exceed the tolerable range of the WCS system, resulting in significant and abrupt changes that exhibit large fluctuations. The self-organization mechanisms of the WCS system cannot assimilate and absorb these disruptive changes. Even minor disturbances may rapidly expand and transmit, breaking down the original structure and governance synergy of the WCS system. Consequently, this can cause a mutation and bifurcation in the system's governance synergy.

2.1.2. The Change Process of the WCS System's Governance Synergy

If the WCS system adopts forceful governance measures, such as enhancing water-resource-utilization efficiency, investing in and maintaining water conservancy facilities, and implementing comprehensive water environmental management, then the system's autonomous resilience can be improved. This, in turn, leads to the decoupling of water resource consumption and pollution from social and economic development. As a result, positive entropy inside the system exceeds the negative entropy obtained from external sources, leading to an imbalance in the coordination of various subsystems. Additionally, the overall coordination of system governance exhibits a tendency to decline and remain constrained by factors such as the wear-and-tear of water conservancy facilities, water environmental pollution, and excessive use of water resources.

The result of bifurcation may be new evolution, collapse, or degeneration (Figure 3). The direction of system governance synergy depends on the implementation of governance measures. If measures such as the integration of resource utilization and environmental protection, the adjustment and optimization of water conservancy planning, the adjustment of economic structure, the investment of irrigation facilities, technical reform, policy formulation, and other measures can be taken to improve the self-recovery of the system, it will promote the positive evolution of the system governance synergy. The specific performance is $dS < 0$ and $d_iS < |d_eS|$. The internal positive entropy of the system is less than the external negative entropy. The development of each subsystem will be coordinated; the overall coordination degree of system governance will reach a new height, forming a new structure, function, and order and then forming a new stable mode, and will continue to develop to a higher level, with a rising differentiation trend. However, if governance and regulation are not in place, it will promote the negative evolution of system governance synergy. The specific performance is $dS > 0$ and $d_iS > |d_eS|$, which indicates that the positive entropy inside the system is greater than the negative entropy obtained from the outside and the coordination of all subsystems is unbalanced, which will tend to decline and degenerate. The degree of coordination will show a downward trend of differentiation, which will continue to be subject to the constraints of water conservancy facilities' wear, water environmental pollution, and the excessive use of water resources.

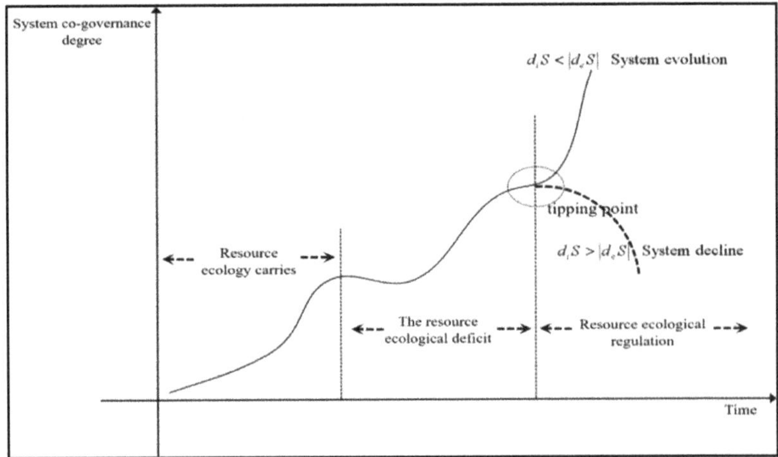

Figure 3. The change process of the WCS system's governance synergy. Note: co-governance refers to collaborative governance.

The constituent subsystems of the WCS system exhibit both opposing and unified characteristics, as well as cooperative and competitive dynamics in collaborative governance. Over time, the co-evolution of the WCS system is boundless and follows an infinite sequence of bifurcation paths. However, due to the geographical conditions, resource allocation, social economy, ecological environment, and other factors, the co-evolution of the WCS system is restrained. The evolution trend of the system's governance synergy generally demonstrates a non-linear change feature resembling the shape of the letter "S" (Figure 3). With appropriate governance and regulation, and by enhancing governance capacity through technical, management, policy, and institutional measures, the WCS system can be promoted to evolve more systematically and improve the level of governance synergy.

3. Materials and Methods

Section 3.1 describes the study area's location, while Section 3.2 outlines the data sources for each indicator representing the various subsystems included in this paper. The specific indicators used to assess the degree of collaborative governance within the WCS system are presented in detail in Section 3.3. Finally, Section 3.4 explains the coupling model and the grey GM (1,1) model employed to evaluate and predict the governance synergy of the WCS system.

3.1. Study Area

The Xiangjiang River Basin is located in Hunan Province, China, covering an area of approximately 94,346 km^2 with longitudes ranging from 109.5° E to 111.1° E and latitudes from 38.2° N to 39.8° N. However, according to Figure 4, the actual area covered is 94,721 km^2. Situated into the radiation zone of the Yangtze River Economic Belt and South China Economic Circle, this basin is home to the Xiangjiang River, a crucial first-level tributary of the Yangtze River and Hunan's mother river. Moreover, this river stretches over 670 kilometers in Hunan and has more than 1300 tributaries across eight cities, namely Yongzhou, Chenzhou, Hengyang, Loudi, Zhuzhou, Xiangtan, Changsha, and Yueyang. With an average rainfall between 1000 and 1800 mm, the Xiangjiang River Basin has a substantial advantage in terms of water resources.

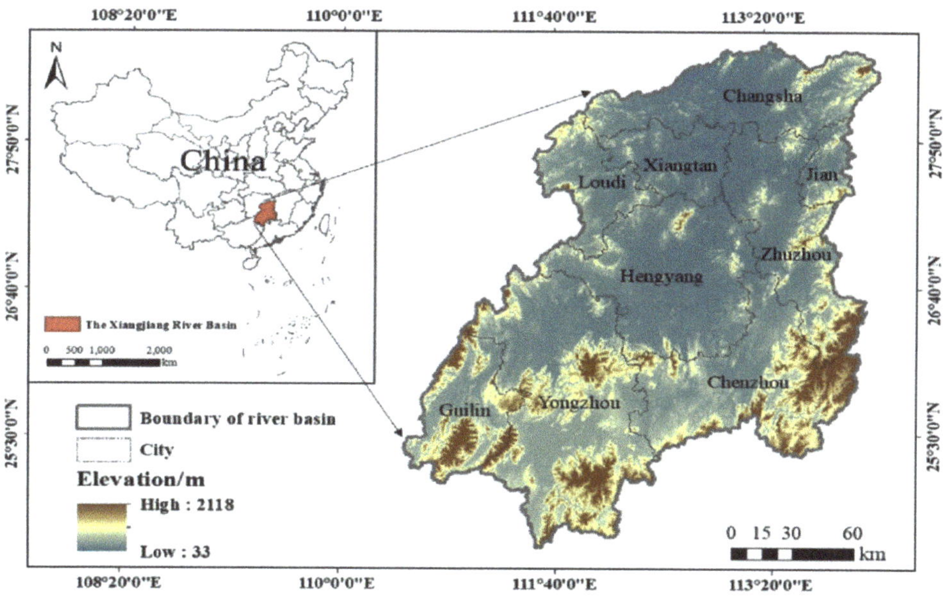

Figure 4. Map of the Xiangjiang River Basin, Hunan Province, China.

This area is densely populated with cities and towns and has a concentrated population, developed economy, rich culture, and convenient transportation. It is the core area and golden zone for Hunan's development, gathering over 70% of the province's large- and medium-sized enterprises. With a total population of 37.74 million by 2019, the Xiangjiang River Basin is the core area of the province's economic and social development. The GDP and industrial value added in 2019 reached approximately 1220.5 billion RMB and 484.2 billion RMB, respectively, accounting for about 76.7% and 82.2% of the total GDP and industrial value added in Hunan Province.

Despite having a solid foundation for development and bright prospects, the Xiangjiang River Basin faces various socio-economic and environmental challenges. As the

traditional extensive development model has not substantially changed, watershed resources and ecological environmental issues have become increasingly prominent in recent years, leading to mounting pressure for sustainable development. Moreover, the uneven distribution of precipitation across time and space, coupled with recurring droughts and floods, poses significant challenges to the basin's development

Currently, the Xiangjiang River Basin lacks an established overall coordination and management system. There is a lack of cooperation mechanisms for upstream and downstream linkage. It is crucial to accelerate the establishment and improvement of a development mechanism that balances ecological environmental protection and construction, considering the macro environment holistically. Urgent demands for changing the current development mode exist, and new institutional arrangements, governance systems, and development paradigms are necessary.

3.2. Data Sources

The panel data of 22 indicators presented in Table 1 were drawn from the Hunan Water Resources Bulletin at http://slt.hunan.gov.cn/slt/xxgk/tjgb/index.html (accessed on 23 July 2021); the Hunan Statistical Yearbook at http://tjj.hunan.gov.cn/hntj/tjsj/tjnj/index.html (accessed on 23 July 2021); the Statistical Bulletin of National Economic and Social Development of Hunan Province at http://www.hunan.gov.cn/hnszf/zfsj/tjgb/202003/t20200319_11815838.html (accessed on 23 July 2021); the China Water Resources Statistics Yearbook at http://www.stats.gov.cn/tjsj/ndsj/ (accessed on 2 October 2020); and the Hunan Ecological Environment Status Bulletin at http://sthjt.hunan.gov.cn/sthjt/xxgk/zdly/hjjc/hjtj/index.html (accessed on 23 July 2021) during the period from 2000 to 2019. Details of the data sources are presented in the Supplementary Material (Table S1).

Table 1. Variable selection and descriptive statistics.

	Subsystems	Indicators	Description	Mean	SD
1	Natural water subsystem	Average annual rainfall	Mean annual precipitation (mm)	0.444	0.188
		Water resource per capita	Per capita water resources (m^3/person)	0.421	0.202
		Total water resource	The total water resources (100 million m^3)	0.441	0.200
2	Water Conservancy Facilities Subsystem	Water conservancy construction investment in agriculture and forestry	Investment in agriculture, forestry and water conservancy (CNY 10,000)	0.454	0.349
		Number of reservoir	Reservoir seat number/individual entries	0.843	0.112
		Reservoir storage capacity	Year-end storage capacity (ten thousand m^3)	0.528	0.289
		Length of embankments	Length of embankments (km)	0.383	0.449
		Effective irrigated area of farmland	Effective irrigated area of farmland (10,000 hm^2)	0.477	0.392
3	Water resource development and utilization subsystem	Water consumption (GDP)	GDP water utilization CNY 10,000 (m^3)	0.678	0.346
		Industrial added value water	Industrial added value water in CNY 10,000 (m^3/t)	0.729	0.267
		Water consumption per unit of grain yield	Per capita water use (m^3)	0.642	0.318
		Comprehensive water consumption per capita	Total supply of water (hundred million m^3)	0.571	0.189
		Total water supply	CNY 10,000 of industrial added value water (m^3)	0.482	0.234
		Water utilization rate	Utilization rate of water resources (%)	0.375	0.186

Table 1. *Cont.*

	Subsystems	Indicators	Description	Mean	SD
4	Ecological Environment subsystem	Green coverage area	Green coverage area (hm^2)	0.494	0.335
		Sewage discharge	Quantity of waste water effluent (10,000 m^3)	0.417	0.273
		Sewage treatment rate	Treatment rate of domestic sewage (%)	0.561	0.372
5	Socio-economic Subsystem	Water quality compliance rate	Water quality standard rate of water functional area (%)	0.676	0.260
		Food production	grain output/t	0.667	0.280
		Per capita disposable income	Per capita disposable income (CNY)	0.417	0.361
		Gross regional product	Gross regional domestic product (CNY 10,000)	0.556	0.297
		Regional fiscal revenue	Regional revenue (CNY 10,000)	0.476	0.386

3.3. Variable Selection and Descriptive Statistics

Table 1 outlines a detailed list of indicators used to measure the degree of collaborative governance within the WCS system. The natural water subsystem focuses on the number of resources and per capita possession, utilizing indicators such as the total water resources, annual rainfall, and per capita water resources. Additionally, the subsystem of water conservancy facilities considers flood control, water storage, and food security, selecting indicators such as investment in agricultural, forestry, and water conservancy construction, dike length; number of reservoirs; and year-end storage capacity. Concerning water consumption and utilization efficiency, the subsystem of water resources development and utilization selects indicators such as water utilized for every CNY 10,000 of industrial added value, water for unit grain output, per capita comprehensive water consumption, and water resources utilization **rate**. Furthermore, the ecological environment subsystem focuses on the pollution and treatment of water resources, utilizing indicators such as sewage discharge, sewage treatment rate, and water quality compliance rate in water function areas. Lastly, the socio-economic subsystem primarily focuses on capital accumulation and utilizes indicators such as regional fiscal revenue, per capita disposable income, and grain output. Overall, there are a total of five subsystems and twenty-two evaluation indicators utilized to measure the degree of collaborative governance within the WCS system.

Accurately estimating the total amount of available water resources in a region can be challenging due to various factors, such as water consumption in the atmospheric cycle, production and daily life, and the level of technological advancement in water resource utilization. Thus, this paper does not provide an exact amount of regional available water resources. Instead, indicators such as CNY 10,000 GDP water consumption, CNY 10,000 industrial added value water consumption, unit grain output water consumption and per capita comprehensive water consumption, among others, found in the water resources development and utilization subsystem can be used to show the current water consumption. Table S3 presents the Xiangjiang River Basin WCS system coordinated governance index information. The implementation and application of the index data from each constituent subsystem within the WCS system rely on the system governance collaborative evaluation model. This method requires calculating the contribution value of each index to the subsystem order degree, determining the subsystem's order degree, and finally obtaining the WCS system's governance synergy degree.

3.4. Methods

Based on the self-organization theory, this paper examines the evolution of collaborative governance in the Xiangjiang River Basin, China. We apply the coupling and gray GM (1,1) model using panel data from the 2000–2019 period to evaluate and predict the governance synergy of five subsystems including; the natural water subsystem, water

conservancy facilities subsystem, water resource development and utilization subsystem, ecological environment subsystem, and socio-economic subsystem.

As the indispensable basic resources and supporting facilities for social and economic development, water resources and water conservancy facilities play a key role in realizing regional sustainable development. However, at present, under the intertwined influence of climate change and resource development, China's water resource shortage, the aging and wear of water conservancy facilities, water ecological degradation, and other problems continue to overlap and become increasingly prominent. This shows the characteristics of systematization and complexity, which have become two of the important problems restricting the sustainable development of the social economy. Therefore, the management of water resources needs to follow the system concept and pay attention to the coordinated management with the water conservancy facilities and social economy. Coupling theory is originally a physical concept, which refers to the phenomenon that two (or more) systems or motion forms affect each other through various interactions. The coupling degree describes the degree of interaction between systems or elements. From the perspective of synergetics, the coupling effect and its degree determine the change trend and direction of the system. Therefore, this paper uses the coupling model to analyze the coordinated governance of water resources, water conservancy facilities, and social economy.

The grey GM (1,1) model used to predict the development synergy of water resources, water conservancy facilities and the socio-economic system can improve the accuracy of the results, but this method is only limited to short-term prediction and is not suitable for long-term prediction. There are similar studies of applying coupling theory and the grey GM (1,1) model in the literature. For example, Nie X and Zhang Z W (2020), Wang Weixin et al. (2020) [40,41] made a quantitative analysis of the coupling relationship between resources and the environment and social and economic development; Xu Hui (2021) [42] and Liu Guofeng (2021) [43] predicted the coupling relationship between resource utilization, the ecological environment, and economic growth and its development trend. The construction of models for the evaluation and prediction of governance-synergy is presented in the following Section 4.2.

3.4.1. Evaluation of the Existing Governance Synergy

In coupling theory, the order parameter index is a key variable used to measure the degree of the coordinated governance of the system; it determines the change process of the system from disorder to order. The degree of coupling is a measure of this synergy. The coupling degree, therefore, is used to construct a system governance collaborative evaluation model [44]. The collaborative governance evaluation model consists of two parts: the degree of order of the subsystems and the degree of subsystem governance synergy.

1. Subsystem Order Degree

In this paper we use i to represent the five constituent subsystems, $i \in [1, 5]$; j represents the order parameter index of each subsystem, $j \in [1, n]$; and = $(_{,,})$ represents the j-th order parameter index of the i-th subsystem. The formula for the order degree S_i of the subsystem is, therefore:

$$S_i = \sum_{j=1}^{n} w_j\, u_{ij}(\delta_{ij}) \qquad (w_j \geq 0 \text{ And } \sum_{j=1}^{n} w_j = 1) \tag{1}$$

w_j in the above formula is the entropy weight of each index,
$u_{ij}(\delta_{ij})$ is the contribution value of each index to the order degree of the subsystem. We then derive the following two equations:

$$w_j = (1 - d_j) / \sum_{j=1}^{n} (1 - d_j), d_j = -\frac{1}{\ln t} \cdot \sum_{t=1}^{h} \left(x_{ij} / \sum_{t=1}^{h} x_{ij} \right) \ln \left(x_{ij} / \sum_{t=1}^{h} x_{ij} \right) \tag{2}$$

$$u_{ij}\left(\delta_{ij}\right)$$
$$= \begin{cases} \left(\delta_{ij} - \beta_{ij}\right)/\left(\alpha_{ij} - \beta_{ij}\right) & \delta_{ij} \text{ is the positive order parameter} \\ \left(\alpha_{ij} - \delta_{ij}\right)/\left(\alpha_{ij} - \beta_{ij}\right) & \delta_{ij} \text{ is the negative order parameter} \end{cases} \quad (3)$$

Whereas,

d_j in the above formula is the information entropy of each index;

t is the number of years evaluated;

x_{ij} is the normalized value of the original data;

α_{ij} is the upper limit of the order parameters;

β_{ij} is the lower limit of the order parameter.

The greater the value of $u_{ij}\left(\delta_{ij}\right)$, the greater the contribution to the order of the subsystem. Avoiding the occurrence of 0 and 1, we enlarge and reduce the extreme values of the order parameter by 1% as the upper limit and lower limit of the critical point.

2. Subsystem Governance Synergy Degree

$$C = \left[\prod_{i=1}^{n} S_i / \left(\frac{1}{n} \sum_{i=1}^{n} S_i \right)^n \right]^k \quad (4)$$

$$D = \sqrt{C \times T}, T = aS_1 + bS_2 + \cdots + eS_5 \quad (5)$$

Whereas,

D is the coordination degree of WCS system governance;

C is the degree of coupling among various subsystems;

i is the number of subsystems;

k is the adjustment coefficient, and $k \geq 2$ ($k = 2$);

T is the coordination index;

$a, b, c, d,$ and e are the undetermined coefficients.

Since the various subsystems influence each other and are indispensable to each other, we adopt Chen's principle to assign weights to all undetermined coefficients as 1/2 [45] (Chen Z, 2020).

3.4.2. Prediction of the Future Governance Synergy

The Xiangjiang River Basin system is in an open state, and each subsystem and the system as a whole can continuously exchange material, energy, and information with the external environment over time. The entire system includes indicators such as reservoirs, domestic water production and consumption, water quality compliance rate, and regional fiscal revenue. These indicators are easy to measure. The system also includes uncertain natural precipitation, total water resources, and other indicators. In this way, a system containing both obvious and uncertain information applies to the gray GM (1,1) model [46]. This article adopts the gray GM (1,1) model for the WCS system's governance synergy prediction.

We set the sequence $x^{(0)}(i) = \{x^0(1), x^0(2), \ldots\ldots, x^0(n)\}$ to be the degree of system governance synergy; $x^{(1)}(k)$ is the original sequence of the $x^0(i)$ accumulatively generated sequence and is denoted as $x^{(1)}(k) = \sum_{i=1}^{k} x^{(0)}(i), k = 1, 2, \ldots, n. \hat{x}^{(1)}(k) = \left(x^{(0)}(1) - \frac{b}{a}\right)e^{-ak} + \frac{b}{a}$ is the gray system prediction equation, and the parameters a and b in the equation are estimated by the least square method:

$$\hat{U} = \begin{pmatrix} a \\ b \end{pmatrix} = \left(B^T B\right)^{-1} B^T Y, Y = \begin{pmatrix} x^{(0)}(2) \\ x^{(0)}(3) \\ \vdots \\ x^{(0)}(n) \end{pmatrix}, B = \begin{pmatrix} -\left(\frac{1}{2}\right)x^{(1)}(2) + x^{(1)}(1) & 1 \\ -\left(\frac{1}{2}\right)x^{(1)}(3) + x^{(1)}(2) & 1 \\ \vdots & \\ -\left(\frac{1}{2}\right)x^{(1)}(k) + x^{(1)}(k-1) & 1 \end{pmatrix} \quad (6)$$

After obtaining the parameters a and b, we can substitute the parameters into $\hat{x}^{(1)}(k) = \left(x^{(0)}(1) - \frac{b}{a}\right)e^{-ak} + \frac{b}{a}$ according to the reduction equation $\hat{x}^{(0)}(k) = \hat{x}^{(1)}(k) - \hat{x}^{(1)}(k-1)$ of the GM (1,1) model, thus we have calculated $\hat{x}^{(0)}(k)$ to obtain the predicted value.

4. Results and Discussion

4.1. The Evaluation of the Existing Governance Synergy

4.1.1. Results of Subsystem Order Degree

The order degree of the five major subsystems over the years in the 2000–2019 period is presented in Figure 5. The overall order of the five subsystems shows a "wave-shaped" upward trend. It can be found that the uncertainty of rainfall has caused huge fluctuations in the orderliness of the natural water subsystem. The largest average rainfall gap over the years is 864 mm. The orderliness of rainy years increased, especially in 2002, 2010, and 2019. The degree of order in drier years declined, with 2007, 2011, and 2018 seen as more typical. This shows that the natural water system is easily affected by rainfall [47]. The amount of rainfall first affects the number of water resources available for surface water and groundwater. Moreover, the amount of rainfall affects social and economic activities, and ultimately fluctuates the degree of coordination of the entire system.

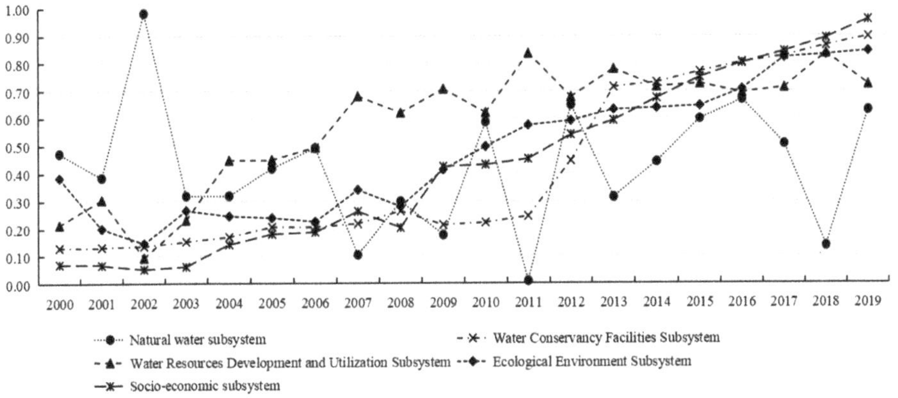

Figure 5. The order degree of each subsystem from 2000 to 2019.

The water conservancy facility subsystem and the socio-economic subsystem contribute significantly to the system's synergy with 27.6%, and 31.9%, respectively. The order degree of the water conservancy facilities subsystem increased significantly from 2011 to 2013. Among them, the number of reservoirs, the length of dykes, and the investment in water conservancy construction increased significantly. The three-year average growth rate of the three indicators exceeded 2%. The order degree of the socio-economic subsystem showed 2008 as the drop point and then increased, mainly because the region suffered low temperatures, rain, snow, and freezing disasters in 2008, which affected social and economic activities.

It is worth noting that the order degree of the two subsystems (the socio-economic subsystem and the water conservancy facilities subsystem) has shown a common upward trend. The analysis suggests that it may be due to the year-on-year improvement of infrastructure such as reservoirs, dikes, and water conservancy construction investment in the basin, which has alleviated the uncertainty of natural precipitation. The water conservancy facility subsystem provides stable water resources for agricultural irrigation, industrial production, and people's everyday lives, guaranteeing the safety of food production in the socio-economic subsystem, increasing per capita disposable income and regional GDP, and ultimately promoting the orderly rise of the socio-economic subsystem [3]. The connection between the socio-economic subsystem and the water conservancy facility subsystem shows that, under the premise of limited natural water resources, the investment and construction of water conservancy facilities are of great significance to socio-economic development and will have a positive impact on the operation of the system. The future regulation of the production and domestic water utilization investment and the maintenance of water conservancy facilities will have a huge impact on the improvement of the synergy of the entire river basin.

The water resource utilization subsystem and the ecological environment subsystem have the second-highest contributions to the WCS system, at 18.32% and 16.53% respectively [48]. Except for the decrease in the orderliness of the two subsystems, the increase in water consumption and sewage discharge of CNY 10,000 increased industrial added value in 2002. In the following years, the government paid more attention to the construction of ecological civilization and environmental protection, as well as the greening construction of water conservation and sewage treatment in the Xiangjiang River Basin, which increased the orderliness of the two subsystems. The reasonable adjustment of the water resources utilization efficiency and sewage treatment rate, protection of water ecological environment, and other measures are, therefore, vital to the coordinated governance of the WCS system in the Xiangjiang River Basin. This can leave other systems with high-quality water and better drive them to coordinate development strategies.

4.1.2. The Contributions of Indicators to the Order of Subsystems

Table 2 presents the contribution value of each order parameter index to the order degree of the subsystem for the years 2000–2019. In general, there is a low contribution of each index to the order degree of the subsystem, showing that there is room for growth. The per capita water resource, investment in agriculture, forestry, and water conservancy construction, length of dikes, water resource utilization, green coverage; sewage treatment volume, and other indicators reflect the low contribution to the construction of water conservancy facilities, water resources utilization, and ecological protection.

Table 2. The contributions of indicators to the order of subsystems over the 2000–2019 period.

δ_{ij} Order Parameter	2000	2003	2006	2009	2012	2015	2018	2019	Mean
δ_{11}	0.473	0.160	0.413	0.252	0.725	0.662	0.299	0.571	0.444
δ_{12}	0.490	0.377	0.518	0.136	0.615	0.552	0.055	0.623	0.421
δ_{13}	0.454	0.376	0.527	0.163	0.624	0.599	0.102	0.680	0.441
δ_{21}	0.012	0.093	0.325	0.194	0.392	0.715	0.909	0.990	0.454
δ_{22}	0.719	0.729	0.739	0.740	0.958	0.961	0.962	0.939	0.843
δ_{23}	0.646	0.650	0.689	0.691	0.988	0.072	0.195	0.290	0.528
δ_{24}	0.003	0.026	0.031	0.053	0.068	0.926	0.974	0.987	0.383
δ_{25}	0.056	0.054	0.085	0.163	0.759	0.820	0.932	0.946	0.477
δ_{31}	0.069	0.207	0.521	0.775	0.916	0.964	0.979	0.999	0.678
δ_{32}	0.463	0.233	0.547	0.746	0.908	0.960	0.999	0.980	0.729
δ_{33}	0.186	0.038	0.717	0.812	0.979	0.895	0.759	0.752	0.642
δ_{34}	0.165	0.449	0.511	0.703	0.594	0.631	0.691	0.826	0.571
δ_{35}	0.220	0.135	0.415	0.400	0.432	0.707	0.858	0.687	0.482
δ_{36}	0.242	0.361	0.275	0.652	0.203	0.276	0.724	0.268	0.375
δ_{41}	0.606	0.061	0.055	0.274	0.423	0.607	0.943	0.982	0.494
δ_{42}	0.381	0.997	0.443	0.521	0.461	0.388	0.134	0.013	0.417
δ_{43}	0.105	0.086	0.176	0.478	0.771	0.909	0.976	0.986	0.561
δ_{44}	0.632	0.811	0.876	0.578	0.744	0.051	0.797	0.921	0.676
δ_{51}	0.505	0.037	0.842	0.890	0.864	0.954	0.641	0.602	0.667
δ_{52}	0.002	0.037	0.107	0.238	0.444	0.629	0.892	0.989	0.417
δ_{53}	0.154	0.204	0.290	0.624	0.545	0.722	0.916	0.990	0.556
δ_{54}	0.001	0.028	0.072	0.408	0.554	0.826	0.933	0.990	0.476

Notes: The description of the order parameters presented in Table S2. For example, "δ_{11}" refers to "Average annual rainfall".

The similar situation of such large changes in indicators is largely due to the influence of natural and human factors. For example, the dike length was 0.003 in 2003 and 0.068

in 2012; but this rose to 0.926 in 2015. This is because the government pays attention to flood control and strengthens the construction of river and lake embankments. Another example is that the grain yield was 0.037 in 2003, because the average annual rainfall was less, which was a typical dry year and affected the grain harvest.

4.1.3. WCS System's Governance Synergy Degree

Looking at the trend over the years, as shown in Figure 6, the WCS system's governance coordination degree of the Xiangjiang River Basin takes 2011 as the critical point, showing a phased "S"-shaped development trend. Specifically, the system governance coordination degree between 2000 and 2011 is between 0.01 and 0.13; from 2012 to 2019, the synergy of system governance has risen to 0.36. However, the overall level is still not high, with an average of 0.1344 in synergy over the past two decades.

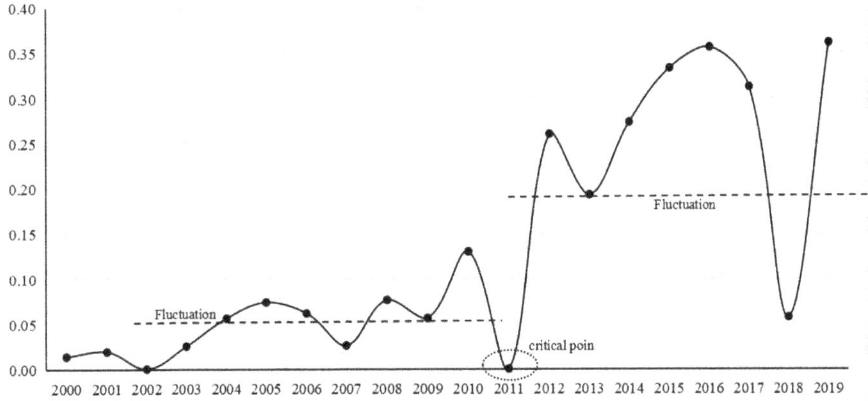

Figure 6. WCS system's governance synergy degree in the Xiangjiang River Basin.

Figure 6 presents the WCS system's governance synergy degree in the Xiangjiang River Basin. It can be seen that the degree of coordination of the system governance suddenly dropped in 2007, 2011, 2018, and at other time points, mainly due to the impact of natural precipitation. The change in rainfall causes a butterfly effect, which leads to a decrease of the reservoir capacity of the water conservancy facility subsystem. The reduction of storage water resources indirectly affects farmland irrigation water, industrial water, and domestic water consumption. Moreover, it ultimately causes a decline in food production, a decrease in per capita disposable income, and a slowdown in the growth of regional GDP and fiscal revenue in the socio-economic subsystem.

On the other hand, the system governance coordination degree steadily increased in 2005, 2012, 2016, and at other time points, especially in the 2011–2012 period where it reached 0.26. The growth in positive sequence parameters such as the river basin green coverage, sewage treatment rate, water conservancy construction investment, per capita disposable income, and regional GDP is responsible for this growth. The green coverage area increased by 81.9%, the sewage treatment rate increased by 11.2%, and the investment in water conservancy construction increased by 66.5%. Moreover, there was a decline in negative sequence parameters such as water consumption per CNY 10,000 of GDP, water per 10,000 yuan of industrial added value, and per capita comprehensive water consumption, where the water consumption for the three indicators decreased by 27.9%, 28.4%, and 0.4% respectively.

Although the overall level of system coordination over the period from 2000 to 2019 is not high, its growth rate is still 34.7%, which reflects that the WCS system of the Xiangjiang

River Basin is well-coordinated and that internal entropy is reduced. The WCS system as a whole evolves to a harmonious and orderly state.

4.1.4. The Governance Synergy Degrees of the Upstream, Midstream, and Downstream Regions of the Xiangjiang River Basin

Figure 7 presents the governance synergy degree of the WCS system in the upstream, midstream, and downstream regions of the Xiangjiang River Basin over the period rang from 2000 to 2019. Low points can be seen in the years 2002, 2002, 2011, and 2018. It develops an "S" shape, and the fluctuation points are mainly affected by rainfall. In accordance with the perspective of regional differences, spatial imbalances have been observed in the regions of 0.1159 upstream, 0.1074 midstream, and 0.1275 downstream. The overall level of governance synergy among the three regions is lower. It can be found that the synergy increased after 2004 and 2007, with an average increase of 31%. The reason for the growth is that Hunan Province is driven by the country's central rise strategy and the radiation of the two-oriented society. The second reason for the growth is due to policy measures taken by the government, such as water resource management assessment measures. Special funds for investment in water ecological construction and protection, water resource allocation guarantees, and social and economic development in the basin are relatively sufficient. This series of comprehensive measures has promoted the continuous increase in the degree of coordination, and the coordination degree of the governance of the upper, middle, and lower reaches of the Xiangjiang River Basin, rendering a WCS system that is still changing into a stable and orderly state.

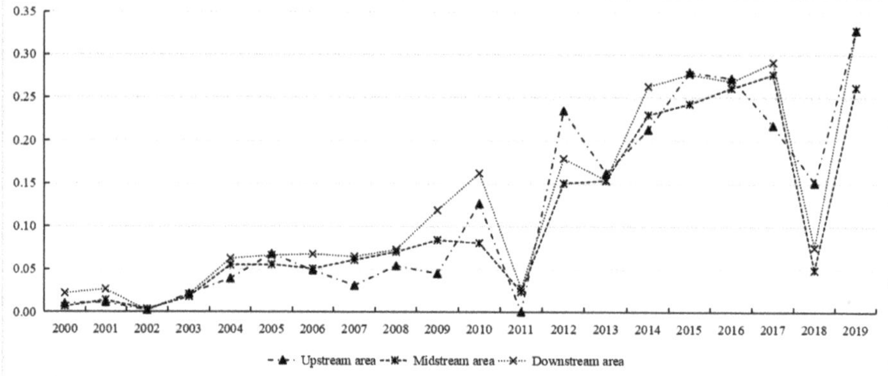

Figure 7. WCS governance synergy degree in the upstream, midstream, and downstream regions of the Xiangjiang River Basin.

4.2. Prediction of Future Governance-Synergy in the WCS System of the Xiangjiang River Basin

Substituting the coordination degree in the WCS system of the Xiangjiang River Basin from 2010 to 2019 into Equation (6), we have estimated parameters ($a= -0.1182, b = 0.0451$) and calculated using the gray GM (1,1) model: the forecast period is 5 years and the coordination degree of the WCS system governance from 2000 to 2024 has been obtained.

The prediction in Figure 8 indicates that the coordination degree of the WCS system governance in the Xiangjiang River Basin will show an overall upward trend from 2020 to 2024. It is predicted that the coordination degree of system governance will be between 0.4701 and 0.5955 from 2020 to 2022, and will be in a medium coordination state. The system governance synergy degree will reach 0.6703 in 2023, entering a highly synergistic state. In 2024. The system governance synergy will further increase, reaching 0.7. Therefore, it is speculated that the WCS system governance coordination degree of the Xiangjiang River Basin will maintain a relatively high marginal growth rate in the period between

2020 and 2024. However, it may be disturbed by natural precipitation and other natural disasters, causing fluctuations in the degree of the coordination of system governance.

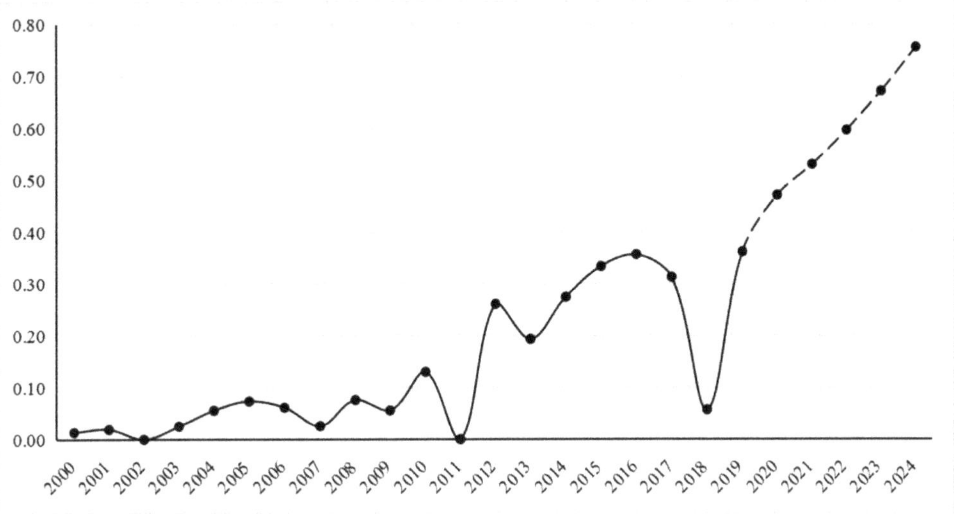

Figure 8. Governance-synergy prediction in the Xiangjiang River Basin.

Research should analyze the content and structure of the WCS system's collaborative governance, analyze the interrelationship of each constituent subsystem and the evolution mechanism of system collaborative governance, and build a model to evaluate the development trend of WCS system governance synergy, which could provide practical guidance and reference for the collaborative development of local water resources, water conservancy facilities and socio-economic systems. Taking the order degrees of the water resources development and utilization subsystem and the ecological environment subsystem in Figure 5 as examples, the order degrees of the two subsystems decreased in 2002, and the contribution of the order degrees of the two subsystems to the system synergy in other years was on the rise. From the perspective of specific indicators, the decrease in the order degreees of the two subsystems in 2002 was due to the large negative impact of the increase in water consumption and sewage discharge of industrial added value. Accordingly, the government should reasonably adjust the utilization efficiency of water resources and the sewage treatment rate, pay more attention to the construction of ecological civilization and environmental protection, pay attention to the greening construction and sewage treatment of water conservation in the basin, promote the orderly rise of the two subsystems, and better promote the coordinated operation of other systems.

5. Conclusions and Recommendation

The analysis revealed several key findings. Firstly, the WCS system's governance showed an S-shaped trend, with subsystems having both cooperative and competitive traits. The evolution process had fluctuations and non-linear features. The evolution process was influenced by factors such as natural precipitation, investment in water conservancy facilities, maintenance, water environment, and water-resource efficiency.

Secondly, from 2000 to 2019, the WCS system governance coordination degree of the Xiangjiang River Basin showed a phased S-shaped change, with 2011 as the critical point. The overall level of coordination was not high, and the upper, middle, and downstream regions had obvious differences. However, the average annual growth rate was up to 34.7%, suggesting a harmonious and orderly development. Thirdly, this study revealed a close relationship between natural precipitation, water conservancy facility development, and changes in the coordination degree of system governance. The utilization of water resources and the ecological environment significantly contributed to the overall system synergy, with contributions of 18.32% and 16.53% respectively. Finally, based on the analysis, the WCS system in the Xiangjiang River Basin is predicted to improve coordination in the next five years (2020–2024). It is expected to reach a highly coordinated stage by 2024.

This study's findings suggest several implications for policymakers. Firstly, the government should prioritize the investment and maintenance of water conservancy infrastructure projects to ensure an efficient water resource allocation and supply system. Additionally, increasing the utilization rate of water resources development is crucial. The allocation of water resources should be determined based on the relationship between socio-economic development and the availability of water resources. An accurate understanding of differences in water use efficiency in the upper, middle, and lower reaches of the river basin could improve the pertinence of water use policies and regulate the matching pattern of productivity development and water resources. Furthermore, regions and industries with low water consumption and utilization rates should be provided with water conservation awareness programs. In contrast, rules and regulations such as water rights transactions should be clarified for regions and industries with high water consumption and utilization rates. By implementing these measures, policymakers can effectively regulate water use and promote sustainable water resource management practices.

To ensure sustainable water resource management, policymakers should prioritize water conservation and ecological protection by supporting water and soil conservation projects, improving sewage treatment facilities, promoting natural restoration in the system, and maintaining the self-purification ability of rivers, lakes, and other water bodies. Additionally, the institutional guarantee system should be strengthened through comprehensive plans, management assessment systems, and responsible departments focusing on the protection of the quantity and quality of water resources, including strengthening water source protection, managing the water-saving system, improving the efficiency of water use, and strictly implementing river quality supervision to prevent pollution. Policymakers must also handle the relationship between trans-regional water conservancy facilities, socio-economic factors, and the water resource environment to provide long-term stable development.

Reasonable adjustments are needed for effective collaborative governance, ensuring moderation to maintain system synergy and dynamic balance. However, external factors can complicate the relationship between governing factors. Therefore, this paper's coordination analysis is a preliminary exploration with limitations; future research should critically investigate the WCS system using multi-level references to support our findings.

Supplementary Materials: The following supporting information can be downloaded at: https://www.mdpi.com/article/10.3390/w15203630/s1, Table S1: Subsystems, indicators and data source. Table S2: Order parameter index. Table S3: Xiangjiang River Basin WCS system coordinated governance index information.

Author Contributions: Conceptualization, J.W., H.L. and A.M.; methodology, J.W.; software, J.W.; validation, J.W., H.L. and A.M.; formal analysis, J.W.; investigation, H.L.; resources, H.L.; data curation, J.W.; writing—original draft preparation, J.W. and A.M.; writing—review and editing, A.M.; visualization, H.L.; supervision, H.L.; project administration, H.L.; funding acquisition, H.L. All authors have read and agreed to the published version of the manuscript.

Funding: This research was funded by Natural Science Foundation of Hunan Province, grant number 2020JJ4378.

Conflicts of Interest: The authors declare no conflict of interest.

References

1. Ahmed, S.S.; Bali, R.; Khan, H.; Mohamed, H.I.; Sharma, S.K. Improved Water Resource Management Framework for Water Sustainability and Security. *Environ. Res.* **2021**, *201*, 111527. [CrossRef]
2. Li, M.; Cao, X.; Liu, D.; Fu, Q.; Li, T.; Shang, R. Sustainable Management of Agricultural Water and Land Resources under Changing Climate and Socio-Economic Conditions: A Multi-Dimensional Optimization Approach. *Agric. Water Manag.* **2022**, *259*, 107235. [CrossRef]
3. Liu, J.; Zang, C.; Tian, S.; Liu, J.; Yang, H.; Jia, S.; You, L.; Liu, B.; Zhang, M. Water Conservancy Projects in China: Achievements, Challenges and Way Forward. *Glob. Environ. Chang.* **2013**, *23*, 633–643. [CrossRef]
4. Dressel, S.; Ericsson, G.; Johansson, M.; Kalén, C.; Pfeffer, S.; Sandström, C. Evaluating the Outcomes of Collaborative Wildlife Governance: The Role of Social-Ecological System Context and Collaboration Dynamics. *Land Use Policy* **2020**, *99*, 105028. [CrossRef]
5. Han, T.; Zhang, C.; Sun, Y.; Hu, X. Study on Environment-Economy-Society Relationship Model of Liaohe River Basin Based on Multi-Agent Simulation. *Ecol. Model.* **2017**, *359*, 135–145. [CrossRef]
6. He, G.; Geng, C.; Zhao, Y.; Wang, J.; Jiang, S.; Zhu, Y.; Wang, Q.; Wang, L.; Mu, X. Food Habit and Climate Change Impacts on Agricultural Water Security during the Peak Population Period in China. *Agric. Water Manag.* **2021**, *258*, 107211. [CrossRef]
7. Klien, M.; Michaud, D. Water Utility Consolidation: Are Economies of Scale Realized? *Util. Policy* **2019**, *61*, 100972. [CrossRef]
8. Salehi, M. Global Water Shortage and Potable Water Safety; Today's Concern and Tomorrow's Crisis. *Environ. Int.* **2022**, *158*, 106936. [CrossRef]
9. Wei, W.; Gao, Y.; Huang, J.; Gao, J. Exploring the Effect of Basin Land Degradation on Lake and Reservoir Water Quality in China. *J. Clean. Prod.* **2020**, *268*, 122249. [CrossRef]
10. Wang, K.; Davies, E.G.; Liu, J. Integrated Water Resources Management and Modeling: A Case Study of Bow River Basin, Canada. *J. Clean. Prod.* **2019**, *240*, 118242. [CrossRef]
11. Shen, D. River Basin Management. In *Water Resources Management of the People's Republic of China*; Global Issues in Water Policy; Springer: Cham, Germany, 2021; Volume 26.
12. Song, J.; Tang, B.; Zhang, J.; Dou, X.; Liu, Q.; Shen, W. System Dynamics Simulation for Optimal Stream Flow Regulations under Consideration of Coordinated Development of Ecology and Socio-Economy in the Weihe River Basin, China. *Ecol. Eng.* **2018**, *124*, 51–68. [CrossRef]
13. Zhang, H.; Jin, G.; Yu, Y. Review of River Basin Water Resource Management in China. *Water* **2018**, *10*, 425. [CrossRef]
14. Li, X.; Zhang, L.; Zheng, Y.; Yang, D.; Wu, F.; Tian, Y.; Han, F.; Gao, B.; Li, H.; Zhang, Y.; et al. Novel Hybrid Coupling of Ecohydrology and Socioeconomy at River Basin Scale: A Watershed System Model for the Heihe River Basin. *Environ. Model. Softw.* **2021**, *141*, 105058. [CrossRef]
15. Reyhani, M.N.; Grundmann, P. Who Influences Whom and How in River-Basin Governance? A Participatory Stakeholder and Social Network Analysis in Zayandeh-Rud Basin, Iran. *Environ. Dev.* **2021**, *40*, 100677. [CrossRef]
16. Gerlak, A.K.; Karambelkar, S.; Ferguson, D.B. Knowledge Governance and Learning: Examining Challenges and Opportunities in the Colorado River Basin. *Environ. Sci. Policy* **2021**, *125*, 219–230. [CrossRef]
17. Huckleberry, J.K.; Potts, M.D. Constraints to Implementing the Food-Energy-Water Nexus Concept: Governance in the Lower Colorado River Basin. *Environ. Sci. Policy* **2019**, *92*, 289–298. [CrossRef]
18. Baudoin, L.; Gittins, J.R. The Ecological Outcomes of Collaborative Governance in Large River Basins: Who Is in the Room and Does It Matter? *J. Environ. Manag.* **2021**, *281*, 111836. [CrossRef]
19. Sullivan, A.; White, D.D.; Hanemann, M. Designing Collaborative Governance: Insights from the Drought Contingency Planning Process for the Lower Colorado River Basin. *Environ. Sci. Policy* **2019**, *91*, 39–49. [CrossRef]
20. Girard, C.; Rinaudo, J.-D.; Pulido-Velazquez, M. Sharing the Cost of River Basin Adaptation Portfolios to Climate Change: Insights from Social Justice and Cooperative Game Theory. *Water Resour. Res.* **2016**, *52*, 7945–7962. [CrossRef]
21. Arfanuzzaman; Rahman, A.A. Sustainable Water Demand Management in the Face of Rapid Urbanization and Ground Water Depletion for Social-Ecological Resilience Building. *Glob. Ecol. Conserv.* **2017**, *10*, 9–22. [CrossRef]
22. Cunha, H.; Loureiro, D.; Sousa, G.; Covas, D.; Alegre, H. A Comprehensive Water Balance Methodology for Collective Irrigation Systems. *Agric. Water Manag.* **2019**, *223*, 105660. [CrossRef]
23. Tiwari, A.K.; Suozzi, E.; Silva, C.; de Maio, M.; Zanetti, M. Role of Integrated Approaches in Water Resources Management: Antofagasta Region, Chile. *Sustainability* **2021**, *13*, 1297. [CrossRef]
24. Li, X.; Yang, Y.; Liu, Y. Research Progress in Man-Land Relationship Evolution and Its Resource-Environment Base in China. *J. Geogr. Sci.* **2017**, *27*, 899–924. [CrossRef]
25. Mao, H.H.; Wang, Y.; Dong, L. Quantitative evaluation of coordinated development of water conservancy and economic society in Haihe River Basin. *Resour. Environ. Arid. Areas* **2011**, *25*, 44–47.
26. Yi, X.B.; Wang, X.J.; Huang, J.L. Evaluation of regional water conservancy socio-economic coordination degree and analysis of spatial difference of comprehensive level. *Water Sav. Irrig.* **2013**, *4*, 57–63.
27. Huang, X.F.; Liu, Z.Z.; Fang, G.H. Evaluation method and application of water conservancy modernization based on cloud model. *Prog. Water Conserv. Hydropower Sci. Technol.* **2017**, *37*, 54–61.
28. Li, Y.L.; Wei, Y.Y.; Li, G.J. Sustainability evaluation of water resources system in Beijing from the perspective of synergy. *China Popul. Resour. Environ.* **2019**, *29*, 71–80.

29. Luo, Z.; Zuo, Q.; Shao, Q.; Ding, X. The Impact of Socio-economic System on the River System in a Heavily Disturbed Basin. *Sci. Total Environ.* **2019**, *660*, 851–864. [CrossRef]
30. Zhang, X.; Li, L.; Su, Z.; Li, H.; Luo, X. Study on Factors Influencing Public Participation in River and Lake Governance in the Context of the River Chief System—Based on the Integrated Model of TPB-NAM. *Water* **2023**, *15*, 275. [CrossRef]
31. Liang, J.Y. Systematic holism of sustainable utilization of regional water resources. *J. Syst. Dialectics* **2005**, 84–88.
32. Zhang, G.P.; Zhu, L.Y. Theory and method of water resources carrying capacity research. *Soil Water Conserv. Res.* **2003**, 148–150.
33. Wang, L.F.; Li, Y.; Zhuang, Y. Evaluation index system and application of water resources carrying capacity. *Soft Sci.* **2007**, 8–10+14.
34. Li, S.H.; Dong, Z.C.; Dong, S.F. Analysis on complex giant system of water resources and its harmony. *Water Conserv. Dev. Res.* **2007**, 10–14.
35. White, R.; Engelen, G. Cellular Automata and Fractal Urban Form: A Cellular Modelling Approach to the Evolution of Urban Land-Use Patterns. *Environ. Plan. A Econ. Space* **1993**, *25*, 1175–1199. [CrossRef]
36. Gilstrap, D.L. Dissipative structures in educational change: Prigogine and the academy. *Int. J. Leadersh. Educ. Theory Pract.* **2007**, *10*, 49–69. [CrossRef]
37. Haken, H.; Portugali, J. Information and Self-Organization. *Entropy* **2017**, *19*, 18.
38. Liu, H.M.; Shi, P.J.; Yang, X.M. Simulation and demonstration of self-organization evolution of human water system. *J. Nat. Resour.* **2014**, 709–718.
39. Wang, Y.M. Entropy change analysis of geographical environment evolution trend. *Acta Geogr. Sin.* **2011**, *66*, 1508–1517.
40. Nie, X.; Zhang, Z.W. Study on time series characteristics of coupling relationship between water resources environment and economic development in Hubei Province. *J. Irrig. Drain.* **2020**, *39*, 138–144.
41. Wang, W.; Xu, J.; Wang, X.; Ji, L.; Qi, C. Spatial and temporal differentiation of the coupling relationship between modern agriculture, regional economy and ecological environment in the Yangtze River Economic Belt. *Agric. Mod. Res.* **2020**, *41*, 64–74.
42. Xu, H.; Wang, Y.; Zhang, Z.; Gao, Y.; Zhang, D. The coupling mechanism of water, energy and grain and the space-time evolution of coordinated development in the Yellow River basin. *Resour. Sci.* **2021**, *43*, 2526–2537.
43. Liu, G.; Ju, W.; Ye, J.; Chu, G. Analysis and prediction of the coordinated development of the coupling of resource utilization, ecological environment and economic growth—Taking the provinces along the Silk Road Economic Belt as an example. *Ecol. Econ.* **2021**, *37*, 191–200.
44. Xu, W.X.; Shu, J.J.; Tang, G.N. Spatiotemporal pattern and dynamic evolution of coordinated development of industrialization, informatization, urbanization and agricultural modernization in China. *Econ. Trends* **2015**, *1*, 76–85.
45. Chen, Z. Construction of index system for coordinated development of five modernizations. *Stat. Decis. Mak.* **2020**, *36*, 26–30.
46. Chen, J.F.; Chen, L.; Liu, L.M. Research on regional "water resources energy food" system security evaluation based on PSR-CGPM. In Proceedings of the 11th China Soft Science Research Association in 2019, Beijing, China, 23–24 November 2019.
47. Gu, J.; Sun, S.; Wang, Y.; Li, X.; Yin, Y.; Sun, J.; Qi, X. Sociohydrology: An Effective Way to Reveal the Coupled Evolution of Human and Water Systems. *Water Resour Manag.* **2021**, *35*, 4995–5010. [CrossRef]
48. Wang, T.; Jian, S.; Wang, J.; Yan, D. Dynamic interaction of water–economic–social–ecological environment complex system under the framework of water resources carrying capacity. *J. Clean. Prod.* **2022**, *368*, 133132. [CrossRef]

Review

Water Resources in Jordan: A Review of Current Challenges and Future Opportunities

Mohammad Al-Addous [1,2,3,*], Mathhar Bdour [1], Mohammad Alnaief [4], Shatha Rabaiah [1] and Norman Schweimanns [3]

1 Department of Energy Engineering, School of Natural Resources Engineering and Management, German Jordanian University, P.O. Box 35247, Amman 11180, Jordan; madher.bdour@gju.edu.jo (M.B.); shathabassil@gmail.com (S.R.)
2 Fraunhofer-Institut für Solare Energiesysteme ISE, 79110 Freiburg, Germany
3 Department of Environmental Process Engineering, Institute of Environmental Technology, Technische Universität Berlin, Secr. KF 2, Straße des 17. Juni 135, 10623 Berlin, Germany; norman.schweimanns@tu-berlin.de
4 School of Applied Medical Sciences, German Jordanian University, P.O. Box 35247, Amman 11180, Jordan; mohammad.alnaief@gju.edu.jo
* Correspondence: mohammad.addous@gju.edu.jo; Tel.: +962-6429-4221

Abstract: Jordan is facing significant challenges related to water scarcity, including overexploitation of groundwater, increasing demand, and wasteful practices. Despite efforts to manage water resources, inadequate planning has resulted in ongoing water security concerns and deteriorating water quantity and quality. To address water stress, Jordan has implemented measures such as desalination, dam construction, and water conservation initiatives. However, water stress remains high, necessitating a comprehensive strategy that includes short-term demand-side interventions and long-term supply-side reforms. Financial and governance challenges hinder the implementation of these measures, requiring private investment and coordination among stakeholders. This paper provides a comprehensive review of Jordan's water resources, analyzing current trends, challenges, and opportunities. The aim is to offer insight into the current situation and propose sustainable management approaches. The findings will be valuable for policymakers, researchers, and stakeholders working towards addressing Jordan's complex water challenges and securing a sustainable water future for its citizens.

Keywords: water resources; management; challenges; opportunities; wastewater; desalination; conventional; nonconventional; harvesting; water sustainability

check for
updates

Citation: Al-Addous, M.; Bdour, M.; Alnaief, M.; Rabaiah, S.; Schweimanns, N. Water Resources in Jordan: A Review of Current Challenges and Future Opportunities. *Water* **2023**, *15*, 3729. https://doi.org/10.3390/w15213729

Academic Editors: Peiyue Li and Jianhua Wu

Received: 14 August 2023
Revised: 15 October 2023
Accepted: 18 October 2023
Published: 25 October 2023

1. Introduction

Acute limitation of freshwater resources emerges as a foremost challenge in the 21st century. This dearth of water resources is anticipated to intensify in the coming years due to various factors, notably the anticipated adverse consequences of climate change. Projections indicate that by the year 2030, approximately 40% of the global populace will experience the ramifications of water scarcity [1,2]. Unfortunately, less than 3% of the global water resources are fresh water, i.e., water with an acceptable salinity level. Less than 1% of the available freshwater can be used directly for industrial, agricultural, and drinking applications [3], as the remaining freshwater is ice, which is a sad fact given that the majority of the water on Earth is saline water from oceans, seas, and deep groundwater [4,5]. Due to population growth, urbanization, industrialization, and improved life quality around the world, especially in the Middle East and North Africa (MENA) region, the demand for freshwater is both undeniable and increasingly unsustainable. The availability of freshwater in MENA is very limited due to low precipitation and high evaporation rates [5]. On the other hand, even though the reduction in water supplies is directly attributed to the low precipitation and high evaporation rates, the reasons behind the increase in water demand are also responsible for the overexploitation of aquifers and the contamination of water supplies, which eventually reduce the available water resources [6].

Jordan, a country in the Middle East region, is one of the countries with the scarcest renewable water resources per capita in the world; it was the second-poorest country in 2017 with only 100 m^3 per capita per year, which could reach almost 80 m^3 in the year 2020 [7]. Jordan has an area of 89,210 km^2, and despite its relatively compact size, its diverse topography and landscape rival those typically found in much larger countries [8]. Based on a physiographic standpoint, the country can be divided into four districts: (1) the ghors (marshes), (2) the highlands, (3) the plains, and (4) the desert. The ghors in the west of the country are further divided into three zones: the Jordan Valley, which starts at Lake Tiberius in the north; the swamps around the Dead Sea; and the Wadi Araba, which stretches southward to the northern shores of the Red Sea (total area: 5000 km^2). The Ghor regions were all discovered to be below sea level. The highlands extend from the north to the south at an elevation of between 600 m and 1600 m above sea level (total area: 5510 km^2). The plains extend from the north to the south along the western borders of the desert (Badiah) (total area: 10,000 km^2). And finally, the desert region (Badiah), an extension of the Middle Eastern desert (total area: 68,700 km^2), is located in the east [8,9]. Jordan's climate is characterized by long, hot, and dry summers and wet, cold winters. The temperature increases towards the south, with the exception of a few southern highlands. Precipitation shifts impressively in the area, primarily due to the country's topography [8,10]. In the Jordan Rift region, the annual mean temperature typically falls within the range of 22 to 25 °C, while in Badia, it fluctuates between 18 and 21 °C. The highlands, in contrast, maintain an average temperature spanning from 14 to 18 °C [11]. Precipitation patterns are notably variable, with the concentration of occurrences during the winter and early spring. More than 93% of the nation receives less than 200 mm of precipitation annually, ranging from 700 mm in the northwest to under 20 mm in the southeastern deserts, as illustrated in Figure 1 [12]. The annual volume of precipitation in Jordan has shown concerning deviations from long-term averages, primarily because of climate change and periods of drought, as detailed in Figure 2. The third national communication report to the UNFCCC anticipates a warmer and drier climate in Jordan by 2100, with potential increases in air temperature ranging from +2.1 °C to +4 °C and potential reductions in annual rainfall between 15% and 35% [13]. The long-term average annual precipitation volume is approximately 8217 million cubic meters. Over a 50-year record, rainfall rates have varied between less than 100 mm and 520 mm. Notably, evaporation accounts for a substantial portion of the total rainfall, ranging from 88% to 93%, while infiltration rates have been observed at approximately 4% to 10% [7,14].

In addition to the previously mentioned reasons for the global water shortage, Jordan, in particular, has experienced the following circumstances: (a) Since 1948, there have been significant influxes of refugees, including more than 1.4 million Syrians since 2010 [15]. (b) Jordan shares 26% of its water resources with neighboring countries [12]. (c) It has a significant amount (roughly 50%) of non-revenue water, of which 70% is a result of illegal use [16]. All these circumstances have combined to put a tremendous amount of pressure on water resources and prevent the implementation of workable management plans.

Water stress is an actual problem in Jordan. Groundwater resources have been over-exploited for decades, and the country's groundwater is estimated to be being used up twice as fast as it can be replenished. Jordan suffers from serious supply shortages, but rapidly increasing demand and a significant increase in waste have been the main causes of exacerbating water shortages in recent years. Due to the interaction of many factors, there is no standalone fix. Jordan is taking steps to reduce its water stress, including attempts to increase water supply through desalination and dam construction and decrease water usage through strengthened public water networks. Despite these efforts, water stress remains at an all-time high. To find a lasting solution to the challenges Jordan faces and ensure a more sustainable water future, both short-term demand-side strategies to decrease and improve water usage efficiency and long-term supply-side reforms to increase water availability are necessary. However, a comprehensive solution to meet all water demands has yet to be found.

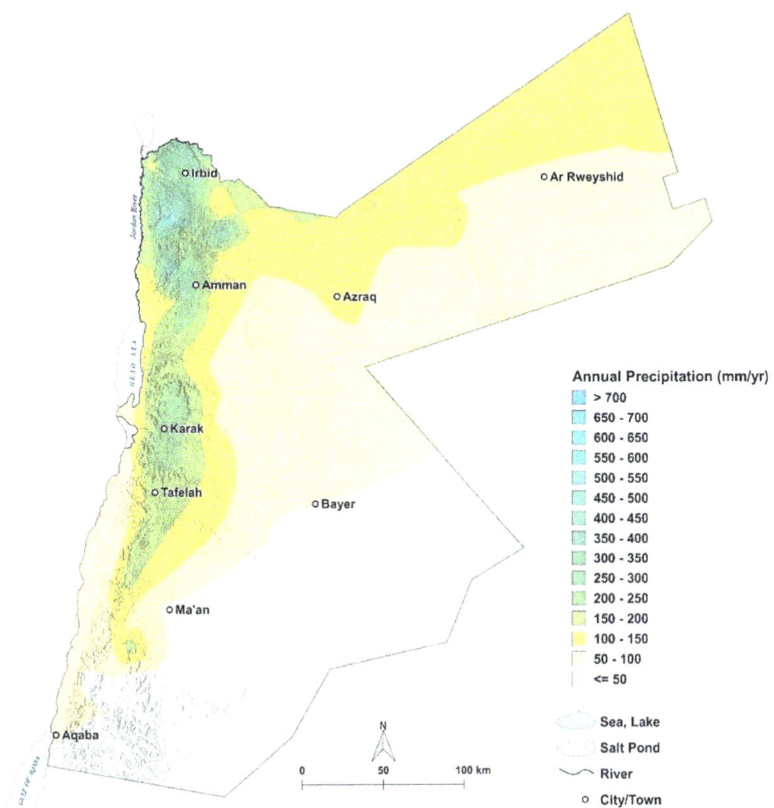

Figure 1. Annual precipitation in Jordan [17].

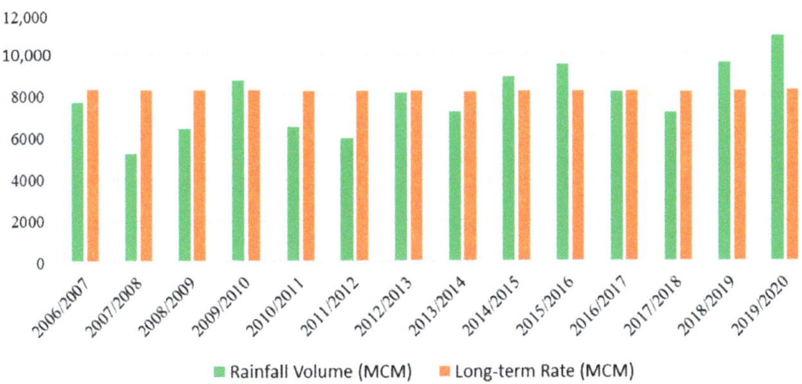

Figure 2. Rainfall volumes in Jordan for water years and long-term rate (2010–2020) [14].

In this paper, we provide a comprehensive review of the current state of water resources in Jordan, including an examination of the available water resources, present and future trends, and the challenges and opportunities faced by the country. The purpose of this review is to provide insight into the current situation and highlight opportunities for a more sustainable future in the management of water resources in Jordan. By examining the available water resources and trends, we aim to identify the challenges faced by water resources in Jordan and explore opportunities for improving the sustainability and management of these resources. Ultimately, this paper provides a resource for policymakers, researchers, and other stakeholders who are working to address the complex water resource challenges faced by Jordan today.

2. Literature Review

Water resource management in Jordan presents a multifaceted challenge due to the nation's limited water resources and the increasing demand [18]. The situation is further complicated by the significant impact on the Jordan River, a vital water source, resulting from actions by neighboring countries and the looming threats of climate change [19]. As a consequence of these challenges, there has been a growing recognition of the need for improved water resource management in Jordan to ensure sustainable water supplies and mitigate the adverse consequences of water scarcity for the environment, economy, and society [20]. Numerous studies have delved into the current challenges and future prospects of water resource management in Jordan [21]. This literature review aims to provide a comprehensive analysis of the present state of water resource management in Jordan, emphasizing the importance of sustainable water supplies and highlighting the adverse effects of water scarcity.

The Water Authority of Jordan (WAJ) is responsible for the operational management of water resources and water supply and wastewater treatment. The Ministry of Water and Irrigation (MWI) plays a central role in water management in Jordan, and the National Water Strategy of Jordan aims to build a resilient sector based on a unified approach and transition towards sustainable management of water and sanitation for all people in Jordan [22]. The strategy covers five pillars and themes: integrated water resources management, water, sewage and sanitation services, water for irrigation, energy and other uses, institutional reform, and sector information management and monitoring [22].

Water law and resource ownership in Jordan are governed by the Water Authority Law of 1988, which was amended by Law No. 22 of 2014. According to Article 25 of the law, all water resources, including surface and groundwater, regional waters, rivers, and internal seas, are considered state-owned property and cannot be used or transferred except in compliance with this law [23]. Nevertheless, the law also facilitates the issuance of licenses to water users and the formation of water user associations. The licensing of water users is provided by the Ministry of Water and Irrigation, which is responsible for overall water resource management in Jordan [24].

The licensing process involves the issuance of permits to water users, which specify the amount of water that can be used and the conditions under which it can be used. The permits are issued for a specific period, and the water users are required to renew them periodically. The permits also specify the fees that the water users are required to pay for the water they use [22]. Water users in Jordan are required to pay for the water they use, and the rates are set by the Water Authority of Jordan based on the cost of water production, treatment, and distribution [25]. The rates are reviewed periodically to ensure that they reflect the actual cost of water production and distribution. The government has also implemented a tariff structure that provides different rates for different categories of water users, such as households, commercial, and industrial users [26].

Leakage in old pipes, administrative errors, illegal water connections, inaccurate metering of the supplied and/or consumed water, and improper billing of water are some of the reasons for non-revenue water use. Illegal use of water in Jordan can occur in various ways, such as unauthorized drilling of wells, illegal water connections, and overpumping of groundwater beyond the permitted limits. The government enforces legal action for illegal water use and has implemented measures to combat illegal water use, such as imposing fines and penalties, confiscating equipment, and disconnecting illegal water connections. The implementation of the legal framework and administrative policies is crucial to ensure equitable and sustainable water use in Jordan [27,28].

The quality of water in Jordan is also a significant concern. Deteriorating water quality is one of the factors exacerbating the country's precarious water situation [29]. The use of treated sewage for irrigation is one strategy to ensure the quality of water resources in Jordan [24]. Furthermore, Jordan is highly vulnerable to the impacts of climate change, which not only diminishes water availability but also worsens water scarcity. The National Water Strategy of Jordan (2016–2025) seeks to enhance water quality and sanitation. In response to the growing water scarcity, nonconventional water supply sources are being considered to address the shortfall. Among these options, water reuse is gaining prominence, especially for agricultural purposes. The government is actively working to improve wastewater treatment for agricultural use [22].

The unequal distribution of water resources in Jordan and persistent water scarcity are among the foremost challenges of water resource management [20]. In spite of the scarcity of water resources in the country, some areas, particularly urban centers, exhibit high water consumption rates. This unequal distribution has led to overextraction of groundwater, resulting in aquifer depletion and aggravating long-term water scarcity [18,30,31]. Climate change compounds these issues, with rising temperatures, reduced precipitation, and increased evaporation rates collectively diminishing water availability and exacerbating water scarcity. Climate change is also expected to heighten the frequency and intensity of droughts and floods, further impacting water resources [32–35].

The adverse effects of water scarcity in Jordan extend to environmental, economic, and societal realms. These include soil degradation, desertification, and loss of biodiversity, with reduced agricultural productivity, higher water costs, and limitations on economic growth adversely affecting the economy. On a societal level, water scarcity leads to health problems, social unrest, and migration [36,37].

Despite these challenges, there are opportunities to enhance water resource management in Jordan. Innovative technologies such as water reuse, desalination, and rainwater harvesting offer promise for augmenting water availability and mitigating scarcity [38–40]. Moreover, the implementation of improved water management policies, including water pricing mechanisms and water conservation measures, can curtail consumption and boost overall efficiency [18,41].

Jordan's water scarcity challenges underscore the pressing need for action. The country serves as a global example of the difficulties posed by climate change and rapid population growth [36,42]. To address these challenges and ensure sustainable water supplies, researchers recommend adopting a nexus approach that considers the interdependence of water, energy, and food [43,44]. They also advocate for a multi-sectoral approach to reform the water sector and highlight the significance of using Geographic Information Systems (GIS) for planning and monitoring freshwater supplies. In the face of these challenges, it is crucial that Jordan continues to innovate and implement comprehensive solutions to ensure a sustainable water future for the country [42].

3. Available Water Resources in Jordan

Sustainable water sources in Jordan are derived from four main resources: surface water, groundwater, desalinated water, and treated wastewater. Freshwater resources in Jordan consist mainly of groundwater and surface water. Desalinated water and treated wastewater are other important nonconventional resources that help to bridge part of the gap between supply and demand, especially in the municipal and agricultural sectors. The different available water resources in Jordan are as follows.

3.1. Conventional Water Resources

Conventional water resources pertain to renewable sources, which encompass natural bodies like rivers, lakes, freshwater wetlands, or subterranean aquifers. In Jordan, this category includes both groundwater and surface water sources. In the year 2019, the cumulative volume derived from these conventional sources reached approximately 906.7 MCM, reflecting a noteworthy variance of 25.2% when compared to the 2018 volume, which was around 678 MCM. The following provides a concise delineation of these resources.

3.1.1. Groundwater

Groundwater, also known as subsurface water, is fresh water found below the surface in rocks and soil. It is also water that is flowing inside aquifers beneath the water table. The storage capacity of subsurface water is much greater than that of surface water, and its turnover is slow. Jordan has 12 main groundwater basins, which are dispersed among 11 hydrological units, as shown in Figure 3 [14]. Groundwater is the main source of water in Jordan, accounting for 60% of all uses and 76% of sources for drinking water [45]. To meet the high water demand, wells have been drilled intensively. According to the Ministry of Water and Irrigation, the number of private and governmental legal operating wells until 2020 was 3208 [14]; the number of pumping wells according to the WIS database was 5160 until 2017 [45]; and they were mainly located in the highly populated northern and central governorates [14]. However, many illegally operated wells are uncovered and backfilled every year. Between 2007 and 2020, the Ministry of Water and Irrigation closed approximately 1548 illegal wells [45]. Unfortunately, the current state of groundwater is critical, where the quantity of overpumping from groundwater is estimated at 200 MCM [14], and most basins have been extracted above safe yields. This has led to a significant decline in the water table within major aquifers. The rate of decline is alarmingly rapid, averaging about 2 m per year, and in severely affected areas, this decline can reach 5 to 20 m per year [14]. As a result, this decline is affecting the quality of groundwater, causing it to exceed the allowable limits set by Jordanian Standard Specification for Drinking Water Quality Guidelines No. 286/2015, which was used as a guide for assessing water quality. Furthermore, this predicament is exacerbated by the escalating pumping costs and the widespread drying up of wells in the central and northern regions [46,47]. Groundwater is considered a major immoderate source; over the past period, the pumping from groundwater expanded from 479 MCM in the year 2006 to almost 508 MCM in the year 2012, while the safe yield of these aquifers is only 275 MCM [48].

3.1.2. Surface Water

Surface water is water in a river, lake, or freshwater wetland. Precipitation naturally replenishes surface water, and evaporation, evapotranspiration, subsurface seepage, and discharge to the seas naturally deplete it. The second most important source of water in Jordan is surface water, which accounts for 31% of all water used across all sectors and is primarily used for drinking and agriculture [49]. There are 15 surface water basins, including transboundary shared basins [17], that discharge into the Dead Sea, Red Sea, and desert mudflats [50]. Table 1 shows the elements of the hydrological water budget for all surface basins in Jordan.

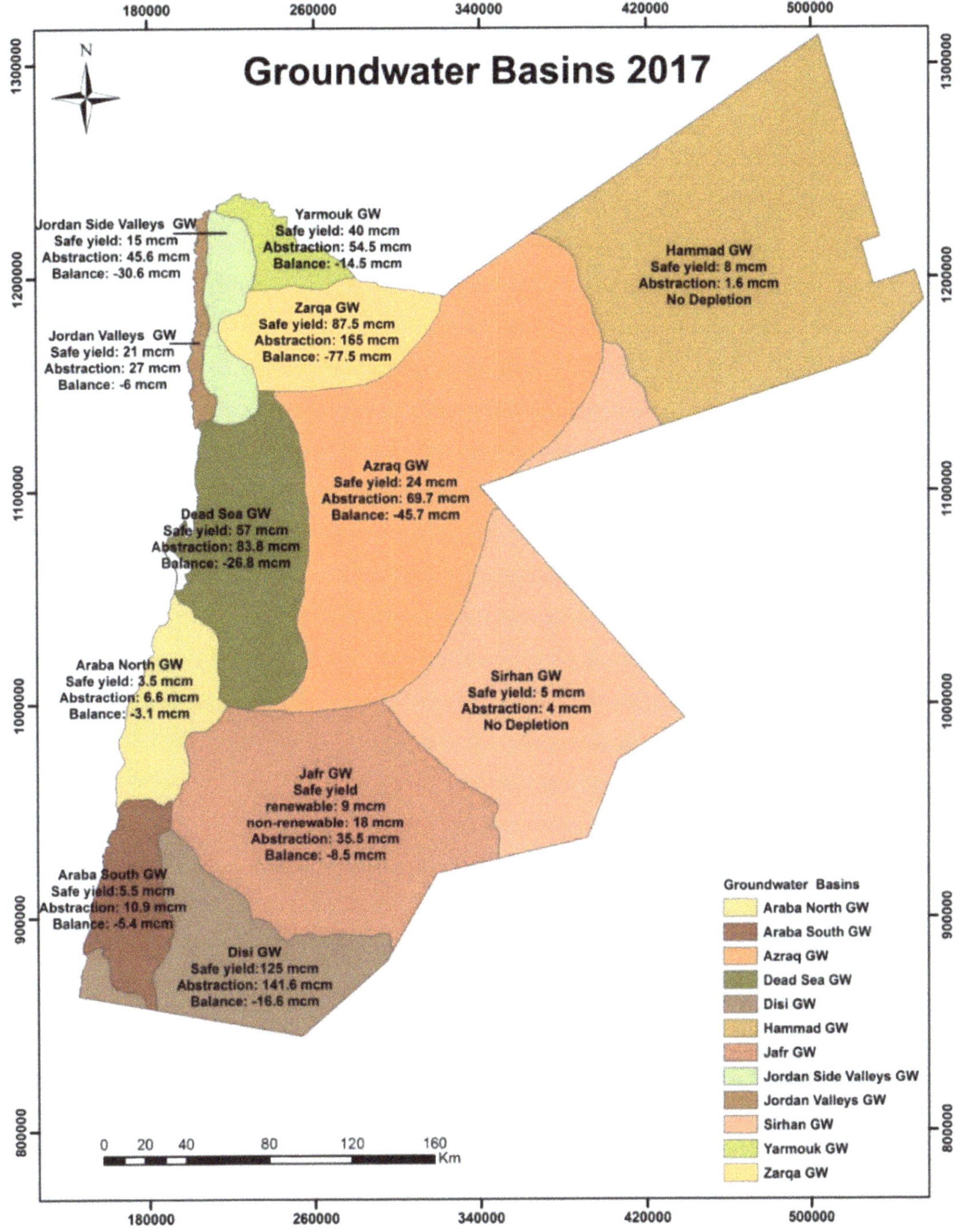

Figure 3. The primary twelve groundwater basins in Jordan [7].

Table 1. Elements of the hydrological water budget for all surface basins in Jordan [49].

Basin Name	Area (km²)	Rainfall Volume (MCM)	Rainfall (mm)	Runoff Coefficient *	Surface Runoff Volume (MCM)	Evaporation (MCM)	Ground Recharge Volume (MCM)
Al-Hamad	17,738.7	1148.9	64.8	1.3	14.9	1119.9	14.0
Al-Sarhan	15,732.9	1101.9	70.0	0.8	8.6	1079.8	13.4
Al-Azraq	12,163.7	1382.8	113.7	1.7	23.4	1310.9	48.5
Al-Mujib	6608.1	1494.9	226.2	4.0	59.6	1329.0	106.3
Al-Hasa	2529.6	308.0	121.7	1.4	4.3	283.0	20.6
Amman-Zarqa	3588.4	1086.0	302.6	4.6	50.0	944.8	91.2
Al-Yarmouk	1362.6	501.5	368.1	7.1	35.6	428.8	37.1
The Dead Sea	1681.1	375.9	223.6	3.8	14.2	345.5	16.2
Northern Side Valleys	964.6	429.5	445.3	3.7	15.7	384.6	29.2
Southern Side Valleys	730.2	262.6	359.6	3.6	9.5	235.3	17.9
Jordan Valley	693.1	287.5	414.9	3.6	10.4	254.2	23.0
North Araba Valley	3010.7	254.2	84.4	1.3	3.2	241.6	9.4
South Araba Valley	5670.2	185.5	32.7	0.9	1.7	180.1	3.7
Al Disi	4234.2	130.9	30.9	0.7	0.9	128.9	1.2
Al Jafr	12,129.6	617.6	50.9	0.8	0.8	604.7	8.0
Total/Average	88,837.5	9567.7	107.7	2.3	2.3	8871.0	439.8

* The runoff coefficient is the percentage of rainfall that flows into the valleys.

The largest surface water supply is the Yarmouk River, which is located to the east of the Jordan Valley Basin. This river accounts for nearly half of Jordan's total surface water resources. Nevertheless, the flow rate of the river has significantly decreased due to the Syrian crisis and substantial water consumption on the Syrian side [50]. The water of the Yarmouk River is of good quality, with total dissolved solids within the range of 400–800 ppm. The second-largest tributary of the Jordan River is the Zarqa River. It is the only river that entirely resides within Jordan's territory [26,51], with a permanent average flow of 92 MCM per year [17]. The river basin is considered the most important surface water basin since it has more than 60% of the population, 80% of the industries, and several agricultural areas [17]. In the last decade, the river water quality has deteriorated because of large discharges of sewerage and industrial effluent. The river is controlled by the King Talal Reservoir, where the ratio of treated wastewater ranges from 45 to 60%. The Zarka River water is used principally for irrigation and stock watering, where sustainable rangeland management in the river basin has increased edible biomass, carbon sequestration, and sediment stabilization [2].

Jordan has several sporadic streams (wadis) spread throughout the surface water basin regions. The principal wadis are listed in Table 2. It is worth mentioning that the base flow includes the WWTP effluent discharge to the wadis, as in the case of Wadi Shuaib, Wadi Karak, and Wadi Fifa [52]. In addition, there are 861 documented springs in Jordan that are categorized as surface water sources [11,34], with an estimated total average annual discharge of 125 MCM/yr based on data from October 2012 to September 2017 [12]. The springs have a significant fluctuating discharge rate that ranges from 1 m³/h up to

1000 m^3/h [12]. The Ministry of Water and Irrigation reports that, out of the 624 springs for which historical data have been collected, 478 have perennial discharge, 16 exhibits seasonal dry behavior, and 130 have recently dried up [12]. To evaluate the status of 591 springs that had no flow measurements between 2015 and 2018 but still had flow measurements from 2013 to 2015, BGR and MWI conducted two national surveys in 2018. The findings showed that 195 of the 591 springs were dry, 12 could not be located in the field, and 361 of the 591 springs were perennial.

Table 2. Details of main wadis in Jordan [53].

Name	Catchment Area (km^2)	Average Rainfall (mm/year)	Base Flow (MCM/year)	Flood Flow (MCM/year)	Total Flow (MCM/year)
Wadi Arab	246	154.0	1.70	4.00	5.70
Wadi Ziglab	100	70.0	7.64	0.38	8.02
Wadi Jurum	23	15.0	8.53	0.29	8.82
Wadi Yabis	122	85.0	2.50	1.49	3.99
Wadi Kufrinja	103	80.0	6.91	1.08	7.99
Wadi Rajib	95	70.0	4.99	1.12	6.11
Wadi Shueib	193	90.0	7.92	1.96	9.88
Wadi Kafrein	159	88.0	12.40	2.80	15.20
Wadi Hisban	88	35.0	3.46	1.23	4.69
Zarqa Ma'in	269	67.0	17.24	2.95	20.19
Wadi Karak	199	43.0	5.89	1.29	7.18
Wadi Mujib and Wala	6727	884.0	31.38	33.62	65.00
Wadi Hasa	2603	334.0	26.26	5.47	31.73
Wadi Feifa	162	30.0	3.91	0.39	4.30
Wadi Kuneizerh	217	38.0	1.43	0.57	2.00
Wadi Dahel	107	20.0	0	0.22	0.22
Wadi Fidan	287	51.0	1.64	0.18	1.82
Wadi Bweirdh	513	125.0	0.80	0.22	1.02
Wadi Musa	165	30.0	0	0.17	0.17
Wadi Hawwar	245	37.0	0	0.28	0.28
Wadi Abu Barqa	141	12.0	0	0.46	0.46
Wadi Rukaya	210	18.0	0	0.34	0.34
Wadi Yutum	2323	73.0	0	0.53	0.53

When taking into account the influence of changing climates, the steady rise in the number of dry springs since 1987 is a sign of ongoing groundwater resource exploitation [45]. However, there were a few reported cases of increasing discharge from some springs, which was attributed to groundwater recharge from leaky water supply networks [12]. The 5-year average spring discharge distribution shows that high recharge areas have high spring discharge, and springs from individual aquifers follow the same discharge variation [45]. In 2016, the share of springs in the drinking and irrigation water supplies was 20.41 MCM and 21 MCM, respectively [8]. There were only 23 springs that provided drinking water [8,45], the majority of which were situated in the northern highlands close to Amman, Zarqa, and Balqa; farther north, in Ajlun, Jarash, and Irbid; and a few springs located in Karak. Although there is a lot of discharge, it is important to note that bacterial pollution prevents many running springs from being used for drinking water [45].

3.2. Nonconventional Water Resources

Different nonconventional water resources are considered potential sources of water supply in Jordan. These include treating wastewater, harvesting water, and desalinating brackish water and seawater. Below is a brief description of these resources.

3.2.1. Treated Wastewater

Approximately 94% of Jordanians have access to fresh water through the public network, and 93% have access to clean sanitation services (63% through the public sewer network and 30% through other safe ways) [12]. As a pioneer in the reuse of treated wastewater, Jordan treats 90% of its annual wastewater production (around 178.2 MCM) at 32 wastewater treatment plants to ensure compliance with irrigation and industrial reuse regulations. This wastewater is then either directly or indirectly used in agriculture [12]. The two tables that follow (Tables 3 and 4) provide an overview of the current state of wastewater treatment plants, their characteristics, and the ultimate use of treated wastewater.

3.2.2. Water Harvesting

In terms of nonconventional water sources, Jordan primarily depends on rainfall. A sizeable portion of Jordan's water budget is made up of the amounts lost to evaporation from soils and temporary open water bodies [49]. Over a large area, rainwater is dispersed and, if properly collected, could significantly increase the nation's water reserves. There are many uses for this water, but agriculture is the first of them. Water harvesting methods can generally be divided into two categories: (1) macro-catchment, floodwater harvesting, and diversion methods and (2) micro-catchment methods, where the catchment area and the cropped area are separate but close to one another [55]. Micro-catchment, as defined by Boers et al. (1986), is a system with a collection area that is less than 100 m in length. On the other hand, macro-catchment methods gather runoff water from hillsides or small, dry watersheds [56]. The Jordanian government has undertaken extensive efforts to maximize the efficient utilization of rainwater resources. This includes the construction of large dams, collectively capable of holding 338.3 million cubic meters of water (as detailed in Table 5), and the implementation of various smaller water harvesting projects such as desert dams, earthen ponds, and concrete ponds (as outlined in Table 6). Jordan faces a significant challenge with water loss due to evaporation, which accounted for a staggering 93% of the total annual rainfall in 2020, as mentioned previously [14]. The loss associated with evaporation from the ten main dams in Jordan on sunny and windy days amounted to about 800 thousand cubic meters in 2015 [57]. In response to this concern, the Jordan Valley Authority (JVA) is contemplating the introduction of a hollow spherical cap system into major dams to curtail water evaporation. It is important to note that water from these dams, if not employed for agricultural purposes, naturally flows downstream to other regions or is preserved in alternative reservoirs. Additionally, there is an estimated collection of 7 million cubic meters of water from households and 15 million cubic meters from desert and rural areas, contributing to water conservation efforts.

Table 3. Characteristics of wastewater treatment plants in Jordan [49,54].

Plant	Designed Capacity (m^3/Day)	Operation Year	Effluent Biochemical Oxygen Demand over 5 Days (BOD$_5$)
Abu Nuseir	4000	1986	1100
Aqaba-Mechanical	12,000	2005	420
Aqaba-Natural	9000	1987	900
Baqa	14,900	1987	800
Ekedar	4000	2005	1500
Fuheis	2400	1997	995
Irbid Center	11,023	1987	800
Jerash-East	9000	1983	1090
Jiza	4000	2008	800
Karak	5500	1988	800
Kufranja	9000	1989	850
Lajoon	1000	2005	1500
Ma'an	5772	1989	700
Madaba	7600	1989	950
Mafraq	6050	1988	825
Mansorah	50	2010	
Meyrad	10,000	2011	800
Mutah and Adnaniyyah	7060	2014	
North Shouna	1200	2015	1200
Ramtha	7400	1987	1000
Salt	7700	1981	1090
Samra	360,000	1984, 2008 [a]	650
Shallaleh	13,750	2014	762
Shobak	350	2010	1850
South Amman	52,000	2015	750
Tafila	7500	1988	1050
Tal Mantah	400	2005	2000
Wadi Arab	21,023	1999	995
Wadi Esseir	4000	1997	780
Wadi Hassan	1600	2001	800
Wadi Mousa	3400	2000	800
Zaatari MBR	1600	2015	1130
Zaatari TF	1726	2015	1130

[a] year of expansion.

Table 4. The use of treated wastewater [49,54].

Plant	Type of Treatment					Treated Wastewater Use										
						Irrigation										
	Activated Sludge	Oxidation Ditch	Trickling Filter	Membrane Bioreactor	Stabilization Ponds	Food	Palm	Nurseries	Windbreakers	Forest Trees	Polo Fields	Olive	Fruit Trees	Ornamental Plants	Landscape	Industrial
Jiza	*					*				*				*		
Karak	*					*				*		*				
Kufranja	*		*			*				*		*				
Lajoon					*	*										
Ma'an	*					*						*				
Madaba	*					*		*				*				
Mafraq					*	*										
Mansorah					*											
Meyrad	*											*	*			
Mutah and Adnaniyyah	*					*										
North Shouna					*											
Ramtha	*					*										
Salt	*											*	*			
Samra	*					*						*				
Shallaleh	*															
Shobak					*											
South Amman	*					*						*				
Zaatari TF			*													
Abu Nuseir	*													*		
Aqaba-Mechanical	*														*	*
Aqaba-Natural					*		*		*						*	
Baqa		*									*	*				
Ekedar					*							*	*			
Fuheis	*					*						*				
Irbid Center																
Jerash-East	*											*	*			
Tafila																
Tal Mantah	*		*													
Wadi Arab	*															
Wadi Esseir		*										*				
Wadi Hassan	*								*			*	*			
Wadi Mousa	*					*						*				
Zaatari MBR				*												

* indicates that the specific type of treatment or wastewater use is applicable to the corresponding location in the table.

Table 5. Dams design capacity, storage, inflows, and outflows in 2020 [14].

Dam	Design Capacity (MCM)	Total Inflows (MCM)	Total Outflows (MCM)	Storage, End of 2020 (MCM)
Wehdeh	110	61.56	53.17	13.75
Wadi Arab	16.8	11.36	9.04	5.23
King Talal	75	156.6	169.05	34.87
Zeqlab	4	0.91	0.57	0.9
Kufranjeh	7.8	16.74	16.21	3.69
Karameh	55	4.1	1.26	23
wadi Shueib	1.4	18.29	18.39	1.42
Kafrain	8.5	18.56	18.28	4.08
Zrqaa maeen	2	1.6	1.73	0.22
Allajon	1	0.8	0.75	0.13
Tanour	16.8	7.15	6.63	2.37
Wala	8.2	20.73	21.15	5.52
Mujeb	29.8	26.2	29.12	7.74
Karak	2	1.94	1.82	0.23
Total	338.3	346.54	347.17	103.15
Percentage of storage from design capacity				30.5%

Table 6. Water harvesting projects (desert dams, earth ponds and concrete ponds) [14].

Water Harvesting Type	Count	Design Capacity (MCM)
Desert dams (constructed)	63	96.55
Concrete ponds	65	0.295
Earth ponds (constructed)	276	25.2
Earth ponds (under construction)	6	0.268
Total	410	122.45

3.2.3. Desalination of Brackish and Seawater

The most promising nonconventional method for increasing the nation's water resources appears to be the use of brackish water, either for direct consumption or after desalination. Numerous sources of brackish groundwater have been identified in diverse regions across the country. Tentative estimates from MWI of stored volumes of brackish groundwater for the major aquifers suggest immense resources, but not all of these quantities will be usable. As such, when referring to statistics about brackish water, the quality, quantity, and location of this resource need to be carefully studied to assess its potential for utilization.

Modern desalination technologies applied to brackish water offer effective alternatives in a variety of circumstances. According to reports from the Jordanian Ministry of Water and Irrigation, the quantity of brackish water, which is extracted from saline layers such as the Zarqa Group layer and includes the water of the Abu Al-Zeghan wells field, reached 3.69 MCM after desalination to be pumped into the water network and used for municipal purposes [49]. As for the desalination of seawater in the Aqaba Governorate, the quantities of water produced after desalination were estimated at 3.12 MCM. Thus, the total amount of nonconventional water for the year 2019 was about 185 MCM, which corresponds to 13% of the total water sources in Jordan [49].

4. Present and Future Trends in Jordan Water Resources

Water law and resource ownership in Jordan are governed by the Water Authority Law of 1988, which was amended by Law No. 22 of 2014. According to Article 25 of the law, all water resources, including surface and groundwater, regional waters, rivers, and internal seas, are considered state-owned property and cannot be used or transferred except in compliance with this law [23]. Nevertheless, the law also facilitates the issuance of licenses to water users and the formation of water user associations. The licensing of water users is provided by the Ministry of Water and Irrigation, which is responsible for overall water resource management in Jordan [24].

The licensing process involves the issuance of permits to water users, which specify the amount of water that can be used and the conditions under which it can be used. The permits are issued for a specific period, and the water users are required to renew them periodically. The permits also specify the fees that the water users are required to pay for the water they use [22]. Water users in Jordan are required to pay for the water they use, and the rates are set by the Water Authority of Jordan based on the cost of water production, treatment, and distribution [25]. The rates are reviewed periodically to ensure that they reflect the actual cost of water production and distribution. The government has also implemented a tariff structure that provides different rates for different categories of water users, such as households, commercial, and industrial users [26].

According to the most recent data from the Ministry of Water and Irrigation's 2019 water budget report, as shown in Table 7, the available water from all resources in 2019 was 1104.8 MCM, distributed as 54.4% groundwater, 30.8% surface water, 14.5% treated wastewater, and 0.3% sea desalinated water, which is used mainly for domestic and agricultural purposes (almost 95%), as shown in Figure 4 [49]. However, a notable concern lies in the largely unregulated extraction of groundwater and the presence of high levels of non-revenue water (NRW) [58,59]. NRW, often resulting from leaks, theft, or metering inaccuracies, refers to water that is produced and transported but not billed to customers. This leads to a critical issue where the estimated water demand may significantly fall short of the actual water usage. To address these issues comprehensively, the water sector in Jordan is not only governed by a robust legal framework but also requires sustained efforts to ensure the efficient management, allocation, and regulation of its vital water resources [16,59].

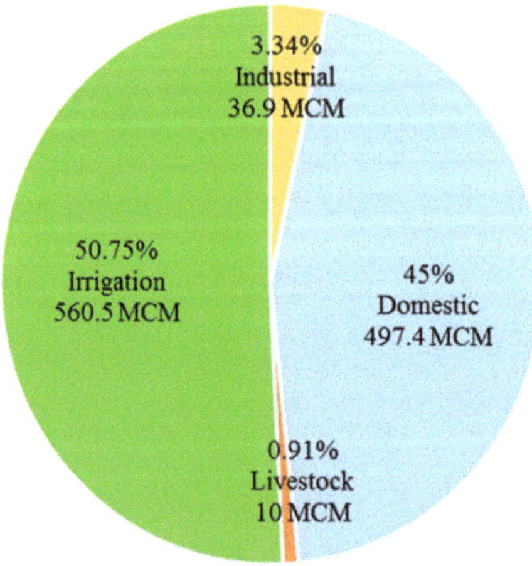

Figure 4. Water uses for 2019 (created based on [49]).

Table 7. Water budget (MCM) in 2019 [49].

Water Source	Domestic	Irrigation	Livestock	Industrial	Total
Surface water	140.54	184.14	8.1	7.66	340.44
Jordan Valley	116.74	141.14	0	7.48	265.36
King Abdullah Canal (KAC)	78	101.71	0	0	179.71
Sothern Ghor and Wadi Araba	38.74	39.43	0	7.48	85.65
Highlands	23.8	43	8.1	0.18	75.08
Springs	21.67	15	0	0.18	36.85
Baseflow and floods	2.13	28	8.1	0	38.23
Treated wastewater (TW)	0	157.4	0	2.5	159.9
TW registered in JV	0	114	0	0	114
TW registered in HL	0	43.4	0	2.5	45.9
Groundwater	355.02	219	1.97	25.41	601.4
Renewable GW	235.14	190.23	1.91	20.7	447.98
Nonrenewable GW	116.18	28.77	0	4.71	149.66
Desalination	3.7	0	0	0	3.7
Sea desalination	1.81	0	0	1.31	3.12
Total	497.37	560.54	10.07	36.88	1104.9

Jordan relies heavily on groundwater, with 54.4% of all uses and 59% of domestic supplies. Figure 5 illustrates that even though renewable aquifers account for the majority of the groundwater supply (74.5%), those aquifers have been extensively exploited above their safe yields, as shown in Table 8, which has led to declining water quality, rising pumping costs, and the drying up of numerous wells in the central and northern regions [46,49]. Every year, many new wells are drilled to replace the dry ones. In addition to the more than 3208 legal wells that are currently in operation, there are a large number of illegal wells as well [14].

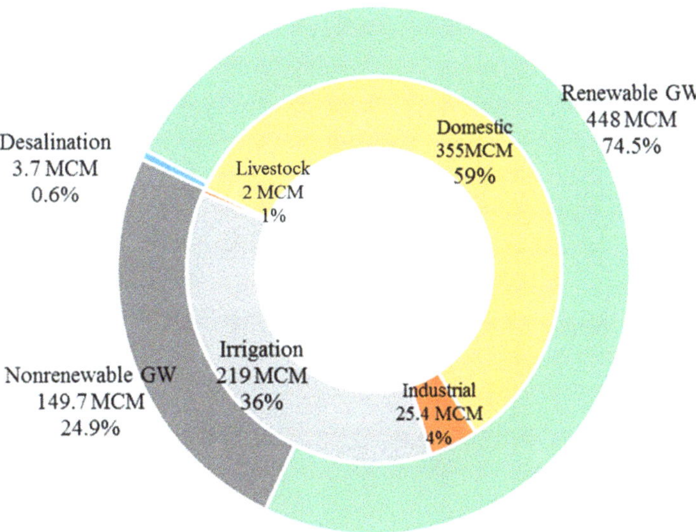

Figure 5. Groundwater sources and sectoral use for 2019 (created based on [49]).

Table 8. Quantities of groundwater extracted from Jordanian water basins and their safe yields in 2019 [49].

| Ground Water Basin | Extraction (MCM) | | | | | | | | Safe Yield (MCM) | Balance (MCM) | Safe Yield % | Number of Wells |
| | Domestic | | Livestock Remote Areas | Industrial | Agriculture | | Recreation | Total | | | | |
	Private	Governmental			Private	Governmental						
Yarmouk	0.75	14.46	0.14	0	17.71	16.65	0	49.71	40	−9.71	124.3	237
Wadi Araba south	0	0	0.05	0	8.14	1.18	0.03	9.4	5.5	−3.9	170.9	57
Wadi Araba North	0.03	1.14	0.8	0	3.95	0.5	0	6.42	3.5	−2.92	183.4	35
Jordan Valley	0.24	9.87	0.06	0.02	14.41	0.47	0	25.07	21	−4.07	119.4	352
Jordan Rift Side Wadis	0.41	40.44	0.01	0	2.93	0.06	0	43.85	15	−28.85	292.3	140
Jafer	0.37	14.42	11.63	0	9.15	0	0	35.57	27 [a]	−8.57	131.7	177
Hammad basin	0.02	1.32	0.03	0	0.2	0	0	1.57	8	6.43	19.6	10
Disi	0	116.18	0	0	25.93	2.84	0	144.95	125 [b]	−19.95	116.0	112
Dead Sea	1.2	44.94	7.33	0.56	25.14	4.57	0.25	83.99	57	−26.99	147.4	467
Azraq basin	1.46	20.81	1.14	0.33	30.17	0	0	53.91	24	−29.91	224.6	574
Amman Zarqa	4.79	81.83	4.22	1.05	71.65	0	0.05	163.59	87.5	−76.09	187.0	1022
Total	9.27	345.41	25.41	1.96	209.38	26.27	0.33	618.03	413.5	−204.53	149.5	3183

[a] The safe yield for the renewable portion of the Jafer is 9 MCM, and its nonrenewable portion is 18 MCM for 50 years (MWI, 2016). [b] The Desi aquifer is nonrenewable, and its safe yield is 125 MCM for 50 years [60].

Like groundwater, surface water is mainly used for domestic and irrigation purposes (95.4%). Approximately (77.9%) of this water originates from the King Abdullah Canal (KAC), Southern Ghor, and Wadi Araba, as shown in Figure 6.

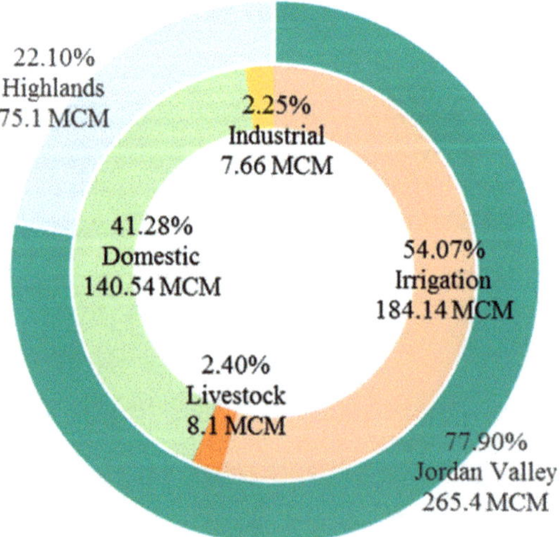

Figure 6. Surface water sources and sectoral uses for 2019 (created based on [49]).

According to reports prepared by the Ministry of Water and Irrigation in recent years, in 2019, Jordan witnessed an improved surface water budget due to increased rainfall and a decline in evaporation rates in contrast to previous years, as shown in Table 9 [49,61]. However, it is important to note that the dams remain significantly below their storage capacity at approximately 31%, as shown in Table 10.

Table 9. Surface water budget 2019 [49].

	MCM	Percentage of Rainfall	Long-Term Average	Percentage to Long-Term Average
Rainfall volume	9567.7		8191	116.8
Evaporation	8871	92.7	7582	117.0
Floods	256.8	2.3	195	131.7
Groundwater recharge	439.8	4.6	419	105.0

Table 10. Storage capacity and water reserves of dams in 2019 [49].

Dam	Design Capacity (MCM)	Storage (MCM)	% of Design Capacity
Wihdah	110	5.3	4.82
Wadi Alarab	16.79	2.84	16.91
Zeglab	3.96	0.56	14.14
Kufranjah	7.8	3.2	41.03
King Talal	75	47.59	63.45
Alkarameh	55	20.16	36.65
Wadi Shueib	1.43	1.51	105.59
Alkafren	8.45	3.81	45.09
Zarqa Maen	2	0.35	17.50
Alwaleh	8.2	5.91	72.07
Almojeb	29.82	10.2	34.21
Altanor	14.7	1.9	12.93
Alajoun	1	0.096	9.60
Alkarak	2	0.112	5.60
Total	338.25	103.6	30.63

The current and projected future water shortages, new technologies, and the significance of sustainable development have all increased awareness of the use of nonconventional water sources. Significant nonconventional resources include treated wastewater and desalinated sea and brackish water [58]. Currently, the reclamation of treated wastewater and the desalination of sea and brackish water are the only nonconventional water resources in Jordan's water budget. Treated wastewater is primarily used for irrigation (98.4%), with a tiny amount used in industry (1.6%). Most of the treated wastewater comes from wastewater treatment plants (WWTPs) registered in the Jordan Valley (71.3%), as shown in Figure 7. Because of the small coastal area in Jordan, seawater desalination has been considered on a limited scale (0.3%) as a source. The desalinated seawater is only used in Aqaba for domestic and industrial applications, as shown in Figure 8.

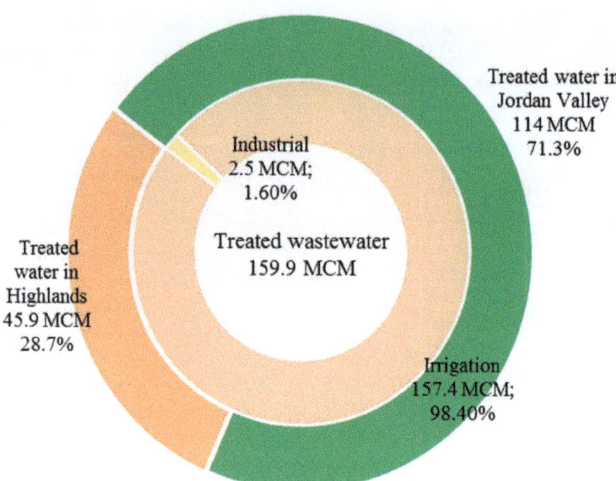

Figure 7. Wastewater use and sectoral applications for 2019 (based on [49]).

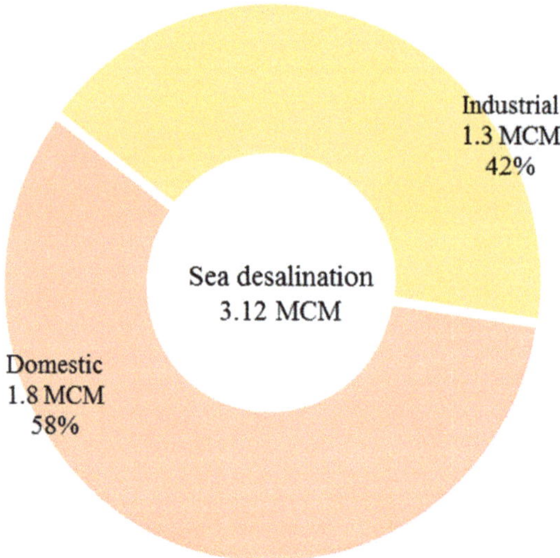

Figure 8. Uses for seawater desalination in 2019 [49].

Figure 9 shows that total and sectoral water use has steadily increased over the last decade, except for the industrial sector, which has slightly decreased. Similarly, the overutilization of resources has increased to meet the increasing demand. The main reasons behind the growing domestic and agricultural water demand could be attributed to government mismanagement of water resources [62], the expanding process of urbanization, the expansion of agriculture activities [45], the inefficient agricultural sector [62], the negative effects of climate change, as well as the influx of refugees [2,58].

Table 11. Future water demand during the period 2020–2025 (MCM) [22].

Year	2020	2021	2022	2023	2024	2025
Municipal, industrial, tourist demands	730	737	746	755	766	778
Irrigation demand	700	700	700	700	700	700
Oil shale and nuclear power demand	25	48	48	48	70	70
Total demand	1455	1485	1493	1503	1536	1548

5. Challenges of Water Resources in Jordan

5.1. Climate Change, Low Precipitation, and Aridity of the Region

Water storage and investment policies are intricately interlinked with the repercussions of climate change on water resources in Jordan [26,51]. The evaluation of these repercussions is imperative for informed decision-making concerning water resource management and infrastructure advancement. In the Jordanian context, these policies assume a pivotal role in alleviating the consequences of declining precipitation, escalating evaporation rates, and mounting aridity [26,51].

The Climate Risk Profile Jordan report supplies forecasts for prospective evapotranspiration and precipitation levels in Jordan under diverse emissions scenarios until the year 2080. The report predicts a substantial reduction in annual precipitation, amounting to a decline of 20–28% in the southern and northeastern regions of Jordan and a decrease of 10–16% in the northwestern and central regions by 2050. Furthermore, the report signifies a surge in air temperatures by up to 4.5 °C by 2080 compared to pre-industrial levels, exacerbating the susceptibility to heatwaves and jeopardizing the populace's capacity to engage in gainful activities [66]. Consequently, strategic investments in water storage and infrastructure become imperative to counteract the anticipated drop in annual rainfall. Such investments can serve to ameliorate the repercussions of decreased precipitation and provide indispensable water reserves to satisfy the burgeoning water demands.

Simultaneously, these policies assume a critical role in the prudent management of groundwater resources, which are undergoing rapid depletion owing to the combined influences of meager precipitation and overextraction [7]. The implementation of sustainable water storage solutions and the adoption of efficacious management protocols have become essential to preserve the quantity and quality of the available groundwater resources.

Moreover, the robust evaporative force intrinsic to Jordan's climatic conditions is underscored by a publication in Scientific Research Publishing. This source elucidates that climate change is progressively exerting an impact on Jordan's water resources, with the potential evaporation force varying from 1800 mm/year in the cooler northwestern zones to 4200 mm/year in the southeastern regions, three and one hundred forty times the amount of average annual precipitation, respectively, as shown in Figure 10 [67]. These challenges are exacerbated by the fact that the once-significant Jordan River has dwindled in significance due to heightened demands from neighboring nations and overextraction practices [68].

The ramifications of climate change extend to the agricultural sector, where reduced precipitation and escalated evaporation contribute to heightened demand for crop irrigation. Furthermore, an escalation in the frequency of extreme weather events, encompassing heavy precipitation and snowstorms, is expected, leading to flash floods and fluctuations in temperature extremes [69]. These alterations will exert far-reaching effects on Jordan's water supply, inducing augmented soil erosion, inundations, and sediment accumulation in rivers, dams, and reservoirs. Additionally, the foreseeable scenario suggests increasing water salinity and variability, potentially resulting in a reduction in major river flows such as the Jordan River by up to 80% [70].

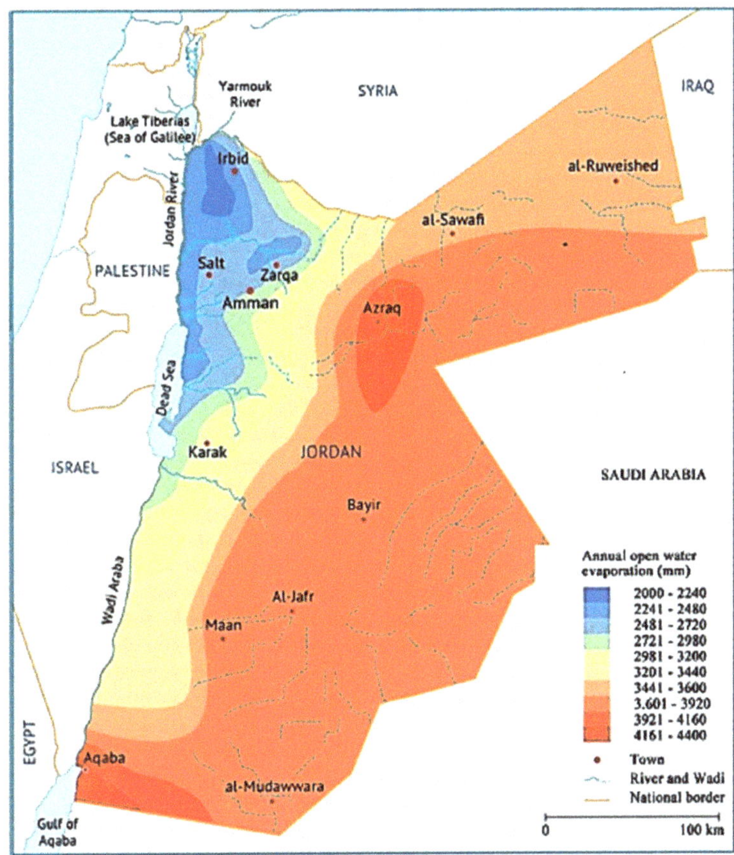

Figure 10. Distribution of potential evaporation across Jordan in mm/year [71].

5.2. Population Growth, and Refugees

Jordan, like many Middle Eastern nations, confronts formidable challenges related to its water resources, driven by population growth and the influx of refugees [72,73]. With an estimated population of 10.5 million for 2021 and a projected surge to 19.04 million by 2050, the nation faces increasing strain on its water supply [54,55]. Hosting a substantial number of refugees has further heightened water demand, especially in refugee camps where resources are limited [74]. To address these challenges, Jordan has been implementing measures such as constructing new water storage facilities and expanding water treatment and desalination plants [37,75], despite these efforts, the challenges of Jordan's water resources remain significant.

The population growth is estimated at 2.1% for Jordanians and 2.2% to 2.9% for other nationalities living in the country [73,76]. The influx of 1.4 million Syrian refugees alone was a major factor in increasing water demand by more than 40%, adding to the burden of the already water-scarce northern and central regions. Despite having the highest rainfall rates and significant water resources, these regions host nearly 90% of the population, and they grapple with rising water supply costs. The refugee crisis compounds Jordan's pre-existing water woes, driven by overextraction, pollution, and inadequate resource management [14]. The rapid population growth pushes the brink of water depletion closer, demanding immediate efforts to ensure water resource sustainability to avert a looming crisis.

5.3. Limited Financial Capacity

Jordan has limited financial capacity to carry out significant projects. Critical projects with low cost–benefit ratios should be the focus of difficult decision-making. Raising fees and taxes could aid in funding the upkeep of current water infrastructure as well as the creation of new projects. However, because of how politically and socially sensitive this is, it is unlikely that there could be any significant rate increases. Jordan relies heavily on external funding sources (e.g., foreign private enterprises, NGOs, foreign aid, and foreign governments). However, foreign funders' priorities do not always coincide with those of the government's goals.

5.4. Public Awareness and Education Campaigns

Jordan is currently grappling with a severe water scarcity crisis where the water demand exceeds the available supply. Nevertheless, a significant portion of the Jordanian population remains largely unaware of the gravity of this issue. Several factors contribute to this lack of awareness, including the normalization of water scarcity, insufficient dissemination of information, a false sense of security resulting from water management efforts, more immediate concerns taking precedence, psychological detachment, and ineffective communication and education campaigns.

To address this awareness gap, the government has embarked on a series of initiatives. These measures encompass media campaigns, the integration of water crisis-related content into educational materials, and the increased involvement of religious leaders in community engagement. These efforts are aimed at cultivating a deeper understanding of the issue and facilitating the effective implementation of sustainable water management strategies.

The ongoing water scarcity challenge in Jordan has created a complex scenario wherein a substantial portion of the population remains uninformed about the critical nature of the problem. While over 98% of Jordanians presently have access to secure water supply services, and sanitation services extend to more than 63% of the population, the government is actively working to raise this coverage to approximately 70% in the near future through initiatives led by the Ministry of Water and Irrigation [8].

Nonetheless, it is imperative to acknowledge the prevalent issues of water supply interruptions and unreliability across Jordan. The statistical figures related to connection rates often do not accurately convey the reality that numerous urban and rural communities routinely experience extended periods without access to piped water, sometimes lasting for weeks.

Confronting these challenges, the government continues to grapple with the task of educating the public about the escalating water crisis and encouraging citizens to adopt water conservation practices. This underscores the pivotal role of education, not only in imparting practical water conservation techniques but also in explaining why water conservation is a matter of paramount significance.

5.5. High Levels of Non-Revenue Water

Jordan's water resources are facing various challenges, with high levels of NRW being one of the major concerns. The levels of NRW in Jordan have reached as high as 60% in some areas, affecting the country's water supply system by reducing the availability of water and increasing scarcity [77]. This also has a negative impact on the economy, as the government invests large amounts in water production and transportation infrastructure but is unable to recover the costs due to high NRW levels. The government has taken measures to reduce NRW, including repairing and replacing aging water distribution networks, installing new metering systems, and educating customers on reducing water waste. However, high NRW levels remain a significant challenge, and the accompanying financial loss for the government is approximately USD 500 million per year [78]. Projects for the reduction in NRW were ongoing through June 2023, with a budget of USD 60 million [29]. In addition, the International Monetary Fund (IMF) has approved a USD 1.3 billion extended arrangement under the Extended Fund Facility for Jordan, which includes efforts to limit losses in the energy and water sectors [79]. The exact amount of effort or money spent on these measures is not specified in the search results, but the 2022 budget for Jordan forecasted USD 15 billion in state spending, including funding for water conservation initiatives [80].

5.6. Managing Shared Water Resources

Jordan confronts a multitude of challenges when it comes to water resource management and allocation with neighboring countries. One of the key predicaments is the uneven distribution of water resources among Jordan and its neighbors, which has created a significant burden on the country's water security. The country has limited water resources and is highly reliant on shared rivers and aquifers with neighboring countries, which account for approximately 26% of the country's total water resources, and most of these resources originate outside the country's borders [14]. Throughout its history, Jordan has faced challenges in accessing its equitable share of these shared resources due to the actions of upstream neighbors who have overexploited these water sources. Here, we will examine the key provisions and agreements that have been made between Jordan and other countries regarding these shared water resources. The Johnston Plan, proposed in 1955, aimed to address the allocation of water in the Jordan River Basin among the riparian states. The plan's proposed allocations were as follows: 55% to Jordan, 26% to Israel, and 9% each to Lebanon and Syria [81–83]. Mediated by the United Nations, the countries agreed to these allocations, but the plan was never ratified. Instead, riparian states pursued unilateral projects, disregarding the outlined allocations. In 1994, Jordan and Israel signed a peace treaty that included an agreement on the distribution of Jordan River water. Despite the Johnston Plan allocating the largest share to Jordan, the treaty stipulated that Israel would transfer only 50 million cubic meters per year (MCM/yr) of water from Lake Tiberias to Jordan. The treaty also encouraged the two countries to cooperate in addressing the ongoing water shortage in the Jordan River Basin by reducing waste, preventing contamination, and developing additional water resources [83]. Jordan and Syria signed their first bilateral agreement regarding the Yarmouk River in 1953. While the agreement did not specify water resource allocations, it addressed irrigation and power generation through the construction of dams on the river. It also led to the establishment of a bilateral commission for dispute resolution. A 1987 treaty modified the agreement, with Jordan committing to cover the costs of the main irrigation dam (Maqarin or Wihdah Dam) and Syria permitting the construction of additional dams on the river system to provide water to Jordan [83–85]. A new commission was established with fewer oversight powers. Disputes arose over Syria's increased construction of dams and groundwater pumping, affecting the flow of the Yarmouk River into Jordan. The agreement primarily covered surface water resources, leading to controversy and misunderstandings. In 2015, Jordan and Saudi Arabia reached an agreement on the shared Disi fossil aquifer, known as Disi on the Jordanian side and Saq on the Saudi Arabian side. The aquifer primarily lies within Saudi Arabia. The agreement prohibited well drilling within a 10 km buffer zone on each side of the border and outlined measures to prevent contamination of the shared resource. The agreement stipulated that the aquifer should be used for domestic and municipal purposes, not for agriculture, reflecting Jordan's efforts to pump water from the Disi aquifer for domestic and municipal use [83]. These agreements and arrangements have played a significant role in shaping Jordan's access to shared water resources. The history of these agreements has been marked by complexities, disputes, and evolving circumstances. Jordan's challenges in securing its water resources have also been influenced by the actions of its upstream neighbors, which have sometimes led to deviations from the intended allocations and usage conditions. As water resources continue to be a critical issue in the region, effective management and cooperation among neighboring countries remain vital for the sustainable use of these shared resources.

5.7. Unsustainable Agricultural Water Use

The government of Jordan is faced with a significant challenge in achieving sustainable water use in the agricultural sector. Agriculture accounts for a substantial 80% of total water withdrawal and more than half of the country's freshwater resources [86,87]. Remarkably, agriculture employs only 3% of the labor force and contributes a mere 5% to the gross domestic product [88]. Despite efforts to improve water efficiency and reduce waste, the high water demands of agriculture have led to the overexploitation of groundwater, resulting in declining water tables and increased salinity levels. To address this issue, the government has promoted the adoption of more efficient drip irrigation systems and implemented water-pricing policies aimed at reducing excessive water use and encouraging efficient practices. However, water stress continues to persist, exerting pressure on the availability of irrigation water, leading to higher costs and reduced crop production available for export [89].

6. Opportunities toward a More Sustainable Future

After analyzing the current situation of the water sector in Jordan from various angles, in addition to studying the adapted projects and measures to ensure the security of the water sector, engineering and social science suggestions are presented to deal with the outstanding water security issues and to reach a more robust and sustainable water sector.

6.1. Enhancing Water Supply

Ensuring a consistent supply of high-quality freshwater is crucial for addressing Jordan's water crisis sustainably. The current usage of freshwater resources is unsustainable and needs immediate attention. Strengthening the water distribution network and minimizing non-revenue water loss should be prioritized to mitigate the further reduction in Jordan's already limited water supply. Investing in wastewater treatment plants, rainwater harvesting, and optimizing virtual water utilization can help to alleviate stress on surface and groundwater resources. Additionally, the establishment of desalination plants is vital for the long-term solution and has the potential to substantially increase the overall water availability. However, the search results also suggest that Jordan has limited financial capacity to implement major projects, and hard choices need to be made to target critical projects with low cost–benefit ratios. Increasing tariffs and taxes could help to finance the maintenance of existing water infrastructure and the development of new projects, but this is politically and socially sensitive, and substantial rate increases would probably not be feasible. Therefore, creative solutions are needed to fight Jordan's water scarcity, and the government should consider expanding cooperation with foreign investors and international donors to address the issue.

6.1.1. Curbing Water Losses through Improved System and Governance

Jordan faces significant challenges in its water management, including high levels of water loss and inadequate governance structures. The government has implemented various initiatives aimed at reducing water losses and strengthening governance in the water sector to achieve a more sustainable future for the country. One of the key initiatives to reduce water losses has been the modernization of the water supply system, including the installation of modern water meters, leak repair, and a centralized billing system [67]. The government has also established the Water Authority of Jordan to regulate the sector and ensure efficient and sustainable water use. The Authority has implemented policies to promote public–private partnerships and investment in the water sector. Despite these efforts, water loss was estimated at 50% in 2020, compared to 43% in 2010, and was divided into more than 50% as administrative losses and less than 50% as technical losses from networks. Consequently, there is a substantial amount of work ahead to enhance the water sector further and maintain progress toward a more sustainable future [14]. This includes the implementation of technology-dependent management tools such as farmers' associations, the control of illegal water tapping, and improved negotiation skills for water disputes. In addition, the water sector must also improve its training programs and prevent both natural and human-caused accidents in water supplies and infrastructure. Laws and regulations should be strengthened to include strict control systems for all significant water infrastructure facilities, including effective control against vandalism and destruction [90]. These efforts, along with continued investment and commitment from all stakeholders, will be necessary to achieve a more sustainable future for Jordan's water sector.

6.1.2. Harnessing Desalination for Sustainable Water Resources

Desalination is a technology that involves removing salt and minerals from the sea or brackish water to make it usable for various purposes. This solution has gained popularity in countries with limited water resources, such as Jordan, due to its reliability and cost-effectiveness. Jordan's desalination sector exhibits substantial potential for expansion, largely owing to the country's coastline and abundant solar energy resources. Desalination offers a dependable supply of fresh water, alleviating the strain on limited freshwater reserves and creating new avenues for economic development. Nonetheless, it is important to acknowledge that the process demands considerable energy, substantial capital, and pervasive costs and may adversely affect the marine environment. Jordan has an opportunity to operate desalination and water transmission systems with renewable energy, given the falling technology costs and increasing use of renewable energy [91], most notably the proposed National Water Carrier Project. However, the National Water Carrier Project, which aims to move water from Aqaba to Amman, must ensure public–private partnership (PPP) financing, limit environmental impacts, and be completed on time and within budget. Additionally, alternative remedies are needed in the interim to offset the output from the desalination plants [26,92].

6.1.3. Harnessing the Potential of Wastewater Treatment and Reutilization

The utilization of wastewater treatment and reuse represents a highly promising strategy with significant potential to contribute to the attainment of sustainability goals in Jordan. Recent years have witnessed an increasing recognition of the importance of harnessing treated wastewater for various purposes, including irrigation, industrial operations, and even potable water supply. Statistical data obtained from the Ministry of Water and Irrigation indicate a notable rise in the overall volume of treated wastewater generated in Jordan, increasing from 150 million cubic meters in 2018 to an impressive 170 million cubic meters in 2020 [93]. This growth can be attributed to several factors, such as the escalating water demand, the decline in conventional water resources, and an enhanced awareness of the significance of water conservation and management. The successful application of treated wastewater for irrigation purposes is particularly noteworthy, with several large-scale projects being implemented across the country. For instance, the projects in Jordan have effectively employed treated wastewater for crop irrigation, including vegetables and fruits, yielding remarkable outcomes [94]. The projects have significantly alleviated pressure on freshwater resources and contributed to mitigating water scarcity in the region. Furthermore, the utilization of treated wastewater in the industrial sector holds great promise, offering diverse benefits such as reduced reliance on potable water, energy conservation, and greenhouse gas emission reduction by eliminating the need for long-distance water transportation. Notably, the incorporation of a wastewater treatment and reuse system by the Jordan Phosphate Mines Company resulted in a remarkable 30% reduction in water consumption [95]. Despite the considerable benefits derived from wastewater treatment and reuse, several challenges need to be addressed. Firstly, substantial investments in infrastructure and technology are crucial to ensure optimal wastewater treatment that aligns with international health and environmental standards. Secondly, both the public and private sectors need to be sensitized to the importance of this approach and encouraged to adopt it extensively. Lastly, societal and cultural obstacles related to the acceptance of treated wastewater for various purposes, especially for potable water supply, must be adequately addressed. The utilization of wastewater treatment and reuse in agriculture is pivotal for bridging the gap between water supply and demand, with research indicating that it can increase water availability by up to 48% [96]. Despite the Jordanian government's efforts to enhance the potential of wastewater treatment plants, only 14.5% of the total water resources are currently considered usable after treatment [14]. There is still considerable room for improvement, particularly in promoting the use of treated water, delivering it to users, and implementing substantial capital investments in sewage treatment facilities. Furthermore, implementing policies and regulations that incentivize the use of treated wastewater and discourage the use of freshwater for non-potable purposes could have a significant impact.

6.1.4. Embracing Rainwater Harvesting for Sustainable Water Supply

Rainwater harvesting (RWH) refers to the collection, storage, and use of rainwater for various purposes, including irrigation, domestic use, and industrial processes. This technology has been in use for centuries, but advancements in recent years have made it more accessible, cost-effective, and energy-efficient. As such, it has become a popular solution for countries with limited water resources, such as Jordan. According to data from the Jordanian Ministry of Water and Irrigation, the number of households practicing RWH in Jordan increased from 10% in 2010 to 30% in 2020 [97]. This growth was due to many factors, including increased awareness of the technology, financial incentives, and the availability of affordable RWH systems. The total amount of RWH in Jordan also increased, reaching over 10 million cubic meters per year. Rainwater harvesting offers several benefits to Jordan's water sector. Firstly, it provides a reliable and consistent source of fresh water that is not subject to the same fluctuations as surface and groundwater resources. This helps to mitigate water stress and reduce the chances of crop failure. Secondly, it reduces the pressure on the country's limited freshwater resources, which are being overexploited due to high demand. Finally, it opens up new opportunities for economic growth as it enables the country to expand its agriculture and industrial sectors. This not only contributes to Jordan's economic prosperity but also fosters job opportunities and enhances overall prosperity. It is crucial to acknowledge that various ecosystems and wildlife rely on the continuous flow of water from springs, dispersed rainwater, and instream flows. Preserving these resources and conserving water is of paramount importance for these ecosystems, given their intrinsic value and their contribution to the tourism sector.

There are several positive models of RWH from other countries that Jordan can learn from. For example, rainwater harvesting is a traditional and reviving technique for collecting water for domestic uses in the United Kingdom [98]. In Australia, RWH is widely used for irrigation and domestic purposes, and the government provides financial incentives to households that install RWH systems [92]. These examples demonstrate the potential of RWH to address water scarcity and the importance of government support and financial incentives in promoting its adoption.

Despite its benefits, RWH also has some drawbacks that must be considered. Firstly, the quality of the collected rainwater may be affected by pollutants in the atmosphere, such as air pollution and dust. Secondly, the cost of installing and maintaining RWH systems can be high, especially in rural areas where access to financing is limited. These factors must be carefully managed to ensure that RWH remains a sustainable solution. The government is expected to establish a financing mechanism under the Green Growth Action Plan 2021–2025 to promote RWH initiatives and raise public awareness [26], but cooperation with various sectors and public acceptance of the technology are still challenges.

6.1.5. Optimizing the Utilization of Virtual Water

The term "virtual water" refers to the amount of water embedded in the production process of goods and services, including water used for crop cultivation, energy generation, and product manufacturing. This concept has gained significant attention as a strategy to address water scarcity and promote sustainable water management. In Jordan, virtual water is increasingly recognized as a potential solution to the country's water crisis and a pathway for economic development. According to data from the Jordan Water Project (JWP), in 2020, the virtual water content of Jordan's exports exceeded 1 billion cubic meters [99]. Key export commodities in Jordan, such as fruits, vegetables, and livestock, have substantial virtual water content, contributing to the country's water footprint. By reducing the virtual water content of these products, Jordan can reduce its water consumption and alleviate strain on its finite freshwater resources. One example of utilizing virtual water to promote sustainability in Jordan is the implementation of water-saving technologies and practices in agriculture. This includes the adoption of efficient irrigation systems like drip irrigation and the cultivation of drought-resistant crops that require less water for growth. By reducing water usage in agriculture, Jordan can conserve its limited water resources while simultaneously enhancing agricultural productivity [100]. Another approach to advancing sustainability in Jordan through virtual water involves promoting industries with lower water consumption and producing goods with lower virtual water content. For instance, the country can encourage the growth of industries specializing in low-virtual water products, such as high-tech electronics. Achieving a balance that allows for water conservation without financial losses is challenging but feasible through a combination of water-efficient practices in agriculture, the promotion of low-virtual water industries, and investments in water-saving technologies. This balance is essential for the long-term sustainability of both Jordan's economy and its water resources [101].

6.2. Optimizing Water Utilization

Enhancing the water supply involves a substantial investment of time and finances. Nevertheless, Jordan has the opportunity to promptly address the issue by curbing highly excessive and inefficient water consumption practices. The current state where ten out of twelve groundwater basins in Jordan are being extracted beyond their sustainable limits calls for immediate action. To mitigate this problem of overexploitation, Jordan can introduce regulations to enhance water efficiency, particularly in agriculture, raise public awareness, and reward and promote more efficient water use and behavior.

6.2.1. Improving Water-Use Efficiency in Agriculture

Improving water-use efficiency in agriculture is crucial for the sustainable development of Jordan, as the country faces increasing demand for food production and limited water resources. Precision irrigation systems, using digital technologies to deliver exactly the amount of water required for crops can significantly reduce water waste. For example, the University of Jordan's Center for Agricultural Research has developed a precision irrigation system that has shown a 25% increase in water-use efficiency [102]. Another effective solution is the promotion of drought-resistant crops, such as the new varieties of wheat and barley developed by the International Center for Agricultural Research in the Dry Areas (ICARDA), which require 30–50% less water compared to traditional crops [103]. Furthermore, employing appropriate crop management strategies like soil conservation, crop rotation, and efficient fertilization can enhance soil health, enabling crops to more efficiently absorb water. This not only reduces the necessity for excessive watering but also mitigates the potential for evaporation and salinization [104]. However, agricultural projects in Jordan are currently hindered by leaky water conveyance systems and outdated irrigation practices [105,106]. Despite the existence of advanced agricultural practices in some areas, the sector as a whole still requires improvement, particularly in terms of the use of metering systems, regular irrigation, soil moisture devices, cameras, telemetry, and leakage detection and monitoring. This will not only make more water available and increase production in both quantity and quality but also create job opportunities for trained workers. Improving the management and practices of the irrigated sector is crucial to reducing water consumption and enhancing agricultural productivity while protecting the rights of others who may benefit from the saved water. More efficient irrigation techniques that can save 30% to 50% of the current water usage are expected to result in a significant increase in agricultural output.

6.2.2. Promoting Public Awareness and Behavioral Change

Public awareness and behavioral change are crucial for creating a more sustainable future for Jordan's water resources. The government launched the National Water Strategy in 2018 to raise public awareness about water conservation and promote a sense of personal responsibility for water use. The strategy includes a nationwide water-saving campaign, which uses various media platforms to reach a wide audience and provide practical water-saving tips [107]. Additionally, the development of water conservation curricula for schools is aimed at fostering a culture of water conservation from an early age [108]. Investment in educational programs that focus on water conservation is crucial to raise public and political awareness of the current water situation. The Green Growth National Plan 2021–2025 is focused on encouraging the public to use more sustainable water alternatives, such as treated wastewater and harvested rainwater. The Ministry of Water and Irrigation has partnered with the Ministry of Education to raise awareness of water scarcity through textbooks, and other initiatives may be developed to increase public awareness in the future [26]. While domestic water use has received the majority of the attention, future campaigns must also focus on the agricultural sector. An environmental assessment can be used to protect the environment and achieve efficient water use by taking into account the effectiveness of water use, its effects on the environment and human health, public awareness, and fair pricing of water in various sectors.

6.3. Creating the Framework

The future of Jordan's water resources is facing obstacles due to the lack of effective coordination among key stakeholders. In order to adequately address the long-term water situation in Jordan, it is crucial to engage in comprehensive collaborative planning and make decisions based on reliable evidence. These decisions should encompass investment and management strategies, taking into account high-quality data and the impact of climate change. Furthermore, the involvement of all relevant parties, such as water users, investors, and donors, is vital. While finance is critical to realize infrastructure projects, fair and inclusive management plans can maximize the utility of existing resources and infrastructures. Given its downstream location, Jordan's water supply is contingent upon cooperation with neighboring countries. The current procedures fail to account for the growing population in Jordan and prove to be ineffective and inadequate. Hence, enhancing cross-border collaboration becomes imperative for discovering enduring solutions to Jordan's water challenges.

6.3.1. Advancing Joint Planning for Sustainable Water Management in Jordan

Advancing joint planning for sustainable water management in Jordan is imperative for ensuring the country's long-term water security and socio-economic development in the face of increasing water stress. Collaborative efforts involving government agencies, local communities, and international organizations can optimize water resources and mitigate the risks associated with water scarcity. By adopting integrated and holistic approaches, considering the interconnectedness of water resources and their impact on agriculture, industry, and domestic use, Jordan can develop sustainable strategies that ensure equitable distribution and minimize conflicts among sectors. Incorporating the private sector is crucial for ensuring financial stability, improving operational efficiency, stimulating innovation, and creating job prospects. Moreover, increased inclusion of marginalized communities bolsters the overall effectiveness of programs. Joint planning facilitates the exchange of knowledge, expertise, and best practices among stakeholders, empowering decision-makers and water managers to make informed choices based on evidence-based practices. Sharing information on water management techniques, innovative technologies, and successful case studies from local and international contexts optimizes resource utilization and minimizes waste. Additionally, joint planning strengthens the resilience of Jordan's water sector by anticipating and preparing for potential disruptions through adaptation measures and infrastructure upgrades. Cooperation on water-related research and development promotes innovation and scalable, sustainable solutions that can benefit other water-stressed regions. Implementing these measures is essential for addressing the challenges within Jordan's water network and achieving sustainable water management goals [26,109].

6.3.2. Enhancing Data Quality and Accessibility

Enhancing the quality and accessibility of data is of paramount importance for the Jordanian Government and the research community. Access to comprehensive data is crucial for developing more informed solutions in the realm of water resource management. At present, the availability of essential hydrological data and related information remains limited and often obscured by paywalls, presenting a significant hindrance to effective planning and decision-making [110].

Hydrological data serve as a linchpin for proactively preparing for extreme events by identifying high-risk areas susceptible to water leakages and other related challenges. Additionally, day-to-day hydrological data play a vital role in optimizing the management and distribution of Jordan's precious water resources, thereby assisting in the identification of leaks and unauthorized water connections [92].

However, it is not only the government and researchers who face difficulties due to this data gap. Academics and researchers working on mapping Jordan's past, present, and future water situation are similarly frustrated by the relative scarcity of available data and resources. This shortage obstructs their efforts to develop the necessary information and data for informed analysis and policymaking. Moreover, the current channels for sharing and disseminating data among key stakeholders, including ministries, are insufficient, further exacerbating the issue [92].

Acknowledging the gravity of this challenge, the government has initiated measures to bolster its data collection capabilities. As indicated in the Green Growth National Action Plan 2021–2025, collaborative efforts between the Ministry of Water and Irrigation (MWI) and the Water Authority of Jordan (WAJ) are underway to implement an updated Geographic Information System (GIS) specifically tailored for newly constructed or rehabilitated water networks. This GIS system is anticipated to efficiently monitor and store data, substantially enhancing the quality and accessibility of information for informed policymaking. Additionally, the plan outlines the implementation of a Joint Work Programme (JWP) model, which is poised to serve as a pivotal strategic planning tool [26,92].

Looking forward, long-term solutions involving advanced and distributed technologies, such as the Internet of Things (IoT) and supervisory control and data acquisition (SCADA) systems, offer promising opportunities for elevating data collection, monitoring, and management processes. For example, remote sensing techniques can be deployed on water network pipes to promptly detect physical leaks. This real-time information can then be relayed to the relevant authorities, significantly reducing response time and improving decision support. Moreover, IoT holds the potential to establish a comprehensive data repository, serving as a robust foundation for evidence-based policy interventions [92,111–113]. The combination of these initiatives will undoubtedly enhance the quality and accessibility of water-related data, empowering stakeholders to make more informed and effective decisions in the field of water resource management.

6.3.3. Promoting Resource Management Amongst Neighboring Regions

Water resource cooperation between neighboring countries is essential for promoting a more sustainable future for water resources in Jordan and the region. By sharing resources, exchanging information, and utilizing new technologies, countries can effectively manage their water resources and improve water security and access. Examples of such cooperation include the Disi Water Conveyance Project in Jordan and the Joint Technical Committee for the Allocation of the Yarmouk River Waters between Jordan, Syria, and Iraq [47,114]. These initiatives have led to increased water supplies, resolved conflicts, and improved cooperation, as well as stimulating economic growth and job creation. Additionally, initiatives such as the Red-Dead Conveyance Project [115] and the Joint Water Commission (JWC) between Jordan, Israel, and the Palestinian Authority have been instrumental in promoting regional cooperation and coordination on water-related matters [116], while the Joint Technical Committee for Water (JTCW), established by the League of Arab States, has helped to build trust and confidence among neighboring countries [117]. Jordan has a unique opportunity to establish a water-energy nexus (WEN) with neighboring countries, particularly Israel and Palestine, to cultivate a mutually beneficial and sustainable partnership. The implementation of the WEN aims to create a regional community centered around desalinated water and solar energy, fostering resource cooperation and enhancing regional stability [43]. With its expansive landmass suitable for large-scale solar power plants, Jordan holds a favorable position to engage in resource-based negotiations. A notable example of collaboration is the water-for-renewable-energy declaration of intent signed in November 2021 between Jordan, Israel, and the United Arab Emirates (UAE). Under this agreement, the UAE plans to construct a solar energy farm in Jordan with an installed capacity of 600 MW of renewable power that will be available for Israel to feed into their grid. In return, Israel would supply Jordan with 200 million cubic meters of desalinated water. This exemplifies the potential for leveraging renewable energy and water resources to forge sustainable partnerships within the region [92,118].

7. Conclusions

Jordan faces a pronounced water scarcity crisis, primarily attributable to its arid region and the increasing variability in climate conditions, particularly with regard to precipitation and rising temperatures. Despite the country's ongoing efforts to manage its limited water resources over recent decades, persistent water security issues prevail due to a lack of proactive decision-making by policymakers. Consequently, the decline in both the quality and quantity of water resources has resulted in social and economic hardships for the population. In order to alleviate water stress, Jordan is implementing a range of measures, including the expansion of water availability through desalination and dam construction, as well as reducing water consumption by enhancing public water supply networks. Nevertheless, water stress levels remain alarmingly high, necessitating the implementation of short-term demand-side interventions and long-term supply-side reforms to ensure a sustainable water future. However, these endeavors encounter financial and governance obstacles, such as the substantial costs associated with supply-side interventions and the limited awareness and incentives for water consumption reduction on the demand side. To overcome these challenges, Jordan must attract private investments by enhancing the business cases and transforming interventions into profitable models. Moreover, significant investments and coordination among domestic stakeholders, neighboring countries, and the international donor community are imperative to raise awareness and foster reduced water consumption. This comprehensive approach holds the potential to secure a sustainable future for water resources in Jordan, ensuring access to clean and safe water for its citizens.

Author Contributions: Conceptualization, M.A.-A., M.B. and S.R.; methodology, M.B. and S.R.; formal analysis, S.R., N.S. and M.A.; investigation, M.A.-A., M.B. and S.R.; resources, M.A.-A., M.B., M.A., N.S. and S.R.; data curation, M.A.-A., M.B., N.S. and S.R.; writing—original draft preparation, M.A.-A., M.B. and S.R.; writing—review and editing, M.A.-A., M.B., M.A., N.S. and S.R.; supervision, M.A.-A. and M.B. All authors have read and agreed to the published version of the manuscript.

Funding: This research received external funding from The Ministry of Higher Education and Scientific Research/Scientific Research and Innovation Support Fund (SRSF), AGREEMed 1/01/2021.

Data Availability Statement: No new data were created or analyzed in this study. Data sharing is not applicable to this article.

Acknowledgments: The authors would like to thank the Deanship of Scientific Research (DSR) at the German Jordanian University, SRSF, the Fraunhofer-Institut für Solare Energiesysteme(ISE), and all partners of AGREEMed project for their continuous support throughout the lifetime of the project.

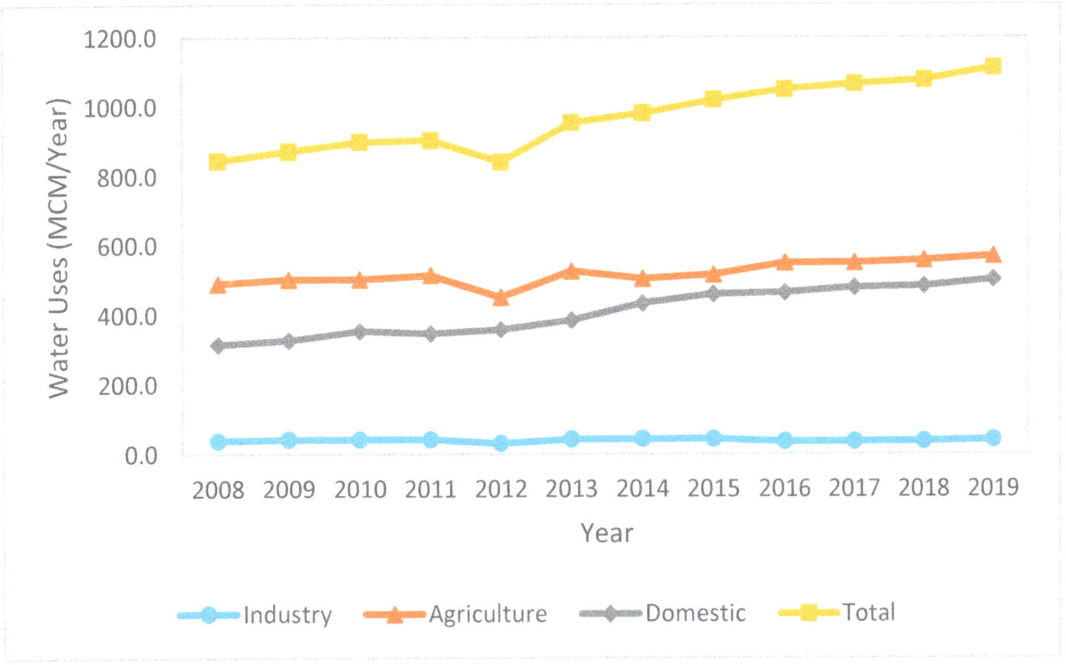

Figure 9. Annual water uses (MCM/year) for different purposes [17,49,61].

The total water demand is expected to increase from 1455 MCM in 2020, according to Jordan's water strategy, to 1548 MCM in 2025, as shown in Table 11. It appears that Jordan is directly overexploiting its water resources by 10% to 100% or more. Jordan's energy mix was projected to consist of 54–62% oil and oil shale through 2020, with a growing demand for energy at a rate of 3% annually [63,64]. Additionally, Jordan has been pursuing nuclear power as a source of energy, with a goal of generating 6% of its electricity from nuclear power by 2020 [65]. The production of oil shale and nuclear power are both water-intensive processes, and the increase in energy demand from these sources is expected to lead to an increase in water demand in Jordan. According to the National Water Strategy of Jordan, power generation from oil shale and nuclear power plants already exceeds 4% of the total surface and groundwater in Jordan, amounting to about 39 MCM in 2023 [9]. Therefore, it is expected that the future water demand from oil shale and nuclear power in Jordan during the period 2020–2025 will increase from 25 MCM in 2020 to 70 MCM in 2025, as shown in Table 11. In terms of additional capital expenditures and the requirement for subsidies, the rising cost of water delivery is adding to the fiscal budget's already heavy load. A tariff increase cannot close the entire difference between the current tariff levels and complete cost recovery. Most water users find full-cost recovery to be prohibitively expensive, particularly low-income residents. Therefore, water subsidies may be reduced. The current average cost of bulk water supply for priority domestic use is USD 0.49 per cubic meter. However, as water shortages worsen and energy costs increase, this cost will rise to USD 1.34–1.55 per cubic meter or even higher because of the difference between demand and supply.

Conflicts of Interest: The authors declare no conflict of interest.

References

1. Mekonnen, M.M.; Hoekstra, A.Y. Four Billion People Facing Severe Water Scarcity. *Sci. Adv.* **2016**, *2*, e1500323. [CrossRef] [PubMed]
2. FAO. *The State of Food and Agriculture 2020: Overcoming Water Challenges in Agriculture. Overcoming Water Challenges in Agriculture (Rome)*; The State of Food and Agriculture (SOFA); FAO: Rome, Italy, 2020; ISBN 978-92-5-133441-6.
3. Khan, M.A.M.; Rehman, S.; Al-Sulaiman, F.A. A Hybrid Renewable Energy System as a Potential Energy Source for Water Desalination Using Reverse Osmosis: A Review. *Renew. Sustain. Energy Rev.* **2018**, *97*, 456–477. [CrossRef]
4. Kalogirou, S.A. Seawater Desalination Using Renewable Energy Sources. *Prog. Energy Combust. Sci.* **2005**, *31*, 242–281. [CrossRef]
5. Taft, H.L. 16—Water Scarcity: Global Challenges for Agriculture. In *Food, Energy, and Water*; Ahuja, S., Ed.; Elsevier: Boston, MA, USA, 2015; pp. 395–429. ISBN 978-0-12-800211-7.
6. Hadadin, N. Dams in Jordan Current and Future Perspective. *Can. J. Pure Appl. Sci.* **2015**, *9*, 3279–3290.
7. MWI. *Jordan Water Sector Facts and Figures 2017*; Ministry of Water and Irrigation: Amman, Jordan, 2018.
8. Al-Kharabsheh, A. Challenges to Sustainable Water Management in Jordan. *Jordan J. Earth Environ. Sci.* **2020**, *11*, 38–48.
9. MoEnv. *National Strategy and Action Plan to Combat Desertification 2015–2020*; Ministry of Environment: Amman, Jordan, 2015.
10. Al-Kharabsheh, A. *Effect of Arab Spring on Water Crises Management in Jordan (Arabic)*; UNESCO: Amman, Jordan, 2013.
11. DoM Department of Meteorology. *Monthly Meteorological Reports*; DoM Department of Meteorology: Amman, Jordan, 2022.
12. MWI. *Water Year Book: Hydrological Year 2018–2019 (Arabic)*; Ministry of Water and Irrigation: Amman, Jordan, 2021.
13. MoEnv. *Jordan's Third National Communication on Climate Change*; Submitted to The United Nations Framework Convention on Climate Change (UNFCCC), Funded by GEF and UNDP; Ministry of Environment: Amman, Jordan, 2014.
14. MWI. *Jordan Water Sector Facts and Figures 2020*; Ministry of Water and Irrigation: Amman, Jordan, 2021.
15. Hamed, T.A.; Bressler, L. Energy Security in Israel and Jordan: The Role of Renewable Energy Sources. *Renew. Energy* **2019**, *135*, 378–389. [CrossRef]
16. Talozi, S.; Al Sakaji, Y.; Altz-Stamm, A. Towards a Water–Energy–Food Nexus Policy: Realizing the Blue and Green Virtual Water of Agriculture in Jordan. *Int. J. Water Resour. Dev.* **2015**, *31*, 461–482. [CrossRef]
17. MWI. *Water Year Book: Hydrological Year 2016–2017*; Ministry of Water and Irrigation: Amman, Jordan, 2018.
18. Al Omari, H. Water Management in Jordan and Its Impact on Water Scarcity. Ph.D. Thesis, Université d'Ottawa/University of Ottawa, Ottawa, ON, Canada, 2020.
19. Amery, H.A.; Wolf, A.T. 1. Water, Geography, and Peace in the Middle East. In *Water in the Middle East*; University of Texas Press: Austin, TX, USA, 2022; pp. 1–18.
20. Salameh, M.T.B.; Alraggad, M.; Harahsheh, S.T. The Water Crisis and the Conflict in the Middle East. *Sustain. Water Resour. Manag.* **2021**, *7*, 69. [CrossRef]
21. Al-Omari, Z.; Alomari, K.; Aljawarneh, N. The Role of Empowerment in Improving Internal Process, Customer Satisfaction, Learning and Growth. *Manag. Sci. Lett.* **2020**, *10*, 841–848. [CrossRef]
22. MWI. *National Water Strategy 2016–2025: Water Demand Management Policy*; Ministry of Water and Irrigation: Amman, Jordan, 2016.
23. Al Naber, M.; Molle, F. Controlling Groundwater over Abstraction: State Policies vs Local Practices in the Jordan Highlands. *Water Policy* **2017**, *19*, 692–708. [CrossRef]
24. Fanack Water Water Challenges in Jordan. Available online: https://water.fanack.com/jordan/water-challenges-in-jordan (accessed on 7 February 2023).
25. Saidan, M.; Al-Addous, M.; Al-Weshah, D.-R.; Obada, I.; Alkasrawi, M.; Barbana, N. Wastewater Reclamation in Major Jordanian Industries: A Viable Component of a Circular Economy. *Water* **2020**, *12*, 1276. [CrossRef]
26. MoEnv. *Green Growth National Action Plan 2021–2025*; Ministry of Environment: Amman, Jordan, 2020.
27. Hussein, H. Lifting the Veil: Unpacking the Discourse of Water Scarcity in Jordan. *Environ. Sci. Policy* **2018**, *89*, 385–392. [CrossRef]
28. El-Naqa, A.; Al-Shayeb, A. Groundwater Protection and Management Strategy in Jordan. *Water Resour Manag.* **2009**, *23*, 2379–2394. [CrossRef]
29. International Trade Administration Jordan—Environment and Water Sector. Available online: https://www.trade.gov/country-commercial-guides/jordan-environment-and-water-sector (accessed on 11 February 2023).
30. Borgomeo, E.; Fawzi, N.A.-M.; Hall, J.W.; Jägerskog, A.; Nicol, A.; Sadoff, C.W.; Salman, M.; Santos, N.; Talhami, M. Tackling the Trickle: Ensuring Sustainable Water Management in the Arab Region. *Earth's Future* **2020**, *8*, e2020EF001495. [CrossRef]
31. Alfarra, A. Water-Energy-Food Nexus in the Arab Region. In *Water, Sustainable Development and the Nexus*; CRC Press: Boca Raton, FL, USA, 2019; ISBN 978-1-315-15590-6.
32. Abdulkadir, T.S.; Okikiola, F.; Bashir, A.; Adeniyi, A.S. Performance Evaluation of Swat-Based Model for the Prediction of Potential and Actual Evapotranspiration. *Jordan J. Civ. Eng.* **2022**, *16*, 1.
33. Abu-Allaban, M.; El-Naqa, A.; Jaber, M.; Hammouri, N. Water Scarcity Impact of Climate Change in Semi-Arid Regions: A Case Study in Mujib Basin, Jordan. *Arab. J. Geosci.* **2015**, *8*, 951–959. [CrossRef]
34. World Health Organization; United Nations. *Framework Convention on Climate Change*; Climate and Health Country Profile 2015; World Health Organization: Geneva, Switzerland, 2015.

35. Qtaishat, T.H.; Al-Karablieh, E.K.; AlAdaileh, H.; El-Habbab, M.S. Drought Management Policies and Institutional Mandate in Jordan. In *Sustainable Energy-Water-Environment Nexus in Deserts: Proceeding of the First International Conference on Sustainable Energy-Water-Environment Nexus in Desert Climates–QEERI, December 2019. At: Doha, Qatar*; Heggy, E., Bermudez, V., Vermeersch, M., Eds.; Springer International Publishing: Cham, Switzerland, 2022; pp. 757–763.
36. Stanford News Jordan's Worsening Water Crisis a Warning for the World. *Stanford News*, 29 March 2021.
37. ReliefWeb Estimated 1.6 Million People in Jordan to Benefit from New Project to Tackle Jordan's Water Crisis and Build Climate Resilience [EN/AR]. Available online: https://reliefweb.int/report/jordan/estimated-16-million-people-jordan-benefit-new-project-tackle-jordans-water-crisis-and-build-climate-resilience-enar (accessed on 12 December 2022).
38. Al-Houri, Z.; Al-Omari, A. Assessment of Rooftop Rainwater Harvesting in Ajloun, Jordan. *J. Water Reuse Desalination* **2021**, *12*, 22–32. [CrossRef]
39. Breulmann, M.; Khurelbaatar, G.; Sanne, M.; van Afferden, M.; Subah, A.; Müller, R.A. Integrated Wastewater Management for the Protection of Vulnerable Water Resources in the North of Jordan. *Sustainability* **2022**, *14*, 3574. [CrossRef]
40. Walschot, M.; Luis, P.; Liégeois, M. The Challenges of Reverse Osmosis Desalination: Solutions in Jordan. *Water Int.* **2020**, *45*, 112–124. [CrossRef]
41. Al-Mefleh, N.K.; AlAyyash, S.M.; Bani Khaled, F.A. Water Management Problems and Solutions in a Residential Community of Al-Mafraq City, Jordan. *Water Supply* **2019**, *19*, 1371–1380. [CrossRef]
42. Abuhalaweh, N. *Student Independent Projects Environmental Studies 2017: Water Scarcity in Jordan: Sustainability Issues and Information Drought*; Grenfell Campus, Memorial University: St. John's, NL, Canada, 2017.
43. Komendantova, N.; Marashdeh, L.; Ekenberg, L.; Danielson, M.; Dettner, F.; Hilpert, S.; Wingenbach, C.; Hassouneh, K.; Al-Salaymeh, A. Water–Energy Nexus: Addressing Stakeholder Preferences in Jordan. *Sustainability* **2020**, *12*, 6168. [CrossRef]
44. Ramirez, C.; Almulla, Y.; Joyce, B.; Huber-Lee, A.; Nerini, F.F. An Assessment of Strategies for Sustainability Priority Challenges in Jordan Using a Water–Energy–Food Nexus Approach. *Discov. Sustain.* **2022**, *3*, 23. [CrossRef]
45. MWI; BGR (Ministry of Water and Irrigation and, Bundesanstalt Fuer Geowissenschaften Und Rohstoffe). *Groundwater Resource Assessment of Jordan 2017*; Ministry of Water and Irrigation: Amman, Jordan, 2019.
46. Al-Karablieh, E.; Salman, A.; Al-Omari, A.; Wolff, H.; Al-Assa'd, T.; Hunaiti, D.; Subah, A. Estimation of the Economic Value of Irrigation Water in Jordan. *J. Agric. Sci. Technol.* **2012**, 487–497.
47. MWI. *Joint Technical Committee for the Allocation of the Yarmouk River Waters*; Ministry of Water and Irrigation: Amman, Jordan, 2018.
48. USAID. *USAID Telling Our Story: Jordan—Bringing Fresh Water to the People. United States Agency for International Development Retrieved 2009-03-12*; USAID: Washington, DC, USA, 2009.
49. MWI. *Water Budget 2019*; Ministry of Water and Irrigation: Amman, Jordan, 2020.
50. Alshibli, F.; Maher, W.; Thompson, R. The Need for a Quantitative Analysis of Risk and Reliability for Formulation of Water Budget in Jordan. *Jordan J. Earth Environ. Sci.* **2017**, *8*, 77–89.
51. Robins, P. *A History of Jordan*; Cambridge University Press: Cambridge, UK, 2019; ISBN 978-1-108-42791-3.
52. WRMD. Water Resources in Jordan: A Primer. In *Government of Newfoundland and Labrador: Water Resources Management Division, Department of Environment and Conservation, Government of Newfoundland and Labrador*; WRMD: Tampa, FL, USA, 2010.
53. JICA. *The Study on Water Resources Management in the Hashemite Kingdom of Jordan Final Report Volume I Main Report Part-A: Water Resources Management Master Plan*; Japan International Cooperation Agency: Tokyo, Japan, 2001.
54. MWI. *Jordan Water Sector Facts and Figures 2015*; Ministry of Water and Irrigation: Amman, Jordan, 2016.
55. Qadir, M.; Sharma, B.R.; Bruggeman, A.; Choukr-Allah, R.; Karajeh, F. Non-Conventional Water Resources and Opportunities for Water Augmentation to Achieve Food Security in Water Scarce Countries. *Agric. Water Manag.* **2007**, *87*, 2–22. [CrossRef]
56. Boers, T.M.; Zondervan, K.; Ben-Asher, J. Micro-Catchment-Water-Harvesting (MCWH) for Arid Zone Development. *Agric. Water Manag.* **1986**, *12*, 21–39. [CrossRef]
57. Husseini, R. 80,000 Cubic Metres of Water from Dams Lost to Evaporation during Hot Spell. Jordan Times. Available online: https://jordantimes.com/news/local/80000-cubic-metres-water-dams-lost-evaporation-during-hot-spell%E2%80%99 (accessed on 15 December 2022).
58. Al-Karablieh, E.; Salman, A. *Water Resources, Use and Management in Jordan-A Focus on Groundwater. IWMI Project Report CEED. Groundwater Governance in the Arab World–Taking Stock and Addressing the Challenges.* 2016. Available online: https://www.researchgate.net/publication/317779966_Water_Resources_Use_and_Management_in_Jordan-A_focus_on_Groundwater_IWMI_Project_Report_CEED_Groundwater_governance_in_the_Arab_World (accessed on 15 December 2022).
59. MWI. *Jordan Water Utilities Monitoring Report 2019*; Ministry of Water and Irrigation: Amman, Jordan, 2019.
60. Schyns, J.F.; Hamaideh, A.; Hoekstra, A.Y.; Mekonnen, M.M.; Schyns, M. Mitigating the Risk of Extreme Water Scarcity and Dependency: The Case of Jordan. *Water* **2015**, *7*, 5705–5730. [CrossRef]
61. MWI. *Water Budget 2018*; Ministry of Water and Irrigation: Amman, Jordan, 2019.
62. Hussein, H.; Natta, A.; Yehya, A.A.K.; Hamadna, B. Syrian Refugees, Water Scarcity, and Dynamic Policies: How Do the New Refugee Discourses Impact Water Governance Debates in Lebanon and Jordan? *Water* **2020**, *12*, 325. [CrossRef]
63. Abu-Rumman, G.; Khdair, A.I.; Khdair, S.I. Current Status and Future Investment Potential in Renewable Energy in Jordan: An Overview. *Heliyon* **2020**, *6*, e03346. [CrossRef]
64. Idris, I. *Economic Situation in Jordan*; Institute of Development Studies: Brighton, UK, 2016.

65. Hickey, S.M.; Malkawi, S.; Khalil, A. Nuclear Power in the Middle East: Financing and Geopolitics in the State Nuclear Power Programs of Turkey, Egypt, Jordan and the United Arab Emirates. *Energy Res. Soc. Sci.* **2021**, *74*, 101961. [CrossRef]
66. Weathering Risk Climate Risk Profile; Jordan; 2022. Available online: https://www.weatheringrisk.org/en/publication/climate-risk-profile-jordan (accessed on 1 March 2023).
67. Salameh, E.; Al-Alami, H. Jordan's Water Sector—Alarming Issues and Future. *GEP* **2021**, *09*, 100–117. [CrossRef]
68. World Bank. *Water Scarcity in Jordan*; World Bank: Washington, DC, USA, 2019.
69. Al Qatarneh, G.N.; Al Smadi, B.; Al-Zboon, K.; Shatanawi, K.M. Impact of Climate Change on Water Resources in Jordan: A Case Study of Azraq Basin. *Appl. Water Sci.* **2018**, *8*, 50. [CrossRef]
70. Namrouqa, H. Floods "only Beginning" of Severe Climate Change Impacts on Jordan. Available online: https://jordantimes.com/news/local/floods-only-beginning-severe-climate-change-impacts-jordan (accessed on 16 January 2023).
71. MWI. Ministry of Water and Irrigation, Jordan, Open Files. Available online: https://www.mwi.gov.jo/Ar/List/%D8%A7%D9%84%D8%AA%D9%82%D8%A7%D8%B1%D9%8A%D8%B1_%D8%A7%D9%84%D8%B3%D9%86%D9%88%D9%8A%D8%A9 (accessed on 26 January 2023).
72. UN. *World Population Prospects 2019*; UN: New York, NY, USA, 2021.
73. DOS. *Estimated Population of Jordan*; Department of Statistics: Amman, Jordan, 2021.
74. WHO. *Access to Water, Sanitation and Hygiene in the Syrian Refugee Response in Jordan*; World Health Organization: Geneva, Switzerland, 2021.
75. CRISIS & ENVIRONMENT. *Jordan Is on the Edge of a Water Disaster—The Home of Jordanians and Arab Refugees Could Run out of Fresh Water in the Next Few Decades; Amman, Jordan.* 2023. Available online: https://crisisandenvironment.com/jordan-is-on-the-edge-of-a-water-disaster-the-home-of-jordanians-and-arab-refugees-could-run-out-of-fresh-water-in-the-next-few-decades/ (accessed on 16 January 2023).
76. Berulmann, M.; Muller, R.A.; Al-Subeh, A.; van Afferden, M. Influx of Syrian Refugees in Jordan—Effects on the Water Sector. In *UFZ with Support from the Ministry of Water and Irrigation*; Helmhltz Centre for Environmental Research: Ammman, Leipzig, 2021.
77. Hussein, H.; Al-Karablieh, A.; Al-Khatib, I. Causes and Mitigation of Non-Revenue Water in the Water Supply System of Jordan. *Water* **2020**, *12*, 1745.
78. USAID. Non-Revenue Water (NRW) Phase I and II Activity, Jordan. Available online: https://www.usaid.gov/infrastructure/results/jordan-non-revenue-water (accessed on 13 February 2023).
79. International Monetary Fund. Jordan: IMF Executive Board Approves US$1.3 Bn Extended Arrangement Under the EFF. Available online: https://www.imf.org/en/News/Articles/2020/03/25/pr20107-jordan-imf-executive-board-approves-us-1-3-bn-extended-arrangement-under-the-eff (accessed on 9 January 2023).
80. Reuters Jordan's Draft 2022 Budget Forecasts $15 Bln in State Spending. Available online: https://www.reuters.com/markets/rates-bonds/jordans-draft-2022-budget-forecasts-15-bln-state-spending-2021-11-28/ (accessed on 28 March 2023).
81. Vajpeyi, D.K. *Water Resource Conflicts and International Security: A Global Perspective*; Lexington Books: Lanham, MD, USA, 2011; ISBN 978-0-7391-7084-7.
82. Katz, D. Basin Management under Conditions of Scarcity: The Transformation of the Jordan River Basin from Regional Water Supplier to Regional Water Importer. *Water* **2022**, *14*, 1605. [CrossRef]
83. Hussein, H. Yarmouk, Jordan, and Disi Basins: Examining the Impact of the Discourse of Water Scarcity in Jordan on Transboundary Water Governance. *Mediterr. Politics* **2019**, *24*, 269–289. [CrossRef]
84. Hussein, H.; Grandi, M. Dynamic Political Contexts and Power Asymmetries: The Cases of the Blue Nile and the Yarmouk Rivers. *Int. Environ. Agreem.* **2017**, *17*, 795–814. [CrossRef]
85. Avisse, N.; Tilmant, A.; Rosenberg, D.; Talozi, S. Quantitative Assessment of Contested Water Uses and Management in the Conflict-Torn Yarmouk River Basin. *J. Water Resour. Plan. Manag.* **2020**, *146*, 05020010. [CrossRef]
86. MWI. *Drip Irrigation Systems in Jordan*; Ministry of Water and Irrigation: Amman, Jordan, 2019.
87. World Bank. *Jordan: Agriculture Sector Review*; World Bank: Washington, DC, USA, 2018.
88. World Bank. *World Bank, National Accounts*; World Bank: Washington, DC, USA, 2020.
89. FAO. *The Future of Food and Agriculture: Trends and Challenges*; FAO: Rome, Italy, 2017.
90. Salameh, E. Water Security Needs What Ought to Be Done to Increase the Future Water Security as a Fundamental Base for Social, Economic and Political Stability—The Case of Jordan. *J. Geosci. Environ. Prot.* **2022**, *10*, 1–17. [CrossRef]
91. Kiwan, S.; Amer, M.A. Renewable Energy Scenarios for Water Desalination and Conveyance: Case of Red Sea-Dead Sea Project. *Renewable Energy Focus* **2023**, *46*, 281–302. [CrossRef]
92. UNICEF Jordan. *Economist Impact Tapped out: The Costs of Water Stress in Jordan*; UNICEF Jordan: Amman, Jordan, 2022.
93. MWI. *Wastewater Treatment and Reuse in Jordan*; Ministry of Water and Irrigation: Amman, Jordan, 2020.
94. Salem, H.S.; Yihdego, Y.; Muhammed, H.H. The Status of Freshwater and Reused Treated Wastewater for Agricultural Irrigation in the Occupied Palestinian Territories. *J. Water Health* **2020**, *19*, 120–158. [CrossRef]
95. Jordan Phosphate Mines Company Wastewater Treatment and Reuse in the Industrial Sector. 2018. Available online: https://www.jpmc.com.jo/En/Pages/Environment (accessed on 22 December 2022).
96. Jaramillo, M.F.; Restrepo, I. Wastewater Reuse in Agriculture: A Review about Its Limitations and Benefits. *Sustainability* **2017**, *9*, 1734. [CrossRef]

97. MWI. *Rainwater Harvesting in Jordan: A Review of Current Trends and Future Opportunities*; Ministry of Water and Irrigation: Amman, Jordan, 2020.

98. Shah, T.; Roy, A.D.; Qureshi, A.S.; Wang, J. Sustaining Asia's Groundwater Boom: An Overview of Issues and Evidence. *Nat. Resour. Forum* **2003**, *27*, 130–141. [CrossRef]

99. JWP. *Virtual Water in Jordan: An Overview*; JWP: Paris, France, 2020.

100. Al-Weshah, R.; Al-Hussaini, S. Virtual Water Trade as a Means of Addressing Water Scarcity in Jordan. *Water* **2018**, *10*, 1083.

101. World Bank. *Water Scarcity in Jordan: Challenges and Opportunities*; World Bank: Washington, DC, USA, 2020.

102. The Jordan Times. *University of Jordan Develops Precision Irrigation System for Small Farmers*; The Jordan Times: Amman, Jordan, 2018.

103. ICARDA. New Drought-Tolerant Crops for Resilient Dryland Livelihoods. 2023. Available online: https://www.icarda.org/media/blog/new-drought-tolerant-crops-resilient-dryland-livelihoods (accessed on 28 March 2023).

104. Jordan News Agency. *Ministry of Agriculture Launches Program to Improve Water-Use Efficiency in Agriculture*; Jordan News Agency: Amman, Jordan, 2020.

105. Sharafan, R.; Kim, S. Agriculture in Jordan: Current State and Future Prospects. *J. Agric. Sci. Technol.* **2012**, *14*, 751–765.

106. Taima, A. Agricultural Water Management in Jordan: Challenges and Opportunities. *J. Water Clim. Chang.* **2015**, *6*, 493–501.

107. MWI. *National Water Strategy*; Ministry of Water and Irrigation: Amman, Jordan, 2018.

108. Nourredine, H.; Barjenbruch, M.; Million, A.; El Amrani, B.; Chakri, N.; Amraoui, F. Linking Urban Water Management, Wastewater Recycling, and Environmental Education: A Case Study on Engaging Youth in Sustainable Water Resource Management in a Public School in Casablanca City, Morocco. *Educ. Sci.* **2023**, *13*, 824. [CrossRef]

109. Engelmann, J.; Al-Saidi, M.; Hamhaber, J. Concretizing Green Growth and Sustainable Business Models in the Water Sector of Jordan. *Resources* **2019**, *8*, 92. [CrossRef]

110. Jordan Water Report. Available online: https://water.fanack.com/jordan/ (accessed on 1 October 2023).

111. Al-Bakri, J.T.; D'Urso, G.; Calera, A.; Abdalhaq, E.; Altarawneh, M.; Margane, A. Remote Sensing for Agricultural Water Management in Jordan. *Remote Sens.* **2023**, *15*, 235. [CrossRef]

112. Aboelnga, H.; Saidan, M.; Al-Weshah, R.; Sturm, M.; Ribbe, L.; Frechen, F.-B. Component Analysis for Optimal Leakage Management in Madaba, Jordan. *J. Water Supply Res. Technol.-Aqua* **2018**, *67*, 384–396. [CrossRef]

113. Al-Bakri, J.T.; D'Urso, G.; Batchelor, C.; Abukhalaf, M.; Alobeiaat, A.; Al-Khreisat, A.; Vallee, D. Remote Sensing-Based Agricultural Water Accounting for the North Jordan Valley. *Water* **2022**, *14*, 1198. [CrossRef]

114. Iseri, Z. *The Impact of Water Politics on Migration in Jordan*; Whittier Scholars Program; Whittier College: St. Whittier, CA, USA, 2023.

115. Karn, R.; Dincer, I.; Rosen, M. Water for Energy and Food Security: A Review of the Red Sea-Dead Sea Conveyance Project. *Renew. Sustain. Energy Rev.* **2019**, *111*, 757–766.

116. JWC. *Joint Water Commission*; Annual Report; JWC: Hillsboro, OR, USA, 2019.

117. League of Arab States. *Joint Technical Committee for Water (JTCW)*; League of Arab States: Cairo, Egypt, 2020.

118. Mahmoud, M. Exploring the Feasibility of the Jordan-Israel Energy and Water Deal. Available online: https://www.mei.edu/publications/exploringfeasibility-jordan-israel-energy-and-water-deal (accessed on 13 June 2023).

Review

Integrated River Basin Management for Sustainable Development: Time for Stronger Action

Minhaz Farid Ahmed [1,*], Mazlin Bin Mokhtar [1,2,*], Chen Kim Lim [1], Izzati Afiqah Binti Che Suza [3], Ku Adriani Ku Ayob [1], Rd. Puteri Khairani Khirotdin [1] and Nuriah Abd Majid [1]

[1] Institute for Environment and Development (LESTARI), Universiti Kebangsaan Malaysia (UKM), Bangi 43600, Selangor, Malaysia; kim@ukm.edu.my (C.K.L.); adriani@ukm.edu.my (K.A.K.A.); rdputeri@ukm.edu.my (R.P.K.K.); nuriah@ukm.edu.my (N.A.M.)
[2] United Nations Sustainable Development Solutions Network (UN SDSN) Asia Headquarters, Sunway University, 5 Jalan Universiti, Petaling Jaya 47500, Selangor, Malaysia
[3] Izzati Suza Enterprise, Ayer Keroh 75450, Melaka, Malaysia; izzatichesuza@gmail.com
* Correspondence: minhaz@ukm.edu.my (M.F.A.); mazlinmokhtar@yahoo.com (M.B.M.); Tel.: +603-8921-3894 (M.F.A.); +603-7491-8622 (M.B.M.)

Abstract: Malaysia has numerous policies, institutions, and experts with foresight and vision for its development. Nevertheless, river basin management has been lacking due to several factors such as insufficient proactive leadership roles of institutions, as well as locally authorized bodies. Reviewing of stakeholders' role in the PENTA-HELIX partnership model (i.e., government, business, academia, community, and NGO) reveals that individuals and institutions with proactive and effective leadership roles along with top-down and bottom-up approaches can create a more strategic policy implementation resulting in better outcomes in river basin management. Local authorities with proactive leadership roles should be encouraged to use a creative and innovative key performance indicator system accompanied by mentoring and training, as well as education, to inspire a passive to active attitude change. A local authority with sound leadership roles can develop proper partnerships with its many stakeholders to improve awareness with more multitasking activities. These can be achieved by motivating all the related stakeholders towards more commitment to creating a sustainable environment. Identifying and recognizing local authorities to manage the rivers will result in more powerful actions in river management. It is essential to ensure quality control and quality assurance at various levels to bring sustainability science at the multi-stakeholders' platforms towards an integrated river basin management to achieve a better living quality for everyone.

Keywords: leadership roles; local authority; river basin; sustainable development

Citation: Farid Ahmed, M.; Mokhtar, M.B.; Lim, C.K.; Suza, I.A.B.C.; Ayob, K.A.K.; Khirotdin, R.P.K.; Majid, N.A. Integrated River Basin Management for Sustainable Development: Time for Stronger Action. *Water* **2023**, *15*, 2497. https://doi.org/10.3390/w15132497

Academic Editors: Athanasios Loukas, Peiyue Li, Steve W. Lyon and Jianhua Wu

Received: 9 May 2023
Revised: 17 June 2023
Accepted: 4 July 2023
Published: 7 July 2023

1. Introduction

Rivers are the main source of drinking water supply in Malaysia. Subsequently, rivers offer approximately 95% to 98% raw water resource primarily utilized for drinking and irrigation systems [1]. Accordingly, the government has developed the National Integrated Water Resources Management Plan, Strategies, and Road Map (Volume 1 and 2) through the Academy of Science Malaysia (ASM) for the administration of the nation's 189 major river basins, which also includes the Langat River Basin in Peninsular Malaysia. Integrated Langat River Basin Management is significant in light of the fact that the Langat River is among the essential primary water sources for drinking for about 33% of the population of the Selangor state, Malaysia [2]. Even though an integrated river basin management (IRBM) could potentially result in safe water supply for the household, the closure of the water treatment plants (WTPs) in the Langat river basin is the unfortunate outcome of contaminations particularly in the form of chemical pollution from both the point and non-point sources of the river [2,3]. Malaysia has sufficient policies, expertise, and institutions alongside the Foresight Institute to deal with water resources. Nevertheless, the fundamental issue related to river contamination is the deficiency in implementing these policies. In Malaysia, for example, managing rivers is the responsibility of the state government even though the federal agency, namely the Department of Irrigation and Drainage (DID), leads the management of the river basin. Surprisingly, the DID is not supported by the constitution of Malaysia in dealing with the river basins particularly in managing pollution [4]. DID was set up as a specialized technical agency to help the federal government for the purpose of flood control and irrigation [5]. The transboundary Langat River Basin shares its borders with the Selangor and Negeri Sembilan states as well as the Federal Territories of Kuala Lumpur and Putrajaya. Due the fact that the mandate given to the DID is inadequate in terms of managing the Langat River, the Selangor state government has established the "Lembaga Urus Air Kuala Lumpur" or the "Selangor Water Management Authority" (LUAS) in 1999 to manage the water bodies within the state including the Langat River. However, difficulties remain in the implementation of those policies for river management, especially pollution management along with the collaboration and cooperation among the federal and state government agencies, particularly due to the absence of agencies like LUAS in the Federal Territories and Negeri Sembilan State.

Therefore, the proactive leadership roles of the seven local authorities (i.e., four in Selangor state, and one in Negeri Sembilan state and in the Federal Territories of Kuala Lumpur and Putrajaya each) in the Langat river basin are crucial due to their mandate in policy implementation using the Local Government Act 1976 [6]. Observing the frequent chemical pollution at the river, it can be assumed that the capacity and capability of the local government officials for river management can be enhanced via multi-disciplinary training from the disaster risk reduction perspective. The customized training can also enhance their real-time decision-making process using technologies such as artificial intelligence (AI) [7], GIS-based risk map [8], etc. especially for the early warning systems at the river basin level [9,10]. Capacity building of local government on hydrological monitoring and early warning system (EWS) can contribute to sustainable and fair water management against water-related disaster risks such as floods and drought. Moreover, the use of AI and GIS-based river basin risk map by local authorities can enhance their holistic and integrated approach to managing Water–Energy–Food (WEF) nexus challenges to meet sustainable river basin management. Watersheds in natural ecosystems, especially forests and wetlands, produce cleaner, purer water than those from agricultural or industrial areas. Some municipalities pay to support the management of protected areas because they provide a cost-effective water supply; others remain virtually unaware that their water comes from a protected area [11]. Such benefits contribute directly to SDG 6, Clean Water and Sanitation, which aims to achieve "universal access to safe and affordable drinking water" and "protect and restore water-related ecosystems". While the Sustainable Development Goals (SDGs) are broadly framed with 17 goals, the goals and their targets inherently connect with each other forming a complex system. Actions supporting one goal may influence progress in other goals, either positively (synergies) or negatively (trade-offs). Therefore, the effective management of the synergies and trade-offs is a prerequisite for ensuring policy coherence, particularly at the river basin level [12].

In September–October 2016, the Sungai Langat and Cheras Mile 11 water treatment plants (WTPs) encountered a few shutdowns on the grounds that the conventional coagulation of the water treatment technology was not able to treat the chemically polluted river water alongside high turbidity [3,13]. There were also several shutdown incidents of WTPs such as Sungai Semenyih WTP, Bukit Tempoi WTP, and as such at the Langat River Basin during the last three years [14–17]. Thus, the government through SPAN (Suruhanjaya Perkhidmatan Air Negara/National Water Services Commission) produced the standard operating procedure (SOP) in managing WTP in relation to odor contamination in raw water to avoid plant shutdowns [18]. Needless to say, without satisfactory safe water, it would be difficult to accomplish many of the universally concurred objectives, for example, the Sustainable Development Goals (SDGs), particularly SDG target 6.1 of having clean water for drinking. Consequently, the capabilities of major agencies in dealing with the Langat basin may be improved through the proactive positions of leadership to connect to all the stakeholders through the methodologies of integrated river basin management (IRBM) in accordance with the integrated water resource management (IWRM), the integrated lake basin management (ILBM), and the integrated coastal zone management (ICZM) [19]. Additionally, the global network of river basin management contended that the engagement of a river basin organization (RBO) relies upon the mandate (geographic inclusion and assignments), authority (formal and informal), and capability (financing and resources) of the state [20]. In any case, the achievement of RBO depends on the interconnectivity among the performance of the RBOs, relationship of the stakeholders, leadership roles, and political cooperation [21]. Accordingly, NGOs could assume an essential function in raising awareness among all the important stakeholders in the management of the river basin. NGOs can carry out environmental awareness projects using courses, campaigns, etc. for the public. However, inadequate coordination has been reported among the stakeholders of the Langat River Basin Management [22].

Subsequently, the execution of water policies at the local level should be guaranteed to lower river contamination and the local government is the most suitable authority to actualize these policies since they have been given the mandate with the Local Government Act 1976 [23] in Malaysia. For instance, the National Agenda on Water Sector Transformation (WST2040) is a Malaysian government initiative to accelerate the implementation of IWRM via the 12th Malaysia Five Year Plan (2021–2025) [24]. Moreover, WST 2040 is an ambitious project of government over the period from 2020 to 2040 in order to contribute to national GDP and to achieve this, WST2040 has highly emphasized the proactive leadership roles of multi-stakeholders, especially of the local government in the developing nation. Local authority with proactive leadership roles will be able to execute these policies and collaborate better with the related agencies and stakeholders. The local authorities are in a position of influencing these approaches from top-down and bottom-up ways to deal with multi-stakeholders' structure to perceive the closest organizations for innovative solutions of the water and wastewater treatment plants and as such. Therefore, this study explored the leadership roles of the local authority in attempting to diminish river contamination to contribute accomplishing the Integrated Langat River Basin Management.

2. Methods

In every form of management, there must be challenges that make policy implementation difficult in management initiatives. River basin management by the local government in Malaysia is also not exempt from the dilemma when various obstacles and challenges often make it difficult for them to implement river basin management activities [25]. This study, based on the thorough analysis of full text, identified a total of eight research articles that fully met the search criteria. Based on the Web of Science (WOS) and SCOPUS databases, articles with the following characteristics were extracted: (i). published between 1970 and 2023, (ii). terms contained in the title, abstract, and keywords were 'river basin management', and 'local government', and (iii). focused on Malaysia. Table 1 summarizes the methodological approaches in river basin management in Malaysia.

Table 1. Various methods of integrated river basin management in Malaysia.

Title (Year)	Objective	Method	Results
Sustainable management of rivers in Malaysia: Involving all stakeholders (2005) [26]	This paper discusses how all stakeholders can contribute by working together in smart-partnerships with government towards effective and sustainable management of rivers in Malaysia.	Literature review to explore sustainable river management in Malaysia.	All stakeholders need to start taking proactive actions, even sacrifices, to manage, protect, conserve, and restore our rivers so that their resources can be sustained for future use.

Table 1. *Cont.*

Title (Year)	Objective	Method	Results
Steady data flow tracks floods (2008) [27]	Aims to understand the flood events, and provided a vital foundation for planning for upcoming floods.	Literature review to explore flood management initiatives in Kuala Lumpur Malaysia.	Data from the SMART project are also helping scientists and officials better understand the local river system.
Perspectives and initiatives on integrated river basin management in Malaysia: A review (2011) [28]	This study attempts to focus on the current situation of water issues in Malaysia in particular on perspectives and initiatives pertaining to Integrated River Basin Management (IRBM).	Literature review was conducted to explore the current status integrated river basin management in Malaysia.	The complex process of decision making for sustainable management of water and river basins needs an integrated and holistic approach involving many stakeholders and disciplines.
Institutional challenges for integrated river basin management in Langat River Basin, Malaysia (2011) [29]	This paper reports on a study of the institutional challenges and factors affecting policy processes and outcomes of integrated river basin management (IRBM) in the Langat River Basin (LRB), Malaysia.	A case study approach using institutional analysis and development (IAD) framework was used, and field observations and interviews with local stakeholders of LRB.	Polycentric institutional arrangements under the Federal administration are likely capable of coordinating and integrating river basin management by extending the scope of an iterative learning through participation of individual stakeholder at the lowest appropriate level.
Resolving Water Disputes via Interstate Co-Operation and Stakeholders' Engagement: A Case Study from Muda River Basin (2017) [30]	This paper highlights the current environmental and political challenges related to water resources in Kedah and Penang.	Qualitative methods are applied to find out solutions to resolve the water disputes and maximise benefit-sharing of water use using a model developed through stakeholders' engagement.	This case study can serve as an important foundation for accessing the negotiations between Kedah and Penang and fostering interactive interstate water co-operation not just in Muda River Basin but other shared watercourses.
Applying a system thinking approach to explore root causes of river pollution: A preliminary study of Pinang river in Penang State, Malaysia (2018) [31]	This study applied a system thinking approach to investigate root causes of pollution in the Pinang River.	Qualitative methods have been applied to explore the system thinking in Pinang River Management.	A causal loop diagram was produced that illustrates the relationship between the local community and the government with regards to improving the water quality of the Pinang River.
Integrating Structural and Non-structural Flood Management Measures for Greater Effectiveness in Flood Loss Reduction in the Kelantan River Basin, Malaysia (2020) [32]	This review explored the flood management in the Kelantan River Basin, Malaysia.	Reviewing government-centric top-down approach focused on flood-control technologies via structural measures.	A combination of structural and non-structural measures is the way forward for Kelantan State as it ensures that government structural measures are effectively supported by public-engaged non-structural measures.
Identification of Water Pollution Sources for Better Langat River Basin Management in Malaysia (2022) [33]	This study explored the pollution sources in the Langat River to suggest an integrated river basin management (IRBM).	Quantitative methods to find the chemical pollutions in Langat River.	The implementation of policies should be effective at the local level for pollution management, especially via the proactive leadership roles of local government for this transboundary Langat River to benefit from IRBM.

Based on the literature review, it is noted that collaboration and cooperation among the stakeholders, i.e., government, non-government, business, academia, and NGO/community sectors, are crucial for successful river basin management towards sustainable development. Therefore, this study, based on the literature review, maps the stakeholders for the Langat river basin management following the modified Penta-helix multi-stakeholder partnership on social innovation framework by Calzada [34]. The Penta-helix multi-stakeholder framework has the advantage of determining the roles and responsibilities of each stakeholder over the triple and quadruple-helix framework model to minimize their overlapping roles [26]. Forss [35] also reported that Penta-helix collaboration works well at the local level in a governance-related model for Penta-helix cooperation when there are proactive leadership roles of each stakeholder especially via citizen-driven processes. Similarly, the Penta-Helix partnership model is useful for natural resources management especially water resources management [36,37] because of its ability to bring sustainable innovations in management initiatives. The literature review also found the challenges faced by the local government in Malaysia for their proactive leadership roles towards river basin management. Man-made disasters and impacts of climate change pose uncertain threats to river basin management when cooperation and collaboration are inadequate among the stakeholders. Therefore, the effective leadership roles of institutions as well as individuals along with their commitment especially by local government are the keys to better coordination of Penta-helix stakeholders for river basin management.

Moreover, information related to the operation of water treatment plants (WTPs) along with effective technological solutions for safe drinking water supply at the household level were collected from the water treatment plant (WTP) authorities i.e., Puncak Niaga Sdn. Bhd., Sungai Semenyih WTPs Authority and Loji Rawatan Air Sg. Labu in the Langat River Basin. Accordingly, the coordinates of the nine drinking water treatment plants (WTPs) were recorded using the GARMIN (GPS, GARMIN, GPSMAP 76CSx, Kansas, MO, USA) machine to produce the map of the WTPs in the Langat Basin using the GIS software.

3. Significance of the Langat River Basin, Malaysia

Approximately 2986 river basins can be found in Malaysia, but only 189 of them are viewed as significant basins dependent on the region of the basin that is >80 km^2 [38]. IRBM has many cross-cutting concerns, thus accomplishing a sound IRBM will connect practically all the Sustainable Development Goals (SDGs) [39]. Langat is the UNESCO HELP (Hydrology for the Environment, Life, and Policy) River Basin in Malaysia [40], and distinct in its attributes because of its flow through three distinct constituencies. Langat River Basin is among the HELP basins from the 91 river basins globally, 26 river basins in the Asia Pacific region, and 3 from the south-east Asia region [41].

3.1. Jurisdiction of Langat River Basin

The Langat river basin ranges around 1815 km^2, and the river's major course is 141 km, located about 40 km east of Kuala Lumpur. It is geographically located from 02°40′152″ N latitude to 3°16′15″ N and 101°19′20″ E to 102°1′10″ E longitude with the most elevated peak at 820.8 m (2691 ft) [42]. Strikingly, about 75% of the catchment area is located on an uneven landscape with a normal slant of 6–9″ and another 25% of the region is under 6″ with a few swamps by the river [41]. The significant tributaries of the Langat River are the Semenyih, Beranang, and Labu rivers; nevertheless, around 40 smaller tributaries are flowing into Langat [43]. A few development projects are being carried out in the transboundary Langat River Basin, which shares the Selangor State (78.14%), Negeri Sembilan State (19.64%), and the Federal Territories of Kuala Lumpur (0.33%) and Putrajaya (1.90%) (Table 2).

Table 2. Percentage area of Langat River Basin among the district councils and states.

State	City/District Council	[1] Area of Langat Basin (%)	[2] Population	[2] Household
Selangor	Majlis Perbandaran Klang	35.39	842,146	201,994
	Majlis Perbandaran Kuala Langat	26.27	220,214	49,798
	Majlis Daerah Hulu Langat	3.39	1,138,198	288,508
	Majlis Perbandaran Sepang	13.09	207,354	49,005
Negeri Sembilan	Majlis Perbandaran Nilai	19.64	200,988	48,430
Federal Territory of Putrajaya	Perbandaran Putrajaya	1.90	68,361	19,511
Federal Territory of Kuala Lumpur	Dewan Bandaraya Kuala Lumpur	0.33	1,588,750	419,187
	Total	100	4,266,011	1,076,433

Note(s): Source: [1] [44]; [2] [45].

The Langat River Basin holds a population of 1,184,917 in the year 2000 with a rate of growth of 7.64% [41]. In 2013, the population grew to 4,266,011 based on the report by the Department of Statistics Malaysia [45]. Langat River is among the basin's primary drinking water sources, it provides drinking water to an incredibly significant number of people in Selangor state including VVIPs residing in Putrajaya, the central federal government city of Malaysia, which also holds the Prime Ministers' Office. Langat River Basin is among the quickest-growing regions in the nation, where a few huge socio-economic development projects have taken place or are in the process of completion [41]. Additionally, the raw water of Langat is utilized for industrial, horticultural, and transportation activities.

3.2. Changes in Land Use Pattern in Langat River Basin

Incidentally, the land use trend has had a major transformation in the Langat River Basin because of rapid development. The development areas around the basin expanded by 23.5% in 2013 contrasted with 2.4% in 1974 (Table 3). Thus, both the farming and forestry areas have reduced in 2013 in comparison to 1974, while the peat swamp and mangrove regions have radically diminished to 9.4% in 2013 compared to 25.7% in 1974. As such, the accessibility to water in the Langat area has been hampered including the loss of biodiversity. Moreover, the fast growth has additionally expanded the chemical and biological contamination of the river which requires treatment prior to drinking.

Table 3. Land use changes of the Langat River Basin, Malaysia during 1974 to 2013.

Land Use Type	1974 (ha)	%	1991 (ha)	%	2001 (ha)	%	2013 (ha)	%
Forest	52,579.7	17.9	50,906.4	17.3	45,071.9	15.4	48,285	16.5
Mangroves and Peat swamp	75,252.6	25.7	37,014.5	12.6	25,630.7	8.7	27,560.8	9.4
Agriculture	155,249	52.9	170,705	58.2	164,841	56.2	142,387.9	48.5
Developed Area	7022.8	2.4	28,510.7	9.7	51,502.8	17.5	69,056.1	23.5
Water body	3267.3	1.1	6401.5	2.2	6207.1	2.1	6009.1	2
Total	293,370.3	100	293,340.5	100	293,253.6	100	293,298.9	100

Note(s): Source: [41].

3.3. Sources of Pollution at Langat River Basin

LUAS [44] has pinpointed the areas of point sources in the Langat River Basin, Malaysia (Figure 1). There are a few industrial areas around the basin in the Nilai industrial area in the state of Negeri Sembilan. Subsequently, the release of illegal effluents in the area is a genuine contamination threat to the river based on the inadequate execution of the Environmental Quality Act 1974 and its revision with the Environmental Quality (Industrial Effluent) Regulations 2009 [2].

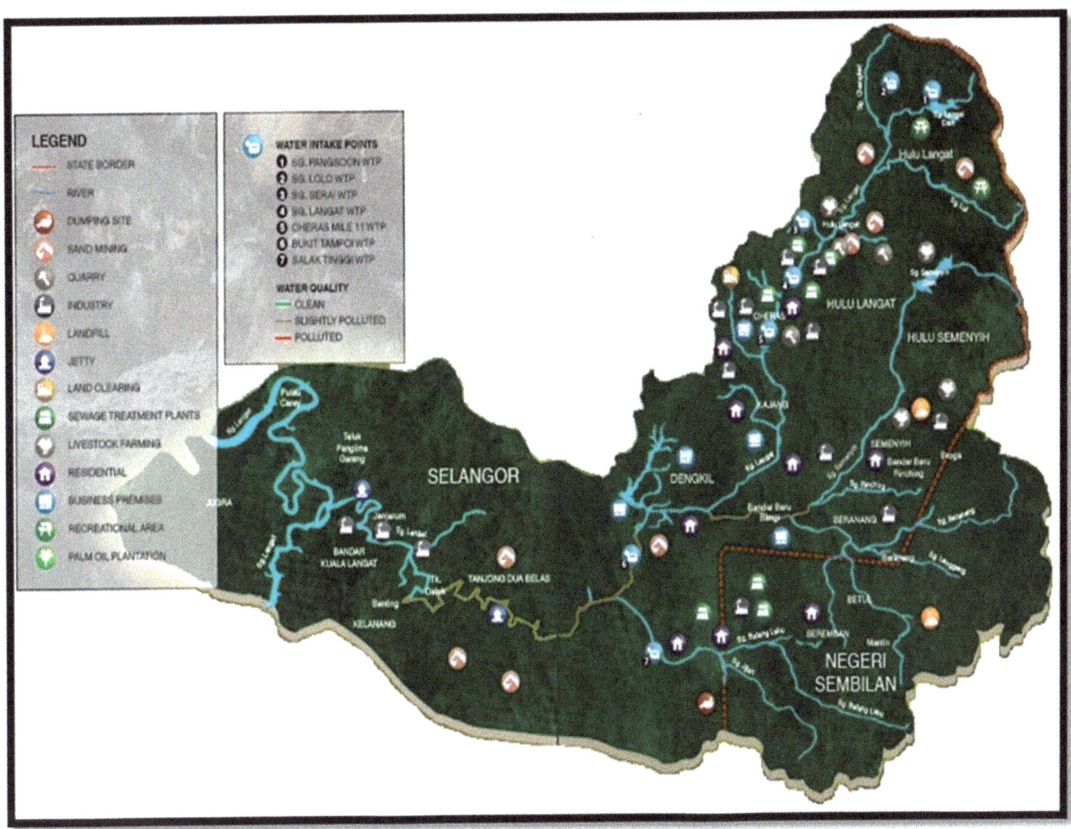

Figure 1. Location of point sources of pollution in the Langat River Basin, Malaysia [44].

The DOE in 2013 detailed that the food industry at 79% is the principal polluter of the Langat River with sewage releases contributing a further 10.8%. Even though industrial waste release decreased to 9.09% in the Langat River in 2013 contrasted with 84.09% in 2002 [42], it still remains a crucial source of chemical contamination in the river in addition to the contamination from the food industry sector. In spite of the fact that sand mining/quarry is liable for just 0.24% contamination in the Langat River [46], there are 86 sand and gravel extraction locations in the Langat Basin out of 198 extraction locations throughout the Selangor state (81 locations) and Negeri Sembilan state (5 locations). Additionally, 43 earth material extraction locations, 21 granite quarries, 2 clay pits, as well as 1 kaolin pit operational site were found in the Langat River Basin [47]. In the meantime, Aris et al. [48] revealed that the mining operations fundamentally increased the predominance of Cu, Sn, Fe, Au, etc. in the surface water of Langat River.

3.4. Impact of Climate Change in Langat River Basin

The contamination of Malaysian rivers based on climatic and anthropogenic issues is a major concern. Numerous studies have already detailed flood occurrence due to heavy rainfall in Malaysia [49–51] and the situation is worsened due to landslides into the riverbank, lack of proper drainage systems, and elevated spring tides [50,52]. Thus, flood is now the most crucial natural disaster as a result of its recurrence and degree in the 189 river basins in Peninsular Malaysia, Sabah, and Sarawak [51]. Flooding has a serious effect on 9% of land in the nation (29,720 km^2) and on 21% of the population (4.915 million) that costs the country about 915 million MYR annually with the extra financial consequential cost of 1.83 billion MYR [51]. Moreover, the frequency and intensity of the recurrence of floods have expanded fundamentally in Malaysia in light of climate change rainfall trends [53].

The quantity of flood occurrences in the Langat River basin has grown from 2006 to 2016. The highest number of floods was recorded at 85 incidents at the river basin in 2015; it was only 36 flood occurrences in 2005 [54]. In addition, 20 dangerous hill slopes have been noted in Selangor, which is in danger of landslides if nothing is carried out to maintain the slopes properly [55]. The Selangor state and Federal Territories Minerals and Geoscience Department reported approximately 1000 hill slopes with potential risk in the Klang Valley [55]. Essentially, the high tides from 21 September to 5 December 2017 at the Langat River Basin were also cause for concern which could result in severe flooding. Residential areas in Selangor, for example, Klang, Kuala Langat, Sepang, Kuala Selangor, and Sabak Bernam were expected to be in danger of flooding with tides as high as 5.5 m to 5.6 m [56]. Additionally, a few landslides were reported from 1999 to 2011 in the Langat River Basin due to the sliding/streaming of soil debris from the hills during overwhelming rainfalls, deforestation, the collapse of river banks, and retrogressive slope downfalls [2], further contributing to the contamination of the Langat River.

4. Shutdown Incidents of Water Treatment Plants (WTPs) at Langat River Basin

The Langat River Basin holds nine drinking water treatment plants (WTPs) (Figure 2) and these WTPs treat the raw water of the Langat River and its tributaries to supply drinking from the basin [57,58].

Figure 2. Water treatment plants at Langat River Basin, Malaysia.

Even though these WTPs supply treated water to Selangor, Kuala Lumpur, and Putrajaya, they were required to shut down a few times during 2006–2019 primarily due to chemical contamination found in the raw water, flood episodes, etc. For example, Sungai Semenyih WTP was closed down multiple times in 2016 as a result of odor contamination from the Nilai and Semenyih industrial zones (Table 4). In addition, floods caused a few shutdown episodes at the Sungai Langat WTP because of higher turbidity in 2012, and the basin, particularly in the greater Kuala Lumpur region, endured the most with its consumable water supply. The WTPs are not able to treat raw water when there is expanded mudflow/turbidity in the river because of floods as there will be a lot of spillover from the substantial downpour [2]. Conversely, the WTPs are also not able to treat raw water when there is a higher concentration of chemicals and circumstances such as drought, anthropogenic operations, etc. In a drought, when there is less water flow, the chemical concentration in the river increases significantly [59].

Table 4. Some shutdowns of water treatment plants (WTP) in Langat River Basin, Malaysia.

Year	WTP	Type of Pollution	Source of Pollution	Affected Area
2019 [9]	Sg. Semenyih	Odor pollution	Private sewage treatment facility in Bandar Mahkota (4th times)	Petaling district, Hulu Langat, Kuala Langat, and Sepang
2016 [1]	Sg. Langat	Odor pollution	Industrial effluent in Semantan river, Pahang mixed in Serai river.	Kuala Lumpur, Petaling Jaya, and Hulu Langat
2016 [1]	Cheras Mile 11	Odor pollution	Industrial effluent in Semantan river, Pahang mixed in Serai river.	Kuala Lumpur, Petaling Jaya, and Hulu Langat
2016 [2,3]	Sg. Semenyih	Odor pollution	Effluent from Nilai and Semenyig industrial area polluting Sg. Buah and Sg. Semenyih	Hulu Langat, Kuala Langat, Sepang, and Petaling
2015 [4]	Sg. Semenyih	Low pH, Manganese, and Ammonia	Leachate from Sanitary Landfill	Bangi and Kajang
2014 [5]	Cheras Mile 11	High Ammonia Concentration	Private Sewage Plant	Hulu and Kuala Langat
2014 [5]	Bukit Tampoi	High Ammonia Concentration	Private Sewage Plant	Hulu and Kuala Langat
2013 [4]	Sg. Semenyih	Bad Smell/Odor, High Turbidity	Leachate from Sanitary Landfill, flood	Bangi and Kajang
2012 [6]	Sg. Langat	High Turbidity	Flood/mudflow	Kuala Lumpur
2012 [6,7]	Salak Tinggi	High Ammonia Nitrogen	Chicken Farm, Industrial Effluent	Hulu Selangor, Sepang
2012 [6]	Cheras Mile 11	Diesel	Quarry	Kuala Lumpur
2012 [4,7]	Sg. Semenyih	High Ammonia Nitrogen (>7.0 mg/L)	Leachate from Sanitary Landfill	Bangi and Kajang
2012 [6]	Cheras Mile 11	High Fluoride (0.25–1.11 mg/L)	Unknown	Kuala Lumpur
2011 [4]	Sg. Semenyih	Diesel	Unknown	Bangi and Kajang
2010 [4]	Sg. Semenyih	High Ammonia Nitrogen (>7.0 mg/L)	Leachate from Sanitary Landfill	Bangi and Kajang
2009 [4]	Sg. Semenyih	Diesel	Unknown	Bangi and Kajang
2009 [8]	Cheras Mile 11	High Ammonia Concentration	Unknown	Cheras and Balakong

Table 4. *Cont.*

Year	WTP	Type of Pollution	Source of Pollution	Affected Area
2009 [8]	Salak Tinggi	High Ammonia Concentration	Effluent from Nilai Industrial Areas	Sepang
2006 [8]	Salak Tinggi	High Ammonia Concentration	Effluent from Nilai Industrial Areas	Sepang
2006 [4]	Sg. Semenyih	High Ammonia, Turbidity, Diesel	Leachate from Sanitary Landfill	Bangi and Kajang

Note(s): Source: [1] [13]; [2] [3]; [3] [60]; [4] [61]; [5] [62]; [6] [63]; [7] [64]; [8] [65]; [9] [66].

Conventional Water Treatment Method in Langat River Basin, Malaysia

The absolute capacity of the 33 WTPs in the state of Selangor is 4476 MLD (million liters per day); however, the 9 WTPs in the Langat Basin contain 1110.80 MLD structured capacity and 1329.4 MLD highest capacity to treat water [67] (Table 5). Subsequently, the structured and maximum capacities are 24.8% and 29.7% contrasted to the absolute capacity of all the WTPs in Selangor and these WTPs offer drinking water to 33% of the Selangor population [2].

Table 5. Description of water treatment plants (WTPs) in Langat River Basin, Malaysia.

WTP	Location	Water Source	Supply Area	Treatment Process	Design Capacity (MLD)	Max. Capacity (MLD)	Filter Performance (Hours)	Water Losses (%)	Water Quality Compliance (%)
Sg. Pangsoon [1]	Batu 24, Kuala Pangsoon, Hulu Langat	Sg Pangsoon	Batu 24, Kuala Pangsoon-Bt. 15, Bukit Kundang, Hulu Langat.	Conventional (Partial and Full WTP)	1.8	1.8 [2]	Continuous Filter	N/A	99.9
Sg. Lolo [1]		Sg. Lolo	Batu 24, Kuala Pangsoon- Bt. 15, Bukit Kundang, Hulu Langat.	Conventional (Partial and Full WTP)	1	3 [2]	N/A	N/A	100
Sg. Serai [1]	Batu 11, Jalan Hulu Langat, Hulu Langat	Sg Serai	Batu 9, Cheras-Bt. 13, Bukit Nanding, Hulu Langat.	Conventional (Full WTP)	1.7	0.9 [2]	8	N/A	99.9
Sg. Langat [1]	Batu 10, Jalan Hulu Langat, Cheras	Sg. Langat	Klang, Petaling Jaya, Kajang, Nanding, Bangi, Beranang, Cheras, Hulu Langat.	Conventional (Full WTP)	386.4	456	56.76	6.45	99.9
Cheras Mile 11 [1]	Batu 11, Jalan Cheras, Kajang	Sg Langat, Sg Raya and Sg Sering	Balakong	Conventional (Full WTP)	27	26.2	72	4.4	100
Bukit Tampoi [1]	Jalan Dengkil-Bukit, Changgan, Dengkil	Sg. Langat	Part of Kuala Langat and Sepang	Conventional; vertical flow	31.5	34.5	56	2.98	100
Salak Tinggi [1]	Jalan Kg Giching, Sepang	Sg. Labu	Salak Tinggi and part of Sepang	Conventional	11.4	5	72	10	99.9

Table 5. *Cont.*

WTP	Location	Water Source	Supply Area	Treatment Process	Design Capacity (MLD)	Max. Capacity (MLD)	Filter Performance (Hours)	Water Losses (%)	Water Quality Compliance (%)
Sg. Labu [3]	Lembah Paya, Salak Tinggi, Sepang	Sg. Labu	Sepang	Conventional	105	120	-	-	-
Sg. Semenyih [4]	Presint 19, Putrajaya	Sg. Semenyig	Kuala and Hulu Langat, Sepang, Putrajaya, Cyberjaya, Seri Kembangan, USJ and Puchong	Conventional	545	682	-	-	-

Note(s): Source: [1] [57]; [2] [63]; [3] [68]; [4] [61].

All the WTPs in the Langat Basin follow the typical coagulation water treatment approach, and the steps include air circulation, substance blending, coagulation, flocculation, filtration, and post-chemical inclusion prior to the supply reaching the end users (Figure 3). Sg. Pangsoon, Sg. Lolo, Sg. Serai, Sg. Langat, Cheras Mile 11, Bukit Tampoi and Salak Tinggi WTPs (Table 5) are managed by Puncak Niaga Sdn. Bhd.; Sg. Labu WTP, which is managed by Konsortium Air Selangor Sdn. Bhd. [68]; and Sg. Semenyih WTP, which is managed by Kumpulan Darul Ehsan Berhad [61], similarly utilize this typical technique for raw water treatment. Out of all the WTPs, the Sg. Semenyih WTP is the most significant plant providing drinking water to 1.5 million people of Selangor and Putrajaya [69].

The WTPs in the Langat River Basin utilize a typical water treatment method that is not capable of expelling trace metals and radionuclides from raw water properly [70]. Thus, a high-pressure reverse osmosis membrane technology could be applied in the WTPs to expel trace metals and a wide range of radionuclides (for example, effectiveness is >90%) from raw water; and the USEPA has recorded this technology as the Best Available Technology (BAT) and the Small System Compliance Technology (SSCT) [71] (Table 6). In addition, worldwide, the average water treatment cost (USD/m^3) for the reverse osmosis approach is the lowest contrasted with other water treatment processes, particularly in the USA, for example, 0.0002–0.0004 USD/m^3 [72], though the average cost of ion trade and lime relaxing is in the scope of 0.08–0.21 USD/m^3 in the USA [73]. Essentially, the average typical water treatment cost in Malaysia is 0.53 USD/m^3 [74].

Table 6. Water treatment technologies approved by the USEPA for radionuclides removal [71].

Treatment Technology	Designation	Customers Served (SSCTs)	Radium (Ra)	Uranium (U)	Alpha (G)	Beta/Photon (B)	Source Water Considerations	Operator Skill Required
Ion Exchange (IX)	BAT and SSCT	25–10,000	✓	✓		✓	All ground waters	Intermediate
Point of Use (POU) IX	SSCT	25–10,000	✓	✓		✓	All ground waters	Basic
Reverse Osmosis (RO)	BAT and SSCT	25–10,000 (Ra, G, B) 501–10,000 (U)	✓	✓	✓	✓	Surface waters usually requiring pre-filtration	Advanced

Table 6. *Cont.*

Treatment Technology	Designation	Customers Served (SSCTs)	Treatment Capabilities				Source Water Considerations	Operator Skill Required
			Radium (Ra)	Uranium (U)	Alpha (G)	Beta/Photon (B)		
POU RO	SSCT	25–10,000	√	√	√	√	Surface waters usually requiring pre-filtration	Basic
Lime Softening	BAT and SSCT	25–10,000 (Ra) 501–10,000 (U)	√	√			All waters	Advanced
Green Sand Filtration	SSCT	25–10,000	√				Typically ground waters	Basic
Co-precipitation with Barium Sulphate	SSCT	25–10,000	√				Ground waters with suitable water quality	Intermediate to Advanced
Electro dialysis/Electro dialysis Reversal	SSCT	25–10,000	√				All ground waters	Basic to Intermediate
Pre-formed Hydrous Manganese Oxide Filtration	SSCT	25–10,000	√				All ground waters	Intermediate
Activated Alumina (AA)	SSCT	25–10,000		√			All ground waters	Advanced
Coagulation/ Filtration	BAT and SSCT	25–10,000		√			Wide range of water qualities	Advanced

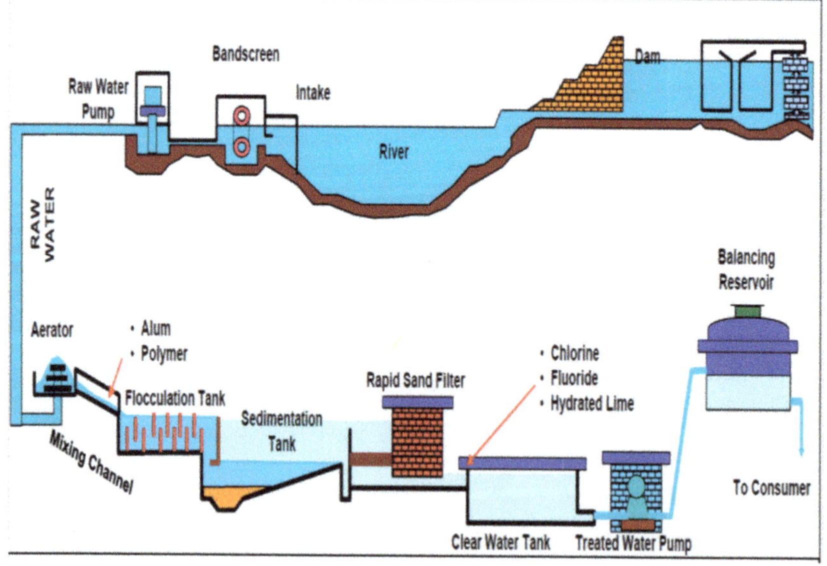

Figure 3. Conventional water treatment plants at Langat Basin, Malaysia [74].

In theory, RO is able to remove about 0.0001 μm molecule size and many of the metal particles measuring 0.0001–0.001 μm [70] in the water such as Al 0.000143 μm [75]. Consequently, the typical water treatment technique at the Langat River Basin plants cannot remove the Al concentration completely. Plant specialists add aluminum sulfate in the treatment of raw water for sterilization due to the frequent transformations of turbidity in raw water in tropical climates. Thus, metal concentration such as Al concentration in the plant's treated water and the household's supply water needs to be resolved to examine the effectiveness of the plants in eliminating the metal particles and the water contamination in the water pipeline distribution process. The RO system needs higher power and its filtrated water yield efficacy is less than other innovations; notwithstanding, with the progression of the membrane technology, the RO cost will also reduce. In 2006, a RO unit cost (300–1000 USD), ion exchange (400–1500), and distillation (300–1200 USD) was a lot more compared to the price in 2017 [72]. In 2017, the RO's unit cost was 150–300 USD, ion exchange was 50 USD, and distillation was 150–250 USD [76].

River contamination is not just an issue in Malaysia; researchers and policymakers globally are considering river water quality since it is the primary source of drinking water. In addition, all around, scientists and policymakers are worried about the nature of the waterway water because at present, it is the significant wellspring of drinking water. Incidentally, the commitment of the stakeholders, for example, GO-NGOs, the private sector, and the community-based organizations in river management has been considered the best methodology.

The entire WTPs in Langat Basin utilize the typical water treatment strategy and the customary molecule filtration technique at the household level can eliminate a molecule size of about 0.5 μm [77], while the Pb particles and other metal particles could be <0.000174 μm [70,78,79]. On the other hand, the particulate elimination efficacy using reverse osmosis (RO) film innovation is 0.0001 μm [79]. In addition, H20 established that the thin film composite RO membrane technology has the most noteworthy metal elimination efficacy from the water supply, for example, Arsenic 94%, Cadmium 98%, Chromium 88%, Lead 99% [70].

5. Policy, Institute, and Expert's Nexus for Integrated River Basin Management

Although strategic policies, plans, institutions, expertise, and visionary organizations exist in Malaysia, the integration and execution of these policies and institutions are not sufficient due to the lack of constitutional support from the government to empower the main authority, the Department of Irrigation and Drainage Malaysia, in managing the water resources. The limitations of river basin management are a major issue given the fact that the majority of the staff only have mono-disciplinary training. Consequently, there are insufficient capacity-building programs of water resource management agencies and multi-stakeholders as well as inadequate participation of the stakeholders. Mokhtar et al. [22] determined the inadequate coordination among the stakeholders for the integrated water resources management in the Langat River Basin.

Inefficient endeavors from the individual level could be a direct result of inadequate training of mid- to lower-ranking officials. Quality assurance (QA) and quality control (QC) are not assured in most of the steps in the management of the Langat River Basin. The stakeholders as well as the individuals are waiting for others to tackle the issues. This individualistic perspective impedes the recognition of the right stakeholders or organizations to deal with the Langat River Basin management and the contamination issue. The differences in political interests between the state and federal government may have additionally been an essential issue in the execution of policies [80].

The lack of cooperation from the stakeholders because of deficient iterative social adapting has just been accounted for in the Langat River Basin Management; in any case, the federal government is equipped for polycentric planning to organize the different stakeholders [30]. The Selangor Water Management Authority (LUAS) is the primary authorized organization of the Selangor State Government to oversee water bodies in

Selangor which includes the Langat River which gives drinking water to about 33% of the populace in the state [42,81]. Langat River Basin shares four distinct constituencies of Selangor state, the Federal Territories of Kuala Lumpur and Putrajaya, and Negeri Sembilan state. As such, river management, especially contamination control, has been a huge challenge. The Department of Environment (DOE) Malaysia screens the quality of the water from the river, which includes the Langat River; nevertheless, there is no particular agency such as LUAS in Negeri Sembilan and Federal Territories. Thus, river management particularly concerning contamination has been crucial since the rivers are located in a particular state or states and the administration falls under the state's responsibility. Consequently, Mokhtar et al. [22] claimed that even though there are numerous policies, institutions, and expertise in Malaysia on river management, LUAS could not coordinate the management of the Langat River Basin among all its stakeholders sufficiently. Thus, the leadership function of the local authority in view of the Local Government Act 1976 can better oversee the rivers in Malaysia including the Langat River Basin through viable execution of water policies such as the national agenda of water sector transformation 2040 (WST2040) and creating workable partnerships with GO-NGOs, businesses, the scholarly world, and civil agencies (Figure 4).

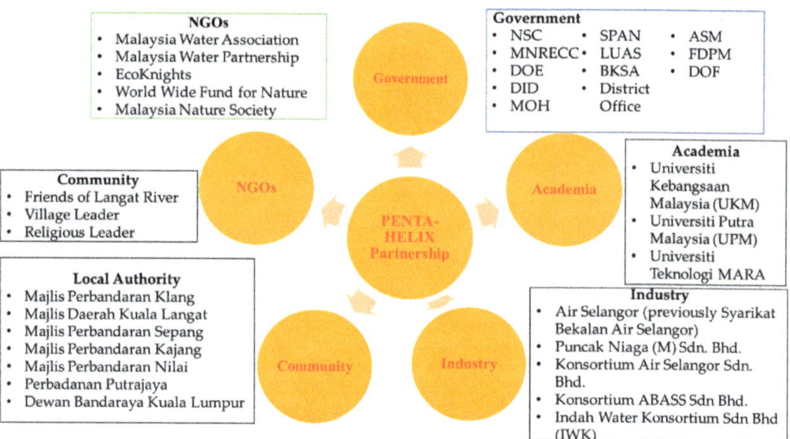

Figure 4. PENTA-HELIX stakeholders at Langat River Basin Malaysia. Note: Government: NSC (National Security Council). MNRECC (Ministry of Natural Resources, Environment and Climate Change). DID (Dept. of Irrigation and Drainage). DOE (Dept. of Environment). MOH (Ministry of Health). SPAN (National Water Services Commission). LUAS (Selangor Water Management Authority). BKSA (Badan Kawal Selia Air Negeri Sembilan). ASM (Academy of Sciences Malaysia). FDPM (Forestry Dept. of Peninsular Malaysia). DOF (Dept. of Fisheries).

The Department of Environment (DOE) Malaysia claimed that 17% of electronics/metal, 50% of paper, 33% of textiles, and 25% of food industries along the Langat River Basin do not comply with the Environmental Quality Act (EQA 1974) as they continue to release effluents into the river [62]. The DOE and MOH (Ministry of Health Malaysia) are accountable for guaranteeing the standard and quality of the river and the treated water. As such, an efficient execution of the Environmental Quality Act (EQA) 1974 and the Environmental Quality (Industrial Effluent) Regulations 2009 in relation to chemical release into the environment is lacking [2].

The Academy of Sciences Malaysia was set up to deliver better policies and to make plans for water and various segments to assist the policymakers as well as the government. Essentially, the Malaysian Industry–Government Group for High Technology (MIGHT) is advancing the public–private partnership to align government policies and strategies. The Malaysian Foresight Institute is additionally helpful in the integration and execution of the government's policies and plans. The Foresight Institute has just utilized suitable techniques to discover new trends in the fields of education, health, technology, etc. to assist in policy-making for the government [82].

The National Integrated Water Resources Management (IWRM) Plan (Volume I and II) set up by the Academy of Science Malaysia is a phenomenal guide to water resource management in Malaysia followed by the National Agenda of Water Sector Transformation 2040 (WST2040) from the year 2020 to 2040. This plan includes water resource management in line with the national economic, food, environment, health, and energy policies as well as advocacy, awareness, and capacity building to accelerate the implementation of IWRM. Moreover, the Integrated Water Resources Management Framework emphasizes aspects of social improvement, monetary advancement, and environmental assurance. Nevertheless, the general model of IWRM concentrates on value by empowering the environment (policies, enactments); enabling environmental sustainability using institutional models (between local and central government, public and private river basin management), and monetary efficacy through management (information, evaluation, assignment, instrument) and a balance of 'water for living and water as an asset' by investing in water infrastructures [83,84]. In addition, the Academy of Science Malaysia announced a 15% increment in consumable water demand (9291 MCM/year) by the year 2050 compared to the 5277 MCM/year water demand in 2010 [85].

Mokhtar et al. [22] recommended polycentric institutional strategies under the federal government for better integration and coordination in the Langat River Basin Management. In addition, polycentric institutional strategies could be successful through the potential utilization of iterative learning procedures. Thus, this methodology could more readily manage the institutional difficulties of versatility and ecosystem-based administration in the Langat River Basin. According to Mokhtar et al. [66], iterative learning would be able to guarantee better cooperation among the stakeholders, even at the lowest suitable level. There should likewise be an iterative learning component inside the inter-organizational network for Integrated Langat River Basin Management as there are a few administration segments on specific rules and agencies (Table 7).

Table 7. IRMB laws and agencies for Langat River Basin; state agency (S), federal agency (F).

Management	Statute	Agency
Pollution control	Environmental Quality Act 1974 Drinking Water Quality Standard Street, Drainage and Building Act 1974 Local Government Act 1976 LUAS Enactment 1999 Water Services Industry Act 2006	D. of Environment (F) D. of Health (F) Local government (S) Local government (S) LUAS (S) Water Commissioner (F)
Catchment area	National Forestry Act 1984 LUAS Enactment 1999 Local Government Act 1976	D. of Forestry (S) LUAS (S) Local government (S)
Land use drainage	Land Conservation Act 1960 Town & Country Planning Act 1976 Local Government Act 1976 Drainage Works Act 1954	Land Office (S) Local government (S) Local Government (S) DID (F)
Flood control	Ministerial Function Act 2008 LUAS Enactment 1999	DID (F) LUAS (S)
Water services	Water Services Industry Act 2006	Water Commissioner (F)

Note(s): Source: [86].

5.1. Raw and Drinking Water Quality of Langat River Basin

The typical coagulation water treatment technique relies upon the physiochemical qualities of raw water. The physiochemical qualities of raw water in the tropical atmosphere are extremely vital in light of its continuous changes because of too much precipitation, unexpected flash floods, dry seasons, landslides, etc. Nevertheless, the experts in water treatment plants mainly rely upon the assurance of physiochemical fixations to blend chemicals in the raw water for the purpose of treatment. Thus, the incessant checking of raw water is extremely fundamental for the typical water treatment technique in light of the regular changes. However, two hour intermittent raw water samplings are not adequate to decide on the physiochemical attributes for treatment purposes. Accordingly, at times, the typical treatment strategy cannot keep up with the treated water quality standard recommended by the World Health Organization (WHO) and the Ministry of Health Malaysia (MOH).

The state authority and concessionary organizations in Malaysia, since the privatization operation of 1987, are accountable for the supply of drinking water at the household level. In spite of the fact that the water is inspected regularly in the drinking water treatment plants before going to the household level through the pipeline, there is potential for natural and chemical pollution in the water supply while being transferred and stored. Thus, local authorities dependent on its mandate can more readily facilitate among the Ministry of Health, Department of Environment, SPAN (water controller), WTP authority, and SYABAS (currently Air Selangor—water supplier) for supervising raw and drinking water at the Langat River Basin. The general water quality index (WQI) of the Langat River and its tributaries demonstrated that a greater part of the tributaries is in Class III [28], which shows that intensive treatment is required before drinking [87].

Despite the fact that the Ministry of Health Malaysia (MOH) does not screen for aluminum (Al) in the river water, it does screen for the degree of Al in treated water and the standard fixed level for Al in drinking water is 200 μg/L [88]. Shockingly, there is just one study on Al concentration in the supplied water from the Langat River Basin, for example, 148 ± 76 μg/L [89]; however, a high Al concentration of 990 ± 1520 μg/L [90], and 210 ± 41.50 μg/L [91] was documented in the drinking water in Johor state, Malaysia. In addition, certain researchers have discovered a relationship between Alzheimer's disease and consuming drinking water with Al over a period of time [92]. Correspondingly, a higher concentration of Pb was documented at 32.5 μg/L in Bandar Sunway [93], basically as a result of corrosion in the piping systems of the old structure. Moreover, the Pb concentration in Bandar Sunway was over the highest limit for the standard of drinking water quality at 10 μg/L as proposed by the Ministry of Health Malaysia (MOH), World Health Organization (WHO), and European Commission (EU), and 15 μg/L as recommended by the United States Environmental Protection Agency (USEPA). Research discoveries of late on water supply systems demonstrate that lead (Pb) is a critical contaminant of grave concern with toxic exposure happening from drinking water [94–96]. In any case, past examinations on arsenic (As), cadmium (Cd), and chromium (Cr) in the drinking water from the Langat River Basin did not cross the limits recommended by the MOH.

Sewage treatment plants, small and medium businesses, residential zones, townships, palm oil plantations, and companies along the downstream of the Langat River Basin are the main sources of Bisphenol A (for example, 1.3 to 215 μg/L) in the raw water. Hence, it might be liable for the endocrine disruption in people in the Langat Basin area, although, in the tap water from Langat River Basin, the concentration of BPA was extremely low, for example, 3.5 to 59.8 μg/L [97]. Correspondingly, individuals are likewise aware that tap water is not totally clean and may comprise microorganisms even though the water is treated in the plants prior to reaching households in the Klang Valley, Malaysia. Likewise, the tap water is chlorinated before reaching the houses. Subsequent to boiling, a concentrated degree of chlorine in the water may represent a huge medical problem, with obscure signs and manifestations [98]. In addition, the Ministry of Health Malaysia in 2002 detailed that in some states in Malaysia, the fluoride concentration in the drinking water crossed the Malaysian drinking water standard because of the synthetic fluoridation of drinking water and it led to high frequencies of dental fluorosis in the individuals [99].

Constructed wetlands in the Langat River Basin would help reduce water contamination from effluent releases by industries. Putrajaya Lake in the Langat River Basin is one of the 32 Ecohydrology Demonstration Sites of UNESCO-IHP Ecohydrology Program (EHP) internationally since 2010 and has been designated as an Operational Demonstration Site. It is the first man-made wetlands in Malaysia and the largest freshwater wetlands in the tropics at 600 hectares [26]. Then again, the Hybrid off River Augmentation System (HORAS) venture was set up for execution by the Selangor state government to satisfy the water needs over the dry seasons. The first HORAS venture (HORAS 600) is presently under development. When it is finished, it will be able to supply 600 million liters of water daily (MLD). The second HORAS venture (HORAS 3000) will be able to provide 3000 million liters of water daily (MLD) [100]. Stormwater could be a significant resource if it is satisfactorily treated [101]. As such, the developed wetlands could be helpful for stormwater maintenance and purification. Likewise, stormwater ought to be treated by the wastewater treatment plant (WWTP), and the management of stormwater ought to consider solid waste management because of its overflow into the water body. Thus, there ought to be appropriate dumping locations for solid waste.

5.2. Global River Basins

There are 263 transboundary river basins globally covering approximately 50% of the Earth and the incredible five river basins including the Congo, Niger, Nile, and Zambezi river basins share boundaries with nine to eleven nations [102]. In addition, thirteen river basins share five to eight riparian countries. Even though 145 countries involving 40% of the global populace [103] are living inside these 263 transboundary river basins, the political limits of twenty-one nations are completely inside these international basins. As such, the Danube River drains through the region of eighteen countries in Europe [86] and it has a practical basin management structure; in any case, the greater part of these transboundary river basins does not have a sufficiently cooperative administration system [103,104].

The International Center for Water Cooperation (ICWC) contended that the harmonious and peaceful settlements among neighboring nations and their growth mainly rely upon transboundary river management [103]. In any case, thirty-seven transboundary water management clashes have been settled since 1948, even though around 295 transboundary river management memorandums were achieved from 1948 to 2015 [105]. Water shortage and a lack of a proper water management structure could have likewise created clashes between the states and regions. UNESCO [106] claimed that 158 river basins out of a total of 263 transboundary river basins do not have a proper framework for river management. Even though 105 river basins as well as river management institutions share more than three riparian states, only 20% of the river basins have multilateral water management arrangements. The modest number of bilateral and multilateral water management arrangements by those involved demonstrate the lack of potential shared advantages because of the absence of strategies, political will, and resources to oversee the shared water resources [106].

Some Best Practices of River Basin Management

Danube River Basin, Europe: Danube River Basin Management by the European Union is among the most commendable river basin management globally as 19 nations share the basin with over 81 million populations of various cultures. The International Commission for the Protection of the Danube River (ICPDR) under the EU is liable for the sustainable and fair utilization of surface and groundwater in the basin. ICPDR and professional management groups, hydrology, and financial task groups manage the development operations at the basin and adhere to the Danube River Protection Convention for the execution of such operations involving the various stakeholders [107].

Murray-Darling River Basin, Australia: Internationally, the Murray-Darling River Basin, Australia (MDBA) is the pioneer river basin authority under the Water Act 2007; the basin shares four states and federal territories in Australia. The Murray-Darling River

Basin Authority (MDBA) is an independent and authorized organization according to the direction of the government. MDBA is liable for sustainable water resource management and development at the basin [108].

6. Proactive Leadership Roles of Local Authority at Langat River Basin

In the Langat River Basin, proactive and successful leadership roles by the local authority are a crucial prerequisite for an integrated and holistic Langat River Basin management, particularly for the safety and availability of the drinking water supply to the household level. To be proactive, four components, namely mandate, financial, human resource, and support are needed for the local authority. At the same, the local authority has the full direction to supervise the rivers and drinking water taken from the Langat Basin under the Local Government Act 1976.

There are additionally money and human resources with the local authority, but the finances are lacking since there is insufficient training of the lowest ranked staff to improve the capacity to perform various tasks and take on a proactive leadership role. Nevertheless, the lack of joint effort among the various stakeholders gives rise to critical issues in managing the Langat Basin by LUAS in collaboration with related organizations, including the DID, DOE, MOH, etc. However, the local authority could do a better job by organizing the various stakeholders' platforms and utilizing the PENTA-HELIX partnership framework by networking with private and public sectors as well as the civil sector to manage and monitor the quality of raw and drinking water.

STEM (science, technology, engineering, mathematics) and SSH (social science and humanities) data on quality assurance (QA) and quality control (QC) will assist in producing a better strategy while refreshing existing policies with the current database. A compelling QA and QC at numerous levels will guarantee the exactness and precision of information and data. Subsequently, individuals will get the best verifiable information which will impact and rouse them to act right and to take the lead in carrying out beneficial work by utilizing their own intelligence. In spite of the fact that the leadership position of the local authority is vital for the administration, nonetheless, without framing a multi-stakeholder platform and without having expertise from various segments, for example, STEM and SSH, river and drinking water management at the Langat Basin will stay fragmented. Thus, the accompanying three-finger, mentor–mentee, multitasking, and mimicking concepts may cause people to be proactive, away from their complacent state, and carry out leadership functions.

6.1. Three-Finger Concept

Individuals generally reprimand others for not completing any task effectively. Thus, the question in this case would be "who do we blame?" for not carrying out the obligations and duties. Furthermore, who will take on the position of leadership in finding solutions related to river basin management and accomplishing the SDGs? At the point when individuals point their index finger at others for not carrying out their responsibilities and obligations effectively, the remaining three fingers—center, ring, and little finger—point at them. Therefore, it implies that individuals ought to carry out their obligations and duties and be fearless to take on a position of leadership to manage tasks and roles within their domain before attempting to blame others for not doing their jobs.

6.2. Mentor–Mentee Concept

Officials and individuals at the local, state, and federal levels ought to be sufficiently courageous to take on the leadership positions. As per the degree of the individual, he/she should create a formal or informal team in the area of environmental management in his/her neighborhood or favored area and get followers to support the mentee so as to stay dynamic in river basin management. The work priority to deal with river management or the environment ought to rely upon the local team/committee's ability and capacity. The team ought to embrace and adapt sustainable river or environment management methodologies for use at the local level and seek help from the closest research/scholarly establishments. The great thinking of the team/committee, particularly the individuals, will motivate others to cooperate as well. Thus, mentors should have successors to proceed with the work development when they retire.

6.3. Multitasking Concept

Individuals or officials should be multitasking to carry out multi-levels of management tasks. Thus, individuals ought to be engaged at the neighborhood-level administration committee in addition to his/her main job. He/she can join NGOs to share his/her capability in improving the welfare of the general public. Subsequently, upon retirement from one's primary job, the individual need not rot at home but rather be an active member at the local level of environmental management. There ought to be preparation, training, and learning in changing the behavior of a person to be agreeable, community-oriented, and committed. Aristotle expresses that virtues are habits and they are manifested in action [109]. Likewise, Durant [110] states that we become what we repeatedly carry out; thus, greatness is not a demonstration but rather a habit. Therefore, the act of performing various tasks linked to environmental management will be changed into an extraordinary habit.

6.4. mimiC Concept

The concept of 'mimiC' depends on an individual's capacity to duplicate the best administration practices (Figure 5). The 'm' in 'mimiC' represents the execution of these management practices with the goal of learning and adapting them to fit in the local culture. The 'i' in mimiC represents the implementation of these practices by the individual and 'm' additionally represents the monitoring of the practices alongside 'i' which represents the improvement of the present practices. Notwithstanding, the capital 'C' represents 'change' and this 'C' represents the person who has the capability and capacity to learn and adapt to the management practice changes.

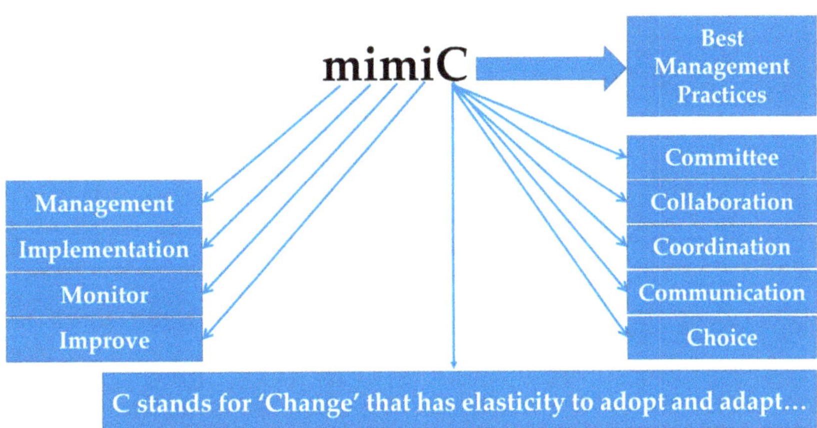

Figure 5. Dynamic perspectives of individual for IRBM.

7. Conclusions and Recommendation

Holistic and integrated Langat River Basin management (ILRBM) depends on the successful implementation of policies at the local level because many beautiful policies already exist in Malaysia along with experts and institutions. Langat is a transboundary river, so the coordination among the local authorities and district offices are very crucial for effective Langat River Basin management due to their mandate in policy implementation as well as revenue generation from the natural resources such as land, water, and forest, respectively. Therefore, the proactive leadership roles of local government officials along with the effective collaboration and cooperation among the Penta-helix multi-stakeholders must exist in real-time decision-making for river basin management. Moreover, the decision-making of local government officials both during the disaster and normal periods should be based on validated data and information from science and technology as well as social science and humanities disciplines; and it should incorporate the technologies such as artificial intelligence for the early warning system. Thus, the Integrated Langat River Basin Management must adhere to the National Agenda of Water Sector Transformation 2040 (WST2040) to accelerate the implementation of the integrated water resources management (IWRM) plan due to its detailed data to speed up the accomplishment of sustainable growth in Malaysia through better execution of water policies.

The capacity building of the local government officials along with other stakeholders can be enhanced using the customized training module produced via the Malaysian National Water Sector Transformation (WST2040) project, and the capacity building will assist them in ensuring quality assurance and quality control at different levels to screen the related stakeholders accountable for water resource management and safe drinking water supply at the household levels. The training alongside creative and innovative key performance indicators (KPIs) will empower them to coordinate with the river polluters, including small and medium enterprises (SMEs), to inspire and bind them legally to eliminate river contamination. Moreover, to prevent shutdowns of water treatment plants (WTPs) and to guarantee safe drinking water supply at the household level, reverse osmosis (RO) water treatment mechanism rather than the present typical coagulation technique would be successful, as based on the United States Environmental Protection Agency (USEPA), it can eliminate contaminants such as radionuclides and metals > 90% from the treated water. On the other hand, the RO filtration technique can be introduced at the house's tap and maintained at a more affordable moderate pond sand filtration at the WTPs, as treated water pollution is apparent in the water pipeline from the WTPs to the households. Subsequently, the introduction of the tow-layer water filtration method at the Langat Basin will be able to guarantee the achievement of the Sustainable Development Goals 2030 (for example, Goal 6.1: By 2030, to accomplish universal and equal access to safe and reasonably priced drinking water for everyone), as well as to reach a healthy living goal for humans. Likewise, safe drinking water will propel Malaysia to achieve its National Transformation to be one of the top 20 countries internationally.

Author Contributions: Conceptualization, M.F.A. and M.B.M.; methodology, M.F.A.; software, M.F.A. and R.P.K.K.; validation, M.F.A., M.B.M., C.K.L. and N.A.M.; formal analysis, M.F.A.; investigation, M.F.A.; resources, M.F.A. and I.A.B.C.S.; data curation, M.F.A. and K.A.K.A.; writing—original draft preparation, M.F.A.; writing—review and editing, M.F.A., M.B.M., K.A.K.A. and C.K.L.; visualization, M.B.M.; supervision, M.F.A.; project administration, M.F.A.; funding acquisition, M.F.A. All authors have read and agreed to the published version of the manuscript.

Funding: This study is supported by Ministry of Higher Education (MOHE) Malaysia Research Grant-FRGS/1/2022/SSl03/UKM/03/1.

Data Availability Statement: The data used in this research can be obtained from the corresponding authors upon rational request.

Acknowledgments: The authors are also grateful to the water treatment plant authorities, i.e., Puncak Niaga (M) Sdn. Bhd.; Konsortium Air Selangor Sdn. Bhd.; and Konsortium ABASS Sdn Bhd. for the provided information.

Conflicts of Interest: The authors declare no conflict of interest.

References

1. Chan, N.W. Managing urban rivers and water quality in Malaysia for sustainable water resources. *Int. J. Water Resour. Dev.* **2012**, *28*, 343–354. [CrossRef]
2. Ahmed, M.F.; Alam, L.; Ta, G.C.; Mohamed, C.A.R.; Mokhtar, M. A review on the chemical pollution of Langat River, Malaysia. *Asian J. Water Environ. Pollut.* **2016**, *13*, 9–15. [CrossRef]

3.   Bernama. Langat, Cheras Water Treatment Plant Shutdown Due to Odour Pollution, Selangor Exco Says, Malaymail Online. 2016. Available online: http://www.themalaymailonline.com/malaysia/article/langat-cheras-water-treatment-plant-shutdown-due-to-odour-pollution-selango (accessed on 27 October 2017).

4.   GOM. *Federal Constitution of Malaysia. Government of Malaysia*; International Law Book Services: Petaling Jaya, Malaysia, 2016.

5.   Khalid, R.M. A Study on Sustainable Water Resources Management within the Malaysian Federalism System. Ph.D. Thesis, Faculty of Law, Universiti Kebangsaan Malaysia (UKM), Bangi, Malaysia, 2014.

6.   Ahmed, M.F.; Lim, C.K.; Mokhtar, M.B.; Khirotdin, R.P.K. Predicting arsenic (As) exposure on human health for better management of drinking water sources. *Int. J. Environ. Res. Public Health* **2021**, *18*, 7997. [CrossRef]

7.   D'Amore, G.; Di Vaio, A.; Balsalobre-Lorente, D.; Boccia, F. Artificial intelligence in the water–energy–food model: A holistic approach towards sustainable development goals. *Sustainability* **2022**, *14*, 867. [CrossRef]

8.   Ziwei, L.; Xiangling, T.; Liju, L.; Yanqi, C.; Xingming, W.; Dishan, Y. GIS-based risk assessment of flood disaster in the Lijiang River Basin. *Sci. Rep.* **2023**, *13*, 6160. [CrossRef] [PubMed]

9.   WMO. *Integrated Water Resources Management and Early Warning System for Climate Change Resilience in the Lake Chad Basin*; World Meteorological Organization (WMO): Geneva, Switzerland, 2022; Available online: https://public.wmo.int/en/media/news/integrated-water-resources-management-and-early-warning-system-climate-change-resilience (accessed on 10 June 2023).

10.   Fernández-Nóvoa, D.; García-Feal, O.; González-Cao, J.; de Gonzalo, C.; Rodríguez-Suárez, J.A.; Ruiz del Portal, C.; Gómez-Gesteira, M. MIDAS: A new integrated flood early warning system for the Miño River. *Water* **2020**, *12*, 2319. [CrossRef]

11.   UNEP. *Protected Planet Report 2020*; United Nations Environment Programme (UNEP): Nairobi, Kenya, 2020; Available online: https://livereport.protectedplanet.net/ (accessed on 10 June 2023).

12.   Zhou, X.; Moinuddin, M.; Renaud, F.; Barrett, B.; Xu, J.; Liang, Q.; Zhao, J.; Xia, X.; Bosher, L.; Huang, S.; et al. Development of an SDG interlinkages analysis model at the river basin scale: A case study in the Luanhe River Basin, China. *Sustain. Sci.* **2022**, *17*, 1405–1433. [CrossRef]

13.   Chan, D. Sungai Langat, Cheras Water Treatment Plants Resume Operations. New Straits Times. 2016. Available online: http://www.nst.com.my/news/2016/10/178979/sungai-langat-cheras-water-treatment-plants-resume-operations (accessed on 14 December 2016).

14.   Bernama. Syabas: Taps to Flow Again for 377,000 Selangor Customers Starting from This Afternoon. *Malaymail Online.* 2019. Available online: https://www.malaymail.com/news/malaysia/2019/06/27/syabas-taps-to-flow-again-for-377000-selangor-customers-starting-from-this/1765898 (accessed on 3 December 2021).

15.   The Straits Times. Malaysia River Pollution Cuts Water Supply to More than 300,000 People. *SPH Media Limited.* 2020. Available online: https://www.straitstimes.com/asia/se-asia/malaysia-river-pollution-leads-to-water-supply-cut-to-more-than-300000-people (accessed on 3 December 2021).

16.   FMT. Water Cut in 463 Areas in Klang Valley Due to Pollution. *FMT Media Sdn Bhd.* 2020. Available online: https://www.freemalaysiatoday.com/category/nation/2021/08/31/water-cut-in-463-areas-in-klang-valley-due-to-pollution/ (accessed on 3 December 2021).

17.   The Star. Sg Semenyih Water Treatment Plant Resumes Operations, Bukit Tampoi Remains Shut. *Star Media Group Berhad.* 2020. Available online: https://www.thestar.com.my/news/nation/2020/10/06/sg-semenyih-water-treatment-plant-resumes-operations-bukit-tampoi-remains-shut (accessed on 3 December 2021).

18.   Awani, A. Ministry Recommends SOP Review for Water Treatment Plant Operators. Astro Awani. 2016. Available online: http://english.astroawani.com/malaysia-news/ministry-recommends-sop-review-water-treatment-plant-operators-120997 (accessed on 14 December 2016).

19.   Chan, N.W.; Raidam, N.; Ao, M.; Foo, K.Y. Water Management in Cities. Chapter 17. In *Sustainable Urban Development Textbook*; Chan, N.W., Imura, H., Nakamura, A., Ao, M., Eds.; Water Watch Penang, Universiti Sains Malaysia: Penang, Malaysia, 2016.

20.   INBO. International Network of Basin Organization. 2017. Available online: http://www.inbo-news.org/spip.php?sommaire&lang=en (accessed on 26 October 2017).

21.   Prayitno, S.B. Secrets of Successful RBOs. Center for River Basin Organizations and Management, Solo, Central Java, Indonesia. 2011. Available online: http://www.inbo-news.org/IMG/pdf/SPS40-Secrets-2.pdf (accessed on 2 October 2017).

22.   Mokhtar, M.B.; Toriman, M.E.H.; Hossain, A.A. Social learning in facing challenges of sustainable development: A case of Langat River Basin, Malaysia. *Res. J. Appl. Sci.* **2010**, *5*, 434–443.

23.   GOM. *Local Government Act 1976*; Government of Malaysia: Putrajaya, Malaysia, 2006.

24.   Unit Perancang Ekonomi. *Water Sector Transformation 2040 (WST2040)*; Bahagian Infrastruktur dan Kemudahan Awam, Unit Perancang Ekonomi, Jabatan Perdana Menteri: Putrajaya, Malaysia, 2022; Available online: https://wst2040.my/ (accessed on 10 January 2023).

25.   Ahmed, M.F.; Mokhtar, M.; Alam, L.; Ta, G.C.; Ern, L.K.; Khalid, R.M. Recognition of Local Authority for Better Management of Drinking Water at the Langat River Basin, Malaysia. *Int. J. Eng. Technol.* **2018**, *7*, 148–154. [CrossRef]

26.   Calzada, I. Democratising smart cities? Penta-helix multistakeholder social innovation framework. *Smart Cities* **2020**, *3*, 1145–1172. [CrossRef]

27.   Chan, N.W. Sustainable management of rivers in Malaysia: Involving all stakeholders. *Int. J. River Basin Manag.* **2005**, *3*, 147–162.

28.   Werblow, S. Steady data flow tracks floods. *Hydro Int.*: HJ Lemmers, The Netherlands. 2009. Available online: https://www.hydro-international.com/content/article/steady-data-flow-tracks-floods (accessed on 5 July 2023).

29. Elfithri, R.; Toriman, M.E.B.; Mokhtar, M.B.; Juahir, H.B. Perspectives and initiatives on integrated river basin management in Malaysia: A review. *Soc. Sci.* **2011**, *6*, 169–176. [CrossRef]

30. Mokhtar, M.B.; Toriman, M.E.H.; Hossain, M.A.A.; Tan, K.W. Institutional challenges for integrated river basin management in Langat River Basin, Malaysia. *Water Environ. J.* **2011**, *25*, 495–503. [CrossRef]

31. Mei, S.L.; Carr, G.; Weng, C.N. Resolving Water Disputes Via Interstate Co-operation And Stakeholders' Engagement: A Case Study from Muda River Basin. *Adv. Sci. Lett.* **2017**, *23*, 9126–9129. [CrossRef]

32. Lai, C.H.; Tan, D.T.; Chan, N.W.; Zakaria, N.A. Applying a system thinking approach to explore root causes of river pollution: A preliminary study of Pinang river in Penang State, Malaysia. In Proceedings of the Proceedings-International Association for Hydro-Environment Engineering and Research (IAHR)-Asia Pacific Division (APD) Congress: Multi-Perspective Water for Sustainable Development, IAHR-APD, Yogyakarta, Indonesia, 2–5 September 2018; Volume 2, pp. 1441–1448.

33. Chan, N.W.; Ghani, A.A.; Samat, N.; Hasan, N.N.N.; Tan, M.L. Integrating Structural and Non-structural Flood Management Measures for Greater Effectiveness in Flood Loss Reduction in the Kelantan River Basin, Malaysia. *Lect. Notes Civ. Eng.* **2020**, *53*, 1151–1162.

34. Ahmed, M.F.; Mokhtar, M.B.; Lim, C.K.; Majid, N.A. Identification of Water Pollution Sources for Better Langat River Basin Management in Malaysia. *Water* **2022**, *14*, 1904. [CrossRef]

35. Muhyi, H.A.; Chan, A.; Sukoco, I.; Herawaty, T. The Penta Helix collaboration model in developing centers of flagship industry in Bandung city. *Rev. Integr. Bus. Econ. Res.* **2017**, *6*, 412–417.

36. Sjögren Forss, K.; Kottorp, A.; Rämgård, M. Collaborating in a penta-helix structure within a community based participatory research programme:'Wrestling with hierarchies and getting caught in isolated downpipes'. *Arch. Public Health* **2021**, *79*, 1–13. [CrossRef]

37. Cabrera-Flores, M.; López-Leyva, J.; Peris-Ortiz, M.; Orozco-Moreno, A.; Francisco-Sánchez, J.; Meza-Arballo, O. A framework of penta-helix model to improve the sustainable competitiveness of the wine industry in Baja California based on innovative natural resource management. *E3S Web Conf.* **2022**, *167*, 06005, EDP Sciences. [CrossRef]

38. DID. *Policy Responses to Attain the Water Quality Target: Malaysia Experience*; Department of Drainage and Irrigation: Kuala Lumpur, Malaysia, 2011. Available online: http://www.wepa-db.net/pdf/1203forum/03.pdf (accessed on 24 September 2017).

39. Mokhtar, M.B.; Lee, K.E.; Sivapalan, S. *Rising to the Challenge: Malaysia's Contribution to the SDGs*; Penerbit Universiti Kebangsaan Malaysia: Bangi, Malaysia, 2017.

40. UNESCO. *HELP-Hydrology for Environment, Life and Policy*; United Nations Educational, Scientific and Cultural Organization (UNESCO): Paris, France, 2010.

41. Elfithri, R. *Establishment of Sustainability Science Demonstration Pilot Project on Restoring and Managing Langat River, Malaysia for Future*; Institute for Environment and Development: Bangi, Malaysia, 2016; Available online: http://jfit-for-science.asia/wp-content/uploads/502_Sustainability-Science-Pilot-Project-Langat-River-Malaysia.pdf (accessed on 24 September 2017).

42. Juahir, H.; Zain, S.M.; Yusoff, M.K.; Hanidza, T.T.; Armi, A.M.; Toriman, M.E.; Mokhtar, M. Spatial water quality assessment of Langat River Basin (Malaysia) using environmetric techniques. *Environ. Monit. Assess.* **2011**, *173*, 625–641. [CrossRef]

43. LUAS. *Sungai Langat, State of the River Report 2015*; Lembaga Urus Air Selangor (LUAS): Shah Alam, Malaysia, 2015; ISBN 2180-2793.

44. DOE. *Study on Pollution Prevention and Water Quality Improvement Program of Sungai Langat*; Shaaban, M.G., Alwan, H.K., Jaafar, M.N., Abdullah, A.R., Ismail, M.Z., Idris, A., Hamidon, A.B., Sharom, A., Zakri, I.M., Othman, P.F., et al., Eds.; Department of Environment (DOE): Kuala Lumpur, Malaysia, 2003.

45. DOS. *Population Distribution by Local Authority Areas and Mukims 2010*; Department of Statistics (DOS): Putrajaya, Malaysia, 2013. Available online: http://newss.statistics.gov.my/newss-portalx/ep/epProductFreeDownloadSearch.seam (accessed on 10 November 2017).

46. DOE. *Malaysia Environmental Quality Report 2013*; Department of Environment (DOE): Putrajaya, Malaysia, 2013.

47. Juahir, H. Water Quality Data Analysis and Modeling of the Langat River Basin. Ph.D. Thesis, University of Malaya, Kuala Lumpur, Malaysia, 2009. Available online: http://repository.um.edu.my/1223/5/Chapter%203%20Revised.pdf (accessed on 20 June 2017).

48. Aris, A.Z.; Lim, W.Y.; Looi, L.J. Natural and Anthropogenic Determinants of Freshwater Ecosystem Deterioration: An Environmental Forensic Study of the Langat River Basin, Malaysia. In *Environmental Management of River Basin Ecosystems*; Ramkumar, M., Ed.; Springer Earth System Sciences: Cham, Switzerland, 2015; pp. 455–476. [CrossRef]

49. Ching, Y.C.; Yatim, B.; Toriman, M.E.; Lee, Y.H.; Surif, S.; Abdullah, M. Impact of Climate Change on Flood Risk in the Muar River Basin of Malaysia. *Disaster Adv.* **2013**, *6*, 11–17.

50. DID. *Flood Management (Vol. 1)*; Department of Irrigation and Drainage (DID): Kuala Lumpur, Malaysia, 2009.

51. Hussaini, H.A. *Flood and Drought Management in Malaysia*; Ministry of Natural Resources and Environment: Kuala Lumpur, Malaysia, 2007.

52. Tam, T.; Ibrahim, A.; Rahman, M.; Mazura, Z. Flood Loss Assessment in the Kota Tinggi. In Proceedings of the IOP Conference Series: Earth and Environmental Science, Kuching, Malaysia, 26–29 August 2013.

53. IPCC. *Climate Change 2014: Impacts, Adaptation, and Vulnerability. Part B: Regional Aspects. Contribution of Working Group II to the Fifth Assessment Report of the Intergovernmental Panel on Climate Change*; Barros, V.R., Field, C.B., Dokken, D.J., Mastrandrea, M.D.,

Mach, K.J., Bilir, T.E., Chatterjee, M., Ebi, K.L., Estrada, Y.O., Genova, R.C., et al., Eds.; Cambridge University Press: Cambridge, UK; New York, NY, USA, 2014.

54. Ahmed, M.F. Risk Based Assessment and Management of Drinking Water in Langat River Basin, Malaysia. Ph.D. Thesis, Universiti Kebangsaan Malaysia (UKM), Bangi, Malaysia, 2019.

55. Arumugam, T. 20 Slopes in Selangor Identified at Critical Risk of Landslide. New Straits Times. 2017. Available online: https://www.nst.com.my/news/nation/2017/09/282057/20-slopes-selangor-identified-critical-risk-landslide (accessed on 24 September 2017).

56. Muthiah, W. Selangor Told to Prepare for Big Waves Due to High Tides over Next Three Months. New Straits Times. 2017. Available online: http://www.thestar.com.my/news/nation/2017/09/20/selangor-told-to-prepare-for-tidal-waves-over-next-three-months/ (accessed on 24 September 2017).

57. Suhaini, I. *Information Related to Water Treatment Plants at Langat River Basin. Interview on 26 January 2015*; Puncak Niaga Sdn. Bhd.: Shah Alam, Malaysia, 2015.

58. Ahmed, M.F.; Mokhtar, M.B. Treated water quality based on conventional method in Langat River Basin, Malaysia. *Environ. Earth Sci.* **2020**, *79*, 1–16. [CrossRef]

59. Rapport, D.J.; Lasley, B.L.; Rolston, D.E.; Nielsen, N.O.; Qualset, C.O.; Damania, A.B. *Managing for Healthy Ecosystems*; CRC Press: New York, NY, USA, 2002.

60. Perumal, E.; Michael, S. Polluters in Their Midst, the Star Online. 2016. Available online: http://www.thestar.com.my/metro/community/2016/11/07/polluters-in-their-midst-residents-say-factories-dotting-the-banks-of-sungai-semenyih-are-the-cause/ (accessed on 20 October 2017).

61. Isnin, M.N. *Shutdown History of Sungai Semenyih Water Treatment Plant (WTP)*; Interview with the Sungai Semenyih WTP Authority: Putrajaya, Malaysia, 2015.

62. Bernama. Ammonia Pollution Found in Sungai Langat, Malaysiakini. 2014. Available online: http://www.malaysiakini.com/news/253179 (accessed on 10 March 2016).

63. SPAN. *Managing Water Pollution for Public Water Supply-Issues and Strategies*; Suruhanjaya Perkhidmatan Air Negara (SPAN): Cyberjaya, Malaysia, 2012. Available online: http://www.mywp.org.my/wp-content/uploads/2012/12/materials/folder1/Paper%2010-Managing%20Water%20Pollution%20for%20Public%20Water%20Supply%20-%20Issues%20and%20Strategies,%20SPAN.pdf (accessed on 20 February 2015).

64. Alkimia, C. Bekalan Air Oleh loji2 Rawatan Air (LRA). 2014. Available online: https://blogkaryait.wordpress.com/2014/05/18/bekalan-air-oleh-loji2-rawatan-air-lra/ (accessed on 15 January 2015).

65. Krishnan, G. Salak Tinggi Water Treatment Plant May be Closed for Good. The Star Online. 2009. Available online: http://www.thestar.com.my/story/?file=%2F2009%2F7%2F15%2Fcentral%2F4317744&sec=central (accessed on 20 February 2015).

66. Zack, J. Water Disruption to Hit More than 370,000 after Pollution Shuts Down Selangor Treatment Plant. The Star Online. 2019. Available online: https://www.thestar.com.my/news/nation/2019/09/28/water-disruption-to-hit-more-than-370000-after-pollution-shuts-down-selangor-treatment-plant (accessed on 20 October 2019).

67. PNSB. Introduction to Puncak Niaga Sdn. Bhd. Shah Alam, Malaysia. 2010. Available online: http://www.jwrc-net.or.jp/aswin/projects-activities/trainee_files/20101004-15_CR04.pdf (accessed on 10 March 2015).

68. Abidin, K.N.B.Z. *Interview with Sg. Labu Water Treatment Plant's Manager*; Loji Rawatan Air Sg. Labu, Kampung Lembah Paya, Salak Tinggi; Sepang: Selangor, Malaysia, 2015.

69. Waseem, S.; Mohsin, I.U. Evolution and fate of haloacetic acids before and after chlorination within the treatment plant using SPE-GC-MS. *Am. J. Anal. Chem.* **2011**, *2*, 522. [CrossRef]

70. H2O. Particle Size Chart: Removal Range by Filtration. H_2O Distributors, USA. 2017. Available online: https://www.h2odistributors.com/pages/info/diagram-particle-sizes.asp (accessed on 20 August 2017).

71. USEPA. *A Regulators' Guide to the Management of Radioactive Residuals from Drinking Water Treatment Technologies*; United States Environmental Protection Agency: Washington, DC, USA, 2005. Available online: https://www.epa.gov/sites/production/files/2015-05/documents/816-r-05-004.pdf (accessed on 10 November 2018).

72. Lesikar, B.J.; Melton, R.; Hare, M.; Hopkins, J.; Dozier, M. *Drinking Water Problems: Radionuclides (Spanish)*; Texas A&M AgriLife Extension Service: TAMU College Station, TX, USA, 2006; Available online: https://twon.tamu.edu/wp-content/uploads/sites/3/2021/06/drinking-water-problems-radionuclides.pdf (accessed on 5 July 2023).

73. Munter, R. Technology for the removal of radionuclides from natural water and waste management: State of the art. *Proc. Est. Acad. Sci.* **2013**, *62*, 122–132. [CrossRef]

74. PASSB. *Performance Report for 2015 (Air Selangor)*; Pengurusan Air Selangor Sdn Bhd.: Kuala Lumpur, Malaysia, 2015.

75. Aquathin. Particle Size Remova; Range by Filtration & Reverse Osmosis. 2017. Available online: http://www.aquathin.com/sites/default/files/particle%20size.pdf (accessed on 22 May 2017).

76. Fixr. Water Treatment System Cost. Fixr. 2017. Available online: https://www.fixr.com/costs/water-purification-system (accessed on 3 October 2017).

77. WHO. Physical Removal Processes: Sedimentation and Filtration. World Health Organization, Switzerland. 2017. Available online: http://www.who.int/water_sanitation_health/dwq/WSH02.07_5.pdf?ua=1 (accessed on 27 October 2017).

78. Anon. Map of Malaysia. 2017. Available online: https://image.slidesharecdn.com/geography-111113043232-phpapp02/95/geography-50-728.jpg?cb=1321159034 (accessed on 7 September 2017).

79. Mark, W.L.; Keung, K.A. Water Treatment and Pathogen Control: Process Efficiency in Achieving Safe Drinking Water. World Health Organization, IWA Publishing, UK. 2004. Available online: http://www.who.int/water_sanitation_health/publications/9241562552/en/ (accessed on 27 October 2017).
80. Khairul, A.M. Jurisdictional and by-Laws Issues Hampering Enforcement of Environmental Quality Act, Says Wan Junaindi. New Straits Times. 2017. Available online: https://www.nst.com.my/news/nation/2017/11/297929/jurisdictional-and-laws-issues-hampering-enforcement-environmental (accessed on 2 November 2017).
81. Alsalahi, M.A.; Latif, M.T.; Ali, M.M.; Magam, S.M.; Wahid, N.B.A.; Khan, M.F.; Suratman, S. Distribution of surfactants along the estuarine area of Selangor River, Malaysia. *Mar. Pollut. Bull.* **2014**, *80*, 344–350. [CrossRef] [PubMed]
82. Shamsudin, N.S. Malaysian Foresight Institute, Good Example to Plan For Country's Future. Singapore News Tribe. 2016. Available online: http://www.singaporenewstribe.com/malaysian-foresight-institute-good-example-to-plan-for-countrys-future/ (accessed on 28 August 2017).
83. Ahmed, M.F.; Mokhtar, M.B.; Lim, C.K.; Hooi, A.W.K.; Lee, K.E. Leadership roles for sustainable development: The case of a Malaysian green hotel. *Sustainability* **2021**, *13*, 10260. [CrossRef]
84. Ahmed, M.F.; Mokhtar, M.B.; Alam, L. Factors influencing people's willingness to participate in sustainable water resources management in Malaysia. *J. Hydrol. Reg. Stud.* **2020**, *31*, 100737. [CrossRef]
85. ASM. *Strategies to Enhance Water Demand Management in Malaysia*; Academy of Sciences Malaysia (ASM): Kuala Lumpur, Malaysia, 2016; ISBN 978-983-2915-27-0.
86. Khalid, R.M.; Mokhtar, M.B.; Jalil, F.; Rahman, A.B.S.; Spray, C. Legal framing for achieving good ecological status for Malaysian rivers: Are there lessons to be learned from the EU Water Framework Directive? *Ecosyst. Serv.* **2017**, *29*, 251–259. [CrossRef]
87. DOE. *Malaysia Environmental Quality Report 2015*; Department of Environment (DOE): Putrajaya, Malaysia, 2015.
88. MOH. *Drinking Water Quality Standard*; Engineering Services Division, Ministry of Health (MOH): Putrajaya, Malaysia, 2010.
89. Ahmed, M.F.; Alam, L.; Ta, G.C.; Mohamed, C.A.R.; Mokhtar, M. A Preliminary Study to Investigate the Trace Metals in Drinking Water Supply Chain. In Proceedings of the In-House Seminar of the Chemical Oceanography Laboratory, Universiti Kebangsaan Malaysia (UKM), Bangi, Malaysia, 2 June 2016.
90. Farizwana, M.R.S.; Mazrura, S.; Fasha, A.Z.; Rohi, G.A. Determination of aluminium and physicochemical parameters in the palm oil estates water supply at Johor, Malaysia. *J. Environ. Public Health* **2010**, *2010*, 615176. [CrossRef]
91. Qaiyum, M.; Shaharudin, M.; Syazwan, A.; Muhaimin, A. Health risk assessment after exposure to aluminium in drinking water between two different villages. *J. Water Resour. Prot.* **2011**, *3*, 268. [CrossRef]
92. Flaten, T.P. Aluminium as a risk factor in Alzheimer's disease, with emphasis on drinking water. *Brain Res. Bull.* **2001**, *55*, 187–196. [CrossRef]
93. Nalatambi, S. Determination of metals in tap water using atomic absorption spectrometry: A case study in Bandar Sunway residential area. *Sunway Acad. J.* **2009**, *6*, 33–46.
94. Sauvé, S.; Desrosiers, M. A review of what is an emerging contaminant. *Chem. Cent. J.* **2014**, *8*, 15. [CrossRef] [PubMed]
95. Deshommes, E.; Tardif, R.; Edwards, M.; Sauvé, S.; Prévost, M. Experimental determination of the oral bioavailability and bioaccessibility of lead particles. *Chem. Cent. J.* **2012**, *6*, 138. [CrossRef] [PubMed]
96. Hu, J.; Ma, Y.; Zhang, L.; Gan, F.; Ho, Y.S. A historical review and bibliometric analysis of research on lead in drinking water field from 1991 to 2007. *Sci. Total Environ.* **2010**, *408*, 1738–1744. [CrossRef] [PubMed]
97. Santhi, V.; Sakai, N.; Ahmad, E.; Mustafa, A. Occurrence of bisphenol A in surface water, drinking water and plasma from Malaysia with exposure assessment from consumption of drinking water. *Sci. Total Environ.* **2012**, *427*, 332–338. [CrossRef]
98. Azlan, A.; Khoo, H.; Idris, M.; Amin, I.; Razman, M.R. Consumption patterns and perception on intake of drinking water in Klang Valley, Malaysia. *Pak. J. Nutr.* **2012**, *11*, 584–590. [CrossRef]
99. Shaharuddin, M.S.; Kamil, Y.M.; Ismail, Y.M.; Firuz, R.M.; Aizat, I.S.; Yunus, A.M. Fluoride concentration in malaysian drinking water. *Am. Eurasian J. Agric. Env. Sci.* **2009**, *6*, 417–420.
100. LUAS. *The 6th General Meeting NARBO, Hybrid Off River Augmentation System (HORAS) Project and Pumping Operation from Alternative Ponds to Selangor River (OPAK) 22–24 FEBRUARY 2017*; Lembaga Urus Air Selangor (LUAS): Shah Alam, Malaysia, 2017; Available online: http://www.narbo.jp/data/01_events/materials_6thgm/2_03.%20HORAS%20LUAS.pdf (accessed on 24 September 2017).
101. Khan, S. *UNESCO Ecohydrology Program Leading the Way to an Interdisciplinary Approach to Sustainable Water Resources Management*; United Nations Educational, Scientific and Cultural Organization (UNESCO): Paris, France, 2017; Available online: http://www.mwa.org.my/events/20130205/1.%20Shahbaz%20Khan_Ecohydrology.pdf (accessed on 24 September 2017).
102. UNDESA. *International Decade for Action 'River for Life' 2005–2015*; United Nations Department of Economic and Social Affairs (UNDESA): New York, NY, USA, 2014; Available online: http://www.un.org/waterforlifedecade/transboundary_waters.shtml (accessed on 26 September 2017).
103. SIWI. *Cooperation over Shared Waters*; Stockholm International Water Institute: Stockholm, Sweden, 2015; Available online: http://www.siwi.org/priority-area/transboundary-water-management/ (accessed on 26 September 2017).
104. UNWATER. *Transboundary Waters*; United Nations: Geneva, Switzerland, 2016; Available online: http://www.unwater.org/water-facts/transboundary-waters/ (accessed on 26 September 2017).

105. UNECE. *Good Practices in Transboundary Water Cooperation*; United Nations Economic Commission for Europe: Geneva, Switzerland, 2015; Available online: http://www.unece.org/fileadmin/DAM/env/water/publications/WAT_Good_practices/2015_PCCP_Flyer_Good_Practices__LIGHT_.pdf (accessed on 26 September 2017).
106. UNESCO. *The United Nations World Water Development Report 2015*; United Nations Educational, Scientific and Cultural Organization: Paris, France, 2015; Available online: http://unesdoc.unesco.org/images/0023/002318/231823E.pdf (accessed on 26 September 2017).
107. ICPDR. *Countries of the Danube River Basin*; International Commission for the Protection of the Danube River: Vienna, Austria, 2017; Available online: https://www.icpdr.org/main/danube-basin/countries-danube-river-basin (accessed on 28 September 2017).
108. MDBA. *River Operations Reports*; Murray-Darling Basin Authority: Canberra City, Australia, 2017. Available online: https://www.mdba.gov.au/ (accessed on 28 September 2017).
109. Sachs, J. Aristotle: Ethics. Internet Encyclopedia of Philosophy. 2017. Available online: http://www.iep.utm.edu/aris-eth/ (accessed on 24 October 2017).
110. Durant, W. *The Story of Philosophy, The Lives and Opinions of Greater Philosopher*; Simon & Schuster, Inc.: New York, NY, USA, 1926; p. 74.

Article

Digitalization of Water Distribution Systems in Small Cities, a Tool for Verification and Hydraulic Analysis: A Case Study of Pamplona, Colombia

Carlos Bonilla [1,2,*], Bruno Brentan [2,3], Idel Montalvo [2,4], David Ayala-Cabrera [2,5] and Joaquín Izquierdo [2]

[1] Department of Civil, Environmental and Chemical Engineering, Faculty of Engineering and Architecture, University of Pamplona, Pamplona 543050, Colombia

[2] Fluing-Institute for Multidisciplinary Mathematics, Universitat Politècnica de València, 46022 Valencia, Spain; brentan@ehr.ufmg.br (B.B.); imontalvo@ingeniousware.net (I.M.); david.ayala-cabrera@ucd.ie (D.A.-C.); jizquier@upv.es (J.I.)

[3] Hydraulic Engineering and Water Resources Department, School of Engineering, Federal University of Minas Gerais, Belo Horizonte 31270-901, Brazil

[4] Business Development, IngeniousWare GmbH, Jollystrasse 11, 76137 Karlsruhe, Germany

[5] Centre for Water Resources Research, School of Civil Engineering, University College Dublin, D04 V1 W8 Dublin, Ireland

[*] Correspondence: carlos.bonilla@unipamplona.edu.co

Abstract: Digitalization in water networks is essential for the future planning of urban development processes in cities and is one of the great challenges faced by small cities regarding water management and the advancement of their infrastructures towards sustainable systems. The main objective of this study is to propose a methodology that allows water utilities with limited budgets to start the path toward the digitalization and construction of the hydraulic model of their water distribution networks. The small city of Pamplona in Colombia was used as a case study. The work explains in detail the challenges faced and the solutions proposed during the digitalization process. The methodology is developed in six phases: an analysis of the cadastre and existing information, the creation and conceptualization of the base hydraulic model, the development of the topography using drones with a limited budget, an analysis of water demand, the development of a digital hydraulic model, and a hydraulic analysis of the system. The product generated is a tool to assess the overall performance of the network and contributes to the advancement of SDG-6, SDG-9, and SDG-11. Finally, this document can be replicated by other cities and companies with similar characteristics (e.g., limited size and budget) and offers an intermediate position on the road to digitalization and the first steps towards the implementation of a digital twin.

Keywords: water distribution network; water management; digitalization; digital hydraulic model; digital elevation model

Citation: Bonilla, C.; Brentan, B.; Montalvo, I.; Ayala-Cabrera, D.; Izquierdo, J. Digitalization of Water Distribution Systems in Small Cities, a Tool for Verification and Hydraulic Analysis: A Case Study of Pamplona, Colombia. *Water* **2023**, *15*, 3824. https://doi.org/10.3390/w15213824

Academic Editors: Peiyue Li and Jianhua Wu

Received: 30 September 2023
Revised: 29 October 2023
Accepted: 30 October 2023
Published: 1 November 2023

1. Introduction

Urban water management carried out by water utilities is based on the implementation of various actions allowing the optimization and sustainability of distribution processes through decision-making [1]. Improvement actions should focus on the sustainability and comprehensive performance of the supply and sanitation networks [2]. To this aim, the Sustainable Development Goals (SDG) set clear objectives and diverse actions to ensure the future availability of water and sanitation to communities through the construction of resilient systems and sustainable cities. These actions aim to improve water management and are mainly framed in three objectives, namely SDG-6, SDG-9, and SDG-11 [3].

In some cases, government entities and water companies develop joint actions to ensure water availability, implementing awareness campaigns in the population about water security and creating responsible water citizens, involving state actors, non-state actors, and citizens, conceptualizing and identifying their roles and responsibilities in water use [4].

Cities and their water utilities face diverse challenges to ensure access to drinking water in a safe environment, such as water scarcity and population growth [5], aging infrastructure [6], and climate change [7], among others. A possible solution to address these problems is found in technological innovation and the digital transformation of water distribution systems (WDSs) [8], because digital models allow for achieving knowledge transfer, resilience during atypical events, and improvements in terms of efficiency [9].

Technology and its advances allow the development of several management actions in WDS amenable to being leveraged via the use of optimization techniques for the planning and management of those systems, thus leading directly to the economic improvement of water companies and respect of the environment. The main management actions are aimed at solving optimization problems in designing WDSs, incorporating new methodologies based on minimizing cost factors, and developing intelligent optimization algorithms [10]. Other actions are based on the digital metering of water consumption, which could allow the detection of leaks at the user level and distribution network level [11]. Likewise, the monitoring and measurement of system parameters (e.g., pressure and water flow) improve decision-making and optimize the resources of water utilities and authorities [12], thus minimizing water leakage in their networks [13,14].

The age and quality of water in the WDS is also an important field of research in which different management and optimization actions based on technology have been proposed. The optimization of the operational management of valves has allowed for the minimization of the age of water in pipes via the implementation of a combination of validated algorithms in networks of different complexities, guaranteeing the supply of water with adequate quality [15]. The analysis of the exposure and vulnerability of users to trihalomethanes (THMs) existing in WDSs, carried out using water quality algorithms, has made it possible to identify critical areas with high THM concentrations with different exposure periods, allowing water companies to design public health strategies to reduce risks [16].

The data available from network operation allows the management and operation of water networks and estimates the real state of their non-monitored elements (e.g., pipes), leading to the building of digital twins (DTs). DTs integrate the data monitored in real-time with optimization algorithms and virtual network models through geographic information systems [17]. An essential requirement to build a DT in a WDS is the digitization of its network and, from this stage, the start of a continuous process of adjustments and learning of the system [18].

To support water management, other digital and technological tools—e.g., unmanned aerial vehicles (drones) and satellite images—have been used to quickly and accurately analyze the behavior of some elements of WDSs. Drones have been used to measure reservoir water levels [19] and analyze water quality in reservoirs or supply sources [20]. Satellite imagery has been used for leak detection in WDSs, successfully identifying leaks and saving operational costs [21]. In this sense, digitalization appears to be a powerful tool for WDS management that translates into economic benefits for water utilities.

The digitalization of WDSs is considered an evolving process that is achieved by managing infrastructure through digital technologies. Currently, the impact of digitalization in the drinking water sector is carried out on a smaller scale compared to that in other sectors [22]. There are different factors of complexity, uncertainty, and dynamism in water supply systems that demand a need to revolutionize the water industry and the adoption of new digital technologies that combine monitoring, supervisory control, and data acquisition (SCADA) systems, decision support in real-time water networks, and information and communication technology (ICT)-based solutions [23].

Currently, research on the digitalization of WDSs seeks to meet socioeconomic, environmental, sustainability, and climate needs to improve their efficiency and productivity [24,25], as well as to generate strategic opportunities to address the challenges associated with the SDGs [26]. On this path to digitization, different technologies, called emerging technologies (ETs), are identified as being employed by water utilities worldwide in the digital transformation of their networks. ET can be grouped into five major groups: cyber-physical systems, the Internet of Things (IoT), big data analytics, artificial intelligence, and cloud computing [23].

Research showing the applicability of ETs in WDSs worldwide can be found in the literature. Cyber-physical systems have been applied in Spain [27,28] and India [29]. IoT has had wider application in countries such as Singapore, South Korea, Malta, and South Africa [23]. Big data analytics have been developed in countries such as Morocco [30,31], India [32], Sri Lanka [33] and Italy [34]. Artificial intelligence has been used in Spain [35,36], Mexico [36], Jerusalem [37], United Kingdom [38] and Jordan [39]. Computation has been implemented in Italy [40], the United States [41], Egypt [42], Spain [43], and India [44]. These investigations provide an overview of digitalization in WDSs and the different approaches adopted in different countries around the world.

As a cornerstone, for the implementation of technological actions in WDSs that promote sustainable water management, water utilities need to digitize their supply systems elements. However, the delay in the digitization (and digitalization) process is directly related to the economic costs associated with its implementation [45]. In many cities, mainly in Latin America, the information available from the WDSs is highly vulnerable and susceptible to losses because the information of their networks (location, diameters, and materials) exists in a physical (not digitized) format and, even in some cases, the information only exists in the memory of the most experienced workers [46]. Some water utilities are unaware of some elements of their WDS and do not have accurate information on their oldest pipes, which makes it challenging to start digitizing their networks [47].

The operation of WDSs in small cities is commonly conducted empirically or experimentally using the applications of the practical know-how of water utility experts. Future planning related to the expansion and construction of new networks is based on factors affecting water demand, such as population growth or the development of new areas in the city according to their needs (e.g., industrial areas) [48]. A city or a water utility without system digitalization does not have sufficient analytical tools to predict/anticipate potential changes that may affect consumers' needs. For example, in the case of a WDS, it may not even have a hydraulic model of its distribution networks. An appropriate (well-calibrated) model allows simulations and test scenarios to be carried out to estimate the effects generated in the existing networks by changes caused by topological modifications or the occurrence of an event, which may compromise the operation of the system. These events include changes such as variations in pressure or flow at different points in the city due to the incorporation of new users of the system, among others. Therefore, building hydraulic models of a WDS becomes an essential step in developing other actions of technological innovation (e.g., the operation optimization of the current network) to improve water management [49].

This work proposes a methodology to advance towards the digitalization and transformation of WDSs. It is mainly focused on networks with reduced size in the context of a limited budget. The proposed methods for creating a hydraulic model of a real WDS, and how to refine (or incorporate new) elements of the system with the assistance of a low-cost technology (in this case drone) are described in detail herein. The challenges faced and solutions adopted during the construction of the hydraulic model are also explained in this paper. This research attempts to become a base statement, serving as a guide that can be replicated by other cities and companies with similar characteristics and concerns with relatively low economic investment. The hydraulic model for the WDS of Pamplona, Colombia, is created. The results obtained in the hydraulic simulations of the created model are analyzed and compared with pressure measurement data at different points in the network. The work carried out provides the city's water company with a fundamental tool to manage and optimize drinking water distribution.

2. Materials and Methods

The methodology proposed in this paper is presented in Figure 1.

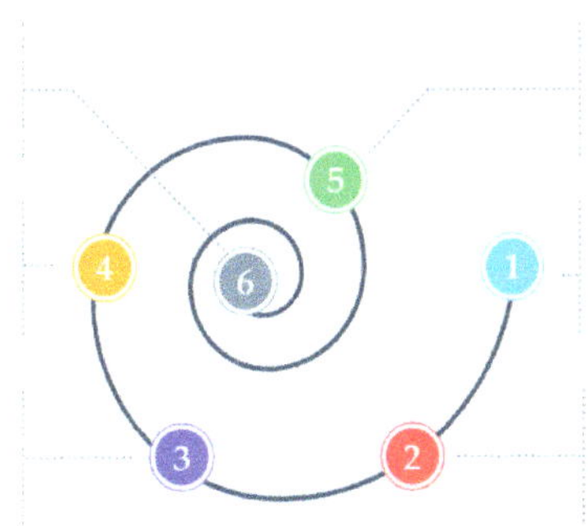

Figure 1. Schematic configuration of the proposed methodology.

In Figure 1, the spiral design of the methodology allows its incorporation at any step of its development. The proposed methodology consists of the following main steps:

Step 1. It begins with collecting and analyzing the existing data related to the land registry of the network infrastructure.

Step 2. The next step is conceptualizing the basic hydraulic model to understand its functioning, such as the direction of the flows and storage volumes.

Step 3. Photogrammetric images are obtained using a drone, which helps create the city's digital elevation model. This digital elevation model is used to obtain the network elevation for both pre-existing and new expansion areas that comprise the city.

Step 4. Existing user consumption data are used in this step to analyze the base demands of the model. The hydraulic model is built using hydraulic computer engines (in this case the EPANET 2.2 software [50–52]) and the model is incorporated into an online platform for monitoring and optimizing water networks (in this paper, WATERing software [53]).

Step 5. The hydraulic analysis of the model is conducted, and the in-field pressure measurements of the water utility are used to compare the obtained results (basic network calibration).

Step 6. Afterward, a preliminary calculation of the non-revenue water index is conducted in this step.

Each step of the proposed methodology is described in detail in the sections below.

2.1. Analysis of Existing Information

Among the main identified problems for building a hydraulic model is the lack of documented information regarding the existing networks and infrastructure of water utilities (this is particularly critical in public institutions) [46]. The lack of standardized processes for storing and documenting public records many times leads to the loss or deterioration of information over time. Therefore, the recorded information is partially reliable and must be verified and updated. In this sense, the first step of the proposed methodology focuses on the collection of the available information from the system and its preliminary digitization. The sources of the information used in this work and essentials of its are as follows.

- Documentary record: In the government offices (technical offices in charge of the infrastructure in each city), physical and digital layouts should be consulted, as well as technical documents related to the land registry of the WDS. Similarly, existing WDS layouts in the water company's technical offices should be found and compiled.

- Urban layouts: The most up-to-date layouts of the city, urban roads, road infrastructure, and land use maps should be obtained from public offices.
- Satellite images: Currently, there is recent, updated, and free-access satellite information that shows the development and urban growth of each community. It is essential to use the existing edited images of the study site and contrast them with the city's physical and digital maps. With these tools, it is possible to identify new neighborhoods (e.g., recent urban developments or unplanned neighborhoods made up of low-income or immigrant families on the outskirts of the city) or non-registered urban settlements in the existing maps. This information should be included and updated.
- Tacit information: Once the existing information is verified and analyzed, field visits should be carried out, preferably with the water utility experts to know the operation of the WDS. The most experienced active personnel in water companies have excellent information in their memory and this information is called tacit information. This type of information is valuable to contrast the information obtained via documentary records, urban layouts, and satellite images with. Fieldwork helps to identify the visible components of the system (valves, reservoirs, reservoirs, pumping stations, hydrants, and sensors) and their physical properties (dimensions, diameters, and materials) to complement the missing information. If necessary, field inspections should be conducted using boreholes to obtain any missing information. Subsequently, this information should be digitized and reflected in an initial layout of the WDS.
- Preliminary digitization: Using the collected information, a preliminary digitization of the WDS is carried out using a computer-aided design (CAD); alternatively, geographic information systems (GIS) can also be used in this activity. It is essential to use several layers in the drawing process to identify and classify the different materials and diameters of the pipes and the existing elements in the network (valves, tanks, and hydrants). This process allows us to roughly understand the general structure of the WDS and to get to know the most important pipes and elements in the network.

2.2. Conceptualization of the Hydraulic Model

It is necessary to understand the WDS's hydraulic operation, identifying the main elements of the network (e.g., main pipes) and the operational hydraulic zones. For the development of the second step of the proposed methodology, it is essential to identify the following.

- Continuity of service: This is determined to know the actual time of service offered by the system during the day, to know if the service is continuous or intermittent, and in the case of the intermittent systems, to identify the service shifts and the areas supplied in the different shifts that may exist. This information is relevant to determining consumption patterns and should be consulted with the water company's experts; if possible, the company's records of pipe damage and repair times should be consulted.
- Pipe information: It is necessary to know the diameter of each pipe, length, type of material, and approximate age. Knowing the age allows for the establishment of an approximate roughness coefficient, which is necessary for the loss equations of the hydraulic model.
- Network fittings: These comprise the location, diameter, and material of strategic valves used by workers to operate the network, as well as pressure regulating valves, need to be identified. Hydrants and other relevant accessories for hydraulic operation should also be identified.
- Storage tanks: It is essential to know the existing tanks, their location, volume, and internal dimensions, as well as the variation in water levels throughout the day. It is also necessary to have an idea of the supply areas of each tank.
- Pumping equipment (if any): The pumping stations that operate in the network must be identified; it is necessary to have clarity on the number of pumps installed, power

characteristics, and models installed to determine the operating curve of each pump and to know the suction and discharge pipes.

- District metered areas: If there are metered district areas in the WDS where flow or pressure data are available, they should be detailed and similarly shown in the network plans and in the hydraulic model to be built. If there is no sectorization, the company's experts should be consulted on how the network is operated, whether it supplies all users continuously and without district zones, or whether there are service shifts in defined areas determined on an experimental basis with valve management. Intermittent service provision should be analyzed based on a thorough knowledge of intermittent water distribution, as recommended [54].
- Monitoring data: The existence of measured data on pressure, flow, and water quality in the network should be investigated. These data can help understand the behavior of the network and will be used to perform the calibration process. In water utilities that do not have this information, it is recommended to implement monitoring campaigns—preferably pressure and flow—at some points of the network and according to the limited budget that can be allocated for these activities.

From all these data, a more approximate idea of the real functioning of the networks will be obtained which is called the "conceptualization" of the hydraulic model.

2.3. Drone Assistance

For building the hydraulic model of a WDS, a detailed topography is needed to obtain topographic elevations of the roads or streets and visible elements of the network. The costs associated with georeferenced topography work increase depending on the area, generally charged per hectare and georeferenced. For some companies with low budgets, the total value of this activity can be costly. In the city of Pamplona, the urban area is approximately 580.30 ha. Using new technologies allows alternatives to obtaining topographic and geospatial information based on photogrammetry with the assistance of unmanned aerial vehicles (drones) to be found. Currently, drones perform high-precision processes to obtain topographies relevant to water management. The general process of drone photogrammetry can be divided into six main activities, according to [55], as follows.

- Flight planning: The areas of interest must be identified, delimiting the general perimeter of the location from which information will be obtained. Mobile applications (e.g., DJI GO o PIX4Dcapture Pro) are used to plan the number of flights to be carried out, called missions, according to the size of the total area. Flights should be made on sunny days; the drone should not be flown on rainy, cloudy, or foggy days, as this affects the quality of the images and the drone's reception signal.
- Configuration of flight parameters: Flight altitude, maximum flight time, and photo capture interval are configured. These parameters depend on aspects such as the height of the buildings or existing infrastructure in the city (telephone towers, electric power antennas, etc.) and the state of the batteries that limits the maximum flight time per mission.
- Image acquisition and processing: Planned flights are carried out to obtain photographs. The information taken by the drone is downloaded to a computer and the quality of images is verified by checking that there are no blurred or distorted photographs due to clouds or any external element around the drone. Subsequently, the images must be processed to obtain the sense and orientation of each picture, which is carried out by analyzing the pixels of each image and the similarity of these pixels in the other photographs to obtain the overall picture of the area. Finally, 3D spatial data are generated, with points containing geographic and elevation information. There are different tools to analyze the photographs taken with a drone. In this case, the Agisoft Metashape photogrammetric application [56] is used to process and orient the images based on the GPS information of the drone and the relative position in each mission. It is necessary to verify that all pictures have information to guarantee the quality of the result. If, due to signal reception problems (e.g., a loss of connection

with GPS satellites or interference with radio or mobile phone signals), information is not obtained from any area, a new mission must be planned for that sector.

- Checkpoints: With the information of the national geodetic network "https://redgeodesica.igac.gov.co/redes/red_geodesica.html (accessed on 17 January 2023)", the existing georeferenced points within the flight area must be located and the elevation of the points and their coordinates, at least, must be obtained, as well as the geodetic coordinate system in which they are located. If these points are unavailable, a high-precision GPS should be used to obtain that information. This information will minimize the error and increase the accuracy of the elevation model obtained.

- Creation of the point cloud: The following process consists of creating the point cloud, which allows for the identification of points with the same pixel information in the photographs and the construction of a first 3D image with the data captured with the drone. From this, the mesh is created, which is the base element with which to obtain elevations of the model; it is recommended to work at a high resolution (a face count parameter equal to 180,000). It is necessary to carry out the process of classification of points to purge elements that are not of interest in the model to be generated.

- Digital elevation model (DEM): Finally, processing is performed to obtain the DEM, contour lines, and orthophoto. These three products can be exported in independent files in shapefile, DWG, or JPG format, among others, so that they can be visualized in software such as AutoCAD, QGIS, or ArcGIS.

Figure 2 summarizes the described process of drone photogrammetry.

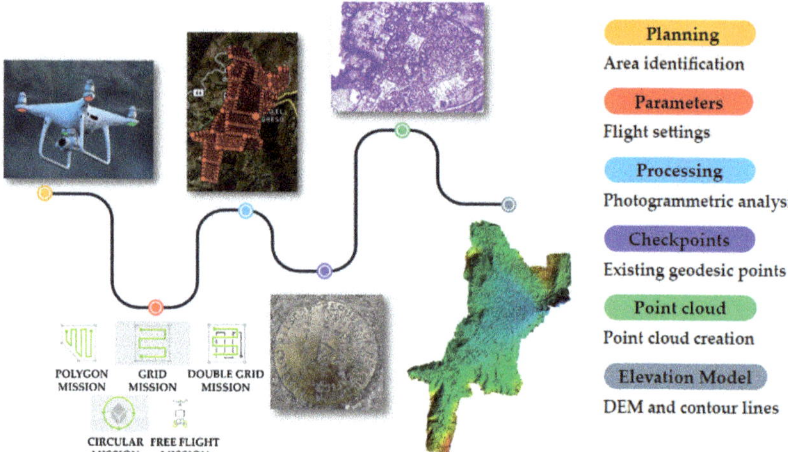

Figure 2. Photogrammetry with drone assistance to update elevation data in a WDS.

2.4. Consumption Analysis and Demand Pattern

Consumption analysis is performed to determine the actual water consumed by the inhabitants of a population. It allows the obtention of an approximate value of the base demand to be used for the hydraulic simulations of the WDS model. To obtain the demands of each node of the hydraulic network, a unit flow is used based on the types of existing users (domestic, commercial, institutional, and industrial) according to the total area supplied by the WDS. Those flows are calculated as described in Equations (1)–(4):

$$\text{Qu}_{\text{RES}} = \frac{\text{CM}_{\text{RES}}}{30 \times 86{,}400 \times \text{A}_{\text{TOTAL}}} \tag{1}$$

$$\text{Qu}_{\text{IND}} = \frac{\text{CM}_{\text{COM}}}{30 \times 86{,}400 \times \text{A}_{\text{TOTAL}}} \tag{2}$$

$$Qu_{IND} = \frac{CM_{IND}}{30 \times 86{,}400 \times A_{TOTAL}} \tag{3}$$

$$Qu_{INST} = \frac{CM_{INST}}{30 \times 86{,}400 \times A_{TOTAL}} \tag{4}$$

Here, QU_{RES} is the unit domestic flow in liters per second per hectare (L/s × Ha), QU_{COM} is the unit commercial flow in L/s × Ha, and QU_{IND} is the unit industrial flow in (L/s × Ha), QU_{INST} is the unit institutional flow in L/s × Ha. CM_{RES}, CM_{COM}, CM_{IND} and CM_{INST} are the monthly average consumptions for each type of user in cubic meters per month (m^3/month × user) and A_{TOTAL} is the total supplied area in hectares (Ha).

The afferent area influencing each node of the WDS hydraulic model is calculated, the number of dwellings associated with each hydraulic node is determined, and the afferent area is estimated from the urban plans mentioned in numeral 2.1. The products of these areas and the estimated unit flows provided by Equations (1)–(4) enable a calculation of the base demand of each node using Equation (5):

$$Q_{Ni} = Qu_{RES} * A_{RES(Ni)} + Qu_{COM} * A_{COM(Ni)} + Qu_{IND}*A_{IND(Ni)} + Qu_{INST} * A_{INST(Ni)}, \tag{5}$$

where Q_{Ni} is the base demand of each node of the hydraulic model in L/s. $A_{RES(Ni)}$, $A_{COM(Ni)}$, $A_{IND(Ni)}$ and $A_{INST(Ni)}$ are the areas afferent to each domestic, commercial, industrial and institutional node, respectively, in hectares (Ha).

To obtain the demand pattern that represents the hourly variation of flows in the WDS, the use of flow data from the flow record at the drinking water treatment plant (DWTP) outlets is proposed. An hourly increase factor (HIF) based on the flow measured at each hour of the day and the average daily flow, the latter being understood as the average of the daily flows, can be calculated as shown in Equation (6):

$$\text{Hourly Increase Factor(HIF)} = \frac{\text{Flow measured each hour}}{\text{Average daily flow}} \tag{6}$$

2.5. Hydraulic Model, Hydraulic Analysis, and Non-Revenue Water Basic Index

For the building of the hydraulic model, the digitized information is converted into DXF format (drawing exchange format). This file, in turn, is converted into INP format to generate the pipes and network nodes according to those described in [57]. The created file can be edited in EPANET and will only have pipe layouts. Lengths, diameters, and roughness must be added. Additionally, reservoirs, tanks, and valves, and the properties of these elements must be inserted.

The base demands in the network nodes are estimated using Equation (5). The demand pattern can also be included in EPANET and the model is configured for extended-period modeling. The changes made are saved to an INP file.

The DEM obtained with the drone and the last INP file modified to include the topographic elevations based on the methodology proposed in [58,59] are used. The physical properties of the hydraulic model are completed and the first hydraulic simulations are performed in EPANET to test the functionality of the network. The WATERing software and the INP file are then used, loaded, and synchronized with an OpenStreetMap satellite base map. A simple graphical interface is obtained once its correct geographical implementation is verified. This software allows water utilities to have an online hydraulic model that offers several analysis options and the possibility of multi-user editing.

With the historical record of the billing and consumption of the users and the measurements of the flow supplied to the WDS by the DWTP for the last three months, the non-revenue water basic index (NRWB) of the system is calculated using Equation (7) described in [60], as follows:

$$NRWB = \left(\frac{SIV - BAC}{SIV}\right) * 100\%, \tag{7}$$

where SIV (system input volume) is the water supplied to the network (m^3/year) and BAC (billed authorized consumption) is the water sold (m^3/year).

2.6. Study Location

The city of Pamplona is located in the state of Norte de Santander in Colombia. Its geographical location is at the coordinates 72°39′ west longitude and 7°23′ north latitude, with an elevation that varies between 2200 and 2600 m above sea level and an estimated population of 51,292 inhabitants for the year 2023. Pamplona is considered a small city and its urban area is approximately 580.30 ha; its main economic activities are education, tourism, and agriculture. According to the different land uses, the city has residential, commercial (e.g., restaurants, stores, cafeterias, and supermarkets), governmental and institutional, and recreational uses (e.g., parks; sports areas) [61]. Figure 3a shows an urban map highlighting the urban perimeter in purple and Figure 3b shows the city's land uses.

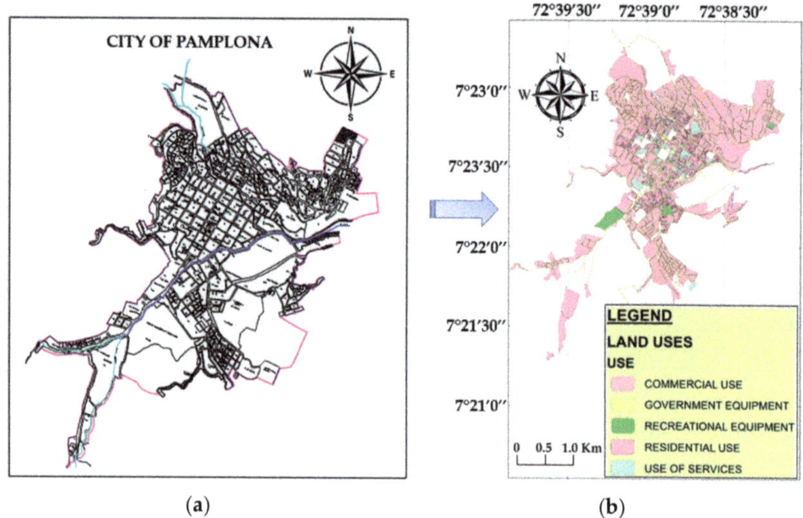

(a)   (b)

Figure 3. Pamplona, Colombia. (**a**) Urban map and (**b**) land uses.

3. Results

3.1. Existing Data and Conceptualization of the Hydraulic Model in Pamplona

In Pamplona, the WDS was not digitalized. There was no hydraulic model that correctly represented the actual behavior of the network and its physical infrastructure. The most relevant existing information consisted of physical plans and the information existing in the memory of the most experienced workers. Physical and PDF (portable document format) layouts provide information on the existing networks in the city up to the year 2016, where the locations of the pipes, materials, and diameters are described, as well as the location of tanks and their storage volumes, valves, pumping stations, and DWTPs. It has been necessary to update information on the cadaster regarding new network zones (installed in recent years in replacement and expansion works).

The updated information up to the year 2021 is in DWG digital format (AutoCAD files) and shapefile format. Using satellite images, new urban settlements were identified that were not registered in the physical or digital plans and currently have water networks installed and in operation. These settlements are observed in the city's northwest, northeast, and southeast areas, as shown in Figure 4.

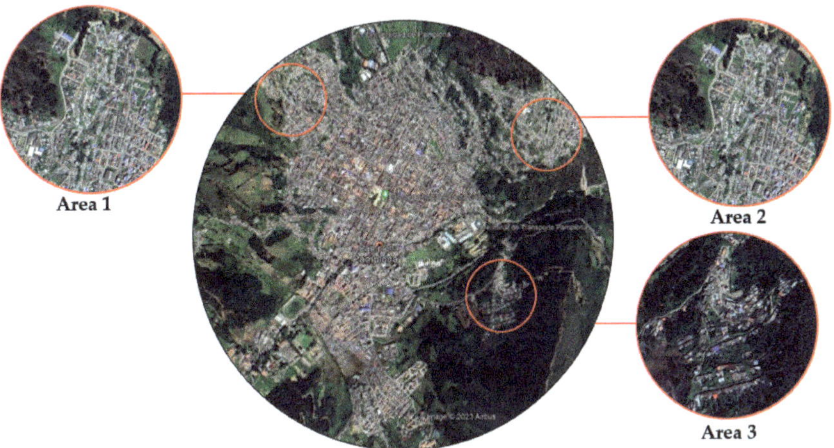

Figure 4. New urban settlements identified in Pamplona.

Due to the incomplete information on the cadaster of the networks installed in recent years, it is necessary to carry out fieldwork to identify and verify some elements and layouts. Figure 5 shows some of the details verified in the field. Figure 5a shows piping and valves and Figure 5b illustrates one pressure-bursting chamber. Once the missing information was confirmed, a preliminary digitization of the WDS was performed.

(a) (b)

Figure 5. Record of field inspections. (**a**) Piping and valves; (**b**) pressure-bursting chamber 4.

The distribution system is operated by the water company EMPOPAMPLONA SA ESP "https://www.empopamplona.com.co/ (accessed on 30 January 2023)". It is a gravity-fed system providing continuous service. Two DWTPs supply water to the piping networks. The Cariongo DWTP has an average operating flow of 104.7 L per second (L/s) and the Monteadentro DWTP has that of 37.8 L/s. The DWTPs are designed as reservoirs and distribute water directly to the network and 12 storage tanks or reservoirs totaling 3530 cubic meters (m^3). The network supplies an area of 3.76 square kilometers (km^2) with an average flow of 142.5 L/s. The water company in the year 2022 had 15,587 users, of which 92% were domestic, 7% commercial, 0.25% industrial, and 0.75% institutional. Figure 6 shows the digitized map of the networks in DWG format.

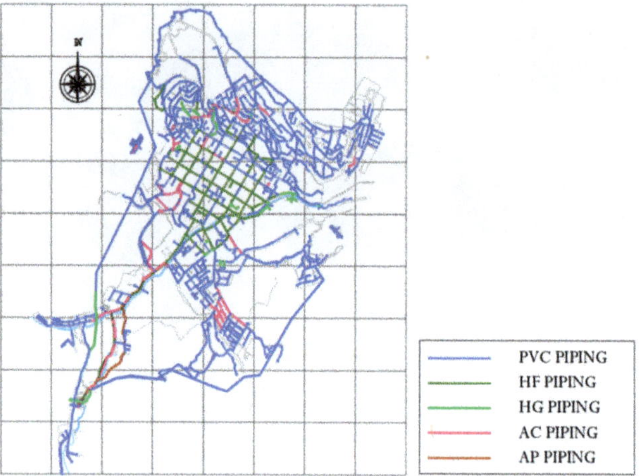

Figure 6. Digitization and updating of the WDS cadaster map.

The main network pipe diameters vary from 150 mm to 400 mm and interconnect to smaller networks with diameters varying from 50 mm and 100 mm. The water network is compounded by polyvinyl chloride (PVC), asbestos cement (AC), cast iron (CI), galvanized steel (GI), and American pipe (AP). There are pressure rupture chambers at some points in the network because there are areas with highly variable topography and differences in elevation between the tanks and the network that can exceed 150 m. There are no pressure-reducing or regulating valves in the WDS. The network lacks hydraulic sectorization, areas with district meters, or a macro measurement of the network's internal flow. It only has a micro measurement for each user.

The water company installed 19 piezometers at different points of the network, which were located after an engineering report, where pressure data are manually recorded at two times of the day, at 9:00 and 16:00 h. These measurements are used to control the hydraulic operation of the WDS. According to existing data, the pressure measured varies between 14 m of the water column (mH$_2$O) and 105 m of mH$_2$O. Isolation valves are distributed along the network and used to close subsystems when pressure problems cause damage and failure. Figure 7 shows the conceptualization scheme of the Pamplona WDS.

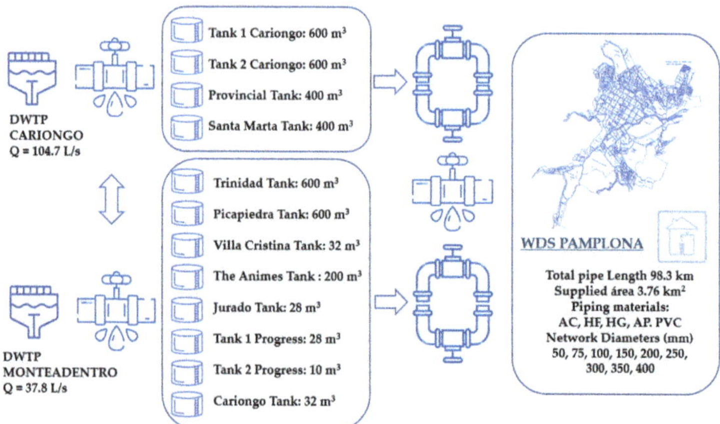

Figure 7. Conceptualization of the WDS of Pamplona.

3.2. Drone Assistance

In this research, a Phantom 4 Pro drone "https://www.dji.com/global/phantom-4-pro (accessed on 2 February 2023)" was used to obtain the digital elevation model to determine the topographic elevations of nodes and elements such as reservoirs and tanks of the network. In this work, 19 missions are planned in "polygonal mission" mode. The minimum flight altitude was selected as 50 m, the maximum height was 150 m above ground level, and the complete transmission range was 7 km. The missions were configured for flight times between 12 and 15 min. These parameters were selected according to the heights of the buildings, the variation of the topography in the city, and the duration of the drone's batteries.

The city of Pamplona has five geodetic control points. This information, included in the Agisoft Metashape software, increased the accuracy of the elevation model. Figure 8a shows the DEM generated with the topographic variation in the city, its minimum height of 2225.8 m above sea level (masl), and its maximum height of 25,652.4 masl. Figure 8b shows the orthophoto of Pamplona.

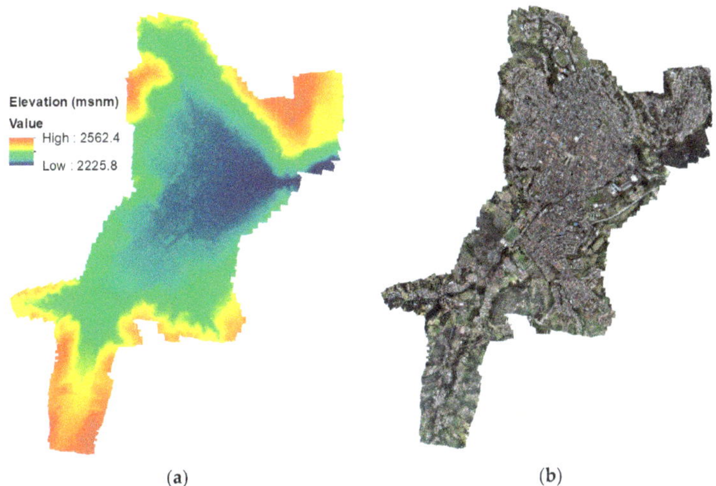

Figure 8. Pamplona, Colombia. (**a**) Digital elevation model; (**b**) orthophoto.

3.3. Consumption Analysis and Demand Pattern

The consumption analysis is based on the water company's billing information. The billing records are analyzed to estimate the average consumption of users and categorized according to the use's type (residential, commercial, institutional, or industrial). This analysis allows the obtention of the average monthly consumption of each user per cubic meter (m^3/month $\times$ user). These values are obtained through a descriptive statistical analysis of the different types of users proposed by [62] and allow the obtention of the values of CM_{RES}, CM_{COM}, CM_{IND}, and CM_{INST}, which are mentioned in Equations (1)–(4). The analysis results show that the average monthly consumption for institutional users was 10.50 m^3, that for industrial users was 10.00 m^3, that for commercial users was 8.25 m^3 for, and that for domestic users was 9.00 m^3.

Another essential element for building a hydraulic model is the hourly demand pattern of the network. For this purpose, the flow values measured in the DWTPs for the last three months are used. The hourly increase factors were calculated using Equation (6) and thus, the demand pattern shown in Figure 9 was obtained.

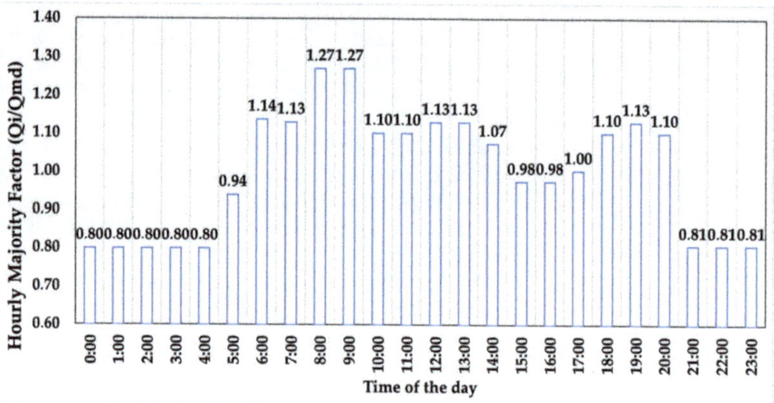

Figure 9. Daily demand pattern of Pamplona WDS.

It is observed that the peak of consumption in the network occurs at 9:00 h and 19:00 h when the population consumes the most significant amount of water. It is also observed that the hours of lower consumption occur between 22:00 h and 4:00 h the following day. During these hours, consumption is relatively low and operations in the DWTPs are reduced to their minimum operation. The HIF was obtained with a value of 1.27. This analysis was performed by averaging the records supplied by the water company. The calculated pattern has a behavior directly related to the habits of the inhabitants of Pamplona.

3.4. Building of the Model for Hydraulic Analysis

The network map shown in Figure 6 is converted into an INP file as indicated in Section 2.5, and an offline model is created and exported to EPANET; some properties such as diameters, roughness, base demand, and demand pattern can be added initially. This file and the DEM generated in Section 3.2 complete the preliminary hydraulic model with the topographic elevations. The first hydraulic simulations are performed in EPANET to test the functionality of the network. The created hydraulic model is loaded and synchronized in WATERing.

Figure 10 shows the WDS network's implementation and general layout in the WATERing software on the Open-Street Map base map.

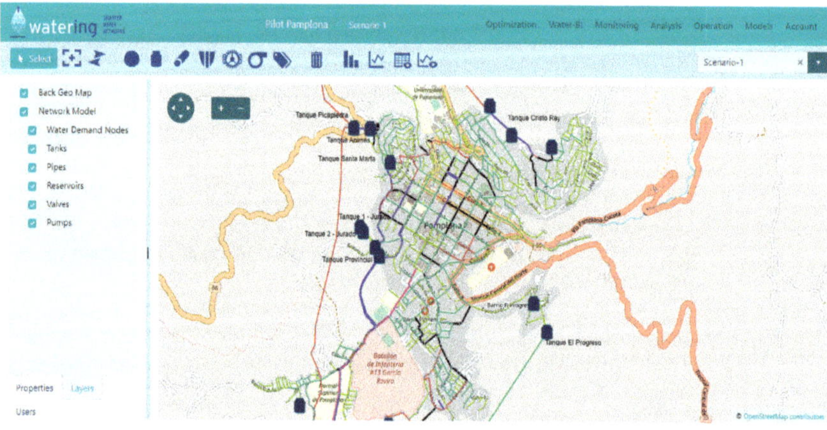

Figure 10. Hydraulic model of the network in the WATERing software platform.

The hydraulic simulation to determine the initial state of analysis of the network was performed in the WATERing software. The hydraulic calculation options were configured and verified that the parameters of all model elements were correctly incorporated. A demand-driven hydraulic analysis (DDA) was performed. WATERing uses the EPANET calculation engine. Once the extended-period simulation was run including the pattern described in Figure 9, the results of the hydraulic behavior of the network were obtained, e.g., the pressure results at the nodes were obtained.

The hydraulic simulation leads to results that are very close to the pressure measurement records. Figure 11a shows the state of the ODS at 0:00 h; at the lowest points of the network (the city valley), the maximum pressures exceed 60 mH$_2$O, and at the highest points, the pressures are between 10 and 15 mH$_2$O. Figure 11b shows the simulation at the time of maximum consumption, at 9:00 h, and the results show low- (pressure less than 15 mH$_2$O) or zero-pressure values in four specific zones of the network. Three of these zones coincide with the areas shown in Figure 4; zero pressures are observed in areas 1 and 2, and in area 3, low pressures are observed. These urban settlements have emerged recently and have grown uncontrollably, having in common that the communities have built their homes in the vicinity and places with topographic heights close to the existing storage tanks in those areas of the city. For this reason, during peak consumption hours, pressures near these tanks are low or non-existent. The inhabitants usually store water at night for daytime consumption.

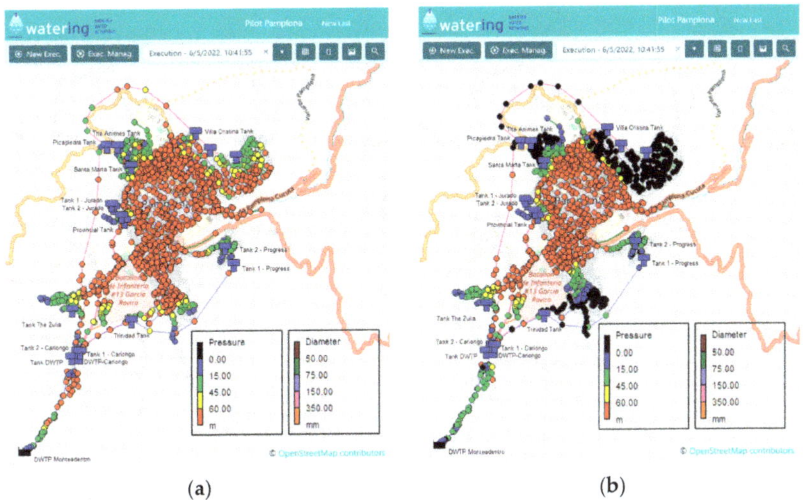

Figure 11. Hydraulic simulation of WDS (**a**) at 0:00 h; (**b**) at 9:00 a.m.

Similarly, Figure 12 shows the contour plot of network pressures at the hour of minimum consumption (0:00 h). It can be seen that the pressure in the city's center exceeds 80 mH2O, and at the extremes (highest points), the pressures are less than 20 mH$_2$O.

The pressure results obtained are close to the average of the measured pressure data; the data corresponding to the last month were used to obtain the average. The differences observed in the model vary between 83 and 123% concerning the data recorded in the field. The root means square error (RMSE) obtained was 4.31 for the measurements for 9:00 h and 5.18 for the data for 16:00 h. The pressure values obtained were compared with the existing data set, and it was observed that there was a significant dispersion between the monitored and simulated values. The absence of flow measurements within the system does not allow the development of a flow-based calibration methodology at this research stage.

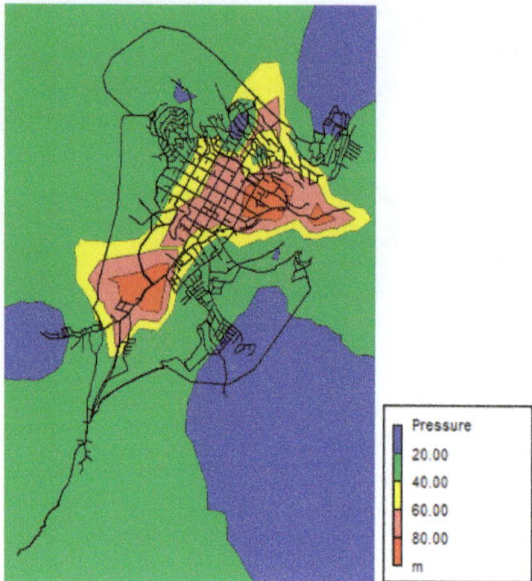

Figure 12. Pressure contour plot of Pamplona WDS at 0:00 h.

For the Pamplona network, based on the existing pressure data measured at some points of the network, which are shown in Figure 13, general calibration was performed, and the results shown in Figure 14 were obtained, where the simulated pressures are observed against the pressures measured in the network.

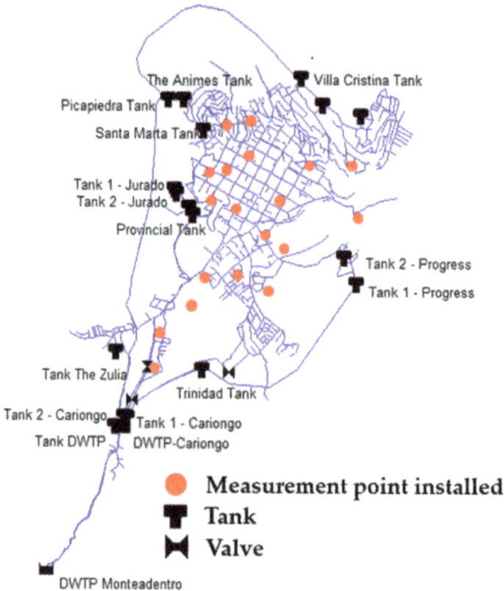

Figure 13. Points in the Pamplona WDS with pressure measurement.

Figure 13 shows the 19 points of the network with pressure measurements, as a reference point the WDS storage tanks and some valves of the network are observed.

Figure 14a shows the results at 9:00 h, and Figure 14b shows the results at 16:00 h. At those hours, pressure records are taken in the network at the 19 points indicated in the graph. The results are for the behavior of the consumption pattern shown in Figure 9; as consumption increases, network pressures decrease.

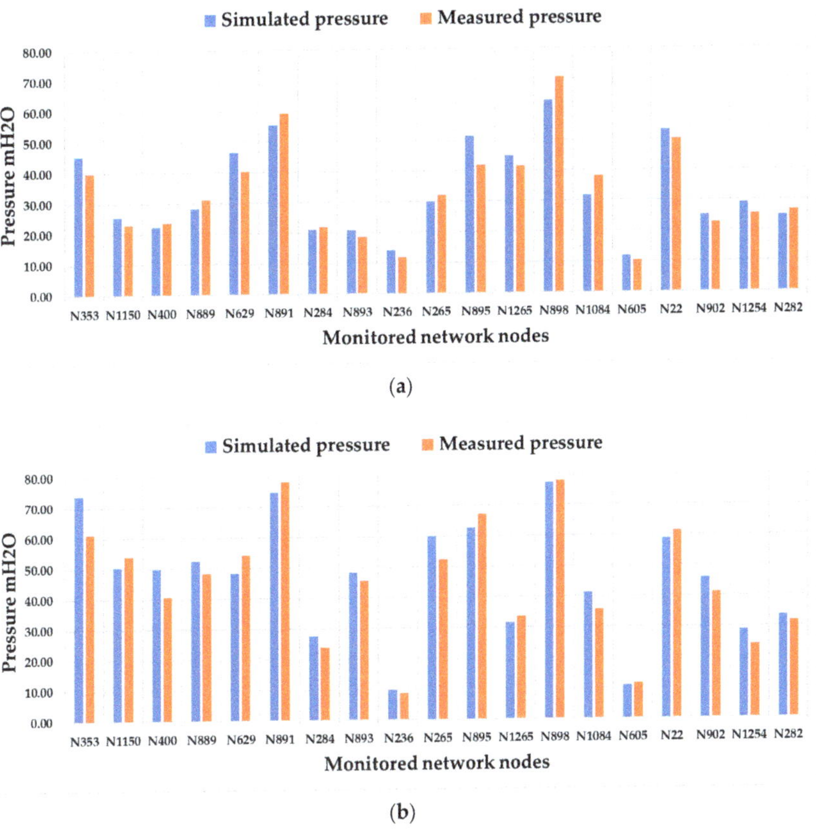

(a)

(b)

Figure 14. Simulated pressures vs. measured pressures (**a**) at 9:00 h; (**b**) at 16:00 h.

Finally, using the flow records at the outlets of the DWTPs for the last three months of the year 2022, a SIV of 382,419.22 m^3 and a BAC of 221,038.31 m^3 were obtained. This represents an NRWB of 42.20% for the WDS of Pamplona. The value obtained is high but typical of Colombian cities. In Colombia, the NRWB values reported by different cities and their water companies vary between 23.90% (Tunja) and 67.00% (Villavicencio). The national average of NRWB between 2007 and 2011 was 45.3% and between 2011 and 2017, it was 43.60% [63].

4. Discussion

The results of this work show an important advance in the path towards digitalization. A digitized network was obtained, and the hydraulic model should be improved by implementing a form of telemetry that allows a comparison of the model results and the pressure and flow values monitored at different points of the network defined by the water company experts or based on optimal sensor location methodologies. Flow and pressure measurements at the sensors should be performed continuously and in constant intervals. To develop a monitoring phase, sensors must be installed to transmit data in real time to the WATERing platform to obtain a time series of data to calibrate the network, analyze hydraulic sectorization scenarios, and identify leaks in the pipes. These recommendations will improve water management in the city of Pamplona and increase the operational efficiency of the existing infrastructure.

The work was designed according to the digitalization needs of water utilities in small cities with low budgets. Describing in detail the necessary steps to obtain the base information required to create the hydraulic model of a WDS is a fundamental step to digitalizing a distribution system. This experience developed in Pamplona teaches other companies and cities the path that will allow them to obtain important results.

The research proposes a guide structured in six steps, taking as a case study a small city in Colombia, which represents to a large extent the typical case of a water company in Latin America and developing countries. A company that has few resources and operates its WDS under the technical decisions made by its engineers was studied. This research explains in detail how to digitize the water network and obtain a useful verification and analysis tool that could be replicated and extended as a guide to companies in other cities and/or countries. The development of traditional actions such as the search for existing data, the conceptualization of the model, and the analysis of water consumption, together with new digital tools such as photogrammetry with drones, cloud computing, and the use of online software to digitize a WDS at low cost are some of the novelties in this research.

5. Conclusions

This research has generated a final product that serves as a fundamental water analysis and management tool for the small city of Pamplona. The hydraulic model created on an online platform can be used from a web browser. The DDA performed on the network yielded good initial results. It is necessary to consider the numerical solution of the network under other approaches such as pressure-driven hydraulic analysis (PDA), as this would allow us to determine if there is a variation in the pressure of the network nodes, specifically in the low-pressure zones evidenced in this study. The results may vary when performing a hydraulic solution based on the assumption of a pressure–demand relationship at the junctions. This tool should be complemented with a set of pressure and flow monitoring measurements of the existing network to determine the analysis option that adequately represents the Pamplona WDS. This would allow an evaluation of the overall performance of the network and contribute to the advancement of SDG-6, SDG-9, and SDG-11. The aim is to ensure the availability of water to the population of Pamplona with a supply system that can be managed and sustained, and that can evaluate resilience scenarios in the face of adverse effects that may arise. The process explained in this work to create a hydraulic model with the assistance of a drone can be applied in small and medium water companies and will allow them to take significant steps towards digital transformation with a low investment cost and few limitations.

Good results can be obtained on the road to water digitalization with the support of current technological elements, such as drones and free-use software, plus the knowledge of the water companies' experts. The creation of hydraulic models of a WDS allows access to network information and the permanent updating of existing elements. It also allows us to permanently evaluate the state and functioning of the network to be able to solve more robust problems using analytical tools. The total investment cost of this research was below USD 1400, not including the cost of the time spent by the researchers, since professionals from the water companies could carry out this work.

The development of this case study shows an important advance in the path to a digital twin; it indicates that there are different options to create value from hydraulic models and the optimal management of drinking water distribution systems.

Author Contributions: Conceptualization and methodology, B.B., I.M. and D.A.-C.; data collection, software, data analysis, and editing, C.B.; writing and preparation of the original draft, C.B., B.B. and J.I.; visualization, I.M. and D.A.-C.; conceptualization, review, and supervision, J.I. All authors have read and accepted the published version of the article. All authors have read and agreed to the published version of the manuscript.

Funding: This research received no external funding.

Data Availability Statement: There is no need for a data statement.

Conflicts of Interest: The authors declare no conflict of interest.

References

1. Makropoulos, C.; Savić, D.A.S.; Nl, C.M. Urban Hydroinformatics: Past, Present and Future. *Water* **2019**, *11*, 1959. [CrossRef]
2. Gómez, M.; Tagle-Zamora, M.; Morales Martínez, D.; Caldera Ortega, J.L.; Mora Rodríguez, A.R.; Delgado-Galván, J.D.J.; Mendoza Gómez, M.; Tagle-Zamora, D.; Morales Martínez, J.L.; Caldera Ortega, A.R.; et al. Water Supply Management Index: Leon, Guanajuato, Mexico. *Water* **2022**, *14*, 919. [CrossRef]
3. Naciones Unidas. *La Agenda 2030 y Los Objetivos de Desarrollo Sostenible: Una Oportunidad Para América Latina y El Caribe. Objetivos, Metas e Indicadores Mundiales*; Naciones Unidas: Santiago, Chile, 2019.
4. Benedict, S.; Hussein, H. An Analysis of Water Awareness Campaign Messaging in the Case of Jordan: Water Conservation for State Security. *Water* **2019**, *11*, 1156. [CrossRef]
5. Ortega-Ballesteros, A.; Manzano-Agugliaro, F.; Perea-Moreno, A.J. Water Utilities Challenges: A Bibliometric Analysis. *Sustainability* **2021**, *13*, 7726. [CrossRef]
6. Snider, B.; McBean, E.A. Watermain Breaks and Data: The Intricate Relationship between Data Availability and Accuracy of Predictions. *Urban Water J.* **2020**, *17*, 163–176. [CrossRef]
7. Pingale, S.M.; Jat, M.K.; Khare, D. Integrated Urban Water Management Modelling under Climate Change Scenarios. *Resour. Conserv. Recycl.* **2014**, *83*, 176–189. [CrossRef]
8. Figueiredo, I.; Esteves, P.; Cabrita, P. Water Wise—A Digital Water Solution for Smart Cities and Water Management Entities. *Procedia Comput. Sci.* **2021**, *181*, 897–904. [CrossRef]
9. Helmbrecht, J. Transformación Digital de Redes Inteligentes: Retos y Riesgos. *Tecnoaqua* **2018**, *32*, 84–86.
10. Shende, S.; Chau, K.W. Design of Water Distribution Systems Using an Intelligent Simple Benchmarking Algorithm with Respect to Cost Optimization and Computational Efficiency. *Water Supply* **2019**, *19*, 1892–1898. [CrossRef]
11. Liu, A.; Mukheibir, P. Digital Metering Feedback and Changes in Water Consumption—A Review. *Resour. Conserv. Recycl.* **2018**, *134*, 136–148. [CrossRef]
12. Bharani Baanu, B.; Jinesh Babu, K.S. Smart Water Grid: A Review and a Suggestion for Water Quality Monitoring. *Water Supply* **2022**, *22*, 1434–1444. [CrossRef]
13. Gupta, A.D.; Kulat, K. Leakage Reduction in Water Distribution System Using Efficient Pressure Management Techniques. Case Study: Nagpur, India. *Water Supply* **2018**, *18*, 2015–2027. [CrossRef]
14. Menapace, A.; Zanfei, A.; Felicetti, M.; Avesani, D.; Righetti, M.; Gargano, R. Burst Detection in Water Distribution Systems: The Issue of Dataset Collection. *Appl. Sci.* **2020**, *10*, 8219. [CrossRef]
15. Quintiliani, C.; Marquez-Calvo, O.; Alfonso, L.; Cristo, C.D.; Leopardi, A.; Solomatine, D.P.; Marinis, G. de Multiobjective Valve Management Optimization Formulations for Water Quality Enhancement in Water Distribution Networks. *J. Water Resour. Plan. Manag.* **2019**, *145*, 04019061. [CrossRef]
16. Quintiliani, C.; Di Cristo, C.; Leopardi, A. Vulnerability Assessment to Trihalomethane Exposure in Water Distribution Systems. *Water* **2018**, *10*, 912. [CrossRef]
17. Ramos, H.M.; Morani, M.C.; Carravetta, A.; Fecarrotta, O.; Adeyeye, K.; López-Jiménez, P.A.; Pérez-Sánchez, M. New Challenges towards Smart Systems' Efficiency by Digital Twin in Water Distribution Networks. *Water* **2022**, *14*, 1304. [CrossRef]
18. Conejos Fuertes, P.; Martínez Alzamora, F.; Hervás Carot, M.; Alonso Campos, J.C. Building and Exploiting a Digital Twin for the Management of Drinking Water Distribution Networks. *Urban Water J.* **2020**, *17*, 704–713. [CrossRef]
19. Erena, M.; Atenza, J.F.; García-Galiano, S.; Domínguez, J.A.; Bernabé, J.M. Use of Drones for the Topo-Bathymetric Monitoring of the Reservoirs of the Segura River Basin. *Water* **2019**, *11*, 445. [CrossRef]
20. Hu, R.Y. Sources and Routes from Terrestrial Exogenous Pollutants Affect Phytoplankton Biomass in Reservoir Bays. *Water Supply* **2021**, *21*, 3913–3931. [CrossRef]
21. Stanford, B.D.; Pochiraju, S.; Yokoyama, T.; Maari, P.; Grijalva, L.; Bukhari, Z. Evaluating Satellite and in Situ Monitoring Technologies for Leak Detection and Response. *AWWA Water Sci.* **2022**, *4*, e1288. [CrossRef]
22. Sadowski, B. Book Review: A Modern Guide to the Digitalization of Infrastructure. *Digit. Policy Regul. Gov.* **2022**, *24*, 220–224. [CrossRef]
23. Adedeji, K.B.; Ponnle, A.A.; Abu-Mahfouz, A.M.; Kurien, A.M. Towards Digitalization of Water Supply Systems for Sustainable Smart City Development—Water 4.0. *Appl. Sci.* **2022**, *12*, 9174. [CrossRef]
24. Balogun, A.L.; Marks, D.; Sharma, R.; Shekhar, H.; Balmes, C.; Maheng, D.; Arshad, A.; Salehi, P. Assessing the Potentials of Digitalization as a Tool for Climate Change Adaptation and Sustainable Development in Urban Centres. *Sustain. Cities Soc.* **2020**, *53*, 101888. [CrossRef]
25. Ceipek, R.; Hautz, J.; Petruzzelli, A.M.; De Massis, A.; Matzler, K. A Motivation and Ability Perspective on Engagement in Emerging Digital Technologies: The Case of Internet of Things Solutions. *Long. Range Plan.* **2021**, *54*, 101991. [CrossRef]
26. Mondejar, M.E.; Avtar, R.; Diaz, H.L.B.; Dubey, R.K.; Esteban, J.; Gómez-Morales, A.; Hallam, B.; Mbungu, N.T.; Okolo, C.C.; Prasad, K.A.; et al. Digitalization to Achieve Sustainable Development Goals: Steps towards a Smart Green Planet. *Sci. Total Environ.* **2021**, *794*, 148539. [CrossRef]

27. Sun, C.; Cembrano, G.; Puig, V.; Meseguer, J. Cyber-Physical Systems for Real-Time Management in the Urban Water Cycle. In Proceedings of the 2018 4th International Workshop on Cyber-Physical Systems for Smart Water Networks, CySWater, Porto, Portugal, 10–13 April 2018; pp. 5–8. [CrossRef]
28. Sun, C.; Puig, V.; Cembrano, G. Real-Time Control of Urban Water Cycle under Cyber-Physical Systems Framework. *Water* **2020**, *12*, 406. [CrossRef]
29. Bhardwaj, J.; Gupta, K.K.; Gupta, R. Towards a Cyber-Physical Era: Soft Computing Framework Based Multi-Sensor Array for Water Quality Monitoring. *Drink. Water Eng. Sci.* **2018**, *11*, 9–17. [CrossRef]
30. Chalh, R.; Bakkoury, Z.; Ouazar, D.; Hasnaoui, M.D. Big Data Open Platform for Water Resources Management. In Proceedings of the 2015 International Conference on Cloud Computing Technologies and Applications, CloudTech, Marrakech, Morocco, 2–4 June 2015. [CrossRef]
31. Moumen, A.; Aghoutane, B.; Lakhrissi, Y.; Essahlaoui, A. Big Data Architecture for Moroccan Water Stakeholders: Proposal and Perception. *Lect. Notes Electr. Eng.* **2022**, *745*, 241–246.
32. Ahirvar, B.P.; Panday, S.; Das, P. Water Indices: Specification, Criteria, and Applications—A Case Study. In *Resilience, Response, and Risk in Water Systems*; Springer: Singapore, 2020; pp. 73–102. [CrossRef]
33. Shayan, M.N.M.; Tushara Chaminda, G.G.; Ellawala, K.C.; Gunawardena, W.B. Evaluation of Water Quality of Community Managed Water Supply Schemes (CMWSS) in Galle District. In *Resilience, Response, and Risk in Water Systems*; Springer: Singapore, 2020; pp. 139–150. [CrossRef]
34. Candelieri, A.; Archetti, F. Smart Water in Urban Distribution Networks: Limited Financial Capacity And Big Data Analytics. *WIT Trans. Built Environ.* **2014**, *139*, 63–73. [CrossRef]
35. Meseguer, J.; Mirats-Tur, J.M.; Cembrano, G.; Puig, V.; Quevedo, J.; Pérez, R.; Sanz, G.; Ibarra, D. A Decision Support System for On-Line Leakage Localization. *Environ. Model. Softw.* **2014**, *60*, 331–345. [CrossRef]
36. Vegas Niño, O.T.; Martínez Alzamora, F.; Tzatchkov, V.G. A Decision Support Tool for Water Supply System Decentralization via Distribution Network Sectorization. *Processes* **2021**, *9*, 642. [CrossRef]
37. Armon, A.; Gutner, S.; Rosenberg, A.; Scolnicov, H. Algorithmic Network Monitoring for a Modern Water Utility: A Case Study in Jerusalem. *Water Sci. Technol.* **2011**, *63*, 233–239. [CrossRef] [PubMed]
38. Wu, Z.Y.; Sage, P. Water Loss Detection via Genetic Algorithm Optimization-Based Model Calibration. In Proceedings of the 8th Annual Water Distribution Systems Analysis Symposium, Cincinnati, OH, USA, 27–30 August 2006; pp. 1–11. [CrossRef]
39. Mamlook, R.; Al-Jayyousi, O. Fuzzy Sets Analysis for Leak Detection in Infrastructure Systems: A Proposed Methodology. *Clean. Technol. Environ. Policy* **2003**, *6*, 26–31. [CrossRef]
40. Candelieri, A. Clustering and Support Vector Regression for Water Demand Forecasting and Anomaly Detection. *Water* **2017**, *9*, 224. [CrossRef]
41. Venkateswaran, P.; Suresh, M.A.; Venkatasubramanian, N. Augmenting In-Situ with Mobile Sensing for Adaptive Monitoring of Water Distribution Networks. In Proceedings of the ICCPS 2019—2019 ACM/IEEE International Conference on Cyber-Physical Systems, Montreal, QC, Canada, 16–18 April 2019; pp. 151–162. [CrossRef]
42. Afifi, M.; Abdelkader, M.F.; Ghoneim, A. An IoT System for Continuous Monitoring and Burst Detection in Intermittent Water Distribution Networks. In Proceedings of the 2018 International Conference on Innovative Trends in Computer Engineering, ITCE, Aswan, Egypt, 19–21 February 2018; pp. 240–247. [CrossRef]
43. Pérez-Padillo, J.; Morillo, J.G.; Ramirez-Faz, J.; Roldán, M.T.; Montesinos, P. Design and Implementation of a Pressure Monitoring System Based on IoT for Water Supply Networks. *Sensors* **2020**, *20*, 4247. [CrossRef]
44. Maroli, A.A.; Narwane, V.S.; Raut, R.D.; Narkhede, B.E. Framework for the Implementation of an Internet of Things (IoT)-Based Water Distribution and Management System. *Clean. Technol. Environ. Policy* **2021**, *23*, 271–283. [CrossRef]
45. Feliciano, J.F.; Arsénio, A.M.; Cassidy, J.; Santos, A.R.; Ganhão, A. Knowledge Management and Operational Capacity in Water Utilities, a Balance between Human Resources and Digital Maturity—The Case of AGS. *Water* **2021**, *13*, 3159. [CrossRef]
46. Feliciano, J.; Almeida, R.; Santos, A.; Ramalho, P.; Ganhão, A.; Covas, D.; Alegre, H. Assessing Human Resources Renovation Needs in Water Utilities. *Water Pract. Technol.* **2016**, *11*, 728–735. [CrossRef]
47. Chirica, Ş.; Luca, M.; Lateş, I. Updating the pipe networks layout plan using modern detection equipment. *Hidrotehnica* **2018**, *64*, 17–25.
48. Sridharan, N.; Pandey, R.U.; Berger, T. Co-Production through Tacit Knowledge for Water Resilience. *Land. Use Policy* **2023**, *126*, 106446. [CrossRef]
49. Areiza, J.; Caraballo, J. *Análisis de Las Pérdidas de Agua En Los Sistemas de Abastecimiento Público, Identificando Sectores y Causas Influyentes En Los Altos Índices de Agua No Contabilizada (IANC) En El Municipio de Turbo Antioquia—10596/28359*; Universidad Nacional Abierta y a Distancia UNAD: Bogotá, Colombia, 2019.
50. Rossman, L.A.; Woo, H.; Tryby, M.; Shang, F.; Janke, R. *Manual Del Usuario de EPANET 2.2*; EPA, US Environmental Protection Agency: Washington, DC, USA, 2002.
51. Avesani, D.; Righetti, M.; Righetti, D.; Bertola, P. The Extension of EPANET Source Code to Simulate Unsteady Flow in Water Distribution Networks with Variable Head Tanks. *J. Hydroinform.* **2012**, *14*, 960–973. [CrossRef]
52. Todini, E.; Pilati, S. A Gradient Method for the Solution of Looped Pipe Networks. In *Computer Applications in Water Supply*; Coulbeck, B., Orr, C.H., Eds.; systems analysis and simulation; John Wiley & Sons: London, UK, 1988; Volume 1, pp. 1–20.

53. Ingeniousware Software for Smarter Water Network Monitoring—WATERing Online. Available online: https://watering.online/watering/ (accessed on 1 February 2023).
54. Klingel, P. Technical Causes and Impacts of Intermittent Water Distribution. *Water Supply* **2012**, *12*, 504–512. [CrossRef]
55. Kaamin, M.; Azhar, S.; Tajudin, A.; Athirah Basri, N.; Rahman, R.A.; Hakimi Mat Nor, A.; Azraie, M.; Kadir, A.; Mokhtar, M.; Luo, P. Unmanned Aerial Vehicle Technology Use in Visual Road Inspection at Ft005, Johor Bahru-Melaka. *Int. J. Nanoelectron. Mater.* **2022**, *15*, 37–48.
56. *Agisoft LLC Agisoft Metashape User Manual Professional Edition*, Version 1.7; Agisoft: St. Petersburg, Russia, 2021.
57. Soriano Olivares, J. Cómo Convertir una Red de Tubería en Autocad a una Red en Epanet. Available online: http://hdl.handle.net/10251/82990 (accessed on 4 February 2023).
58. Tapiero, D.I.S.; Valencia, M.M. SIG Aplicado a La Optimización Del Tiempo de Diseño En Redes de Distribución de Agua Potable. *Ing. Hidráulica Ambient.* **2021**, *42*, 68–80.
59. Vegas, O.T.; Santos, R.; Delgado, X.; Fernando, M.A.; Rodriguez, J. Herramienta para convertir un modelo de datos de una red de suministro de agua en formato shape y epanet. In Proceedings of the 5tas Jornadas México—11vas Jornadas Latinoamericanas y del Caribe de gvSIG, Guanajuato, Mexico, 15–16 August 2019; pp. 1–9.
60. Ociepa, E.; Mrowiec, M.; Deska, I. Analysis of Water Losses and Assessment of Initiatives Aimed at Their Reduction in Selected Water Supply Systems. *Water* **2019**, *11*, 1037. [CrossRef]
61. Alcaldía de Pamplona. Plan Básico de Ordenamiento Territorial Municipio de Pamplona Norte de Santander 2015. 2015. Available online: https://pamplona-nortedesantander.gov.co/Transparencia/PlaneacionGestionyControl (accessed on 15 February 2023).
62. Bonilla Granados, C.A.; Tarazona Tobo, L.V.; Caicedo Calderón, A.D. Statistical Analysis of Residential Drinking Water Consumption in Toledo, Colombia. *BISTUA Rev. Fac. Cienc. Básicas* **2022**, *20*, 70–75. [CrossRef]
63. Herrera, D.B.; Ávila, E.M.; Mejía, C.A.Z. Analysis of non-revenue water in the urban supply system of the municipality of Facatativá, Colombia. *Tecnura* **2020**, *24*, 84–98. [CrossRef]

 *water*

Article

Development of an Optimal Water Allocation Model for Reservoir System Operation

Eunkyung Lee [1], Jungwon Ji [1], Seonmi Lee [1], Jeongin Yoon [1], Sooyeon Yi [2] and Jaeeung Yi [1,*]

1 Department of Civil Systems Engineering, Ajou University, 206 Worldcup-ro Yeongtong-gu, Suwon 16499, Republic of Korea; oplk100@ajou.ac.kr (E.L.); log58@ajou.ac.kr (J.J.); sunki7070@ajou.ac.kr (S.L.); pinkashley@ajou.ac.kr (J.Y.)
2 Landscape Architecture & Environmental Planning, University of California, Berkeley, CA 94720, USA; sooyeon@berkeley.edu
* Correspondence: jeyi@ajou.ac.kr; Tel.: +82-31-219-2507

Abstract: Allocating adequate water supplies under the increasing frequency and severity of droughts is a challenge. This study develops an optimal reservoir system operation method to allocate water supplies from upstream reservoirs to meet the downstream water requirements; validates the proposed optimization model through the system operation of upstream reservoirs; and proposes new water supply policies that incorporate a transformed hydropower reservoir with an add-on water supply function and two multipurpose reservoirs. We use linear programming to develop an optimal water allocation model. This model provides an operational strategy for managing upstream reservoirs with different storage capacities. By integrating the effective storage ratio of each reservoir into the allocation estimation, the model ensures an optimal distribution of downstream water requirements. The results indicated well-balanced, effective storage ratios among the Chungju, Soyanggang, and Hwacheon Reservoirs across varying hydrological conditions. Specifically, during drought years, the average effective storage rates were 20.5%, 20.6%, and 19.07%, respectively. In normal years, these figures, respectively, were 59.3%, 68.6%, and 52.4%, while in wet years, the rates stood at 64.08%, 62.90%, and 54.61%. This study enriches the reservoir operation literature by offering adaptable solutions for collaborative reservoir management and presents efficient strategies for reservoir operations.

Keywords: water allocation model; reservoir optimization; effective storage ratio; linear programming; water resources management; Han River Basin

Citation: Lee, E.; Ji, J.; Lee, S.; Yoon, J.; Yi, S.; Yi, J. Development of an Optimal Water Allocation Model for Reservoir System Operation. *Water* **2023**, *15*, 3555. https://doi.org/10.3390/w15203555

Academic Editors: Peiyue Li, Jianhua Wu and Achim A. Beylich

Received: 11 September 2023
Revised: 27 September 2023
Accepted: 10 October 2023
Published: 12 October 2023

1. Introduction

Climate change is significantly altering hydrologic systems, intensifying precipitation in wetter regions while exacerbating drought in drier areas [1,2]. Extreme events like droughts pose severe challenges to both water supply and demand [3]. South Korea has an average annual precipitation of 1331.7 mm, which is well above the global rainfall average of 884 mm [4]. South Korea experiences four distinct seasons, resulting in substantial changes in precipitation and temperature. Approximately 68% of its annual precipitation occurs during the flood season from June to September, and precipitation varies greatly temporally and spatially [5]. With 65% of its terrain being mountainous with slopes exceeding 20%, the country experiences rapid run-off into rivers [6]. To manage its water resources, South Korea relies extensively on reservoirs throughout the country [7]. In the event of a water deficit during the dry season, reservoirs supply the stored water during the flood season to prevent drought. However, challenges arise during periods of insufficient rainfall throughout the year. The country experiences long-term drought cycles of 5–7 years and short-term cycles of 2–3 years [8]. Localized droughts have become more frequent in recent years, contributing to increased water scarcity. Due to a lack of precipitation throughout the summer months, a prolonged drought occurred from 2014 to 2017. The drought during this period is considered the most severe in the Han River Basin to date, and long-term droughts are anticipated to worsen as a result of climate change [9,10].

Preparing for drought often involves increasing storage capacity, commonly achieved through the construction of new reservoirs [11]. However, these construction projects come with notable environmental and social drawbacks [12]. Social conflicts can emerge between upstream communities, potentially affected by submersion from a new reservoir, and downstream communities, who may experience alterations in flow regimes, water quality, and ecosystems [13]. Also, changing reservoir operations can affect the natural streamflow regime and impact the aquatic ecosystem biodiversity. The construction of reservoirs, as exemplified by the Lake Powell River Reservoir in the Colorado River, USA, modifies the natural river flow [14]. The building and functioning of the Three Gorges Dam have substantially changed the hydrological patterns downstream on the Yangtze River, impacting environmental conditions, biodiversity, landscape structure, and human development [15]. Recognizing these trade-offs, the subsequent sections of this study focus on optimizing reservoir operations to balance these competing interests. To address water scarcity, the South Korean government has investigated ways to secure additional water supplies, for example, by adding on a water supply function to the existing hydropower reservoirs in the Han River Basin. The Soyanggang and Chungju Reservoirs supply water to Seoul, the capital city of South Korea. In April 2020, the Korean government decided to add a water supply function to the Hwacheon Reservoir, which originally served as a hydropower reservoir, to better meet the increasing water demand upon severe droughts [16].

Numerous countries leverage existing reservoirs to fulfill their water requirements. An initiative movement for the multifunctional use of existing hydropower reservoirs emerged prominently during the 6th World Water Forum in Marseille, France, in 2012 led by Électricité de France and the World Water Council [17]. Many countries, like Peru, Costa Rica, the United States, France, Cameroon, Niger, India, Nepal, China, and Austria, have already added water supply functions to the existing hydropower reservoirs [18]. For example, the Serre-Ponçon Reservoir in France primarily served as hydropower generation but now supports secondary functions like water supply and local tourism [19]. Extensive research has also been conducted on the multipurpose usage of hydropower reservoirs for water demand and ecosystem preservation. As another example, the Keswick Reservoir, a hydropower reservoir in California, USA, added a water supply function to better meet the domestic and ecosystem water needs by developing the optimal reservoir operation rules [20].

Although initially constructed for hydropower generation, the Hwacheon Reservoir has substantial storage capacity and is a desirable candidate to potentially have a water supply function. In the past, the Hwacheon Reservoir irregularly released water during some drought events. However, these operations have often proceeded without a systematic framework. The Hwacheon Reservoir in South Korea stands as the first example of a transformed multipurpose reservoir that encompasses not only power generation but also water supply. However, the purposes of water supply and hydropower generation conflict with each other occasionally. This is because the timing of hydropower generation is not always compatible with the timing of water supply.

Several studies have aimed to estimate the volume of water that the Hwacheon Reservoir could provide with a 95% reliability for water supply. This initiative led to the development of a Hwacheon Reservoir rule curve specifically for drought conditions [21]. Another study proposed estimating the water supply capacity for an individual reservoir by incorporating operational aspects of hydropower reservoirs into a reservoir operation model [22,23]. Another study used a data-driven model for predicting inflows into hydropower reservoirs and assessed the water supply capabilities of hydropower reservoirs [24]. Prior research has predominantly focused on operation of a single reservoir, leaving a gap in the literature regarding integrated operations with other reservoirs for effective water resources management. In reviewing the existing literature, several gaps become evident: While there are instances in other countries of hydropower reservoirs being converted for multipurpose use, there is an absence of research focusing on the cooperative operation of such transformed reservoirs with other types of reservoirs. Previous research has largely centered on securing water supply through the cooperative management of multipurpose and water-only reservoirs. Studies have also been conducted on reservoir simulation considering emergency storage and water supply volumes within the same basin. Notably, there is a lack of studies applying optimization techniques to efficiently allocate water through the cooperative operation of existing multipurpose reservoirs initially designed for water supply with hydropower reservoirs newly repurposed for water supply. These gaps in the literature underscore the novelty and significance of the present study, which aims to address these unexplored areas.

To bridge this gap, the present study aims to achieve multiple objectives to enhance water resource management in the Han River Basin through a more integrated approach. Specifically, this study seeks to (a) develop an optimal reservoir system operation method to allocate water supplies from multiple upstream reservoirs to meet the downstream water requirements; (b) validate the proposed optimization model through the system operation of three upstream reservoirs using historical inflow data; and (c) propose new water supply policies that incorporate a transformed hydropower reservoir with an add-on water supply function and two multipurpose reservoirs. This ensures an optimal downstream water supply while accounting for the unique characteristics and status of the upstream reservoirs.

The novel contributions of this study are multifold: Unlike traditional studies that focus solely on multipurpose reservoirs, our work extends the functionality of a hydropower reservoir by incorporating a water supply component. Our research addresses a gap in current methodologies by developing rules for water allocation, particularly when only a limited number of reservoirs are available to meet downstream demand. The paper presents a case study that transforms an existing hydropower reservoir to serve additional functions, offering a cost-effective and environmentally sustainable alternative to building new infrastructure. This study offers a collaborative framework for cases like South Korea, where multipurpose reservoirs and their operating institutions differ, thereby necessitating cooperation for optimal reservoir management. This holds true even if the same institution operates all the reservoirs, enabling rational operations to achieve the best results in non-flood water supply seasons. In the context of South Korea, the transformation of existing multipurpose reservoirs by adding hydropower functions for water supply enables the securing of additional water volumes, thereby contributing to the optimal integrated operation of the three reservoirs to prevent water supply shortages during non-flood seasons. These contributions bring new perspectives to the field of water resources management and offer practical solutions for enhanced system operation.

2. Water Allocation Methods for Multireservoir Systems

Optimization models serve as invaluable tools in water resources management for designing optimal system configurations or operational measures [25,26]. Managing reservoirs is complex, requiring a balance of diverse operational objectives for optimal operation [27]. Numerous studies have employed optimization models to develop and assess reservoir operations. Various optimization techniques, such as linear programming (LP) [28,29], nonlinear programming (NLP) [30,31], dynamic programming (DP) [32,33], and heuristics algorithms (HA) [34–36], have been explored for over four decades. These models aim to either maximize or minimize reservoir objectives while satisfying operational constraints [37]. A crucial step in model development is the careful formulation of the objective function and constraints. The model in this study employs LP, a commonly used technique in the water resources field for its simplicity and capability to find the global optimum [38].

The water level within a reservoir fluctuates from a low water level to either a normal high or a restricted level during flood seasons. The release of water is based not just on downstream demand, but also on reservoir storage conditions [39–41]. Reservoir system operation considers conditions of the individual reservoir such as inflows, storages, and demands, enabling more effective water resources management. Consequently, the topic of reservoir system operations in the basin has long been a subject of ongoing research, as it provides a more stable water supply and better flood management compared to single reservoir operations [42]. Earlier studies have developed guidelines for the operation of single-purpose reservoirs arranged in series or parallel. Further research has categorized reservoir system operation into those focused on water supply and flood control, leading to more efficient operation plans depending on the objective [43–45]. Specifically, various approaches have been investigated to determine the supply allocation from multiple reservoirs to meet downstream demand [46].

2.1. Method 1: Dry Season Allocation Using Predicted Inflow and Available Reservoir Storage

Effective storage in a reservoir is the storage capacity between the low water level and normal high water level. Available storage is the storage capacity between the current water level and the normal high water level, which is reserved space for additional water storage (Equation (1)). The first step is to determine the proportion of available storage in each reservoir relative to the total available storage across all reservoirs (Equation (2)). The second step is to estimate the predicted inflow for the dry season (from the present to the onset of the flood season) for each reservoir. The third step is to assess the proportion of each reservoir's predicted inflow to the total predicted inflow for the dry season (Equation (3)). The final step involves calculating each reservoir's allocation, which is the ratio of its available storage to the predicted inflow during the dry season (Equation (4)).

This method is only applicable to the period preceding the flood season. In South Korea, the annual cycle is divided into a flood season (21 June–20 September) and a dry season (21 September–20 June of the following year). The water allocation for each reservoir is determined to secure available storage. This allocation ensures a consistent ratio of inflow in the dry season to available storage for all reservoirs (Equation (5)).

The total demand is equal to the sum of the allocation amounts for the reservoirs in a given month (Equation (6)). In Equation (5), setting $\sum_{i=1}^{n}[SN_{ii} - (S_{i,t-1} + i_{i,t})]$ equal to N_t and integrating with Equation (6) yields Equation (7). The allocation amount for each reservoir is expressed in Equation 8.

This approach tends to minimize reservoir release during the dry season because it only considers relative available storage. However, multipurpose reservoirs in South Korea need to supply more than the contracted amount of water. This method does not account for the minimum release requirements for each reservoir and is uncertain when estimating the allocation amount in flood season.

$$v_{i,t} = SN_i - (S_{i,t-1} + i_{i,t} - x_{i,t}) \tag{1}$$

where $v_{i,t}$ is the available storage in reservoir i for the period t; SN_i is the reservoir storage at the normal high water level during the dry season and the restricted water level during flood season; $S_{i,t}$ is the storage of reservoir i for the period t; $i_{i,t}$ is the inflow of reservoir i for the period t; and $x_{i,t}$ is the allocation from reservoir i for the period t.

$$P_{i,t} = \frac{v_{i,t}}{\sum_{i=1}^{n} v_{i,t}} = \frac{SN_i - (S_{i,t-1} + i_{i,t} - x_{i,t})}{\sum_{i=1}^{n}[SN_i - (S_{i,t-1} + i_{i,t} - x_{i,t})]} \tag{2}$$

$$\alpha_{i,t} = \frac{ir_{i,t}}{\sum_{i=1}^{n} ir_{i,t}} \tag{3}$$

where $ir_{i,t}$ is the predicted inflow into reservoir i during the remainder before the flood season.

$$P_{i,t} = \alpha_{i,t} \tag{4}$$

where $P_{i,t}$ is the ratio of free volume in reservoir i to the total free volume in the system, and $\alpha_{i,t}$ is the proportion of the period t's demand allocated to reservoir i.

$$\alpha_{i,t} = \frac{SN_i - (S_{i,t-1} + i_{i,t} - x_{i,t})}{\sum_{i=1}^{n}[SN_i - (S_{i,t-1} + i_{i,t} - x_{i,t})]} \tag{5}$$

$$\sum_{i=1}^{n} x_{i,t} = D_t \tag{6}$$

where D_t is the total demand.

$$SN_i - (S_{i,t-1} + i_{i,t} - x_{i,t}) = \alpha_{i,t}(N_t + D_t) \tag{7}$$

where N_t is the reservoir's available storage, excluding the portion occupied by the current water storage from full storage capacity.

$$x_{i,t} = \alpha_{i,t}(N + D) - SN_i + S_{i,t-1} + i_{i,t} \tag{8}$$

2.2. Method 2: Allocation Based on Storage Ratio

This method estimates the allocation amount for this month using the storage ratio at the end of the preceding month for each reservoir (Equation (9)). A higher storage ratio results in a larger allocation amount. This method only considers storage, and it does not consider future reservoir condition, capacity, and inflow. Thus, this method is not applicable in real reservoir operation.

$$x_{i,t} = \left[\frac{S_{i,t-1}}{\sum_{i=1}^{n} S_{i,t-1}} \right] \times D_t \tag{9}$$

2.3. Method 3: Allocation Considering Storage and Inflow

This method considers both storage and predicted inflow to determine the water allocation amount (Equation (10)). The allocation amount is estimated based on the ratio of available water, which is computed by adding reservoir storage for the previous month and the predicted inflow for the current month. This approach is suitable for reservoirs with similar capacities but not for those with significant differences in reservoir capacities. This method is impractical, as it relies only on the volume of available water while neglecting the hydrologic conditions and capacity of each reservoir.

$$x_{i,t} = \left[\frac{S_{i,t-1} + i_{i,t}}{\sum_{i=1}^{n}(S_{i,t-1} + i_{i,t})} \right] \times D_t \tag{10}$$

2.4. Method 4: Allocation Using Effective Storage Ratio

The method incorporates the effective storage ratio of each reservoir in allocation estimation (Equation (11)). The effective storage is achieved by subtracting the storage at the low water level from the storage at the normal high water level. The current effective storage is calculated by subtracting the storage at the low water level from the storage at the current water level. The effective storage ratio is the current effective storage divided by the effective storage. This method can estimate the allocation amount for those reservoirs with different storage capacities (Equation (12)). Thus, this study adopts this method, as it considers both reservoirs with different capacities and the current reservoir condition.

$$R_{i,t} = \left[\frac{S_{i,t} - SL_i}{SN_i - SL_i} \right] \times 100 \tag{11}$$

$$x_{i,t} = \left[\frac{R_i}{\sum_{i=1}^{n} R_i} \right] \times D_t \tag{12}$$

where $R_{i,t}$ is the effective reservoir storage rate in reservoir i for the period t, and SL_i is the reservoir storage at the low water level in reservoir i.

2.5. Optimization Model for Equitable Water Allocation in Multireservoir Systems

In this study, we develop an optimization model for water allocation using the effective storage ratio method. This method provides an optimal operational strategy for downstream water requirements. By integrating the effective storage ratio of each reservoir into the allocation estimation, the model ensures an optimal distribution of downstream water requirements while considering the different reservoir capacities and the current reservoir condition. The objective of the optimization model is to maximize the sum of the minimum monthly storage ratios for each reservoir throughout its operation period (Equation (13)). This objective function is different from its traditional objective function, which maximizes the sum of the monthly storage for all reservoirs. Such a traditional approach can lead to an operation where a single reservoir supplies all downstream demands, increasing the storage of other reservoirs. This study uses an objective function that maintains a similar effective storage ratio for all reservoirs.

$$\max \sum_{t=1}^{12} R_{i_{min},t} \tag{13}$$

where $R_{i_{min},t}$ is the water storage rate for reservoir i with a minimum R among reservoirs in month t.

Constraints of the optimization models are as follows. Equation (14) represents the water balance equation in a reservoir. The sum of all allocations from upstream reservoirs should exceed the downstream water requirement (Equation (15)). For each reservoir, the effective storage ratio is estimated using (Equation (16)) and is required to surpass the minimum storage ratio for all reservoirs (Equation (17)). The storage for each reservoir should remain between the storage for the low water level and the storage for the normal high water level (Equation (18)). The allocation amount from each reservoir should exceed the planned water supply or the minimum instream flow (Equation (19)). The planned water supply is the predetermined volumes of water released from the multipurpose reservoirs. All variables are constrained to be positive (Equation (20)).

$$S_{i,t} = S_{i,t-1} + i_{i,t} - x_{i,t} - w_{i,t} \tag{14}$$

where $w_{i,t}$ is the discharge excluding the allocation of reservoir i in month t.

$$x_{1,t} + x_{2,t} + \cdots + x_{n,t} \geq D_t \tag{15}$$

$$R_{i,t} = \left[\frac{S_{i,t} - SL_i}{SN_i - SL_i} \right] \times 100 \tag{16}$$

$$R_{i,t} \geq R_{i_{min},t} \tag{17}$$

$$SL_i \leq S_{i,t} \leq SN_i \tag{18}$$

$$x_{i,t} \geq IF_i \text{ or } WSP_{i,t} \tag{19}$$

IF_i represents the stream maintenance flow of reservoir i; $WSP_{i,t}$ is the planned water supply of reservoir i for month t; and n is the number of reservoirs.

$$S_{i,t}, i_{i,t}, x_{i,t}, w_{i,t}, SL_i, SN_i, R_{i,t}, R_{min,t}, IF_i, WSP_{i,t} \geq 0 \tag{20}$$

3. Application

3.1. Study Area

The capital area of South Korea, encompassing Seoul, Incheon Metropolitan City, and Gyeonggi Province, covers an area of 11,856 km^2, or 11.8% of the nation's total land area of 100,210 km^2. This region has approximately 26 million people, which accounts for 50.5% of the national population [47]. Due to its high population density, the capital area is sensitive to water supply shortage. With four distinct seasons with the most precipitation in the summer, reservoirs are crucial for storing water during flood season for later use in the dry season.

The Han River system includes three multipurpose reservoirs (Soyanggang, Chungju, and Hoengseong), seven hydropower reservoirs (Hwacheon, Chuncheon, Uiam, Cheongpyeong, Goesan, Paldang, and Doam), and one flood control reservoir (Pyeonghwa) (Figure 1 and Table 1). The three multipurpose reservoirs release predetermined volumes of water (Table 2). The Hoengseong Reservoir, completed in 2000, is designed for water scarcity mitigation and flood control in the Seomgang River Basin, a Han River tributary. The Soyanggang and Chungju Reservoirs supply water and control floods in the capital area. Soyanggang Reservoir is the largest reservoir and was completed in 1973 with a basin area of 2703 km^2. The Chungju Reservoir, built in 1985, has a basin area of 6648 km^2 with an average annual inflow of 154.5 CMS—approximately 2.78 times greater than the inflow for Soyanggang (55.5 CMS). The effective storage of Hoengseong Reservoir is relatively small and only accounts for 3.00% and 3.16% of the storage capacities of Soyanggang and Chungju, respectively. The Hoengseong Reservoir is a substantially smaller multipurpose reservoir that only accounts for 8.92% of the capacity of the Hwacheon Reservoir.

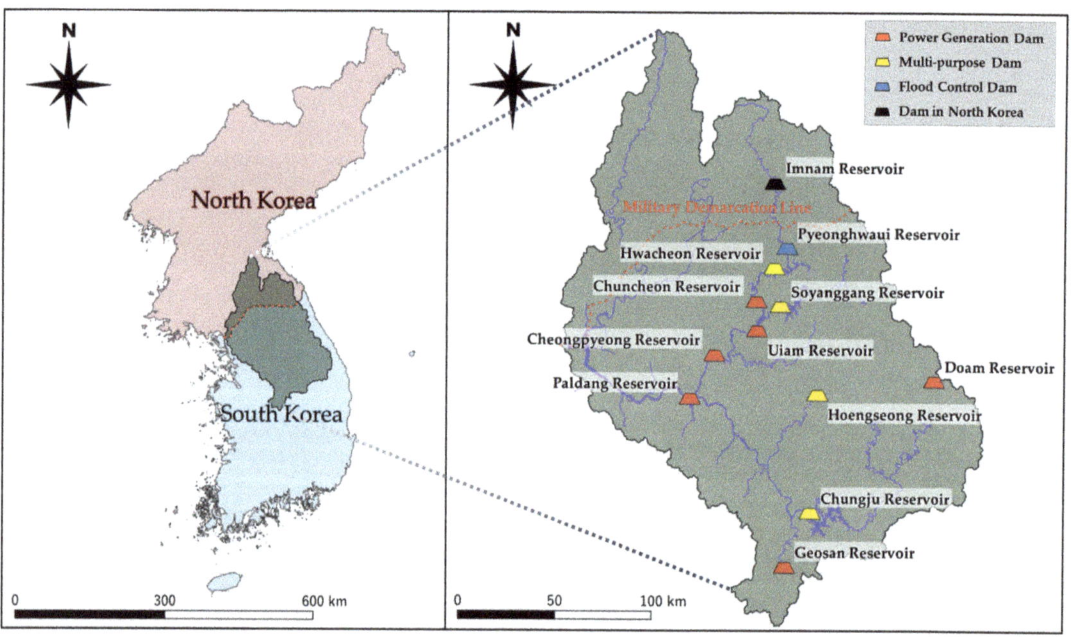

Figure 1. A study area map of Han River Basin.

Table 1. Comparative analysis of storage capacities (million cubic meters (MCM)) and water levels (EL.m) in multipurpose and Hwacheon reservoirs.

Reservoir	Total Storage	Conservation Storage	Flood Water Level	Normal High Water Level	Restricted Water Level	Low Water Level
	MCM	MCM	EL.m	EL.m	EL.m	EL.m
Chungju	2750	1789	145	141	138	110
Soyanggang	2900	1900	198	193.5	190.3	150
Hoengseong	86.9	73.4	180	180	178.2	160
Paldang	244	18	27	25.5	na	25
Hwacheon	974.2	572.8	183	181	175	156.8

Table 2. Monthly planned water supply (WS) of the multipurpose reservoirs. Multipurpose reservoirs serve as domestic (D), industrial (I), agricultural (A), and instream (IS) water supply (CMS). A* is the water intake within the reservoir, and B** is the water intake from downstream of the reservoir.

Reservoir		Soyanggang			Chungju			Hoengseong		
Types of WS		D and I	Ag	IS	D and I	Ag	IS	D and I	Ag	IS
January	A*	-	-	-	-	-	-	2.3	-	-
	B**	38.1	-	8.1	86.6	-	10.6	-	-	1.8
February	A*	-	-	-	-	-	-	2.3	-	-
	B**	38.1	-	8.1	86.6	-	10.6	-	-	1.5
March	A*	-	0.4	-	-	-	-	2.3	-	-
	B**	38.1	-	8.1	86.6	-	10.6	-	-	0.9
April	A*	-	1	-	-	-	-	2.3	0.1	-
	B**	38.1	-	8.1	86.6	9.1	10.6	-	0.3	0.5
May	A*	-	1	-	-	-	-	2.3	0.1	-
	B**	38.1	-	8.1	86.6	21.8	10.6	-	1	1.8
June	A*	-	1	-	-	-	-	2.3	0.1	-
	B**	38.1	-	8.1	86.6	28	10.6	-	1	2.1
July	A*	-	1	-	-	-	-	2.3	0.1	-
	B**	38.1	-	8.1	86.6	18	10.6	-	1	0.5
August	A*	-	1	-	-	-	-	2.3	0.1	-
	B**	38.1	-	8.1	86.6	23.7	10.6	-	1	-
September	A*	-	1	-	-	-	-	2.3	0.1	-
	B**	38.1	-	8.1	86.6	10.6	10.6	-	0.7	0.4
October	A*	-	0.3	-	-	-	-	2.3	0.1	-
	B**	38.1	-	8.1	86.6	8	10.6	-	0.2	0.2
November	A*	-	-	-	-	-	-	2.3	0.1	-
	B**	38.1	-	8.1	86.6	-	10.6	-	0.1	1.2
December	A*	-	-	-	-	-	-	2.3	-	-
	B**	38.1	-	8.1	86.6	-	10.6	-	-	1.3
Mean	A*	-	0.4	-	-	-	-	2.3	0.1	-
	B**	38.1	-	8.1	86.6	10	10.6	-	0.4	1

Reservoir management is divided among different agencies. For example, K-water manages the multipurpose reservoirs and the Korea Hydro & Nuclear Power Company operates hydropower reservoirs. While flood control and hydropower reservoirs serve specific functions, multipurpose reservoirs are versatile, catering to water supply, flood control, and energy generation. The importance of securing a consistent water supply for the capital area was highlighted during the severe drought from 2015 to 2018, prompting the government to explore expanding the roles of multipurpose reservoirs within the Han River system. In April 2020, a pilot project commenced to convert the Hwacheon Reservoir, the largest hydropower reservoir in the Han River system, into a multipurpose reservoir. This marks the first conversion of a hydropower reservoir to a multipurpose reservoir in South Korea. Since then, the Hwacheon Reservoir constantly releases 22.2 CMS.

This study aims to develop an optimization model for the coordinated operation of the Hwacheon Reservoir and the Chungju and Soyanggang Reservoirs. Among the hydropower reservoirs, this study considers the Hwacheon and Paldang Reservoirs, while other hydropower reservoirs are excluded because they are run-off reservoirs. Among the multipurpose reservoirs, this study considers the Soyanggang and Chungju Reservoirs. The Hoengseong Reservoir is excluded because it is a local reservoir with a small capacity. The multipurpose and hydropower reservoirs in the Han River system are operated jointly only during emergencies, such as drought or floods, under the supervision of the Han River Flood Control Center. For example, if the Chungju, Soyanggang, and Hwacheon Reservoirs are operated by a single agency, then this agency can consider the specific conditions of each reservoir for all.

3.2. Data Collection

North and South Korea share the North Han River. The Imnam Reservoir in North Korea is located upstream on the North Han River, while the Hwacheon Reservoir in South Korea is downstream of Imnam Reservoir. Built in 2004, the Imnam Reservoir diverts inflow into the East Sea year-round for hydropower generation. During the flood season, Imnam Reservoir releases water via its spillway, leading to downstream flooding events in South Korea. Conversely, the release from the reservoir diminishes during the dry season. Since the construction of the Imnam Reservoir, the average inflow into the Hwacheon Reservoir has notably decreased from 90.1 CMS (based on average data from 1967 to 2003) to 51.3 CMS (based on average data from 2004 to 2022), representing a 57.0% reduction (Figure 2).

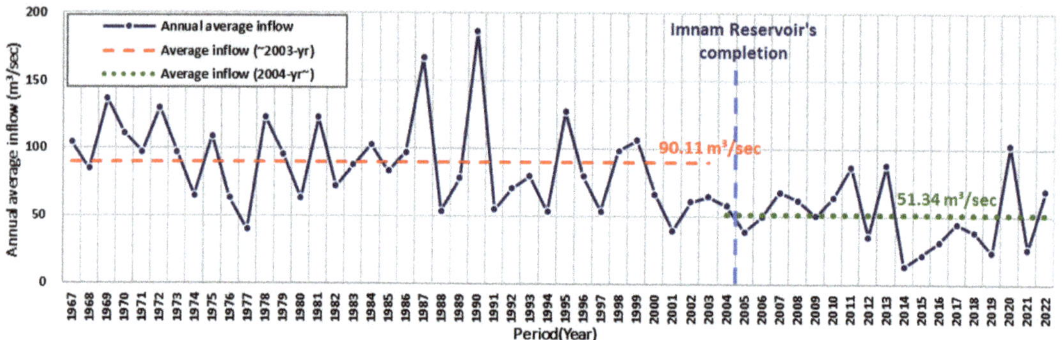

Figure 2. Annual average inflow to Hwacheon Reservoir, 1967–2022: comparison of pre- and post-2004 periods.

This study used historical inflow data, focusing on the period after 2004 when the inflow of the Hwacheon reservoir was significantly reduced upon the completion of Imnam Reservoir. We built and tested the model for every single year from 2004 to 2022. However, instead of presenting the results for a total of 19 years' worth of data, we identified three representative years for dry, normal, and wet conditions. Between 2004 and 2022, the average inflows into the Chungju, Soyanggang, and Hwacheon Reservoirs were 150.1 CMS, 67.5 CMS, and 51.3 CMS, respectively (Table 3). The year 2015 was selected as the dry year because it experienced an extended drought and had the lowest average annual inflow. The year 2018 was selected as a normal year, as its annual average inflow value was closest to the average inflow value. The year 2020 was chosen as a wet year due to its highest average annual inflow. We applied the monthly inflow data for 2015, 2018, and 2020 to develop the model and to assess the feasibility of system operation for the three reservoirs. All input data came from the Han River Flood Control Office (http://www.wamis.go.kr/, accessed on 13 April 2023).

Table 3. Average annual inflow (CMS) for Chungju, Soyanggang, and Hwacheon Reservoirs from 2004 to 2022.

Year	Chungju Reservoir (CMS)	Soyanggang Reservoir (CMS)	Hwacheon Reservoir (CMS)
2004	214.0	81.1	58.0
2005	175.4	63.1	39.0
2006	244.7	95.8	49.9
2007	212.2	76.2	68.1
2008	96.3	58.6	62.3
2009	128.0	75.3	50.8

Table 3. *Cont.*

Year	Chungju Reservoir (CMS)	Soyanggang Reservoir (CMS)	Hwacheon Reservoir (CMS)
2010	169.0	74.8	64.4
2011	283.1	105.2	86.9
2012	159.7	56.0	35.5
2013	144.8	75.1	88.5
2014	73.5	29.2	13.3
2015	55.6	33.8	21.7
2016	91.7	52.1	31.4
2017	108.6	62.6	45.1
2018	160.7	66.8	38.5
2019	74.7	36.4	23.8
2020	193.9	107.0	102.9
2021	108.6	45.3	26.1
2022	167.8	87.3	69.3
Average	150.1	67.5	51.3

4. Results

This study developed a model to allocate appropriate water supplies from the Chungju, Hwacheon, and Soyanggang Reservoirs to meet the required discharge for Paldang Reservoir (124 CMS and 138 CMS for the nursery and transplantation season). The model incorporated the specialized discharge requirements for Paldang Reservoir during the nursery and transplantation period from 27 May to 10 June, with rates set at 126.3 CMS for May and 128.7 CMS for June. The effectiveness of the model was evaluated using actual inflow data from three representative years: a dry year (2015), a normal year (2018), and a wet year (2020).

4.1. Model Evaluation for Dry Year (2015)

Supplementary Tables S1 (Chungju Reservoir), S2 (Soyanggang Reservoir) and S3 (Hwacheon Reservoir) present the outcomes of applying the optimization model during the drought year of 2015. These tables delineate the optimal water allocation for each reservoir with the spillway discharge. These tables are represented in Figure 3a–c.

In December 2014, the initial water levels for the three reservoirs (end-of-month water levels) were as follows: 126.2 EL.m for Chungju Reservoir, 165.8 EL.m for Soyanggang Reservoir, and 165.2 EL.m for Hwacheon Reservoir. The continuing drought since 2014 resulted in low initial water levels for reservoir operations in 2015. Given the modest inflows and low water levels during the dry season of 2015, the model requirement for each reservoir to release more than the planned water supply was unsatisfied. Thus, only instream flows were discharged during the dry season.

In June, the reservoirs recorded their lowest water levels: 114.5 EL.m for Chungju Reservoir, 156.3 EL.m for Soyanggang Reservoir, and 159.1 EL.m for Hwacheon Reservoir (Tables 4–6). Due to their different basin areas and locations, these reservoirs showed significant differences in average annual inflows. Hence, the average annual inflows for the Chungju, Soyanggang, and Hwacheon Reservoirs were 55.4 CMS, 33.6 CMS, and 21.5 CMS, respectively. Their average annual allocation amount for historical data was 65.8 CMS (52.8%), 36.1 CMS (29.0%), and 22.7 CMS (18.2%), respectively. Among the three reservoirs, the Chungju reservoir had larger initial storage than the other two and received a large initial allocation. As the effective storage ratios of the three reservoirs converged over time, the patterns of their low water levels evolved in a largely similar manner (Figure 3).

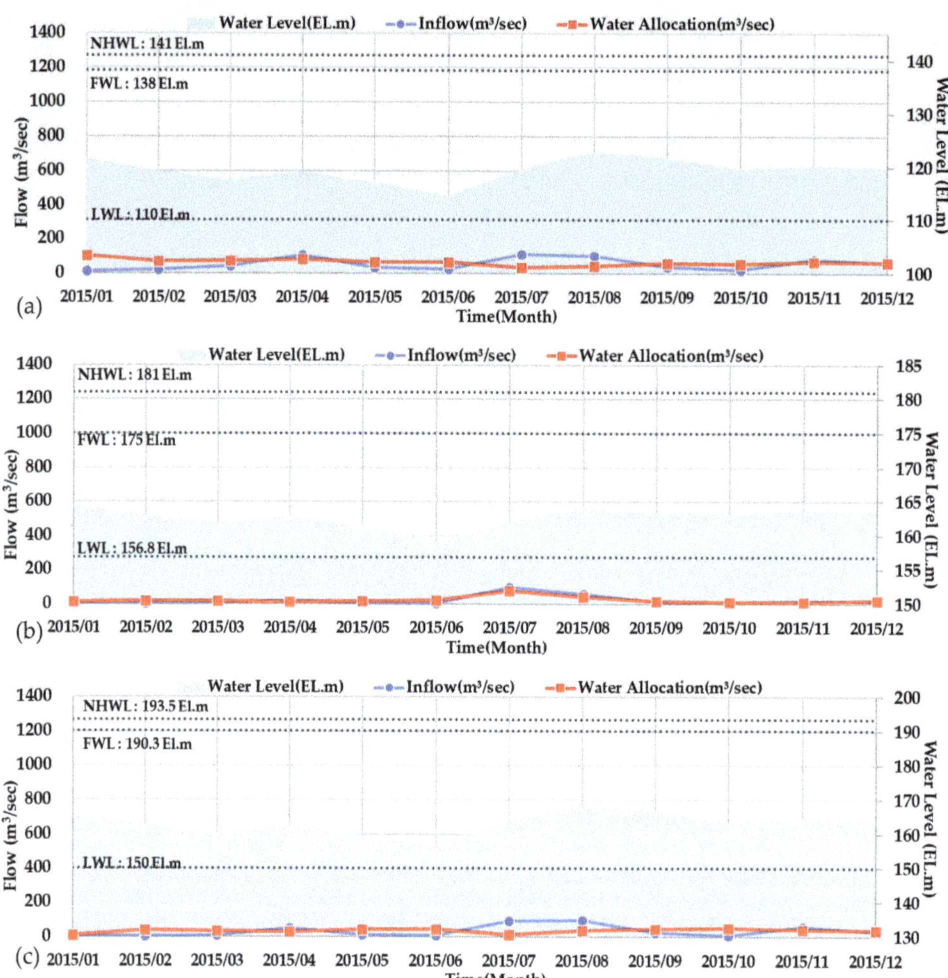

Figure 3. The results from (**a**) Supplementary Table S1, (**b**) Supplementary Table S2, and (**c**) Supplementary Table S3 are represented in the graphs. In this figure, the water level is depicted as a sky-blue shaded area, inflow is shown as a blue solid line, and water allocation is represented in red.

4.2. Model Evaluation for Normal Year (2018)

Supplementary Tables S4 (Chungju Reservoir), S5 (Soyanggang Reservoir) and S6 (Hwacheon Reservoir) show the results of applying the optimization model during the normal year of 2018, as represented in Figure 4a–c. The initial water levels (end-of-month water levels in December 2017) in December 2017 for the Chungju, Soyanggang, and Hwacheon Reservoirs were 128.47 EL.m, 180.97 EL.m, and 169.39 EL.m, respectively. Thus, the planned water supply was met for each reservoir in 2018. The lowest water levels for the three reservoirs occurred in March, recorded as 119.1 EL.m for Chungju Reservoir, 173.0 EL.m for Soyanggang Reservoir, and 165.0 EL.m for Hwacheon Reservoir. These levels were maintained to mitigate dry season impacts, and sufficient inflows were observed up to the flood season in June. The average annual inflows for 2018 were 159.7 CMS for Chungju Reservoir, 66.1 CMS for Soyanggang Reservoir, and 38.1 CMS for Hwacheon Reservoir. The average annual allocation amounts were 137.3 CMS (59.1%), 53.0 CMS (25.4%), and

34.8 CMS (15.6%). No spillway releases were required, as the maximum releases for the Chungju, Soyanggang, and Hwacheon Reservoirs were 775.0 CMS, 250.0 CMS, and 185.0 CMS, respectively. While Soyanggang Reservoir had the highest effective storage ratio at the beginning of the year, Chungju Reservoir had the highest effective storage ratio by the end of the year.

Table 4. Optimization model-determined water allocation and effective reservoir storage ratios for the three reservoirs in the dry year of 2015.

2015-MM	Model-Derived Water Allocation (CMS)			Model-Derived Effective Reservoir Storage Ratio (%)			Total Water Supply (CMS)
	Chungju	Soyanggang	Hwacheon	Chungju	Soyanggang	Hwacheon	
01	104.3	8.1	11.6	26.0	24.7	24.8	124.0
02	67.6	40.3	16.1	19.9	19.9	20.0	124.0
03	72.6	36.6	14.8	15.8	15.8	15.9	124.0
04	82.1	29.7	12.2	19.2	19.2	19.3	124.0
05	66.9	43.5	15.9	14.5	14.5	14.6	126.3
06	65.1	44.1	19.5	8.8	9.1	7.2	128.7
07	33.7	12.6	77.8	20.7	21.6	16.7	124.0
08	44.4	37.0	42.7	29.4	30.5	23.7	124.0
09	59.8	46.5	17.8	26.3	27.4	21.2	124.0
10	58.6	53.0	12.5	20.7	20.7	20.8	124.0
11	67.8	44.1	12.1	22.7	22.7	22.8	124.0
12	66.7	37.6	19.7	21.6	21.6	21.7	124.0

Table 5. Optimization model-determined water allocation and effective reservoir storage ratios for the three reservoirs in the normal year of 2018.

2018-MM	Model-Derived Water Allocation (CMS)			Model-Derived Effective Reservoir Storage Ratio (%)			Total Water Supply (CMS)
	Chungju	Soyanggang	Hwacheon	Chungju	Soyanggang	Hwacheon	
01	97.2	78.3	22.2	33.9	49.6	37.7	197.7
02	97.2	46.2	22.2	21.7	43.6	32.0	165.6
03	97.2	46.2	22.2	19.2	41.0	27.1	165.6
04	106.3	46.2	22.2	29.8	46.9	32.2	174.7
05	119.0	46.2	22.2	49.0	63.4	63.1	187.4
06	125.2	46.2	94.1	37.3	59.2	30.0	265.5
07	115.2	46.2	47.9	72.8	75.8	68.9	209.3
08	192.5	78.2	67.7	85.8	89.3	68.9	338.4
09	397.5	63.7	29.9	85.8	89.3	68.9	491.1
10	105.2	46.2	22.2	93.6	90.1	66.5	173.6
11	97.2	46.2	22.2	94.2	90.2	68.9	165.6
12	97.2	46.2	22.2	88.6	85.3	64.5	165.6

Table 6. Optimization model-determined water allocation and effective reservoir storage ratios for the three reservoirs in the wet year of 2020.

2020-MM	Model-Derived Water Allocation (CMS)			Model-Derived Effective Reservoir Storage Ratio (%)			Total Water Supply (CMS)
	Chungju	Soyanggang	Hwacheon	Chungju	Soyanggang	Hwacheon	
01	97.2	46.2	22.2	68.0	49.7	53.3	165.6
02	97.2	46.2	22.2	63.1	46.2	50.3	165.6
03	97.2	46.2	22.2	61.1	44.9	48.1	165.6

Table 6. *Cont.*

2020-MM	Model-Derived Water Allocation (CMS)			Model-Derived Effective Reservoir Storage Ratio (%)			Total Water Supply (CMS)
	Chungju	Soyanggang	Hwacheon	Chungju	Soyanggang	Hwacheon	
04	106.3	46.2	22.2	52.9	42.9	44.1	174.7
05	119.0	46.2	31.1	49.6	47.3	47.4	196.3
06	125.2	46.2	22.2	35.7	44.4	44.9	193.6
07	115.2	46.2	26.5	68.1	61.0	47.2	187.9
08	778.0	250.0	185.0	85.8	89.3	68.9	1213.0
09	458.8	250.0	185.0	85.8	89.3	68.9	893.8
10	105.2	46.2	22.2	77.0	85.0	65.4	173.6
11	97.2	46.2	22.2	67.0	80.5	61.8	165.6
12	97.2	46.2	24.4	54.9	74.4	55.0	167.8

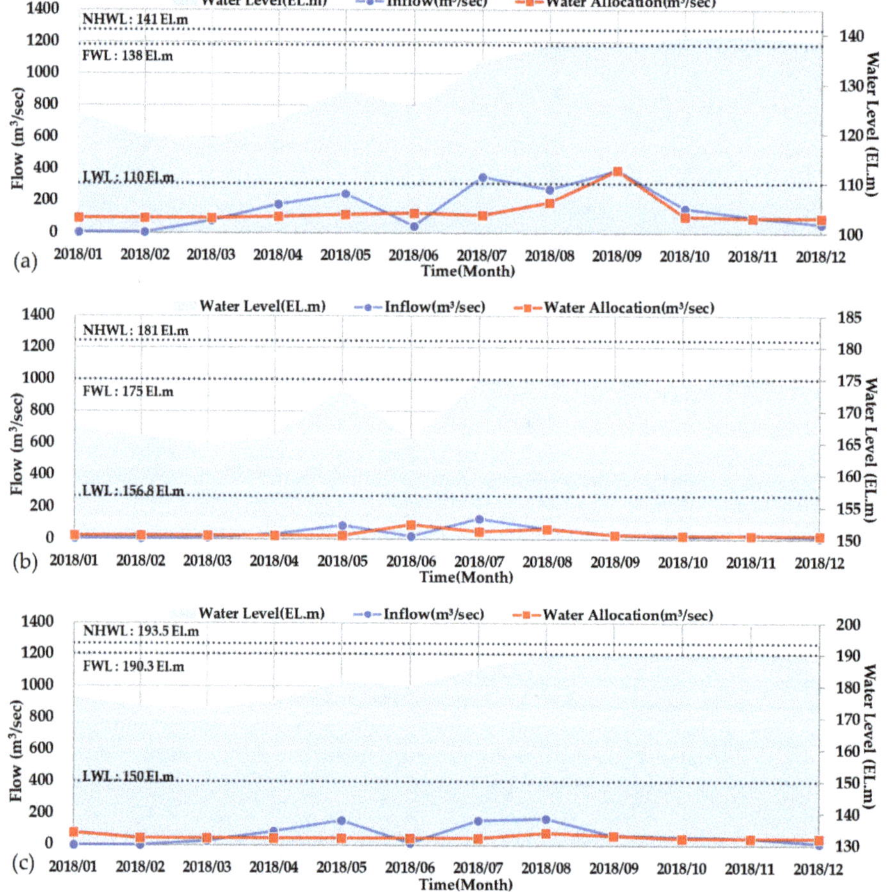

Figure 4. The results from (**a**) Supplementary Table S4, (**b**) Supplementary Table S5, and (**c**) Supplementary Table S6 are represented in the graphs.

4.3. Model Evaluation for Wet Year (2020)

Supplementary Tables S7 (Chungju Reservoir), S8 (Soyanggang Reservoir) and S9 (Hwacheon Reservoir) present the results during the wet year of 2020, as illustrated in

Figure 5a–c. South Korea experienced extreme floods in 2020. In 2019, the Han River Basin experienced lower-than-normal inflows, leading to minimal releases from the three reservoirs due to concerns about a prolonged drought. Consequently, water levels were high at the end of 2019. The initial water levels (December 2019 month-end water levels) for the Chungju, Soyanggang, and Hwacheon Reservoirs were 134.4 EL.m, 176.6 EL.m, and 171.6 EL.m, respectively. In 2020, reservoirs had the planned water supply for each reservoir and the spillway releases. The average annual inflows for the Chungju, Soyanggang, and Hwacheon Reservoirs in 2020 were 192.6 CMS, 106.4 CMS, and 102.3 CMS, respectively. The average annual allocation amounts were 191.1 CMS (60.1%), 80.2 CMS (26.2%), and 49.3 CMS (13.7%). Although significant flooding occurred in 2020, a substantial drop in inflow occurred after the flood season, resulting in a decrease in the effective storage ratio.

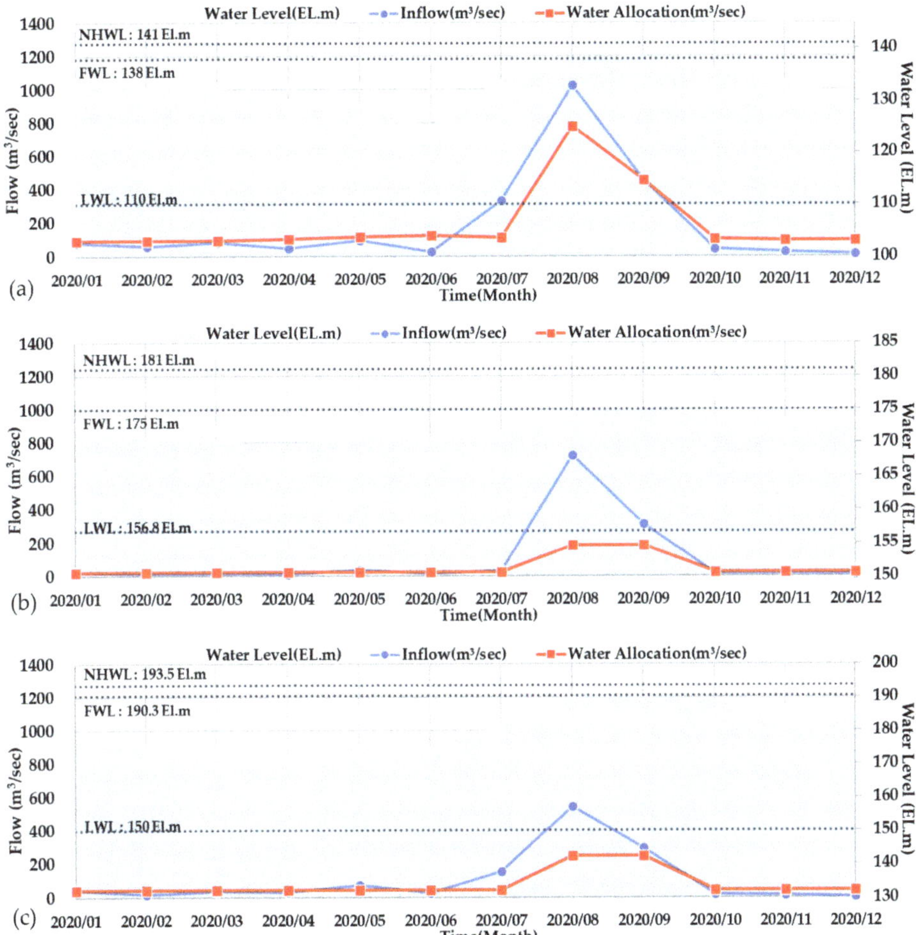

Figure 5. The results from (**a**) Supplementary Table S7, (**b**) Supplementary Table S8, and (**c**) Supplementary Table S9 are represented in the graphs.

5. Discussion

Globally, constructing new reservoirs is becoming increasingly challenging, necessitating the exploration of methods to repurpose existing reservoirs for multiple uses. In South Korea, water supply functions have been newly added to existing hydropower reservoirs.

This approach is expected to contribute to the country's sustainable development goals. The following subsections delve into the specifics of optimization and performance assessment under varying hydrological conditions. We also engage in a comparative analysis with prior research in the field, further establishing the relevance and contributions of the present study to the literature on reservoir operation.

5.1. Optimization and Performance Assessment in Dry Year (2015)

Table 4 presents the optimized water allocation and effective storage ratios for the three reservoirs during the dry year of 2015. Table 7 compares the monthly water allocation based on historical data with optimization results for three reservoirs in the dry year. Figure 6a depicts the optimized water supply of 2015. Figure 7a displays the historical monthly water allocation ratios for the three reservoirs, while Figure 7b illustrates the corresponding ratios determined with the optimization model.

Table 7. Comparison of historical data and optimization results for monthly water allocation in three reservoirs for the dry year of 2015.

	Historical Data								Optimization Results							
2015-MM	**Chungju Reservoir**		**Soyanggang Reservoir**		**Hwacheon Reservoir**		**Total**		**Chungju Reservoir**		**Soyanggang Reservoir**		**Hwacheon Reservoir**		**Total**	
	CMS	**%**	**CMS**	**%**	**CMS**	**%**	**CMS**	**%**	**CMS**	**%**	**CMS**	**%**	**CMS**	**%**	**CMS**	**%**
01	79.1	59.0	45.1	33.6	9.9	7.4	134.0	100	104.3	84.2	8.1	6.5	11.6	9.3	124.0	100
02	78.9	8.9	44.9	33.5	10.2	7.6	134.1	100	67.6	54.5	40.3	32.5	16.1	13.0	124.0	100
03	85.8	67.6	33.1	26.1	8.1	6.4	127.1	100	72.6	58.5	36.6	29.5	14.8	12.0	124.0	100
04	85.6	67.7	28.2	22.3	12.7	10.0	126.5	100	82.1	66.2	29.7	24.0	12.2	9.9	124.0	100
05	85.1	53.7	41.5	26.1	32.1	20.2	158.6	100	66.9	53.0	43.5	34.4	15.9	12.6	126.3	100
06	32.7	31.3	44.3	42.3	27.6	26.4	104.6	100	65.1	50.6	44.1	34.3	19.5	15.1	128.7	100
07	16.5	28.0	4.8	8.2	37.5	63.8	58.8	100	33.7	27.2	12.6	10.2	77.8	62.7	124.0	100
08	25.1	52.1	4.9	10.2	18.2	37.7	48.2	100	44.4	35.8	37.0	29.8	42.7	34.4	124.0	100
09	29.0	29.5	4.7	4.7	64.6	65.8	98.3	100	59.8	48.2	46.5	37.5	17.8	14.3	124.0	100
10	29.7	30.0	36.1	36.4	33.3	33.6	99.1	100	58.6	47.2	53.0	42.7	12.5	10.1	124.0	100
11	16.4	56.9	11.9	41.5	0.5	1.6	28.7	100	67.8	54.7	44.1	35.6	12.1	9.8	124.0	100
12	14.6	60.7	4.9	20.4	4.5	18.9	24.0	100	66.7	53.8	37.6	30.4	19.7	15.9	124.0	100
average	48.2	49.6	25.4	25.5	21.6	25.0	95.2	100	65.8	52.8	36.1	29.0	22.7	18.3	124.6	100

In a dry year like 2015 with low initial water levels and modest inflows, it is challenging to satisfy the planned water supply. Thus, operations focused on maintaining at least the instream flow while optimizing each reservoir's effective storage ratio. Thus, the storage ratios of all three reservoirs converged, and their total supply exceeded Paldang Reservoir's required discharge. Given that the reservoirs are managed by different agencies, adopting a balanced approach based on an effective storage ratio method is practical, especially under basin-wide severe drought conditions. In periods of low inflow, it is observed that optimal results were well-achieved in accordance with the objective, as there was no occasion for the reservoir levels to reach the normal high water level, unlike in normal or flood seasons. Slight differences occurred in the monthly effective storage ratios of the three reservoirs, influenced by the watershed-specific inflow rates. In June 2015, the effective storage ratios dropped below 10%, marking a critical state; however, the subsequent reservoir operation raised low water levels across all three reservoirs. Notably, the combined release from the three reservoirs was identical to the requirements of the Paldang Reservoir's discharge. By exceeding instream flows, the model aimed to increase storage ratios while minimizing releases from the reservoir with the lowest storage ratio, thereby achieving comparable effective storage ratios over time.

According to the actual and historical operations, water was mainly supplied from the Chungju and Soyanggang Reservoirs, as the water supply function was inactivated in Hwacheon Reservoir in 2015. Except for the period from January to May, the supply from the Chungju and Soyanggang Reservoirs was not sufficient to satisfy the required discharge of Paldang Reservoir due to the continuous drought. After June 2015, the combined discharge of the three reservoirs was smaller than Paldang Reservoir's required discharge.

During the months from July to October 2015, the Hwacheon Reservoir released more than the other two reservoirs, thereby assisting with water supply for the Seoul area.

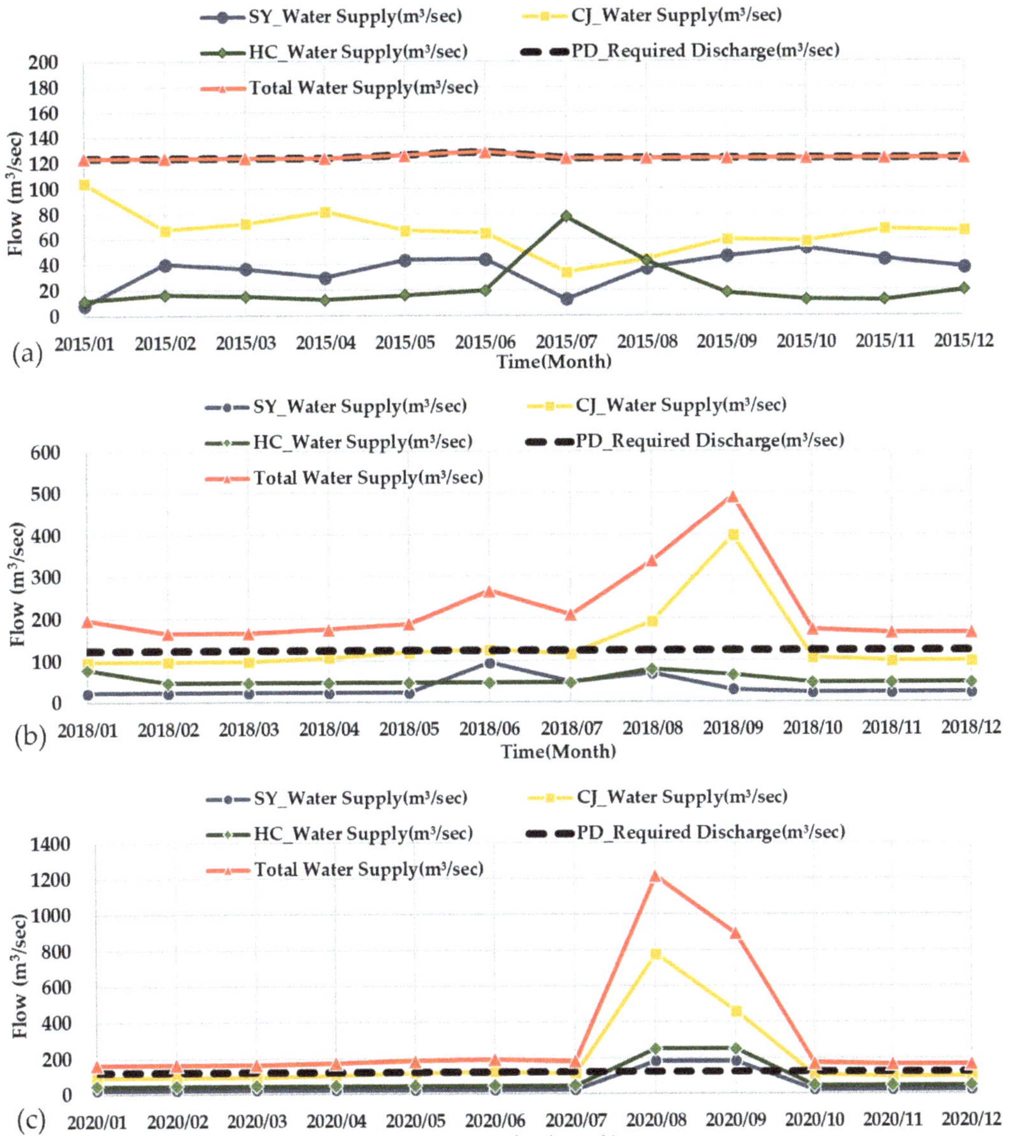

Figure 6. Optimized water supply for the years (**a**) 2015, (**b**) 2018, and (**c**) 2020 in correspondence with Tables 4–6.

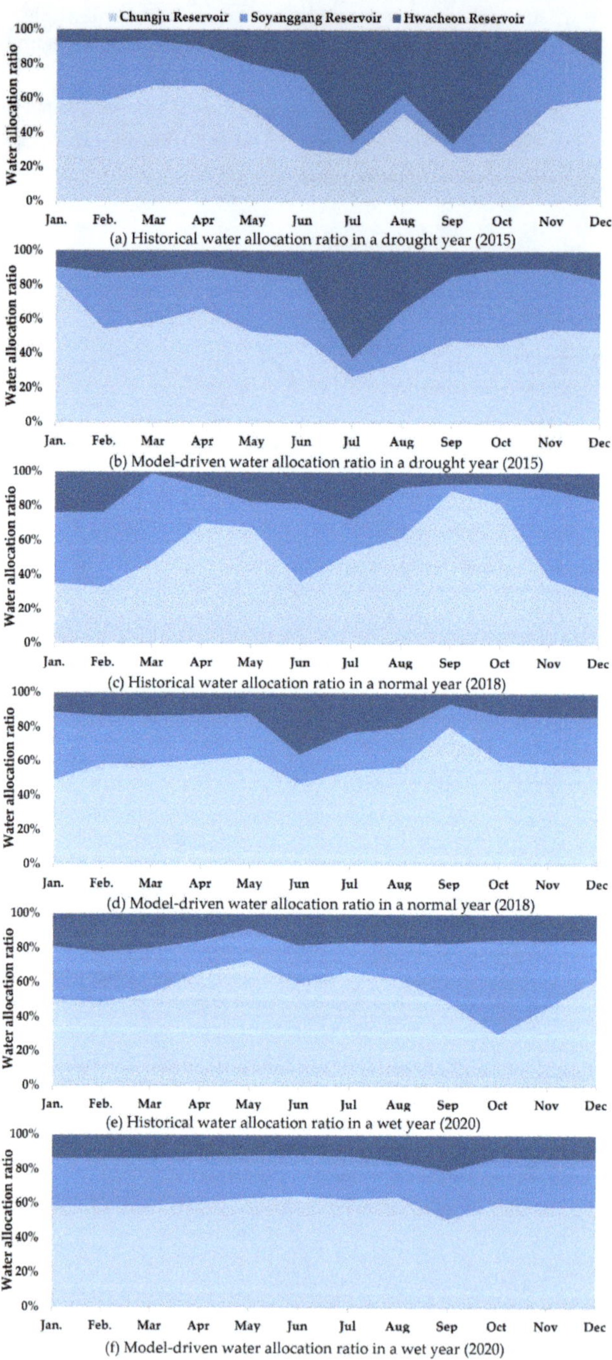

Figure 7. Subfigures (**a**,**c**,**e**) presenting the historical data on monthly water allocation ratios for three reservoirs in the years 2015, 2018, and 2020, respectively. Subfigures (**b**,**d**,**f**) displaying the monthly water allocation ratios for these reservoirs in the same years, as determined with the optimization model.

According to the optimization model results, the total allocation amount of the three reservoirs could supply the required discharge of each reservoir through the joint operation. Additionally, there was a tendency for stable allocation among the three reservoirs consistently during the dry season, as shown by a significant reduction in differences in the water allocation ratios of the three reservoirs (Figure 7a,b).

5.2. Optimization and Performance Assessment in Normal Year (2018)

Table 5 shows the optimized water allocation and effective storage ratios for the three reservoirs in the normal year. Table 8 compares the monthly water allocation based on historical data with optimization results for the three reservoirs in the normal year. Figure 6b shows the optimized water supply for the normal year. Figure 7c presents the historical data on monthly water allocation ratios for the three reservoirs, while Figure 7d provides the corresponding ratios estimated with the optimization model.

Table 8. Comparison of historical data and optimization results for monthly water allocation in three reservoirs for the normal year of 2018.

| | Historical Data | | | | | | | | Optimization Results | | | | | | | |
| | Chungju Reservoir | | Soyanggang Reservoir | | Hwacheon Reservoir | | Total | | Chungju Reservoir | | Soyanggang Reservoir | | Hwacheon Reservoir | | Total | |
2018-MM	CMS	%	CMS	%	CMS	%	CMS	%	CMS	%	CMS	%	CMS	%	CMS	%
01	52.0	35.3	60.0	40.8	35.2	23.9	147.2	100	97.2	49.2	78.3	39.6	22.2	11.2	197.7	100
02	52.0	33.1	68.1	43.3	37.0	23.6	157.1	100	97.2	58.7	46.2	27.9	22.2	13.4	165.6	100
03	52.0	47.5	56.6	51.7	0.9	0.9	109.5	100	97.2	58.7	46.2	27.9	22.2	13.4	165.6	100
04	107.1	70.1	32.9	21.5	12.8	8.4	152.7	100	106.3	60.9	46.2	26.5	22.2	12.7	174.7	100
05	343.3	67.8	75.5	14.9	87.3	17.3	506.1	100	119.0	63.5	46.2	24.7	22.2	11.9	187.4	100
06	128.5	36.5	158.9	45.2	64.5	18.3	351.9	100	125.2	47.2	46.2	17.4	94.1	35.5	265.5	100
07	177.3	53.4	64.4	19.4	90.3	27.2	331.9	100	115.2	55.0	46.2	22.1	47.9	22.9	209.3	100
08	99.1	62.0	46.4	29.1	14.2	8.9	159.7	100	192.5	56.9	78.2	23.1	67.7	20.0	338.4	100
09	284.7	89.7	11.3	3.6	21.5	6.8	317.4	100	397.5	80.9	63.7	13.0	29.9	6.1	491.1	100
10	118.2	82.0	16.3	11.3	9.6	6.7	144.1	100	105.2	60.6	46.2	26.6	22.2	12.8	173.6	100
11	57.7	38.3	78.4	52.1	14.6	9.7	150.6	100	97.2	58.7	46.2	27.9	22.2	13.4	165.6	100
12	51.8	28.6	101.0	55.7	28.6	15.8	181.4	100	97.2	58.7	46.2	27.9	22.2	13.4	165.6	100
average	52.0	35.3	60.0	40.8	35.2	23.9	147.2	100	137.3	59.1	53.0	25.4	34.8	15.6	225.0	100

During a normal year, a reservoir releases water based on the planned water supply. However, the model aimed to increase the lowest effective storage ratio for the three reservoirs. Spillways opened throughout the flood season to maintain reliable reservoir operation. The reservoir operation model focused on the dry season rather than the flood season, focusing on the optimization model as it aimed to build a long-term reservoir operation plan. The model executed spillway releases during the flood season to enhance the effective storage ratio during the next dry season. Each reservoir exceeded its planned water supply, and at the same time, the sum of the three reservoir releases surpassed Paldang Reservoir's required discharge (Table 5, Figure 6b). Due to different inflows and planned water supplies across the three reservoirs, a uniform storage ratio was more challenging to achieve. By implementing reservoir operation strategies aimed at increasing the water level in the reservoir with the lowest effective storage ratio, the differences in effective storage ratios were substantially reduced. The model aimed to raise the water level of the reservoir with the minimum water level considering the water availability. This dropped the differences in the effective storage ratio from over 15% to approximately less than 1%.

According to the actual and historical operations, in 2018 right after the extreme drought spanning from 2014 to 2017, the Chungju and Soyanggang Reservoirs could not meet their planned water supply. It was only after incorporating the water allocation from Hwacheon Reservoir that the combined allocation exceeded the anticipated discharge of the Paldang Reservoir. The Hwacheon Reservoir, still functioning as hydropower generation in 2018, had release fluctuations in monthly water allocation during the dry season, which lasted from October to May.

The optimization model released more water than the monthly planned water supply, thereby ensuring that the sum of water allocations from all three reservoirs satisfied the discharge requirements of the Paldang Reservoir. This setup resulted in uniform water allocation ratios among the three reservoirs during the dry season (Figure 7c,d).

5.3. Optimization and Performance Assessment in Wet Year (2020)

Table 6 outlines the optimized water allocation and effective storage ratios for the three reservoirs in the wet year. Table 9 compares the monthly water allocation based on historical data with optimization results for the three reservoirs in the wet year. Figure 6c shows the optimized water supply. Figure 7e presents the historical data on monthly water allocation ratios for the three reservoirs, while Figure 7f provides the corresponding ratios as estimated with the optimization model.

Table 9. Comparison of historical data and optimization results for monthly water allocation in three reservoirs for the wet year of 2020.

| | Historical Data | | | | | | | | Optimization Results | | | | | | | |
| | Chungju Reservoir | | Soyanggang Reservoir | | Hwacheon Reservoir | | Total | | Chungju Reservoir | | Soyanggang Reservoir | | Hwacheon Reservoir | | Total | |
2020-MM	CMS	%	CMS	%	CMS	%	CMS	%	CMS	%	CMS	%	CMS	%	CMS	%
01	74.1	50.2	45.3	30.7	28.2	19.1	147.6	100	97.2	58.7	46.2	27.9	22.2	13.4	165.6	100
02	74.7	48.8	44.4	29.0	33.9	22.2	153.0	100	97.2	58.7	46.2	27.9	22.2	13.4	165.6	100
03	75.0	53.9	36.3	26.1	27.9	20.0	139.2	100	97.2	58.7	46.2	27.9	22.2	13.4	165.6	100
04	107.6	66.1	29.7	18.2	25.7	15.7	162.9	100	106.3	60.9	46.2	26.5	22.2	12.7	174.7	100
05	121.7	73.3	29.5	17.8	14.8	8.9	166.0	100	119.0	63.5	46.2	24.7	22.2	11.9	187.4	100
06	109.6	60.0	39.7	21.7	33.5	18.3	182.8	100	125.2	64.7	46.2	23.9	22.2	11.5	193.6	100
07	124.3	66.5	31.8	17.0	30.8	16.5	186.8	100	115.2	62.8	46.2	25.2	22.2	12.1	183.6	100
08	541.7	62.3	185.1	21.3	143.3	16.5	870.0	100	778.0	64.1	250.0	20.6	185.0	15.3	1213	100
09	286.5	49.3	196.4	33.8	98.2	16.9	581.2	100	458.8	51.3	250.0	28.0	185.0	20.7	893.8	100
10	46.4	30.5	83.2	54.7	22.7	14.9	152.3	100	105.2	60.6	46.2	26.6	22.2	12.8	173.6	100
11	69.6	46.3	58.3	38.7	22.6	15.0	150.5	100	97.2	58.7	46.2	27.9	22.2	13.4	165.6	100
12	94.9	62.0	35.4	23.2	22.7	14.8	152.9	100	97.2	58.7	46.2	27.9	22.2	13.4	165.6	100
average	74.1	50.2	45.3	30.7	28.2	19.1	147.6	100	191.1	60.1	80.2	26.2	49.3	13.7	320.6	100

In a wet year, water releases are based on the planned water supply of each reservoir, like the normal year, and at the same time, the lowest effective storage ratios of the three reservoirs are increased. To ensure reliable reservoir operation, the reservoirs opened spillways and released water throughout the flood season. Notably, these spillway releases were more substantial in wet years compared to normal or dry years. To establish a long-term reservoir operation plan, the developed optimization model was primarily focused on operation for the dry season rather than for the flood season. In the dry season, the reservoirs released the minimum required supply to improve the effective storage ratio. Each reservoir released water above its planned water supply, and the sum of supplies from the three reservoirs exceeded the required discharge of the Paldang Reservoir (Table 6, Figure 6c). The operations aimed to maximize the storage ratio of each reservoir, similar to the approach in the normal year of 2018. In 2020, exceeding the normal inflows from January to the flood season led to greater water supply, hydropower generation, and spillway releases. The high monthly inflows resulted in significant fluctuations in the effective storage ratio throughout 2020.

Hwacheon Reservoir commenced its pilot operation in April 2020. Hwacheon Reservoir has consistently released 22.2 CMS from April to the present, leading to a more balanced water allocation ratio for all three reservoirs. However, there are instances, such as in October 2020, when the allocation ratio of Soyanggang Reservoir surpasses that of Chungju Reservoir. This happens because water releases are based on individual reservoir standards rather than a system operation among the three reservoirs. However, the model results ensured a relatively stable water allocation ratio by considering the storage ratios of all three reservoirs in determining the allocation amount (Figure 7e,f).

5.4. Comparative Analysis with Previous Studies in Reservoir Operation

To contextualize the contributions of the present study, it is instructive to compare its methodologies and findings with those of previous studies in the field of reservoir operation and water allocation. One study used two-stage stochastic linear programming to optimize reservoir operations in the Han River Basin, focusing solely on the multipurpose reservoirs of Chungju and Soyanggang [48]. In the first stage, the model determined reservoir storage, while in the second stage, it set water supply and environmental flow rates based on actual demand. The optimization aimed to minimize discrepancies between the target and actual reservoir storage and any water supply and environmental flow shortages. The study also utilizes a Hedging Rule to adjust planned release rates based on current reservoir storage, thereby enhancing the model's applicability and relevance.

In contrast to this study, which relied on an artificial Hedging Rule, our research formulated an optimal water allocation model that incorporated the effective storage ratio of each reservoir into the allocation calculations. This approach ensured a more balanced and efficient distribution of downstream water requirements. We also broadened the scope by incorporating a transformed hydropower reservoir with an add-on water supply function and two multipurpose reservoirs. This offered a versatile and comprehensive solution for reservoir management, particularly in regions where collaboration between different types of reservoirs is essential for optimal performance.

Another study focused on the coordinated operation of multipurpose reservoirs (Soyanggang, Chungju, and Hoengseong), a water supply-only reservoir (Gwangdong), and a large-scale hydropower reservoir (Hwacheon) within the Han River Basin [49]. The paper employed a five-level Hedging Rule and used mixed integer linear programming (MILP) to develop a reservoir operation model. The model aimed to approximate the actual storage volume of existing reservoirs to target storage volumes while maximizing water supply rates and ensuring maximum river maintenance flows. The paper focused on phased water supply reductions to optimize water supply during drought conditions.

As opposed to the approach taken in this research, which employed a five-level Hedging Rule and MILP for optimizing water allocation primarily during drought conditions, our research focused on a broader range of hydrological scenarios. Our research aimed for efficient water allocation from the Soyanggang, Chungju, and Hwacheon Reservoirs, which are responsible for water supply in the metropolitan area. Additionally, our work introduced a transformed hydropower reservoir into the system, offering a more versatile and sustainable solution for water resources management.

6. Conclusions

In summary, the proposed method optimized water supply capacity across multiple upstream and downstream reservoirs and focused on rational operational strategies for downstream water requirements. Validated against historical inflow data, the model effectively balanced the unique operational needs of upstream hydropower and other multipurpose reservoirs while ensuring optimal downstream water supply.

When comparing the monthly storage ratios of each reservoir, the model aimed to minimize differences in storage ratios while determining the allocation. During the dry year (2015), the effective storage ratios of the three reservoirs were operated almost equally each month. The actual annual average allocation ratios for the Chungju, Soyanggang, and Hwacheon Reservoirs were 49.6%, 25.5%, and 25.0%, respectively, while the model-estimated ratios were 52.8%, 29.0%, and 18.3%. The higher allocation for the Chungju and Soyanggang Reservoirs compared to actual data was due to their higher effective storage ratios and greater inflows. In the normal year (2018), the actual annual average allocation ratios for the Chungju, Soyanggang, and Hwacheon Reservoirs were 35.3%, 40.8%, and 23.9%, respectively. Soyanggang Reservoir had more discharge than Chungju Reservoir because there were periods when Chungju Reservoir did not supply the basic planned amount. However, considering the effective storage ratio and the planned water supply, the estimated allocation ratios changed to 59.1%, 25.4%, and 15.6%, respectively, increasing the proportion allocated to Chungju Reservoir. In the wet year (2020), the actual annual average allocation ratios for the Chungju, Soyanggang, and Hwacheon Reservoirs, respectively, were 50.2%, 30.7%, and 19.1%, while the model-estimated ratios were 60.1%, 26.2%, and 13.7%. The Chungju Reservoir, with its larger catchment area and higher inflow, had a higher estimated proportion of the allocation.

This study delivers several key findings that significantly contribute to the field of water resources management. First, the optimization model effectively allocated water among three critical reservoirs—Chungju, Soyanggang, and Hwacheon—across diverse hydrological scenarios, including dry, normal, and wet years. It should be noted that, in addition to the case studies presented for these specific hydrological conditions, the model was also rigorously tested for every single year from 2004 to 2022, yielding consistently satisfactory results. This extensive validation across a wide range of inflow conditions underscores the model's robustness and reliability. Second, the model adeptly balanced effective storage ratios among these reservoirs, enhancing drought preparedness and water resource management. Third, historical data served as a robust validation mechanism, confirming the model's effectiveness in appropriately distributing downstream water supply. Specifically, during a dry year such as 2015, the model excelled in water management even under low inflow conditions. In a normal year like 2018, the model not only maintained the storage ratios for each reservoir but also exceeded the required discharge for Paldang Reservoir. Additionally, the model demonstrated adaptability in a wet year, like 2020, by effectively managing increased inflows. The pilot operation of Hwacheon Reservoir further solidified the model's capability, resulting in more balanced water allocation across the reservoirs. Moreover, the model was multiobjective, serving various functions, such as maintaining instream flows and preparing for extreme hydrological events. These key findings not only have academic implications but also offer practical utility in real-world reservoir and water resources management.

While the study's results underscore the model's efficiency in water allocation and flood control, it is crucial to acknowledge the inherent trade-offs in reservoir operation. Environmental impacts can be mitigated through ongoing monitoring and adaptive management strategies, while stakeholder engagement can help resolve social conflicts. By recognizing and proactively managing these trade-offs, the proposed model aligns with broader sustainability objectives. This integrated approach contributes significantly to the sustainable operation of reservoir systems, thus benefiting both the environment and the communities they serve.

This study presents a comprehensive model for reservoir management but has limitations and uncertainties. The model's scope was restricted by its exclusion of various hydrological factors, like tributary inflows and water losses, and by assumptions regarding discharge capacities of the reservoirs involved. While validated against historical data, the model was not tested in real-world settings and did not extend to extreme hydrological scenarios. One limitation of our model is that it did not address the potential conflicts between water supply and hydropower generation. Specifically, the optimal timing for each may differ, and adding a water supply function to hydropower reservoirs could adversely affect hydropower output. This impact was not examined in the current study. The limitations and uncertainties of the proposed methodology include the following: The optimization model was designed to meet the objective function within specified constraints and may yield infeasible solutions when not all conditions can be satisfied. This is particularly true during extreme drought conditions, where the planned water supply may be insufficient. To address this, we modified the constraints to maintain instream flow during droughts. Additionally, the use of linear-programming techniques limited the model's applicability to nonlinear equations, which is why hydropower generation was excluded from this study.

While the current study provided model results based on predefined scenarios such as dry, normal, and wet years, it is crucial to highlight the model's flexibility in allowing allocation amounts to be estimated based on user-defined inflow periods. Future research perspectives include capitalizing on this flexibility by testing the model's adaptability and performance under various inflow conditions. Another significant aspect to consider is the integration of other hydrological factors. The model was designed with the assumption that the sum of the allocations from the three reservoirs would exceed each reservoir's required discharge. However, it did not account for additional factors, like tributary inflows, water withdrawals, losses, and return flows. Given that the Han River system is South Korea's largest and maintains higher inflow levels even during extended droughts, it is imperative for future research to integrate the allocation model with a water balance model for more precise inflow estimates into the Paldang Reservoir. To further improve model performance, upcoming research should focus on developing a simulation model using the rule curves of the three reservoirs and advanced reservoir allocation techniques. Such an approach is anticipated to yield a more realistic operational plan closely aligned with real-world conditions. The optimally generated plan could then be incorporated into this new simulation model. Lastly, future studies could also benefit from generating long-term inflow data to evaluate the model's robustness over extended timeframes. This would enable the model to be tested against more severe drought and flood scenarios, providing valuable insights into its long-term applicability for water resources management.

In summary, the model presented in this study offered a robust framework for the rational operation of three reservoirs, Chungju, Soyanggang, and Hwacheon, which are crucial for the water supply in Seoul. By enabling these reservoirs to be operated at comparable storage ratios, the model can serve as a valuable guide for proposing rational operational plans to stakeholders. Furthermore, its versatility allows for the examination of long-term preparedness strategies against extreme events, like droughts and floods, through preliminary analyses. This is especially pertinent during dry years when water supply is highly vulnerable, making the model an objective indicator in such scenarios. Importantly, the model not only considers the storage ratio between reservoirs but also takes into account the planned water supply for each, delivering optimal outcomes for both joint and individual operations. These findings underline the effectiveness of the proposed method in achieving a balanced and reliable water supply, and they offer valuable insights for future research and practical applications in water resources management.

Supplementary Materials: The following supporting information can be downloaded at: https://www.mdpi.com/article/10.3390/w15203555/s1; Table S1. Optimization results for Chungju Reservoir during drought year based on initial water level of 126.2 EL.m; Table S2. Optimization results for Soyanggang Reservoir during drought year based on initial water level of 165.2 EL.m; Table S3. Optimization results for Hwacheon Reservoir during drought year based on initial water level of 165.8 EL.m; Table S4. Optimization results for Chungju Reservoir during drought year based on initial water level of 128.5 EL.m; Table S5. Optimization results for Soyanggang Reservoir during drought year based on initial water level of 181.0 EL.m; Table S6. Optimization results for Hwacheon Reservoir during drought year based on initial water level of 169.4 EL.m; Table S7. Optimization results for Chungju Reservoir during drought year based on initial water level of 134.4 EL.m; Table S8. Optimization results for Soyanggang Reservoir during drought year based on initial water level of 176.6 EL.m; Table S9. Optimization results for Hwacheon Reservoir during drought year based on initial water level of 171.6 EL.m.

Author Contributions: Conceptualization, E.L. and J.Y. (Jaeeung Yi); methodology, E.L. and J.Y. (Jaeeung Yi); software, E.L.; validation, E.L.; formal analysis, E.L.; investigation, E.L.; resources, E.L., J.J., S.L., J.Y. (Jeongin Yoon), S.Y. and J.Y. (Jaeeung Yi); data curation, E.L.; writing—original draft preparation, E.L., J.J. and J.Y. (Jaeeung Yi); writing—review and editing, S.Y. and J.Y. (Jaeeung Yi); visualization, E.L.; supervision, J.Y. (Jaeeung Yi); project administration, E.L. and J.Y. (Jaeeung Yi); funding acquisition, J.Y. (Jaeeung Yi). All authors have read and agreed to the published version of the manuscript.

Funding: This work was supported by the Korea Environment Industry & Technology Institute (KEITI) through the Water Management Project for Drought funded by the Korea Ministry of Environment (MOE) (2022003610004).

Data Availability Statement: Not applicable.

Conflicts of Interest: The authors declare no conflict of interest.

References

1. Sillmann, J.; Roeckner, E. Indices for extreme events in projections of anthropogenic climate change. *Clim. Chang.* **2008**, *86*, 83–104. [CrossRef]
2. He, X.; Pan, M.; Wei, Z.; Wood, E.F.; Sheffield, J. A Global Drought and Flood Catalogue from 1950 to 2016. *Bull. Am. Meteorol. Soc.* **2020**, *101*, E508–E535. [CrossRef]
3. Yi, S.; Kondolf, G.M.; Sandoval-Solis, S.; Dale, L. Application of machine learning-based energy use forecasting for inter-basin water transfer project. *Water Resour. Manag.* **2022**, *36*, 5675–5694. [CrossRef]
4. Choi, Y.; Ahn, J.; Ji, J.; Lee, E.; Yi, J. Effects of inter-basin water transfer project operation for emergency water supply. *Water Resour. Manag.* **2020**, *34*, 2535–2548. [CrossRef]
5. Choi, J.; Kim, H.; Wang, B. Interdecadal variation of Changma (Korean summer monsoon rainy season) retreat date in Korea. *Int. J. Climatol.* **2020**, *40*, 1348–1360. [CrossRef]
6. Cho, Y.; Woo, S.; Noh, J.; Lee, E. Application of the weather radar-based quantitative precipitation estimations for flood runoff simulation in a dam watershed. *J. Korea Water Resour. Assoc.* **2020**, *53*, 155–166.
7. Kondolf, M.; Yi, J. Dam Renovation to Prolong Reservoir Life and Mitigate Dam Impacts. *Water* **2022**, *14*, 1464. [CrossRef]
8. Gwon, Y.; Son, K.; Lee, K.; Choi, G. Development and Utility Evaluation of a Multi-Composite Water Balance Model. *J. Korean Soc. Hazard Mitig.* **2020**, *20*, 239–250. [CrossRef]
9. Kim, G.J.; Kim, Y.-O.; Reed, P.M. Improving the Robustness of Reservoir Operations with Stochastic Dynamic Programming. *J. Water Resour. Plan. Manag.* **2021**, *147*, 04021030. [CrossRef]
10. King, A.D.; Pitman, A.J.; Henley, B.J.; Ukkola, A.M.; Brown, J.R. The role of climate variability in Australian drought. *Nat. Clim. Chang.* **2020**, *10*, 177–179. [CrossRef]
11. Di Baldassarre, G.; Wanders, N.; AghaKouchak, A.; Kuil, L.; Rangecroft, S.; Veldkamp, T.I.E.; Garcia, M.; van Oel, P.R.; Breinl, K.; Van Loon, A.F. Water shortages worsened by reservoir effects. *Nat. Sustain.* **2018**, *1*, 617–622. [CrossRef]
12. Karami, S.; Karami, E. Sustainability assessment of dams. *Environ. Dev. Sustain.* **2020**, *22*, 2919–2940. [CrossRef]
13. Jamali, A.A.; Tabatabaee, R.; Randhir, T.O. Ecotourism and socioeconomic strategies for Khansar River watershed of Iran. *Environ. Dev. Sustain.* **2021**, *23*, 17077–17093. [CrossRef]
14. Benenati, E.P.; Shannon, J.P.; Blinn, D.W.; Wilson, K.P.; Hueftle, S.J. Reservoir–river linkages: Lake Powell and the Colorado River, Arizona. *J. N. Am. Benthol. Soc.* **2000**, *19*, 742–755. [CrossRef]
15. Zhang, X.; Dong, Z.; Gupta, H.; Wu, G.; Li, D. Impact of the Three Gorges Dam on the Hydrology and Ecology of the Yangtze River. *Water* **2016**, *8*, 590. [CrossRef]
16. Lee, E.; Lee, S.; Ji, J.; Yi, J.; Jung, S. Estimating the water supply capacity of Hwacheon reservoir for multi-purpose utilization. *J. Korea Water Resour. Assoc.* **2022**, *55*, 437–446.
17. *A Study on the Optimal Management of Existing Reservoirs to Improve the Water Resource's System Management*; Ministry of Environment and Ministry of Trade I and E: Sejong, Republic of Korea, 2019.
18. Branche, E. The multipurpose water uses of hydropower reservoir: The SHARE concept. *Comptes Rendus Phys.* **2017**, *18*, 469–478. [CrossRef]
19. Branche, E. *The Durance-Verdon River Basin in France: The Role of Infrastructures and Governance for Adaptation to Climate Change*; Springer: Berlin, Germany, 2016; pp. 129–155.
20. Zarri, L.J.; Danner, E.M.; Daniels, M.E.; Palkovacs, E.P. Managing hydropower dam releases for water users and imperiled fishes with contrasting thermal habitat requirements. *J. Appl. Ecol.* **2019**, *56*, 2423–2430. [CrossRef]
21. Choi, Y.; Lee, E.; Ji, J.; Ahn, J.; Kim, T.; Yi, J. Development and Evaluation of the Hydropower Reservoir Rule Curve for a Sustainable Water Supply. *Sustainability* **2020**, *12*, 9641. [CrossRef]
22. Jeong, G.; Choi, J.; Kang, D.; Ahn, J.; Kim, T. Evaluation of hydropower dam water supply capacity (I): Individual and integrated operation of hydropower dams in Bukhan river. *J. Korea Water Resour. Assoc.* **2022**, *55*, 505–513.
23. Jeong, G.; Kang, D.; Kim, T. Evaluation of hydropower dam water supply capacity (III): Development and application of drought operation rule for hydropower dams in Han river. *J. Korea Water Resour. Assoc.* **2022**, *55*, 531–543.
24. Jeong, G.; Kang, D.; Kim, D.H.; Lee, S.O.; Kim, T. Evaluation of hydropower dam water supply capacity (II): Estimation of water supply yield range of hydropower dams considering probabilistic inflow. *J. Korea Water Resour. Assoc.* **2022**, *55*, 515–529.
25. Janga Reddy, M.; Nagesh Kumar, D. Evolutionary algorithms, swarm intelligence methods, and their applications in water resources engineering: A state-of-the-art review. *H₂Open J.* **2020**, *3*, 135–188. [CrossRef]
26. Yang, L.; Bai, X.; Khanna, N.Z.; Yi, S.; Hu, Y.; Deng, J.; Gao, H.; Tuo, L.; Xiang, S.; Zhou, N. Water evaluation and planning (Weap) model application for exploring the water deficit at catchment level in beijing. *Desalin. Water Treat.* **2018**, *118*, 12–25. [CrossRef]
27. Fayaed, S.S.; El-Shafie, A.; Jaafar, O. Reservoir-system simulation and optimization techniques. *Stoch. Environ. Res. Risk Assess.* **2013**, *27*, 1751–1772. [CrossRef]
28. Datta, B.; Houck, M.H. A Stochastic Optimization Model for Real-Time Operation of Reservoirs Using Uncertain Forecasts. *Water Resour. Res.* **1984**, *20*, 1039–1046. [CrossRef]
29. Datta, B.; Burges, S.J. Short-Term, Single, Multiple-Purpose Reservoir Operation: Importance of Loss Functions and Forecast Errors. *Water Resour. Res.* **1984**, *20*, 1167–1176. [CrossRef]
30. Huock, M.H.; Cohon, J.L. Sequential explicitly stochastic linear programing models: A proposed method for design and management of multipurpose reservoir systems. *Water Resour. Res.* **1978**, *14*, 161–169. [CrossRef]

31. Mariño, M.A.; Mohammadi, B. Reservoir Operation by Linear and Dynamic Programming. *J. Water Resour. Plan. Manag.* **1983**, *109*, 303–319. [CrossRef]
32. Cuevas-Velásquez, V.; Sordo-Ward, A.; García-Palacios, J.H.; Bianucci, P.; Garrote, L. Probabilistic Model for Real-Time Flood Operation of a Dam Based on a Deterministic Optimization Model. *Water* **2020**, *12*, 3206. [CrossRef]
33. Oron, G.; Mehrez, A.; Rabinowitz, G. Forecasting in Optimizing Dual System for Energy Generation and Irrigation. *J. Water Resour. Plan. Manag.* **1991**, *117*, 287–300. [CrossRef]
34. Bras, R.L.; Buchanan, R.; Curry, K.C. Real time adaptive closed loop control of reservoirs with the High Aswan Dam as a case study. *Water Resour. Res.* **1983**, *19*, 33–52. [CrossRef]
35. Stedinger, J.R.; Sule, B.F.; Loucks, D.P. Stochastic dynamic programming models for reservoir operation optimization. *Water Resour. Res.* **1984**, *20*, 1499–1505. [CrossRef]
36. Bellman, R. Dynamic Programming. *Science* **1966**, *153*, 34–37. [CrossRef] [PubMed]
37. Thaeer Hammid, A.; Awad, O.I.; Sulaiman, M.H.; Gunasekaran, S.S.; Mostafa, S.A.; Manoj Kumar, N.; Khalaf, B.A.; Al-Jawhar, Y.A.; Abdulhasan, R.A. A Review of Optimization Algorithms in Solving Hydro Generation Scheduling Problems. *Energies* **2020**, *13*, 2787. [CrossRef]
38. Haddad, O.B.; Afshar, A.; Mariño, M.A. Honey-Bees Mating Optimization (HBMO) Algorithm: A New Heuristic Approach for Water Resources Optimization. *Water Resour. Manag.* **2006**, *20*, 661–680. [CrossRef]
39. Ngamsert, R.; Techarungruengsakul, R.; Kaewplang, S.; Hormwichian, R.; Prasanchum, H.; Sivanpheng, O.; Kangrang, A. Optimizing Solution in Decision Supporting System for River Basin Management Consisting of a Reservoir System. *Water* **2023**, *15*, 2510. [CrossRef]
40. Jothiprakash, V.; Shanthi, G. Single Reservoir Operating Policies Using Genetic Algorithm. *Water Resour. Manag.* **2006**, *20*, 917–929. [CrossRef]
41. Tatano, H.; Okada, N.; Kawai, H. Optimal operation model of a single reservoir with drought duration explicitly concerned. *Stoch. Hydrol. Hydraul.* **1992**, *6*, 123–134. [CrossRef]
42. Lund, J.R.; Guzman, J. Derived Operating Rules for Reservoirs in Series or in Parallel. *J. Water Resour. Plan. Manag.* **1999**, *125*, 143–153. [CrossRef]
43. Zeng, X.; Hu, T.; Xiong, L.; Cao, Z.; Xu, C. Derivation of operation rules for reservoirs in parallel with joint water demand. *Water Resour. Res.* **2015**, *51*, 9539–9563. [CrossRef]
44. Kang, L.; Zhang, S.; Ding, Y.; He, X. Extraction and Preference Ordering of Multireservoir Water Supply Rules in Dry Years. *Water* **2016**, *8*, 28. [CrossRef]
45. Meng, W.; Wan, W.; Zhao, J.; Wang, Z. Optimal Operation Rules for Parallel Reservoir Systems with Distributed Water Demands. *J. Water Resour. Plan. Manag.* **2022**, *148*, 04022020. [CrossRef]
46. Houck, M.H. Optimizing Reservoir Resources: Including a New Model for Reservoir Reliability. *Eos Trans. Am. Geophys. Union* **2002**, *83*, 285–286. [CrossRef]
47. Abrishamchi, A.; Dashti, M.; Tajrishy, M. Development of a Multi-Reservoir Flood Control Optimization Model: Application to the Karkheh River Basin, Iran. In *World Environmental and Water Resources Congress 2011*; American Society of Civil Engineers: Reston, VA, USA, 2011; pp. 3048–3057.
48. Chung, G.-H.; Ryu, G.-H.; Kim, J.-H. Optimization of Multi-reservoir Operation considering Water Demand Uncertainty in the Han River Basin. *J. Korean Soc. Hazard Mitig.* **2010**, *10*, 89–102.
49. Ryu, G.-H.; Chung, G.-H.; Lee, J.-H.; Kim, J.-H. Optimization of Multi-reservoir Operation with a Hedging Rule: Case Study of the Han River Basin. *J. Korea Water Resour. Assoc.* **2009**, *42*, 643–657. [CrossRef]

Article

The Base Value of the Ecological Compensation Standard in Transboundary River Basins: A Case Study of the Lancang–Mekong River Basin

Yue Zhao [1], Fang Li [2,*], Yanping Chen [1], Xiangnan Chen [3] and Xia Xu [4]

[1] Business School, Hohai University, Nanjing 211100, China; zh_yyyyy@hhu.edu.cn (Y.Z.); chenyp@hhu.edu.cn (Y.C.)
[2] Business School, Changzhou University, Changzhou 213159, China
[3] Business School, Jiangsu Open University, Nanjing 211100, China; cxn1128@126.com
[4] Architectural Engineering School, Tongling University, Tongling 244000, China; 027627@tlu.edu.cn
* Correspondence: lf@cczu.edu.cn

Abstract: The ecological compensation standard in transboundary river basins should be determined by the basin countries through negotiation on the basis of the base value of the ecological compensation standard. This paper calculated the base value range of the ecological compensation standard, determining the upper limit based on the spillover value of ecosystem services for the ecosystem-service-consuming country and the lower limit according to the cost of ecological protection for the ecosystem-service-supplying country. The final range was determined by integrating this with the willingness to pay and the actual effort in each basin country. Taking, for example, the Lancang–Mekong River basin, the results indicate that the spillover value of ecosystem services in Laos, China and Myanmar was positive and these three countries were ecosystem-service-supplying countries, while in Cambodia, Vietnam and Thailand it was negative and these three countries were ecosystem-service-consuming countries. Among the ecosystem-service-supplying countries, the cost of ecological protection of them was in descending order of Laos, China and Myanmar, which was related to their own level of economic development. Considering the adjustment coefficient for the payment of ecosystem service value and the cost-sharing coefficient of each basin country, the feasible range for the base value of the ecological compensation standard was determined to be $[2.47, 229.67] \times 10^8$ \$, which provided the basis for the negotiation on the determination of the ECS. In addition, implementation suggestions were proposed from three aspects: establishing a basin-information-sharing mechanism and platform, establishing an integrated management organization for transboundary river basins, and strengthening and improving the coordination and supervision model of ecological compensation.

Keywords: transboundary river basins; base value; ecological compensation standard; spillover value; cost; Lancang–Mekong River Basin

Citation: Zhao, Y.; Li, F.; Chen, Y.; Chen, X.; Xu, X. The Base Value of the Ecological Compensation Standard in Transboundary River Basins: A Case Study of the Lancang–Mekong River Basin. *Water* **2023**, *15*, 2809. https://doi.org/10.3390/w15152809

Academic Editors: Peiyue Li and Jianhua Wu

Received: 5 June 2023
Revised: 27 July 2023
Accepted: 1 August 2023
Published: 3 August 2023

1. Introduction

Globally, there are 310 transboundary river basins (TBRBs) covering 150 countries and regions, and the basins cover 47.1% of the global land area, with about 52% of the global population living in TBRBs [1]. As the global shortage of freshwater resources intensifies and the population grows rapidly, the utilization and demand for water resources in TBRBs by basin countries (BCs) are increasing, leading to increasingly serious water resource problems such as water shortage [2], water environment pollution [3], and water ecology damage [4], and, thus, resulting in increasingly prominent conflicts among BCs. In the era of global integration, countries in the world are increasingly connected and dependent. It has gradually become a rational choice for BCs to solve contradictions and conflicts through cooperation. In TBRBs, due to the involvement of numerous BCs, each of which is an independent sovereign state, it is not possible to handle basin-related matters through mandatory means. Therefore, resolving the issues of ecological environment protection and benefit allocation in transboundary river basins through negotiation has become an effective means [5]. For example, the United States and Canada conducted full negotiations on the distribution of interests in the Columbia River, and finally signed an agreement in 1976 to clarify the rights and obligations of both parties, which became a successful case in the world to reasonably solve the distribution of interests in transboundary water resources [6]. Moreover, in the treatment of the allocation for the water resource in the Colorado River basin, the negotiation of various stakeholders played a key role in improving the water resource allocation model and resolving conflicts [7]. In addition, due to the mobility of water resources, TBRBs show a strong characteristic of basin integrity and form a water community with a shared future. This characteristic determines that each BC should prioritize the overall interests and promote sustainable development when dealing with ecological environmental protection issues in the basin.

Ecological compensation, as a means of environmental economic management, can effectively coordinate the relationship between resource exploitation and ecological environmental protection [8]. It can also effectively adjust the unequal distribution of benefits caused by resource exploitation and utilization, and alleviate the conflicts among BCs. It is an important means to propel the sustainable development of TBRBs forward. In the process of TBRB development, some BCs have ceded part of their own benefits in order to make the overall benefits of the TBRB maximum, and some BCs have enjoyed the overall benefits too much. In order to achieve sustainable development of TBRBs, beneficiary countries need to compensate the countries that transfer interests, thereby reducing conflicts and contradictions caused by uneven distribution of basin interests. For example, in the Elbe River basin, Germany compensated 9 million marks to the Czech Republic in 2000 for the construction of urban sewage treatment plants at the border of the Czech Republic and Germany to ensure the health and the stability of the water environment [9]; in the Lesotho highlands of Orange River, the upstream country, Lesotho, built dam facilities, and South Africa bore most of its construction costs to ensure that it could obtain water from upstream of Lesotho [10]; during the construction process of the Gabčíkovo-Nagymaros hydroelectric station in the Danube River, Hungary, undertook a portion of the engineering construction on the territory of Czechoslovakia to ensure equal sharing of costs and equal distribution of power-generation benefits [11]. These compensation measures have reduced the occurrence of conflicts and contradictions between different basin countries.

When carrying out ecological compensation activities, it is crucial to determine the ecological compensation standard (ECS) [8], which affects whether compensation activities can be effectively carried out. The quantity and magnitude of the ECS are influenced by factors such as the loss or gain of stakeholders, the international situation, national policy and so on. Because of the differences in the perspectives and main factors involved in different scholars' research, although there has been more research on the ECS, a unified standard or calculation method has not yet been formed. At present, scholars mostly determine the ECS from the perspective of investment and income [12]. The methods based on the investment perspective mainly include the cost of ecological protection (CEP) method [13,14], while the methods based on the income perspective mainly include the ecosystem service value (ESV) method [15,16] and the willingness to pay method [17,18]. The CEP method mainly includes the method of the direct cost and the opportunity cost [19]. When using the CEP method for calculation, the key lies in the selection of calculation indicators; that is, the coverage of the CEP. When using the ESV method, due to the large amount of the ESV and its tendency to be overestimated, there may be a significant deviation between the conclusions obtained and the actual situation [20]. When using this, scientific measurement can be made by deducting the consumption of ecosystem service value (CESV) by oneself. The willingness to pay method ensures the acceptance and recognition of compensation by relevant stakeholders through considering willingness to pay in the calculation of the ESV. However, during the investigation process, information asymmetry may occur, which may not match the actual willingness to pay [21]. Therefore, more objective methods need to be used to measure this.

In summary, scholars have generally recognized the importance of determining the ECS in TBRBs. Although the research methods are different, scholars are in agreement in the pursuit of fairness and rationality in the formulation of the ECS. However, currently, the determination of the ECS, both domestically and internationally, is based on one or several methods, with the author choosing the appropriate results or finding the mean or median as the ECS, lacking universality. In TBRBs, due to the involvement of numerous BCs, determining an ECS that can be accepted by all BCs should not be done solely through one method. In addition, each BC should be given sufficient negotiation autonomy to comprehensively determine the ECS in TBRBs, thereby improving the acceptability and feasibility of the ECS. In view of the role of various stakeholders in the ecological environment of TBRBs, the BC is taken as the basic research unit and can be subdivided into an ecosystem-service-supplying country (ESSC) and an ecosystem-service-consuming country (ESCC). During the negotiation, the ESSCs and the ESCCs are taken as two together, to conduct one-on-one negotiation, and the negotiation space of both sides should be determined first. The base value of the ECS in TBRBs is the theoretical value of the ECS formulated objectively and reasonably considering the input cost of the ESSC and the benefit degree of the ESCC, which can provide the basis for the negotiation among the BCs.

Therefore, in order to promote the healthy and stable development of TBRBs, reduce interest disputes, and provide a theoretical basis for negotiation among various BCs; this paper attempted to design a feasible range of the base value of the ECS in TBRBs. Taking the BC as the research subject, the BC was divided into two main bodies: the ESCC and the ESSC. From the perspective of ecological beneficiaries, the upper limit of the base value of the ECS was determined on the basis of the spillover value of ecosystem services (SVES) of the ESCC. From the perspective of ecological protectors, the lower limit of the base value of the ECS was determined according to the CEP of the ESSC. The feasible range for the base value of the ECS in TBRBs was determined based on the actual water consumption and the willingness to pay of each BC. The general idea of this paper is shown in Figure 1.

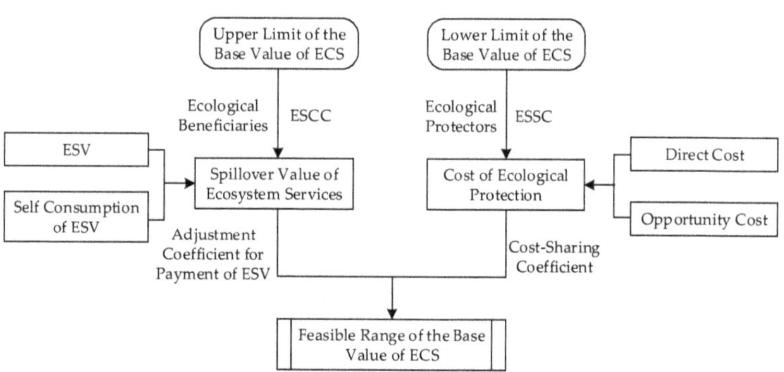

Figure 1. General idea of this paper.

2. Study Area and Data Resource

2.1. Study Area

The Lancang–Mekong River is the only transboundary river in Asia that connects six countries with one river. It originates in Qinghai, China and enters Vietnam through Myanmar, Laos, Thailand, and Cambodia [22]. The Lancang–Mekong River basin (LMRB) runs through the entire of southeast Asia and is linked to the economic and social development of various countries. The LMRB is rich in ecological resources, which are exploited and utilized by each BC based on its own development needs. However, due to the difference in economic development level, the exploitation and utilization degree and efficiency of basin resources are also different. In the process of resource utilization, the BCs have caused different degrees of impact on the basin's ecological environment. In addition, the differences in the willingness and ability of the basin's ecological protection result in the unbalanced relationship between the ecological environment and economic interests among the BCs. At present, although some transboundary cooperation has been carried out among Lancang–Mekong countries, interest conflicts in resource utilization and ecological protection occur from time-to-time. These conflicts pose a certain threat and challenge to the water security and long-term development of the LMRB. Therefore, ecological compensation can be used to solve this problem in the LMRB. The research area is shown in Figure 2 [23].

Figure 2. Geographical distribution of LMRB.

2.2. Data Resource

The basic data involved in the determination of the base value of the ECS in the LMRB are mainly used to calculate the SVES and CEP of BCs. Considering the time development and data availability, this paper took 2019 as the research benchmark year. When confirming different land use types and areas of the BCs in the LMRB, the classification standards made by the Food and Agriculture Organization of the United Nations (FAO) and the Mekong

River Commission are adopted [24]. In this paper, land use in the LMRB was divided into seven types: farmland, grassland, forestland, water area, wetland, construction land and unused land. The land use data were sourced from the National Qinghai Tibet Plateau Scientific Data Center with a resolution of 10 m [25], and various land use areas were obtained through visual interpretation by ArcGIS. Specific data are shown in Table 1.

Table 1. Area of different land use types in LMRB. Unit: 10^4 km^2.

Category	Farmland	Grassland	Forestland	Water Area	Wetland	Construction Land	Unused Land
China	0.56	8.77	6.82	0.09	0.01	0.08	0.14
Myanmar	0.12	0.75	1.30	0.01	0.00	0.01	0.00
Laos	2.60	6.59	11.28	0.25	0.02	0.08	0.00
Thailand	11.32	3.36	3.44	0.27	0.08	0.52	0.00
Cambodia	6.89	1.35	6.45	0.77	0.75	0.11	0.00
Vietnam	2.01	1.07	2.01	0.07	0.00	0.08	0.00

The equivalent factor of the ESV was modified based on the Xie Gaudi Edition [26] and according to the actual land use situation in the LMRB. Among them, in the farmland ecosystem, the LMRB is mainly planted with rice, which has high value in food production and soil and water conservation [27]; in the forestland ecosystem, the LMRB mainly grows tropical rainforest with various values such as climate regulation and soil erosion prevention [28], and its NPP is relatively high [29,30]; in the construction land ecosystem, some construction land contains public green spaces, which have certain ESV such as environmental greening and soil conservation [31–33]. Taking the practical characteristics of the LMRB into account, the equivalent factor values for ecosystems of grassland, water area and unused land were determined on the basis of secondary classification of shrubland [34], water system, and bare land [31]. In addition, the precipitation in the LMRB is relatively abundant [35], and corresponding adjustments were made according to the adjustment rules of precipitation spatiotemporal adjustment factors [26,31]. The equivalent value per unit area in the LMRB determined in this paper is shown in Table 2.

Table 2. Ecosystem service equivalent value per unit area in LMRB.

Classification	Supply Service			Regulating Service				Support Service			Cultural Service
	FP	MP	WS	GR	CR	DE	HR	SC	NCM	BD	AL
Farmland	1.36	0.09	−0.88	1.11	0.57	0.17	8.16	1.30	0.19	0.21	0.09
Grassland	0.38	0.56	0.93	1.97	5.21	1.72	11.46	2.40	0.18	2.18	0.96
Forestland	0.42	0.96	1.48	3.15	9.43	2.80	20.62	3.84	0.29	3.49	1.54
Water Area	0.80	0.23	24.87	0.77	2.29	5.55	306.72	0.93	0.07	2.55	1.89
Wetland	0.51	0.50	7.77	1.90	3.60	3.60	72.69	2.31	0.18	7.87	4.73
Construction Land	0.01	0.03	0.41	0.75	2.01	0.97	5.08	0.92	0.07	0.89	0.69
Unused Land	0	0	0	0.02	0	0.10	0.09	0.02	0	0.02	0.01

Notes: FP: Food Production; MP: Material Production; WS: Water Supply; GR: Gas Regulation; CR: Climate Regulation; DE: Decontamination Environment; HR: Hydrologic Regulation; SC: Soil Conservation; NCM: Nutrients Cycle Maintenance; BD: Biological Diversity; AL: Aesthetics Landscape.

The output, price, planting area and agricultural output value of major food crops were from the FAO database [36]. The population, water resources and socio-economic data of the basin were from the Transboundary Waters Assessment Programme (TWAP) database [37], Transboundary Freshwater Dispute Database (TFDD) [38], Mekong River Commission [39], World Bank database [40] and the existing literature [32,41,42]. Among them, when calculating the direct cost of ecological protection, China uses data from the Yunnan Province (including Baoshan, Pu'er, Lincang, Xishuangbanna, Dali, Nujiang, and Diqing), sourced from the Yunnan Provincial Statistical Yearbook [43].

3. Materials and Methods

3.1. Upper Limit of the Base Value—SVES

3.1.1. Determination Ideas of the SVES

From the perspective of ecological beneficiaries, this paper determines the upper limit of the base value of the ECS according to the SVES of ESCCs. In this paper, the SVES is the surplus value, which is the remaining part after subtracting the ecological value used by residents for production and life based on the ESV of the basin. Therefore, the measurement of the SVES should contain two sections: the first is the measurement of the ESV, and the second is the measurement of the CESV by the BCs themselves. In the process of calculation, it is scientific and reasonable to use the SVES as the upper limit, taking into account the elimination of self-consumption based on the ESV. In the application of methods, it is comparable to obtain the SVES on the basis of the equivalent factor method, combining with the ecological footprint and ecological carrying capacity.

Based on the above analysis, a calculation model of the SVES in TBRBs is obtained, and it serves as the determination basis of the upper limit of the base value of the ECS in TBRBs. The specific calculation is shown in Equation (1).

$$SE_i = FE_i - CE_I \tag{1}$$

where i = 1, 2, ..., n is the BC; SE_i is the SVES in BC_i(\$); FE_i is the ESV in BC_i(\$); CE_i is the CESV in BC_i(\$). When $SE_i > 0$, it indicates that the CESV of BC_i is less than the ESV it owns, and its ESV status belongs to a surplus state. BC_i is the ESSC. When $SE_i = 0$, it indicates that the CESV of BC_i is equal to the ESV it owns, and its ESV status is in an equilibrium state. BC_i is neither an ESSC nor an ESCC. When $SE_i < 0$, it indicates that the CESV of BC_i is greater than the ESV it owns, and its ESV status belongs to a deficit state. BC_i is the ESCC.

3.1.2. Determination of ESV

This paper calculates the total ESV of each BC using the equivalent factor method. When calculating, the main consideration is the land use area of various ecosystems, equivalent factor of various ecosystem service functions, and economic value corresponding to each equivalent factor. The specific calculation is shown in Equation (2).

$$FE_i = \sum_{j=1}^{a} \sum_{p=1}^{b} \alpha_{jp} \cdot \beta_i \cdot A_i \tag{2}$$

where j, j = 1, 2, ···, a is different ecosystem types; p, p = 1, 2, ···, b is different ecosystem service functions; α_{jp} is the equivalent factor value corresponding to the p-th ecosystem service function in the j-th ecosystem; β_i is the economic value (\$/hm^2) corresponding to the equivalent factor of BC i; A_i is the land use area (hm^2) of various ecosystems in BC i; the other symbols are the same as above.

(1)   Determination of equivalent factor value (α_{jp})

The determination of equivalent factor value α_{jp} is mainly based on the basic equivalent table revised by Xie Gaodi et al., in 2015 [26], and actual modifications are made based on the specific situation of the research basin. When making corrections, adjustments can be made based on net primary productivity (NPP), differences in precipitation, etc. [26].

(2)   Determination of economic value corresponding to equivalent factors (β_i)

The economic value corresponding to each equivalent factor is the economic value generated by the national average grain production per hectare of farmland under natural conditions. In farmland ecosystems, food value is produced by natural factors and human factors, so it is difficult to accurately measure the specific amount of food value under the action of natural factors. Therefore, according to relevant research [44–46], the economic value corresponding to each equivalent factor is determined as 1/7 of the national average grain yield market value for that year. The economic value of grain crops is mainly determined by calculating the economic value generated by rice, wheat and corn, as shown in Equation (3).

$$\beta_i = \frac{1}{7}\sum_{m=1}^{3}\frac{Q_{im}\cdot P_{im}}{S_{im}} \tag{3}$$

where m, $m = 1, 2, 3$ represents rice, wheat and corn, respectively, in the farmland ecosystem; Q_{im} is the annual average yield (kt) of grain crop m in the farmland ecosystem of BC i; P_{im} is the average annual price ($/kt) of grain crop m in the farmland ecosystem of BC i; S_{im} is the average annual planting area (hm^2) of grain crop m in the farmland ecosystem of BC i; the other symbols are the same as above.

3.1.3. Determination of CESV

This paper calculates the CESV according to the ecological consumption coefficient, which is the ratio of ecological consumption and ecological supply of each BC. After the ecological consumption coefficient is defined, the CESV can be obtained in consideration of the ESV. The specific calculation is shown in Equation (4).

$$CE_i = FE_i \cdot \theta_i \tag{4}$$

where θ_i is the ecological consumption coefficient of BC i; the other symbols are the same as above.

The specific calculation of θ_i is shown in Equation (5).

$$\theta_i = \frac{EC_i}{ES_i} \tag{5}$$

where EC_i is the ecological consumption of BC$_i$(hm^2); ES_i is the ecological supply of BC$_i$ (hm^2); the other symbols are the same as above.

(1)   Determination of ecological consumption of each BC (EC_i)

The ecological consumption of each BC can be determined through the ecological footprint, which refers to the biological productive land area needed for the conversion of waste generated by product production and consumption during the development process [47]. It usually includes six types: farmland, grassland, forestland, waters, construction land and fossil energy land. The specific calculation of the ecological footprint is shown in Equations (6) and (7).

$$EC_i = N_i \cdot ec_i \tag{6}$$

$$ec_i = \sum_{k=1}^{6}\sum_{s=1}^{n} r_k \cdot \frac{C_{iks}}{GP_{ks}} \tag{7}$$

where N_i is the population of BC i; ec_i is the ecological footprint per capita of BC$_i$ (hm^2/person); k, $k = 1, 2, 3, 4, 5, 6$ represents six land types: farmland, grassland, forestland, waters, construction land and fossil energy land; r_k is the equilibrium factor of various types of land; s, $s = 1, 2, \ldots, n$ represents the goods produced by various types of land; C_{iks} is the per capita output of the s-th commodity produced on the k-th land in the BC$_i$ (kg/person); GP_{ks} is the global average production of commodity s produced by the k type of land (kg/hm^2); the other symbols are the same as above.

(2)   Determination of ecological supply of each BC (ES_i)

The ecological supply of each BC can be determined by ecological carrying capacity, which refers to the total bioproductive area provided by a region [48]. It usually includes six land types: farmland, grassland, forestland, waters, construction land and fossil energy land. Generally speaking, when measuring ecological carrying capacity, 12% of the area needs to be subtracted for biodiversity conservation, in order to maintain regional sustainable development [16]. The specific calculation of ecological carrying capacity is shown in Equation (8).

$$ES_i = 1 - 12\% \sum_{k=1}^{6} A_{ik} \cdot r_k \cdot x_k \tag{8}$$

where A_{ik} is the utilization area (hm^2) of the k-th type of land in BC$_i$; x_k is the yield factor of each type of land; the other symbols are the same as above.

3.2. Lower Limit of the Base Value—CEP

3.2.1. Determination Ideas of CEP

From the perspective of ecological protector, the lower limit of the base value of the ECS is determined according to the CEP of ESSCs. The CEP in this paper refers to the total investment of each BC in the governance and protection of the ecological environment within the TBRB, as well as the loss caused by the development opportunity given up, including direct cost and opportunity cost [49]. The positive SVES exists in ESSCs, which has a positive effect on other regions, indicating that the ESSCs have paid a lot of funds and efforts to manage and maintain the ecological environment of the basin, and made certain sacrifices in their own economic development. Due to the systematic and holistic nature of TBRBs, ESCCs have excessively enjoyed the benefits of basins, posing a certain degree of threat and damage to the rights that ESSCs should have. According to the principles of fair and reasonable utilization as well as equal rights and responsibilities, from the perspective of sustainable development, it is reasonable for ESCCs, as beneficiaries, to compensate ESSCs to a certain extent for their efforts.

Therefore, the calculation of the CEP in a TBRB should contain two sections: the first is the calculation of direct cost of the CEP, and the second is the calculation of opportunity cost of the CEP. In the process of calculation, it should be fully combined with the characteristics of the basin and the collection of relevant data, so as to sort out and calculate the CEP invested by the ESSCs.

Based on the above analysis, a calculation model of the CEP in a TBRB is obtained, and it serves as the determination basis of the lower limit of the base value of the ECS in the TBRB. The specific calculation is shown in Equation (9).

$$PC_i = DPC_i + OPC_i \tag{9}$$

where i, $i = 1, 2, \cdots, n$ is the ESSCs; PC_i is the CEP of ESSCs ($); DPC_i is the direct cost for ecological protection of ESSCs ($); OPC_i is the opportunity cost for ecological protection of ESSCs ($).

3.2.2. Determination of Direct Cost for Ecological Protection in ESSCs

(1)　Accounting Scope

The direct cost is the direct display of the funds invested by ESSCs in basin ecological environment protection work. Due to the differences in the actual situation of development and protection work in different basins, and the high demand for related data to determine the direct cost, there is no fixed standard for the accounting scope of direct cost. However, in general, for the sake of protecting the ecological environment of the basin and providing sustainable ESV, it is necessary for ESSCs to implement corresponding protection measures according to various land use types, including the construction of related protection projects and the management of related environmental problems. Thus, when assessing the accounting scope of direct cost for ecological protection in ESSCs, in this paper, it is divided into four categories: forestry and grassland construction cost (C_{Li}), water environment management and protection cost (C_{Wi}), wetland protection cost (C_{Si}) and biodiversity protection cost (C_{Di}). The accounting scope of direct cost for ecological protection in ESSCs is shown in Table 3.

Table 3. Accounting scope of direct cost for ecological protection in ESSCs.

Direct Cost	Index	Index Interpretation
Forestry and grassland construction cost (C_{Li})	Cost of forest protection	Investment in reducing deforestation, artificial afforestation, closed mountain afforestation etc.
	Cost of returning farmland to forest or grassland	Investment in returning sloping farmland to forests and grasslands, afforestation in barren mountains and wasteland etc.
	Cost of natural ecological protection	Investment in the construction and management of ecological function protection zones, ecological restoration, resource development supervision etc.
	Construction and management costs of nature reserves	Investment in infrastructure construction, daily maintenance, management operations etc.
Water environment management and protection cost (C_{Wi})	Cost of water conservancy project construction	Investment in the water facilities construction, operation and maintenance of water engineering etc.
	Cost of water pollution control	Investment in point and non-point source pollution etc.
	Cost of water quality monitoring	Investment in the construction and operation management of water quality monitoring stations, scientific research etc.
	Cost of saving water	Investment in water-saving projects, renovation and upgrading of water-saving facilities, innovation of technologies etc.
	Cost of soil and water conservation	Investment in regional comprehensive governance, related engineering construction etc.
Wetland protection cost (C_{Si})	Wetland protection cost	Investment in the construction of wetland protection areas, returning farmland to wetlands, restoring degraded wetlands etc.
Biodiversity protection cost (C_{Di})	Cost of plant and animal protection	Investment in the renovation and restoration of animal and plant habitats, as well as pilot projects in national parks

(2)   Accounting Method

On the basis of the above analysis, the specific calculation of the direct cost for ecological protection in ESSCs is shown in Equation (10).

$$DPC_i = C_{Li} + C_{Wi} + C_{Si} + C_{Di} \tag{10}$$

where C_{Li} is the forestry and grassland construction cost of $ESSC_i$; C_{Wi} is the water environment management and protection cost of $ESSC_i$; C_{Si} is the wetland protection cost of $ESSC_i$; C_{Di} is the biodiversity protection cost of $ESSC_i$; the other symbols are the same as above.

3.2.3. Determination of Opportunity Cost for Ecological Protection in ESSCs

The opportunity cost is an indirect reflection of the development value sacrificed by ESSCs in ecological environment protection work in TBRBs. In order to maintain the overall healthy and long-term development of the basin environment, ESSCs have to restrict the exploitation and utilization of some natural resources; thus, losing the benefits obtained from utilizing these resources. When measuring the opportunity cost, it is difficult to obtain sufficient and accurate actual data as there is no unified measurement standard and method, and the TBRB situation is quite complex. As a consequence, this paper adopts the empirical comparison method to measure it [50]; that is, the difference between the economic development level of the ESSC and the neighboring region. The development opportunity cost lost by the ESSC due to ecological protection is calculated in detail as shown in Equations (11) and (12).

$$OPC_i = (G_s - G_i) \cdot N_i \cdot \sigma_i \tag{11}$$

$$\sigma_i = \frac{R_{Ai}}{R_{Ti}} \times 100\% \tag{12}$$

where G_s is the per capita disposable income (\$/person) in the vicinity of $ESSC_i$; G_i is the per capita disposable income (\$/person) in $ESSC_i$; σ_i is the regulatory factor, that is, the proportion of the total agricultural product of the basin in the total basin product of $ESSC_i$; R_{Ai} is the total agricultural product of the basin in $ESSC_i$ (\$); R_{Ti} is the total basin product of $ESSC_i$ (\$); the other symbols are the same as above.

3.3. Determination of the Range for the Base Value of the ECS in TBRBs

On the basis of the analysis of Sections 3.1 and 3.2, this paper takes the SVES of ESCCs as the upper limit, and the CEP of ESSCs as the lower limit; thus, forming the selection range for the base value of the ECS in the TBRB. On this basis, it provides a basis for negotiation on the final value of the ECS between the ESSC and ESCC in the TBRB. Since the ESSCs and ESCCs are taken as two together to conduct one-to-one negotiation in this paper, the final range of the base value is shown in Equation (13).

$$\gamma \cdot \sum_{i=1}^{n} PC_i \leq S_0 \leq \sum_{i=1}^{n} l_i \cdot |SE_i| \tag{13}$$

where γ is the cost-sharing coefficient; S_0 is the base value of the ECS in the TBRB (\$); l_i is the adjustment coefficient for payment of ESV for the BC i; the other symbols are the same as above.

The cost-sharing coefficient (γ) is mainly determined according to the utilization of water resources in the TBRB by ESSCs and ESCCs; that is, the CEP should be shared according to the proportion of the water consumption in ESCCs to the entire basin. The specific calculation of γ is shown in Equation (14).

$$\gamma = \frac{\sum_{i=1}^{n} W_{ci}}{\sum_{i=1}^{n} W_i} \tag{14}$$

where W_{ci} is the total water use (m^3) of basin water resources for the $ESCC_i$, including agricultural, industrial and domestic water use; W_i is the total water use (m^3) of basin water resources for the BC_i; the other symbols are the same as above.

The adjustment coefficient for payment of the ESV (l_i) is mainly determined according to the economic development level of each BC. Li Jinchang et al. proposed a method to get the value of willingness to pay on the basis of combining of the S-shaped Pearl growth curve model and the Engel coefficient [51], and applied it with significant results. Since then, many scholars have continuously used this model to get the adjustment coefficient for payment of the ESV [52], and its specific calculation is shown in Equation (15).

$$l_i = \frac{1}{1 + e^{-t_i}}, t_i = \frac{1}{E_i^k} - 3 \tag{15}$$

where the image of l_i is called the Pearl Growth Curve, the range of which is from 0 to 1; E_i^k is the Engel coefficient for the year k of each BC; the other symbols are the same as above.

4. Results and Discussion

4.1. SVES of BCs in the LMRB

4.1.1. ESV of BCs in the LMRB

According to Equations (2) and (3), on the basis of clarifying the area of different land use types and the equivalent factor value in each BC, the ESV of each BC can be obtained by combining the economic value of main food crops. The specific calculation results are shown in Table 4.

Table 4. ESV of BCs in LMRB. Unit: 10^8 \$.

Category	Farmland	Grassland	Forestland	Water Area	Wetland	Construction Land	Unused Land	Total
China	6.30	221.59	296.17	28.94	0.93	0.91	0.03	554.87
Myanmar	1.32	18.53	55.02	1.83	0.00 *	0.09	0.00	76.79
Laos	37.03	211.84	622.67	97.76	2.18	1.13	0.00	972.61
Thailand	115.08	77.27	135.77	75.90	7.11	5.05	0.00	416.18
Cambodia	144.05	63.89	523.12	451.80	134.83	2.19	0.00	1319.88
Vietnam	42.66	51.27	165.79	40.56	0.26	1.66	0.00	302.20
Total	346.44	644.39	1798.54	696.79	145.31	11.03	0.03	3642.53

Note: The actual value of 0.00 * is 0.00257359.

The Table 4 indicated that the overall ESV in the LMRB was 3642.53 × 10^8 \$. From the perspective of BCs, Cambodia had the highest ESV, which was 1319.88 × 10^8 \$, accounting for 36.23% of the total basin. It was followed by Laos, China, Thailand, Vietnam and Myanmar, which accounted for 26.70%, 15.23%, 11.43%, 8.30% and 2.11%, respectively. The ESV in Myanmar was the lowest in the basin: only 76.79 × 10^8 \$. In terms of land use type, the ESV of forestland was the highest in the entire ESV of each BC, among which Myanmar and Laos accounted for 71.65% and 64.02%. The proportion of construction land and unused land was the lowest. Except China, there was no unused land in the other five countries. The proportion of ESV of different land use types in the LMRB is shown in Figure 3.

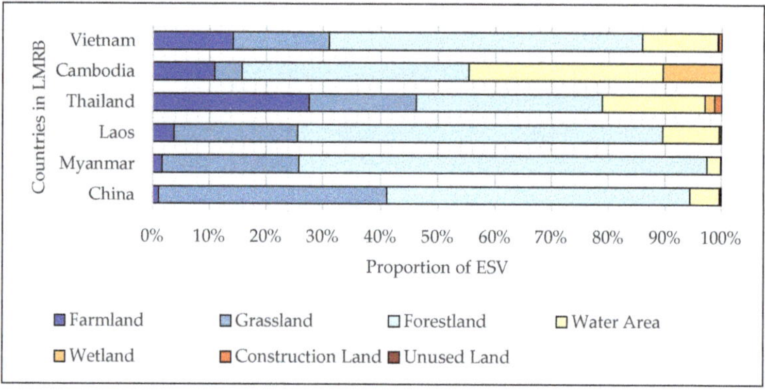

Figure 3. The proportion of ESV of different land use types in LMRB.

4.1.2. CESV of BCs in the LMRB

According to Equations (5)–(8), the ecological consumption coefficient of each BC can be obtained by combining the equilibrium factor and yield factor of various land types, based on clarifying the population of each BC, the area of different land use types, and the main production commodity yield of each land type. Among them, the equilibrium factors for farmland, grassland, forestland, water area, construction land and land for energy and fuel are 2.10, 0.47, 1.33, 0.36, 2.18 and 1.35, respectively [32]. The yield factors are 1.65, 0.20, 0.91, 0.99, 1.65 and 0, respectively [32]. The specific calculation results are as shown in Table 5.

According to Table 5, the ecological consumption coefficient of China, Myanmar and Laos was less than 1, and that of Thailand, Cambodia and Vietnam was greater than 1, indicating that the ecological supply of China, Myanmar and Laos exceeded the ecological consumption, while the ecological consumption of Thailand, Cambodia and Vietnam exceeded the ecological supply. The ecological consumption coefficient of Thailand reached 1.6, which was higher than that of Cambodia and Vietnam. In terms of geographical location, China, Myanmar and Laos lie in the upstream and middle stream of the LMRB, while

Thailand, Cambodia and Vietnam lie in the downstream. The ecological consumption level in the downstream was significantly higher than that in the upstream and middle stream.

Table 5. Ecological consumption coefficient of BCs in LMRB.

Category	Ecological Footprint (10^4 hm^2)	Ecological Carrying Capacity (10^4 hm^2)	Ecological Consumption Coefficient
China	1328.70	1641.66	0.81
Myanmar	115.13	184.41	0.62
Laos	1425.80	2083.29	0.68
Thailand	6430.64	4017.26	1.60
Cambodia	2172.48	2022.20	1.07
Vietnam	1148.37	864.20	1.33

From the perspective of ecological footprint, Thailand was the highest, which was 6430.64×10^4 hm^2, accounting for about 50.95% of the whole basin, and the rest were Cambodia, Laos, China, Vietnam and Myanmar, in turn. From the perspective of ecological carrying capacity, Thailand still was the highest, which was 4017.26×10^4 hm^2, accounting for 37.15% of the whole basin, and the rest were Laos, Cambodia, China, Vietnam and Myanmar, in turn. From a geographical location point of view, both aspects generally showed a trend of gradual growth from the up-reaches to the low-reaches of the basin, which was closely affected by the geographical distribution, basin area, economic conditions, natural resource endowment conditions and other factors of the BCs.

According to Equation (4), combined with the ESV shown in Table 4 and the ecological consumption coefficient shown in Table 5, the CESV of each BC can be obtained. Specifically, the CESV of China, Myanmar, Laos, Thailand, Cambodia and Vietnam was 449.09×10^8 \$, 47.94×10^8 \$, 665.65×10^8 \$, 666.20×10^8 \$, 1417.96×10^8 \$ and 401.57×10^8 \$, respectively.

4.1.3. SVES of BCs in the LMRB

According to Equation (1), the SVES of each BC in the LMRB can be obtained by combining the ESV of each BC and its own CESV. Specifically, the SVES in China, Myanmar, Laos, Thailand, Cambodia and Vietnam was 105.78×10^8 \$, 28.85×10^8 \$, 306.96×10^8 \$, -250.02×10^8 \$, -98.08×10^8 \$ and -99.37×10^8 \$, respectively.

The results indicated that the SVES in China, Myanmar and Laos was positive, resulting in positive ecological spillover, indicating that these three countries could offer ESV to other BCs after deducting their own CESV. They were ESSCs and should receive compensation. The SVES in Thailand, Cambodia and Vietnam was negative, showing that they had excessively consumed the ESV and destroyed the balance of nature of the river basin in the process of development. They were ESCCs and should compensate.

In real life, Thailand, Cambodia and Vietnam lie in the middle and lower reaches and have relatively high levels of utilization of basin resources, consuming a large amount of water and ecological resources in their respective development processes. This also explains why these three countries are ESCCs. Thailand mainly diverts the Mekong River for agricultural irrigation and daily life. The area of Thailand in the LMRB accounts for nearly 1/4 of the entire basin. It has formulated multiple Mekong River diversion plans, introducing massive water resources into the country for self-use, which has seriously affected downstream water use [42]. Cambodia is a typical agricultural country, with more than 85% of its land area located in the LMRB. Meanwhile, 60% of the water supply of Tonlé Sap in Cambodia comes from the Mekong River. Its economic development has a high demand for river water resources [41]. The lower Mekong Delta is home to high-quality rice from around the world, with Cambodia accounting for about 1/5 of the area and the remaining area belonging to Vietnam. Here, 90% of Vietnam's exported rice grows. Therefore, it needs a large amount of irrigation and ecological water [53]. China and Laos, which lay in the upstream, have put much energy and resources into basin ecological environment protection and water conservancy engineering construction, which can reduce the losses caused by special water conditions in downstream countries during drought periods and floods. For instance, in 2016, China implemented emergency water replenishment for the downstream of the Mekong River through sluice opening, which effectively relieved the drought in Vietnam [54]. Laos has built dams upstream to alleviate water scarcity and flood disasters in downstream countries during dry periods [32]. As a result, in the light of the principle of beneficiary compensates, it is fair for the ESCCs to compensate the ESSCs.

4.2. CEP of BCs in the LMRB

According to the analysis in Section 4.1.3, China, Myanmar and Laos were the ESSCs. Therefore, the CEPs of these three countries need to be calculated to determine the lower limit of the base value of ESV in LMRB. According to Equations (9)–(12), combined with the direct and opportunity cost in China (Yunnan Province), Myanmar and Laos, the CEP of ESSCs in LMRB can be obtained. The specific results are shown in Table 6.

Table 6. CEP of BCs in LMRB. Unit: 10^4 \$.

Category	Direct Cost	Opportunity Cost	CEP
China	4954.40	4155.27	9109.67
Myanmar	216.73	1062.93	1279.66
Laos	2148.06	15,839.28	17,987.34
Total	7319.19	21,057.48	28,376.67

As seen from Table 6, the entire CEP of ESSCs in the LMRB was $28{,}376.67 \times 10^4$ \$, in which the direct cost was 7319.19×10^4 \$, accounting for 25.79% of the total cost. The opportunity cost was $21{,}057.48 \times 10^4$ \$, accounting for 74.21% of the total cost. From the perspective of BCs, China had the highest direct cost, accounting for 67.69% of the total direct cost, while Laos had the highest opportunity cost, accounting for 75.22% of the total opportunity cost.

The CEP of ESSCs in the LMRB is closely related to its own level of economic development. The higher the level of economic development, the higher the awareness and ability of ecological protection, and the more funds can be directly invested in ecological protection for the basin. Myanmar and Laos, due to their relatively backward economy and small population size, are unable to fully undertake the obligations of resource development and ecological environment protection, resulting in significantly smaller direct cost investments. Although Myanmar lies in the upstream of the LMRB, only about 2% of the basin coverage area is within Myanmar's borders. Therefore, Myanmar has low enthusiasm for water resource development in the LMRB, and focuses more on forest environmental protection in the basin [4], resulting in higher opportunity costs. Most of Laos' national territory is located in the LMRB. The water production of the LMRB in Laos accounts for 35% of the total basin volume, ranking first among the six countries in the basin [42]. Although Laos has great potential for hydropower development, its utilization level is relatively low, resulting in higher opportunity costs. China has the highest level of economic development among the three countries, a strong enthusiasm for ecological protection, and has invested the most in costs. For example, China actively carries out the construction of Sanjiangyuan National Park to improve the water conservation and grassland coverage of the Lancang River; it actively promotes the construction of Potatso National Park; and constantly improves the value of forest ecosystem services. In addition, because of the fact that China's utilization of water resources in the LMRB is mainly based on hydropower and shipping, which are classified as non-consumable water, the actual consumption of water in China is relatively low [22]. So the opportunity cost is also high.

4.3. Range for the Base Value of the ECS in the LMRB

Combined with the analysis in Section 3.3 and the judgment above, the SVES in Thailand, Cambodia and Vietnam was taken as the upper limit, and the CEP in China, Myanmar and Laos was taken as the lower limit, when determining the feasible range for the base value of the ECS in the LMRB. The final confirmation of the range was combined with the cost-sharing coefficient and adjustment coefficient for payment of the ESV.

When determining the cost-sharing coefficient, it is necessary to consider the proportion of water resource consumption in Thailand, Cambodia and Vietnam in the total basin water consumption. The water resources utilization of countries in the LMRB is shown in Table 7. When determining the adjustment coefficient for payment of the ESV, it is necessary to consider the Engel coefficients of Thailand, Cambodia and Vietnam. According to statistics from the US Department of Agriculture's Bureau of Economic Research, they are 0.276, 0.446 and 0.411, respectively.

Table 7. Water resources utilization of countries in LMRB. Unit: 10^8 m^3.

Category	Agricultural Water	Industrial Water	Domestic Water	Total
China	21.43	2.15	4.10	27.68
Myanmar	1.36	0.03	0.10	1.49
Laos	39.44	0.20	2.39	42.03
Thailand	98.09	1.40	11.23	110.72
Cambodia	89.54	0.20	5.20	94.94
Vietnam	259.14	1.22	5.45	265.81
Total	509.00	5.20	28.47	542.67

To sum up, according to Equations (13)–(15), the range for the base value of the ECS in the LMRB can be obtained, as shown in Table 8.

Table 8. Range for the base value of ECS in LMRB.

	Upper Limit of the Base Value				Lower Limit of the Base Value			
	SVES	ACP	APV	Total		CEP	CSC	Total
Thailand	250.02	0.65	162.51		China	0.91		
Cambodia	98.08	0.32	31.39	229.67	Myanmar	0.13	0.87	2.47
Vietnam	99.37	0.36	35.77		Laos	1.80		
Range for the base value of ECS in LMRB: [2.47, 229.67] $\times 10^8$ \$								

Notes: ACP: adjustment coefficient for payment of ESV; APV: actual payment value; CSC: cost-sharing coefficient.

As seen from Table 8, the feasible range for the base value of the ECS in the LMRB was [2.47, 229.67] $\times 10^8$ \$ considering the actual development level of each BC, the willingness to pay and the actual water consumption. That is to say, the theoretical space for the negotiation between the ESSCs and ESCCs in the LMRB on the ECS was [2.47, 229.67] $\times 10^8$ \$, within which the two sides can conduct full negotiation in order to work out a fair and reasonable scheme for the ECS in the LMRB.

The ECS based on the SVES in the LMRB may exceed the actual tolerance range of ESCCs, as the ESV is huge and difficult to estimate, and often several times the economic value. The research method in this paper provided a way of thinking. However, the ECS was still influenced by factors such as the geographical location of each BC and the acquisition of relevant data, which can only be used as the theoretical upper limit of the ECS. The ECS based on the CEP is in accordance with the ecological environment governance and protection costs and opportunity costs of ESSCs. Influenced by factors such as economic and social development level, people's living standards, and government policies, coupled with high requirements for data accuracy, it was easy to be underestimated during accounting and can only be used as the theoretical lower limit of the ECS. Therefore, the final ECS must be fully negotiated and determined by each BC.

Due to the particularity of TBRBs, the current practice of ecological compensation in TBRBs mainly relies on negotiated compensation between two countries. For example, Canada agreed to build three reservoirs on the upstream of the Columbia River to regulate the runoff, so as to meet the flood control requirements in the lower reaches of America. And, America paid Canada for the construction of reservoirs and shared the benefits of power generation [55]. However, theoretical research on ecological compensation in TBRBs pays more attention to the theoretical framework of compensation mechanisms, and there is less research on the setting of the ECS. Yu Jiawen et al. explored the ECS for the Mekong River Basin through the ESV method, and obtained the actual compensation amounts that Thailand, Vietnam and Cambodia need to pay separately as the ecological compensation objects [32]. The base value of the ECS discussed in this paper is a range of intervals, and from the perspective of the overall basin. The compensation subjects and objects are discussed as two entities. In this paper, the setting of the base value of the ECS offers quantitative support for the development of the ecological compensation activities in TBRBs. Meanwhile, it provides a theoretical basis for the negotiation among BCs, which helps to clarify the ecological protection responsibilities of the BCs, reduces the occurrence of basin disputes, and promotes the sustainable development of TBRBs. In terms of the final ECS for TBRBs and the amount of compensation each BC should bear and accept, respectively, this paper believes that the final decision can be made by the BCs themselves through full negotiation according to the actual economic and social development of each BC, combined with their benefit degree, ability to pay, fund use efficiency and other aspects. In terms of specific compensation methods, this paper believes that various methods such as financial compensation, ecological financing, water rights trading, political coordination, unimpeded trade, technical support and industrial assistance can be adopted through negotiation by various BCs to promote the long-term movement of ecological compensation activities.

5. Conclusions

For the reason of reducing the conflicts caused by the inharmonious relationship between resource exploitation and ecological environmental protection of BCs, and realize the sustainable development of TBRBs, this paper divided BCs into ESSCs and ESCCs in view of the overall basin. The paper took them as two together to conduct one-to-one negotiation on setting the ECS of the basin, and discussed the basis of the two parties' negotiation—the base value of ECS. Firstly, a calculation model for the SVES of BCs in TBRB was constructed. The equivalent factor method was used to estimate the ESV of each BC, and the CESV of each BC was measured by the ecological consumption coefficient. On this basis, the SVES of each BC was obtained. The upper limit of the base value of the ECS was determined based on the SVES in ESCCs. Secondly, a calculation model of the CEP in the TBRB was constructed, and the lower limit of the base value of the ECS was determined according to the CEP in ESSCs. Finally, the feasible range for the base value of the ECS in the TBRB was determined by combining the adjustment coefficient for payment of the ESV and cost-sharing coefficient of each BC.

This paper took the LMRB as an example to conduct a case study, and the main conclusions were as follows.

(1)　The SVES in LMRB was Laos, China, Myanmar, Cambodia, Vietnam and Thailand, in descending order. Among them, Laos, China and Myanmar had positive SVES, and they were ESSCs. Cambodia, Vietnam and Thailand had negative SVES, and they were ESCCs.

(2)　The CEP of ESSCs in the LMRB was in descending order of Laos, China and Myanmar. Among them, China had the highest direct cost and Laos had the highest opportunity cost. The CEP of ESSCs in the LMRB is closely related to its own level of economic development. The higher the level of economic development, the higher the awareness and ability of ecological protection, and the more funds can be directly invested in ecological protection for the basin.

(3)   Based on the adjustment coefficient for payment of the ESV and the cost-sharing coefficient of each BC, the feasible range for the base value of the ECS in the LMRB was determined to be $[2.47, 229.67] \times 10^8$ \$, which provided the basis for the negotiation between the ESSCs and the ESCCs on the determination of the ECS.

Ecological compensation for TBRBs is a complex activity involving many stakeholders. In the specific implementation process, it involves various influencing factors and prerequisites such as communication and cooperation between various BCs, selection and operation of negotiation platforms, fair and effective determination of the ECS, and so on. Consequently, in the practical application of compensation, there is a relatively large difficulty in coordination. To successfully achieve the goal of ecological compensation in TBRBs, it is necessary to fully leverage the negotiation autonomy of each BC, and establish and improve a guarantee mechanism for conducting ecological compensation activities in the TBRB. Firstly, establish a basin-information-sharing mechanism and platform to regularly share information related to ecological resources, such as water and forest resources in the basin; secondly, establish an integrated management organization for the TBRB, and jointly formulate rules for ecological environment protection and ecological compensation activities, especially for the formulation of rules of procedure in special situations; thirdly, strengthen and improve the coordination and supervision model of ecological compensation in the TBRB, establish a platform for basin coordination and supervision, standardize the cooperation and coordination procedures and improve the regulatory system of ecological compensation.

However, because of the difficulty in obtaining basin information and related data, this paper still has certain limitations and can be further explored in future research. Firstly, the variability and development trend of the ESV in each BC can be considered, and exploration can be conducted from the perspective of value increment; secondly, when modifying parameters such as equivalent factor values, differential corrections can be made based on the individual situation of each BC; thirdly, when setting the base value of the ECS, the impact of cross-sectional water quality monitoring data on the determination of the ECS can be further explored. In the future, BCs need to strengthen the concept of integrated development, adhere to interest sharing and mutually beneficial cooperation, establish an information-sharing mechanism, and assist in the long-term development of the basin.

Author Contributions: Conceptualization, Y.Z.; methodology, Y.Z. and F.L.; software, F.L. and X.C.; validation, Y.Z. and X.C.; formal analysis, Y.Z. and X.X.; investigation, Y.C.; resources, F.L. and Y.C.; data curation, X.C. and X.X.; writing—original draft preparation, Y.Z. and F.L.; writing—review and editing, Y.C., X.C. and X.X.; funding acquisition, Y.C. and X.X. All authors have read and agreed to the published version of the manuscript.

Funding: This research was funded by the Major Projects of National Social Science Fund of the People's Republic of China (Project No. 17ZDA064), General Projects of the National Social Science Fund of the People's Republic of China (Project No. 21BGL289), Anhui Provincial Education Department Humanities Key Fund (No. SK2021A0652), Tongling College Talent Fund (2021tlxyr15).

Data Availability Statement: The data supporting the findings of this paper are available from the corresponding author upon reasonable request.

Acknowledgments: We thank all the members of the Major Projects of National Social Science Fund of the People's Republic of China (Project No. 17ZDA064), as well as the editors and reviewers for their valuable comments and suggestions.

Conflicts of Interest: The authors declare no conflict of interest.

Abbreviations

TBRB	Transboundary river basins
BC	Basin country
ECS	Ecological compensation standard
CEP	Cost of ecological protection
ESV	Ecosystem service value
CSEV	Consumption of ecosystem service value
ESSC	Ecosystem service supplying country
ESCC	Ecosystem service consuming country
SVES	Spillover value of ecosystem services
NPP	Net primary productivity
LMRB	Lancang-Mekong River Basin

References

1. Mccracken, M.; Wolf, A.T. Updating the Register of International River Basins of the world. *Int. J. Water Resour. Dev.* **2019**, *35*, 732–782. [CrossRef]
2. UNESCO. *United Nations World Water Development Report 2020: Water and Climate Change*; UNESCO: Paris, France, 2020.
3. He, D.; Liu, H.; Feng, Y.; Ni, G.; Kong, L.; Long, A.; Zhang, C. Perspective on theories and methods study of transboundary water resources under the global change. *Adv. Water Resour.* **2016**, *27*, 928–934.
4. Wen, Y.D. A Study on the Allocation of the Water Resources of Lancang-Mekong River. Ph.D. Thesis, Wuhan University, Wuhan, China, 2016.
5. Wang, Z.J.; He, Q.E. Impact of international rivers on national security. *J. Econ. Water Resour.* **2013**, *31*, 23–26+76.
6. Zhang, C.C.; Fan, Y.F. Study on the unbalanced benefits of transboundary water resources and conflict prevention measures. *J. Bound. Ocean Stud.* **2020**, *5*, 80–90.
7. Ge, Y.P.; Zhang, H. Evaluation and reference of Colorado River water resources allocation model. *J. Econ. Water Resour.* **2022**, *40*, 55–60+93.
8. Mu, G.L.; Wang, Y.J.; Li, L.; Ma, J.L.; Wang, J.G.; Tang, H.L. Development and application of the dynamic calculation model for proposing a water source eco-compensation standard. *China Environ. Sci.* **2018**, *38*, 2658–2664.
9. Zhang, H.O.; Song, Y.; Hao, M.L. Experience of international watershed ecological compensation (WEC) and research progress of WEC mechanism in China. *Environ. Dev.* **2020**, *32*, 232–233+235.
10. He, Y.M. Modes and its development of the equitable and reasonable use of international river water resources. *Resour. Sci.* **2012**, *34*, 229–241.
11. Hu, W.J.; Chen, J.W.; Zhang, C.C. Practices of the international cooperation across the Danube River basin and their inspirations. *Resour. Environ. Yangtze Basin* **2010**, *19*, 739–745.
12. Li, J.; Zhang, D. Study on Ecological Compensation Mechanism for Water Source—A Case Study on Heihe Reservoir of X'an. *Environ. Sci. Manage.* **2014**, *39*, 155–158. [CrossRef]
13. Thu, T.P.; Campbell, B.M.; Garnett, S. Lessons for pro-poor payments for environmental services: An analysis of projects in Vietnam. *Asia Pac. J. Public Adm.* **2009**, *31*, 117–133.
14. Wunder, S. The efficiency of payments for environmental services in tropical conservation. *Conserv. Biol.* **2007**, *21*, 48–58. [CrossRef]
15. Muenzel, D.; Martino, S. Assessing the feasibility of carbon payments and Payments for Ecosystem Services to reduce livestock grazing pressure on saltmarshes. *J. Environ. Manag.* **2018**, *225*, 46–61. [CrossRef] [PubMed]
16. Fu, R.M.; Miao, X.L. A new financial transfer payment system in ecological function areas in China: Based on the spillover ecological value measured by the expansion emergy analysis. *Econ. Res. J.* **2015**, *50*, 47–61.
17. Ren, Y.S.; Lu, L.; Zhang, H.M.; Chen, H.F.; Zhu, D.C. Residents' willingness to pay for ecosystem services and its influencing factors: A study of the Xin'an River basin. *J. Clean. Prod.* **2020**, *268*, 122301. [CrossRef]
18. Khan, I.; Zhao, M. Water resource management and public preferences for water ecosystem services: A choice experiment approach for inland river basin management. *Sci. Total Environ.* **2019**, *646*, 821–831. [CrossRef]
19. Chen, Y.P.; Cheng, Y.X. The eco-compensation tactics of basin with seriously disrupted water environment. *J. Huazhong Agric. Univ.* **2018**, *646*, 121–128.
20. Wang, Y.Q.; Li, G.P. The evaluation of the watershed ecological compensation standard of ecosystem service value: A case of Weihe watershed up-stream. *Acta Ecol. Sin.* **2019**, *39*, 108–116.
21. Jiang, Y.Q.; Chen, K. A review of researches on payment for watershed ecosystem services. *Ecol. Econ.* **2016**, *32*, 175–180.
22. Liu, Y.L.; Zhao, Z.X.; Sun, Z.L.; Wang, G.Q.; Jin, J.L.; Wang, G.X.; Bao, Z.X.; Liu, C.S.; He, R.M. Multi-objective Water Resources Allocation in Trans-boundary Rivers Based on the Concept of Water Benefit-sharing: A Case in the Lancang-Mekong River. *Sci. Geol. Sin.* **2019**, *39*, 387–393.
23. Li, F.; Wu, F.P.; Chen, L.X.; Xu, X.; Zhao, Y. Transboundary river water resource allocation based on weighted bankruptcy game model. *Sci. Geol. Sin.* **2021**, *41*, 728–736.

24. Latham, J.S.; He, C.; Alinovi, L.; DiGregorio, A.; Kalensky, Z. *FAO Methodologies for Land Cover Classification and Mapping*; Springer: New York, NY, USA, 2002.
25. Wang, Z.F.; Liu, J.G.; Li, J.B.; Meng, Y.; Pokhrel, Y.; Zhang, H.S. Basin-scale high-resolution extraction of drainage networks using 10-m Sentinel-2 imagery. *Remote Sens. Environ.* **2021**, *255*, 112281. [CrossRef]
26. Xie, G.D.; Zhang, C.X.; Zhang, L.M.; Chen, W.H.; Li, S.M. Improvement of the Evaluation Method for Ecosystem Service Value Based on Per Unit Area. *Nat. Resour. Res.* **2015**, *30*, 1243–1254.
27. Dugan, P.J.; Barlow, C.; Agostinho, A.A.; Baran, E.; Cada, G.F.; Chen, D.Q.; Cowx, L.G.; Ferguson, J.W.; Jutagate, T.T.; Mallen-Cooper, M.; et al. Fish Migration, Dams, and Loss of Ecosystem Services in the Mekong Basin. *Ambio* **2010**, *39*, 344–348. [CrossRef] [PubMed]
28. Delgado-Aguilar, M.J.; Konold, W.; Schmitt, C.B. Community mapping of ecosystem services in tropical rainforest of Ecuador. *Ecol. Indic.* **2017**, *73*, 460–471.
29. Liu, X.J. Main Land Vegetation Productivity and Their Relationship with Climatic Factors in China. Master's Thesis, Shanxi University, Taiyuan, China, 2019.
30. Gu, L.; Yue, C.R.; Zhang, G.F.; Zhao, X.; Jin, J. Temporal and Spatial Analysis of Vegetation NPP in the Greater Mekong Subregion Based on Google Earth Engine Platform from 2001 to 2019. *J. West China For. Sci.* **2021**, *50*, 132–139.
31. Zhou, Y.J. A Quantitative Study on Environmental Carrying Capacity and Ecological Compensation Standard in Urban Environmental Planning. Ph.D. Thesis, Huazhong University of Science and Technology, Wuhan, China, 2017.
32. Yu, J.W.; Long, A.H.; Deng, X.Y.; Liu, Y.D.; He, X.L.; Zhang, J. Ecosystem services and benefit compensation mechanism in the Mekong River Basin. *Trans. Chin. Soc. Agric. Eng.* **2020**, *36*, 280–290.
33. Hu, H.B.; Liu, H.Y.; Hao, J.F.; An, J. Spatio-temporal variation in the value of ecosystem services and its response to land use intensity in an urbanized watershed. *Acta Ecol. Sin.* **2013**, *33*, 2565–2576.
34. Su, F.J. The Analysis of Temporal and Spatial Variation Charateristics of Vegetation ner Primary Productivity and Its Influencing Factors in Lancang-Mekong River Basin from 2001 to 2020. Master's Thesis, Yunnan Normal University, Yunnan, China, 2022.
35. Li, M.C. Study of NDVI and Its Relationship with Precipitation over the Lancang-Mekong River Basin from 2000 to 2017. Master's Thesis, Yunnan Normal University, Yunnan, China, 2019.
36. Food and Agriculture Organization (FAO) Database. Available online: http://www.fao.org/faostat/en/#data (accessed on 5 April 2023).
37. Transboundary Waters Assessment Programme (TWAP) Database. Available online: http://twap-rivers.org/#home (accessed on 5 April 2023).
38. Transboundary Freshwater Dispute Database (TFDD). Available online: https://transboundarywaters.science.oregonstate.edu (accessed on 5 April 2023).
39. Mekong River Commission. Available online: http://ffw.mrcmekong.org/weekly_report/ (accessed on 5 April 2023).
40. World Bank. Available online: https://data.worldbank.org (accessed on 5 April 2023).
41. Li, F. A Two-Level Allocation Model of Trans-Boundary Water Resources from the Perspective of Cooperation. Ph.D. Thesis, Hohai University, Nanjing, China, 2021.
42. Zhang, Z.F. The construction of Water Quantity Trading Mechanism of International Rivers Based on Water Resources Conflict. Ph.D. Thesis, Hohai University, Nanjing, China, 2020.
43. Yunnan Provincial Bureau of Statistics. *Statistical Yearbook of Yunnan Province*; China Statistics Press: Beijing, China, 2018–2022.
44. Xie, G.D.; Lu, C.X.; Leng, Y.F.; Zheng, D.; Li, S.C. Ecological assets valuation of the Tibetan Plateau. *J. Nat. Resour.* **2003**, *18*, 189–196.
45. Xie, G.D.; Zhang, C.X.; Zhang, C.S.; Xiao, Y.; Lu, C.X. The value of ecosystem services in China. *Resour. Sci.* **2015**, *37*, 1740–1746.
46. Yang, S.C.; Liu, W.W. Evaluating the influence of hydropower development on ecosystem service value based on value equivalence: Taking Gannan Jiudianxia as example. *J. Cent. South Univ.* **2018**, *24*, 78–85.
47. Yang, Y.; Meng, G.F. A bibliometric analysis of comparative research on the evolution of international and Chinese ecological footprint research hotspots and frontiers since 2000. *Ecol. Indic.* **2019**, *102*, 650–665. [CrossRef]
48. Guan, D.J.; Jiang, Y.N.; Yan, L.Y.; Zhou, J.; He, X.J.; Yin, B.L.; Zhou, L.L. Calculation of ecological compensation amount in Yangtze River Basin based on ecological footprint. *Acta Ecol. Sin.* **2022**, *42*, 1–15.
49. Zhang, H.N.; Ge, Y.X.; Jie, Y.M. Research on Basin Ecological Compensation Mechanism of Main Functional Areas. *Mod. Econ. Res.* **2017**, 83–87. [CrossRef]
50. Gao, H.Z.; Liu, H.; Xu, F.R.; Zhang, C.L.; Li, X. Calculation of trans-provincial water ecological compensation standard and allocation of funds based on entropy weight method: Case study of Dongjiang River Basin. *J. Econ. Water Resour.* **2021**, *39*, 72–76+80.
51. Li, J.C.; Jiang, W.L. *Ecological Axiology*; Chongqing University Press: Chongqing, China, 1999.
52. Dai, M.; Liu, Y.N.; Chen, L.J. The study on quantitative standard of eco-compensation under major function-oriented zone planning and opportunity cost. *J. Nat. Resour.* **2013**, *28*, 1310–1317.
53. Piao, J.Y.; Li, Z.F. Water cooperation governance: New issues of the regional relationship construction in the Lancang-Mekong River Basin. *Southeast Asian Stud.* **2013**, *5*, 27–35.

54. Li, Y.Q.; Li, Z.P.; Dai, M.L.; Wei, L.Y. Effect evaluation of emergency water supplement from cascade reservoirs on Lancang River to Mekong River in 2016. *Yangtze River* **2017**, *48*, 56–60.
55. Huang, X.S.; Zheng, R. Compensation principle of the beneficiaries of the transboundary rivers. *Resour. Environ. Yangtze Basin* **2012**, *21*, 1402–1408.

Article

Groundwater Management for Agricultural Purposes Using Fuzzy Logic Technique in an Arid Region

Amjad Al-Rashidi *, Chidambaram Sabarathinam, Dhanu Radha Samayamanthula, Bedour Alsabti and Tariq Rashid

Water Resources Development and Management Program, Water Research Center, Kuwait Institute for Scientific Research, Safat 13109, Kuwait; csabarathinam@kisr.edu.kw (C.S.); vdhanuradha@kisr.edu.kw (D.R.S.); bsabti@kisr.edu.kw (B.A.); trashed@kisr.edu.kw (T.R.)
* Correspondence: arashidi@kisr.edu.kw

Abstract: The study aimed to determine groundwater's suitability for irrigation and cattle rearing in Kuwait. In this regard, groundwater samples were collected from Umm Al Aish (UA) and adjoining Rawdhatain (RA) water wellfields to develop groundwater suitability maps for irrigation purposes using the fuzzy logic technique in ArcGIS. RA was dominated by Na-Cl, Na-Ca, and Ca-SO$_4$ water types, whereas UA was dominated by the Ca-Mg water type. Due to the influence of the temperature and pCO$_2$, the carbonates were inferred to be more susceptible to precipitation in the soil than the sulfates. The ternary plots for both regions revealed that the samples' suitability ranged from good to unsuitable. Spatial maps of nine significant parameters governing the irrigation suitability of water were mapped and integrated using the fuzzy membership values for both regions. The final suitability map derived by overlaying all the considered parameters indicated that 8% of the RA region was categorized as excellent, while UA showed only 5%. Samples situated in the study areas showed an excellent to very satisfactory range for livestock consumption. Developing a monitoring system along with innovative water resource management systems is essential in maintaining the fertility of the soil and existing groundwater reserves.

Keywords: groundwater suitability; irrigation; fuzzy logic; GIS; water quality; water management; Kuwait; arid region

check for updates

Citation: Al-Rashidi, A.; Sabarathinam, C.; Samayamanthula, D.R.; Alsabti, B.; Rashid, T. Groundwater Management for Agricultural Purposes Using Fuzzy Logic Technique in an Arid Region. *Water* 2023, *15*, 2674. https://doi.org/10.3390/w15142674

Academic Editor: Saseendran S. Anapalli

Received: 13 June 2023
Revised: 20 July 2023
Accepted: 23 July 2023
Published: 24 July 2023

1. Introduction

The lack of freshwater availability is the primary challenge in the world's driest regions. Water quality is quickly degrading because of the combined effects of the dry climate and global warming. Climate change has an immediate impact on the amount and quality of groundwater/surface water in arid and semi-arid regions [1,2]. Studies on long-term climate change, adopting modeling techniques using land use–land cover and meteorological parameters, have also yielded similar observations [3]. Additionally, the water quality is declining due to the rising demand for human consumption and agricultural purposes. These needs cannot be satisfied by existing water sources, where water quality is also crucial for farming in many deserts and semi-arid areas [4]. In arid regions such as Kuwait, evaporation rates are higher due to increased temperatures, lesser rainfall, and a lack of surface water bodies, and the dependency on groundwater has increased. Groundwater dependence has not only resulted in increased salinity but has also depleted the freshwater resources in arid regions [5], thus affecting its utility for agricultural purposes. Hence, agriculture in arid regions mainly focuses on the availability of sustainable freshwater resources. Therefore, the need for water in an arid region is heterogeneous; it varies with utility, availability, and proximity. However, due to increased rates of evaporation and poor precipitation, groundwater often becomes the main source for irrigation, despite its inferior quality. Because of this, the use of low- to medium-quality groundwater for irrigation is becoming increasingly important [6]. The increased use of agrochemicals, variability of land use, changes in climate, increase in population, etc., are a few major factors affecting the groundwater quantity and quality, especially in arid regions [7]. Further, due to the presence of certain ions at unsafe/harmful levels, groundwater that is not fit for industrial or drinking purposes may be fit for irrigation purposes [8,9].

As intensive crop production and irrigated agriculture [10] are rapidly increasing the water demand, the current groundwater requirements for crop production cannot keep pace with the demand [11]. This is because farmers, specifically in arid regions, are forced to use brackish to saline groundwater for irrigation purposes, which has a high concentration of dissolved salts. In most cases, this leads to crop failure and the development of saline or sodic soils, requiring expensive remediation to restore their productivity. In fact, studies have indicated that the extensive practice of pumping brackish groundwater for agricultural purposes has led to enhanced conductivity values and a drop in groundwater levels, thus affecting the suitability of the region's future irrigation practices [12,13]. Hence, the irrigation suitability should be assessed when the low-quality water consumption rises.

Kuwait has three agricultural regions: Abdally, Wafra, and Kabd. Brackish groundwater is used for agriculture in these agricultural regions, with treatment in certain farms. The brackish groundwater serves as a substitute for irrigation due to the absence of fresh groundwater resources in Kuwait. This practice of brackish groundwater utilization for agriculture has been reported globally for food safety [14]. Since brackish groundwater serves as a source, the type and variety of crops cultivated are limited due to the higher salinity [15]. Nevertheless, when brackish/saline water is used in an innovative manner, it may contribute to the production of a variety of salt-tolerant crops. Therefore, brackish groundwater should be used with caution considering the chemical constituents and their concentration levels in water [16]. The treated water also results in the generation of brines, which affect the environment; hence, the cultivation of plants tolerant to the salinity of groundwater in arid regions is necessary [17].

In Kuwait, the Rawdhatain field (RA) and Umm Al Aish field (UA) are the only known groundwater fields with exploitable freshwater lenses. About 40 years ago, the groundwater in RA and UA was reported to range from 205 mg/L to 700 mg/L of TDS [18]. Later, the RA and UA freshwater lenses were protected with the study of the local conditions, which included the catchment boundary, size, lithology, rainfall rates, and drainage patterns [19]. Paleoenvironmental studies indicated that the wadis formed under varied environmental conditions in the Pleistocene transported rainfall runoff and infiltrated to form the freshwater lenses [19]. The wide catchment area and higher percolation rates have led to enormous volumes of water to recharge the lenses, despite the high temperatures of the region. The long-term assessment of groundwater in the RA and UA regions indicated that a small number of wells around UA were strongly contaminated with hydrocarbons [20], and these could reach parts of the RA field if no necessary action was taken to prevent this contaminated plume migration [21]; this was also later confirmed by groundwater modeling studies [22].

Al-Rashed [23] found that the Sabriya Oil Fields' seawater pits infiltrated to the groundwater and thus seawater ingression impaired the groundwater by increasing its salinity. Hence, these studies suggest that the groundwater of the region needs to be protected from the infiltration of contaminants. Further, Mukhopadhyay [24] reported that the infiltration of surface runoff and the leaching of salts and hydrocarbons from the surface soil resulted in high groundwater TDS levels in Northern Kuwait.

It is worth noting that regions with different groundwater quality are to be mapped both spatially and temporally in lateral and vertical dimensions to ascertain the use and manage the resources strategically. A geographic information system (GIS) is a powerful tool for the monitoring and management of groundwater resources at a local or regional level, the analysis of water quality, and to provide tangible solutions [25,26]. The basic chemical parameters in determining the groundwater's characteristics and its appropriateness for irrigation purposes include the electrical conductivity (EC), sodium percentage (Na%), sodium adsorption ratio (SAR), Kelly's ratio (KR), magnesium hazard (MAR), residual sodium carbonate (RSC), permeability index (PI), potential salinity (PS), and soluble sodium percentage (SSP). Many authors have investigated the irrigation suitability of groundwater (ISGW) based on some of the above-mentioned parameters [27–30]. Several studies have used fuzzy logic to assess land suitability for yield prediction [31], regions with promising irrigation water quality [32,33], and regions with fresh groundwater resources suitable for drinking purposes [34]. Apart from spatiotemporal variation studies, a ternary plot has also been used to integrate four different water quality parameters for the assessment of irrigation suitability [30].

A bibliometric review on the Scopus database using the keywords (groundwater, irrigation, fuzzy, and GIS) identified 25 research articles. The retrieved file was checked for duplicate research articles and similar words were merged to derive a meaningful representation in the plot (Figure 1). The network-based visualization map of the co-occurrence of keywords in the selected articles, was obtained with the VOSviewer software [35]. Later, filtering the articles with a minimum of five occurrences resulted in 14 frequently used keywords.

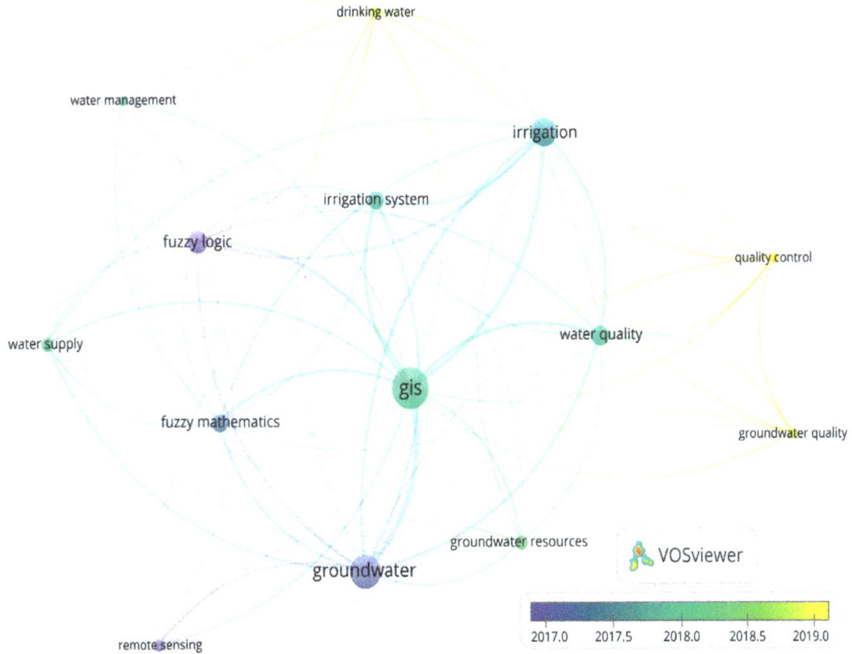

Figure 1. A network-based visualization map derived from a bibliometric review on the Scopus database representing the frequency of keywords and their linkages. The size of the circles and the thickness of the lines are proportional to the number of uses and the frequency of the linkages, respectively. The colors reflect the chronology of the usage and their linkages.

Irrigation-related groundwater studies using fuzzy logic techniques initially focused on groundwater using remote sensing techniques. Later, studies concentrated more on irrigation, GIS, groundwater resources, water supply, water management, and water quality. Recently, more emphasis has been given to drinking water, groundwater quality, and quality control. Hence, it could be inferred that fuzzy logic techniques have been very recently adopted in studying the irrigation suitability of groundwater, as the oldest work in the database was reported by Dixon [36] for groundwater vulnerability assessment, followed by studies with an emphasis on irrigation systems and GIS. Recently, more studies on fuzzy logic have aimed to unravel the ISGW and focused on drinking water purposes by involving more related parameters through the analytic hierarchical process (AHP) [32,37,38].

Although the freshwater reserves are being preserved in the RA and UA wellfield regions, the quality of the resources is reported to be deteriorating [19,39]. Further, there are no detailed studies investigating the quality and suitability of brackish and fresh groundwater lenses for irrigation purposes. Thus, studying the hydrogeochemical characteristics of the region and identifying the wells/zones for ISGW is essential for the development and application of groundwater resources. Hydrochemical characterization associated with hydrological inferences is needed to evaluate and manage the groundwater resources [40]. Thus, the formation of policies and governance regarding the risk of damaging the potential aquifers and available freshwater resources are actions that urgently need to be executed. Hence, water management considering the future needs of the region is critical in arid regions for strategic planning.

Study Aims and Objectives

Water is one of the main focuses of the United Nations Sustainable Development Goals (SDG), particularly for SDG 6 (clean water and sanitation), and it also relates to agriculture, i.e., SDG 13 (climate change) and SDG 15 (life on land). Thus, the current study considered the pursuit of these SDGs by using a fuzzy GIS technique to develop a map of groundwater quality and ISGW. The study also attempted to identify areas suitable for the purpose of groundwater irrigation and cattle rearing, along with optimizing the groundwater utilization in the study area.

2. Materials and Methods

2.1. Study Area

Kuwait is an arid region, receiving average rainfall varying from 110 to 120 mm [41]. During the country's dry hot summers (May to September), the temperature reaches 45 to 50 °C, while, in the winter (December to February), the temperature can fall below 10 °C [42]. The study area's RW and UA fields are located north of Kuwait (Figure 2) and are bounded by latitudes 29°46′0″ to 29°46′0″ N and longitudes 47°38′0″ to 47°52′0″ E. The RA and UA regions, adjoining each other, are reported to have the only known economically viable freshwater lenses in Kuwait [24]. The RA and UA regions cover an area of 53 km^2 and 44.5 km^2, respectively. The Rawdhatain Oilfield lies to the southeast of the Rawdhatain groundwater field, whereas a portion of the Sabriya Oilfield is situated NE of the Umm Al Aish groundwater field.

The tertiary sediment sequence (Paleogene and Neogene) of Kuwait has been divided into two main groups [43]. The formation of the Neogene Aquifer System, which is also referred to as the Kuwait Group Aquifer, occurred during the Pleistocene pluvial. It is noteworthy that, during this phase, Wadi Ar-Rimah and Wadi al Batin [44] were the mainstream channels. The Dibdibba Formation (Upper Miocene to Pliocene), Fars Formation (Middle Miocene), and Ghar Formation (Lower Miocene) are the three formations that make up the Neogene Aquifer System. Moreover, the oldest group that is found to be overlain by the Kuwait Group is the Hasa Group, where it consists of the Dammam Formation, Umm Er Radhuma Formation, and Rus Formation [45]. Additionally, the Hassa Group, which is made up of Paleogene rocks, and the Neogene Formations, which are overlain by Quaternary sediments, are separated by a disconformity layer [44]. The Dibdibba Formation is characterized by the presence of fresh groundwater lenses in the northern area of Kuwait. Further, the Dibdibba Formation is generally defined by two main units. The first unit is the Pliocene–Pleistocene (upper unit), composed of gravel, sand, and gypsiferous cement. The second unit is the Miocene–Pliocene (lower unit), which contains pebble-sized sandstone cemented with chalky carbonates [22].

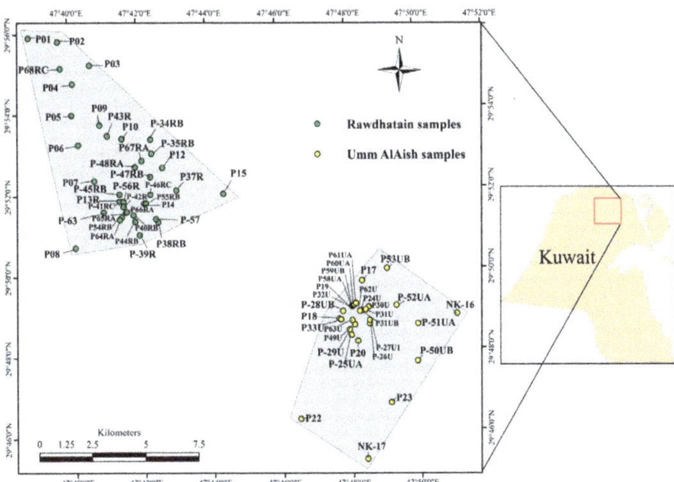

Figure 2. Map of sampling locations in the studied groundwater fields of Rawdhatain (RA) and Umm Al Aish (UA) regions. The northwestern region in the figure indicates the RA wellfield and the southeastern region represents the UA wellfield.

In general, the Neogene Aquifer System is considered an unconfined aquifer, although, in deeper layers of the aquifer, confined conditions may exist. The Neogene Aquifer System was recharged during the cold–humid period based on isotopic data, and it dates back to 30,000 years ago [22,44]. An analysis of more than 75 pumping tests carried out by the Ministry of Electricity and Water (MEW) in the study area for aquifer evaluation suggests that the transmissivity ranges between eighty thousand and twenty-six thousand gallons per day per foot, with a storage coefficient of 5.1×10^{-6} to 0.13×10^{-4} [46]. Hence, the transmissivity values that have been documented in Kuwait from the southwest to the northeast are 1.15×10^{-4} m^2/s and 1.73×10^{-2} m^2/s, respectively [22]. In addition, the hydraulic parameters of the same system vary widely according to the saturated thickness and lithological differences, ranging from 0.24 to 21 m per day, with an average value of 7.40 m per day [47].

For the studied wells in RA, the water level ranges from 29.4 to 38 mbgl, and the well depth ranges from 31 to 76 m. In the UA wells, the water level ranges from 17.3 to 29.1 mbgl, with a well depth of 22.4 to 57 m, covering only the top part of the Kuwait Group in both fields. The water level ranges from 2.4 to 8.4 m (amsl) in RA and from 6.2 to 16.5 m (amsl) in UA, reflecting the change in topography, where it has been recorded that the groundwater in Kuwait flows from SW to NE [48].

2.2. Sample Collection and Analysis

Thirty UA groundwater wells and thirty-eight RA groundwater wells were chosen for sampling. Using disposable Teflon bailers, a total of 68 samples were collected from the monitoring wells. All the samples were transported to the Water Research Center (WRC), Kuwait Institute for Scientific Research (KISR), and filtered before analysis. Temperature, electrical conductivity (EC), and pH analysis was conducted onsite and in the WRC laboratory by means of calibrated meters of a portable type. The analysis of total dissolved solids (TDS), cations of Ca^{2+}, K^+, Mg^{2+}, Na^+ type, and anions of HCO_3^-, Cl^-, Br^-, F^-, NO_3^-, and SO_4^{2-} were determined using the Standard Methods for the Examination of Water and Wastewater (SMEWW) and American Society for Testing and Materials (ASTM) standard methods. All the major ions (except HCO_3) were analyzed with an ion chromatograph; calibration, duplication, standard checks, and certified reference materials were applied to the analytes as part of the quality control (QC) and quality assurance (QA). The HCO_3 was determined by the sulfuric acid titration method.

Ternary plots were developed for both the sampling locations using the SAR, EC, Na%, and PI values in AQUACHEM. The values of the parameters used for the study were calculated using the CHIDAM software [49], considering the analytical data of groundwater from both the wellfields. The saturation indexes (SIs) of carbonate and sulfate minerals, and their variation with respect to a temperature change from 5 to 50 °C, were determined in the PHREEQC software [50]. Based on the literature review summarized in a recent study [30], nine typical parameters, namely EC, Na% [51], SAR [52], KR [53], MAR [54], RSC [52], PI [55], PS [56], and SSP [57], were selected in the current study to validate the irrigation suitability of the groundwater. The recommendations of the Food and Agriculture Organization (FAO) [58] were considered to study the groundwater's suitability for livestock.

2.3. Geospatial Analysis Using Fuzzy Membership

Fuzzy logic membership [59] is a technique that enables the semantic descriptions of experts to be transformed into a numerical spatial model that forecasts the suitability zones of a certain parameter. Using a scale of 0 to 1, the fuzzy membership technique evaluates the input data depending on how likely they are to be part of a detailed set [60].

Any of the functions and operators of the Spatial Analyst extension tool in ArcGIS can be used to change the input data, reclassifying them to a 0 to 1 scale. The values of fuzzy membership for the current study were obtained by adopting the fuzzy linear membership function. The spatial maps of the fuzzy membership were developed using the inverse distance weighted (IDW) interpolation method. Using the Fuzzy Overlay tool, all the fuzzy membership maps were merged into a single integrated map for the studied regions. The ArcGIS software was also used to plot the groundwater level, ground elevation, and depth to water level along with the flow direction of the groundwater. The fuzzy gamma 0.9 operator was used in this study to overlay the maps, due to its ability to vary the amount of decreasing and increasing effects [34,61]. The schematic approach of fuzzy GIS adopted for the development of the map for ISGW is represented in Figure 3.

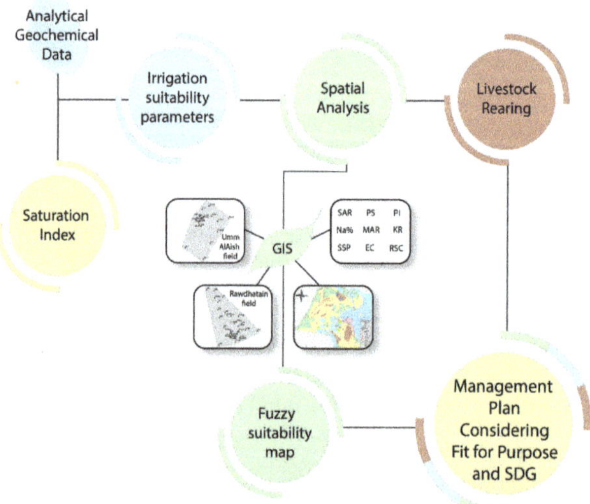

Figure 3. Flowchart depicting the methodology used to derive a groundwater suitability map for irrigation purposes using fuzzy GIS method and agricultural management plan.

3. Results and Discussion

3.1. Hydrochemistry

The maximum, minimum, and average values of all the chemical constituents of the groundwater samples for the RA and UA fields are listed in Table 1.

Table 1. The results of physical and chemical analyses of the groundwater samples (mg/L is the unit for all parameters except electrical conductivity (EC) (μS/cm) and temperature ($^\circ$C)).

Parameter	Rawdhatain (RA), n = 38			Umm Al Aish (UA), n = 30		
	Max	Min	Avg	Max	Min	Avg
Ca^{2+}	639	14.0	201	2802	35.3	349
Mg^{2+}	121	1.10	30.0	365	1.10	48.8
Na^+	2932	21.2	502	6828	16.0	666
K^+	22.3	0.30	5.00	39.8	3.00	11.2
Cl^-	2385	16.0	514	15,244	30.0	1191
HCO_3^-	318	19.0	131	420	13.0	168
NO_3^-	113	1.50	38.1	94.3	1.20	23.0
PO_4^{3-}	1.10	<0.01	0.10	1.70	<0.01	0.20
SO_4^{2-}	5042	21.0	857	4039	72.1	761
TDS	9040	328	2196	27,821	397	3078
EC	14,130	512	3222	43,470	621	4809
pH	9.20	6.80	7.70	8.80	6.90	7.60
Temperature	32.0	26.0	29.4	31.0	28.0	30.0

The pH of the RA samples ranged from 6.80 to 9.20, while that for UA ranged from 6.91 to 8.80. Although the RA samples tended to be more alkaline, both the study areas had a similar average pH, with 7.60 and 7.70. Groundwater tends to be alkaline, owing to the presence of HCO_3 percolated through rainfall runoff and aided by eroded soil [62]. The TDS of the RA and UA fields ranged from 328 to 9040 mg/L and 397 to 27,821 mg/L, respectively. There are a multitude of causes of the high TDS concentrations in groundwater, but the marine influence on geological formation and sabkhas is the most plausible explanation (the Arabic word sabkha describes both coastal and inland salt flats with the same meaning). Sabkha soil typically contains four to six times the amount of salt found in the seawater of the region [63]. Moreover, a dry climate, increased irrigation, and less rainfall recharge also contribute to elevated TDS levels [64]. The salinity also increases when water infiltrates a salty surface to reach the water table.

3.2. Mechanism Controlling the Type and Chemistry of the Groundwater

Earlier studies on the hydrochemical classification of UA groundwater performed by Robinson and Al-Ruwaih [18] identified the following four major groundwater groups: group 1, calcium bicarbonate water type in which Na > Cl; group 2, calcium chloride water type in which Na > Cl; group 3, calcium sulfate water type in which Na > Cl; group 4, calcium sulfate water type, with Cl > Na. In the present study, however, the southern and eastern parts of the RA and UA fields indicated the predominance of the Na water type, whereas the western and the northern areas were represented by the Ca water type (Figure 4). RA is mostly dominated in the east by Na-Cl, and in the north and south by Na-Ca and Ca-SO$_4$ water types. However, Na-Ca is the dominant water type throughout the UA area, except for the southwestern part, where it is dominated by the Ca-Mg water type.

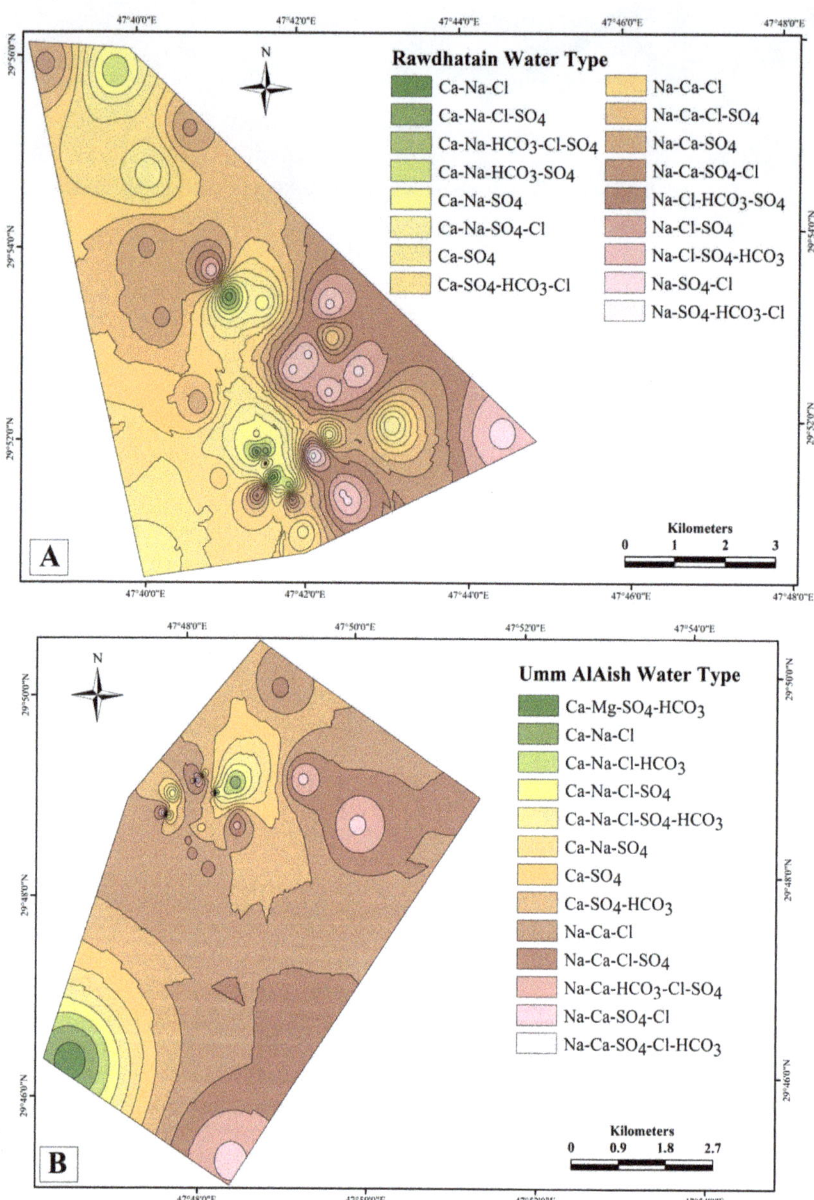

Figure 4. Spatial map showing the different water types of (**A**) RA and (**B**) UA fields, where the eastern part is dominated by Na-Ca and Na-Cl water types in both wellfields.

By exchanging Ca^{2+} for Na^+, the water is softened [65]. A 2001 study by Hidalgo and Cruz-Sanjulián [66] stated that the cation exchange mechanisms are typically indicated by the $Na-HCO_3$ water type [67]. The term "exchange waters" is used to describe bodies of water in which the HCO_3^- concentration is greater than that of the alkaline-earth cations $(Ca^{2+} + Mg^{2+})$ [67]. As a result of the exchange interaction with the exchange sites, the Na^+ ions together with excess bicarbonate ions are released into the groundwater.

A correlation was found between the Na-SO$_4$, Ca-Cl, and Ca-SO$_4$, water types and samples with NO$_3^-$ pollution. Groundwater samples of the Na-Ca-Cl-SO$_4$ and Ca-SO$_4$-Cl type were mainly due to interactions between rock and water (gypsum dissolution), and the exchange of ions. Both evaporation and the ion exchange process influenced the facies, as shown by the Na-Ca-SO$_4$-Cl and Na-Ca-Cl-SO$_4$ types.

3.3. Saturation Index

Compounds with values exceeding the solubility limits are most likely to precipitate, especially sulfates and carbonates. An increase in ions has a direct relationship with the saturation of minerals. However, the mixed precipitation of salts has not been determined with a definite pathway due to the non-availability of thermodynamic data for mixed or co-precipitation condition [68,69]. Authors have indicated that the presence of minor amounts of one compound may hinder the precipitation kinetics of another compound [68,69]. The solid form of a compound/mineral may contribute ions to the salt formation by the process of dissolution or behave similarly to a germination seed or as an adsorbent [70,71].

In general, the irrigated water salinity is related to biomass production and evapotranspiration. The yield of a crop is noted to be higher when there is a decrease in salinity and an increase in evapotranspiration [72,73]. The pores between the soil particles are generally clogged due to the precipitation of salts in these spaces, thereby increasing the bulk density. Thus, the soil salinity and the crop yield are mainly influenced by salt formation [74,75]. This situation is common in arid regions such as Kuwait, with increased rates of evaporation and irrigation with brackish water. The prolonged practice of using brackish groundwater for irrigation increases the salinization of the soil, thus affecting the permeability and crop yields. In this regard, the SI of a mineral provides the probability of salt precipitation on the soil based on the groundwater composition. Hence, the saturation states of carbonates (calcite, aragonite, and dolomite) and sulfates (gypsum and anhydrite) were determined for the analyzed samples.

The minerals of carbonate, aragonite, calcite, and dolomite, along with sulfate minerals like anhydrite and gypsum, were studied for variations in the SI values with respect to annual temperature fluctuations in Kuwait (5–50 °C) (Figure 5). The SI of anhydrite in UA (Figure 5A) at 5 °C varied from >−2.5 to <0.4, but an increase in temperature to 50 °C in the groundwater tends to increase the SI to range from >−2.0 to saturation around zero values. However, no significant variation was noted with the gypsum composition. All carbonate minerals reflected an increase in SI values with an increase in temperature from 5 to 50 °C. The lowest SI value for aragonite at 5 °C was noted to be −0.3, and, at 50 °C, it was calculated as 0.01. Similarly, the value of aragonite increased from 0.6 to 0.8 with an increase in temperature from 5 to 50 °C. Relatively, a greater increase in the SI of calcite was observed for a similar increase in temperature. The highest increase in the SI with regard to a rise in temperature was noticed for dolomite, tending towards supersaturation at 50 °C. Hence, there is a greater probability of the precipitation of dolomite at higher temperatures, followed by calcite and aragonite. However, at low temperatures, calcite tends to precipitate first, followed by aragonite and then dolomite. The SIs of sulfate minerals, irrespective of temperature variation, are noted to shift between saturated and undersaturated states, so they are less likely to be precipitated in soil. Groundwater samples from RA also showed a similar trend (Figure 5B) for all the studied minerals, but with relatively smaller ranges of SIs at both higher and lower temperatures compared to UA.

The partial pressure of carbon dioxide (pCO$_2$) governs the formation of carbonate salts along with the chemical composition of the groundwater and the temperature. The samples from the RA and UA fields had log pCO$_2$ values above the atmospheric value ($10^{-3.5}$) [76] of pCO$_2$ computed for groundwater samples, ranging from $10^{-4.46}$ to $10^{-1.48}$ and from $10^{-4.8}$ to $10^{-1.61}$, respectively. The decomposition of organic matter and plant root respiration in the top soil leads to an increase in the pCO$_2$ of the infiltrating water compared to that of the atmosphere ($10^{-3.5}$). Thus, the soil carbon dioxide may easily enter the groundwater system [77]. The pCO$_2$ value in a closed system is $>10^{-2.5}$ [78,79] and it governs the saturation states of carbonate minerals in groundwater. The pCO_2 of the Kuwaiti rainwater studied for a period of five years (2018–2023) ranged from $10^{-3.81}$ to $10^{-2.29}$, with an average of $10^{-2.86}$ in 2018; it ranged from $10^{-3.50}$ to $10^{-2.05}$ with an average value of $10^{-2.86}$ in 2019 [42], and from $10^{-4.29}$ to $10^{-2.09}$ with an average value of $10^{-3.15}$ in 2022, reflecting a higher value than the reported atmospheric pCO_2 ($10^{-3.5}$), indicating that carbonates remained in the solution due to the regional atmospheric pCO_2 levels, in certain samples from the study area. However, at higher atmospheric temperatures and under the variation in the thermodynamic nature of the solution, carbonates are forced to precipitate.

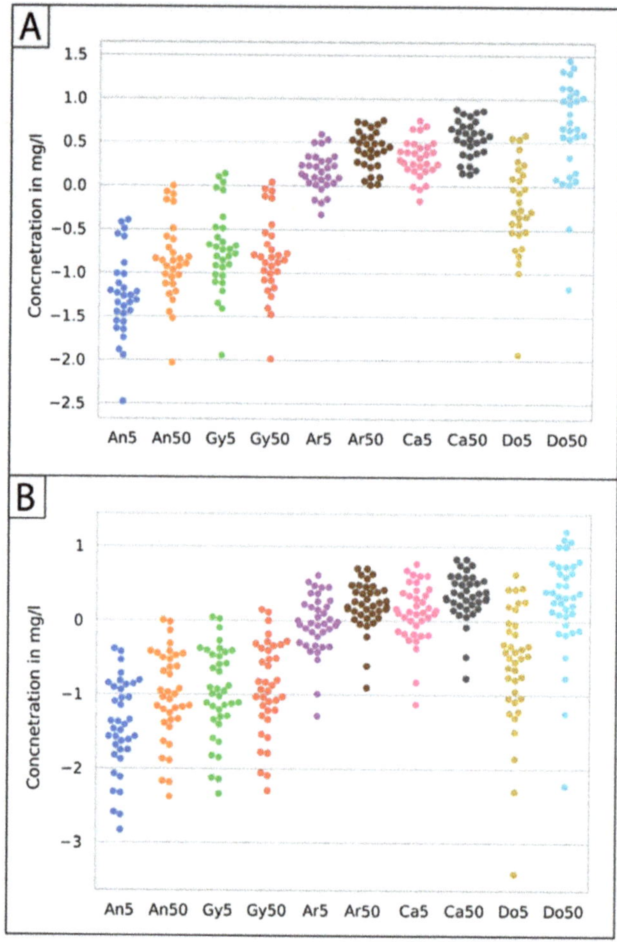

Figure 5. The variations in saturation index (SI) values with respect to annual temperature (5–50 °C) for sulfate (An = anhydrite, Gy = gypsum) and carbonate (Ar = aragonite, Ca = calcite, Do = dolomite) minerals in (**A**) UA and (**B**) RA.

The undersaturation of calcite, dolomite, and aragonite was observed in the samples due to the higher atmospheric pressure of the dissolved CO_2, irrespective of the wellfield. The dissolution of carbonates is prominent under pCO_2-rich conditions and the reverse reaction precipitates $CaCO_3$ when CO_2 escapes from the solution, either due to variations in pressure or variations in temperature, leading to the lowering of the pH. The pCO_2, a significant factor in weathering, favors carbonate weathering, which is responsible for the higher HCO_3 concentration [80]. As a result of a rise in HCO_3^-, Mg^{2+}, and Ca^{2+}

concentrations due to evaporation and mineral breakdown, groundwater becomes slightly supersaturated with regard to dolomite and calcite [80].

Due to the presence of secondary minerals inside the host rock, such as calcite and dolomite, the geochemistry of groundwater is more complicated. Cation exchange between groundwater and clay minerals and the dissolution and precipitation of secondary carbonate minerals impact the groundwater chemistry. High levels of bicarbonate in some waters might cause toxicity due to a lack of iron [81]. $CaCO_3$ precipitation from these groundwaters reduces the dissolved Ca^{2+} concentrations, boosting the SAR and thereby the soil exchangeable Na^+ [81]. In the study, the groundwater was subsaturated with gypsum and anhydrite, which is consistent with the lack of evaporites in the area, except for the presence of gypsum in the Lower Fars and Ghar Formation [82]. The disintegration of the gypsiferous formation of the aquifer is the source of the salinity [83]. The precipitation of gypsum is rather straightforward as it depends on the Ca^{2+} and SO_4^{2-} concentrations at a given pressure and temperature. The concentration of Ca^{2+} may be altered by the removal of this ion during the formation of carbonates, as it leads to the common ion effect [84].

3.4. Irrigation Suitability Plot

In general, if the EC of groundwater ranges from 1000 to 2000 μS/cm, it is likely to be of good quality. Apart from EC, parameters that indicate its suitability for agricultural purposes are Na%, SAR, and PI. The samples of RA represented four categories, reflecting a few samples with low suitability for agricultural purposes. Similarly, samples of UA were also represented in all four categories, but only one sample was observed in the unsuitable category (Figure 6) as the EC was >10,000 μS/cm. It was also noted that although the EC was higher in a few samples, the samples fell under the moderate, good, and doubtful categories. Moreover, it was observed that samples with EC ranging from 2000 to 5000 μS/cm and 5000 to 10,000 μS/cm were classified as good to moderate, which could be due to their higher Ca^{2+} and Mg^{2+}. Similarly, samples falling into the doubtful to unsuitable categories had higher Na% [30].

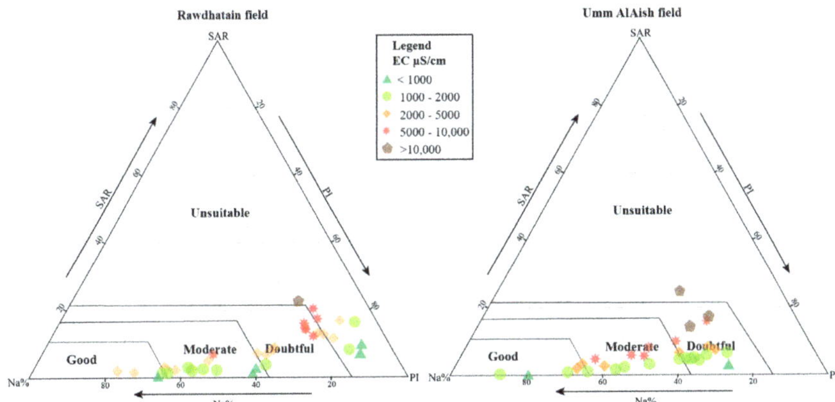

Figure 6. Ternary plot of RA and UA groundwater samples to determine suitability for irrigation purposes using EC, sodium percentage (Na%), sodium adsorption ratio (SAR), and permeability (PI).

3.5. Fuzzy GIS Maps and Irrigation Groundwater Quality Indices

The ISGW parameters used for the classification and determination of the fuzzy membership values were determined based on previous studies [30] (Table 2). In RA, based on the EC classification [85], only 7.8% of the samples were categorized as good, 36.8% were permissible, 36.8% were doubtful, and 18.4% were unsuitable. The EC map showed high fuzzy membership values in the southern part, which were identified as unsuitable, with small patches of lower values reflecting good suitability in the north (Figure S1).

Table 2. Summary of the classification of fuzzy membership values adopted for the determination of irrigation water quality indices.

Parameter	Result	Rawdhatain		Result	Umm Al Aish	
		Classification	Fuzzy Membership Classification		Classification	Fuzzy Membership Classification
EC	512–750	Good	<0.017	<750	Good	<0.003
	750–2250	Permissible	0.017–0.128	750–2250	Permissible	0.003–0.037
	>2250	Doubtful	0.128–1	2250–5000	Doubtful	0.037–0.108
				>5000	Unsuitable	0.108–1
KR	<1	Safe	<0.064	<1	Safe	<0.381
	>1	Unsafe	0.064–1	>1	Unsafe	>0.381
Na%	<20	Excellent	<0.021	<20	Excellent	<0.200
	20–40	Good	0.021–0.296	20–40	Good	0.200–0.512
	40–60	Permissible	0.296–0.560	40–60	Permissible	0.512–0.821
	60–80	Doubtful	0.560–0.836	60–80	Doubtful	0.821–1
	>80	Unsuitable	0.836–1			
PI	51–75	Good	0.860–1	51–75	Good	0–0.552
	>75	Suitable	<0.860	>75	Suitable	>0.552
PS	1.3–3	Suitable	<0.013	<3	Suitable	<0.003
	3–15	Moderate	0.013–0.114	3–15	Moderate	0.003–0.027
	15–20		0.114–0.157	15–20		0.027–1
	20–120	Unsuitable	0.157–1	>20	Unsuitable	
SAR	0.6–10	Excellent	<0.325	<10	Excellent	<0.301
	10–18	Good	0.325–0.592	10–18	Good	0.301–0.554
	18–26	Fair	0.592–0.916	18–26	Fair	0.554–0.792
	26–29	Poor	0.915–1	>26	Poor	>0.792
SSP	<20	Excellent	<0.037	<20	Excellent	<0.210
	20–40	Good	0.037–0.306	20–40	Good	0.210–0.517
	40–60	Permissible	0.306–0.570	40–60	Permissible	0.517–0.823
	60–80	Doubtful	0.507–0.834	60–80	Doubtful	0.823–1
	>80	Unsuitable	0.834–1			
MAR	4.4–18	Safe	0	0.7–14	Safe	0
	18–31			14–27		
RSC	<1.25	Good	0	<0.2	Good	0

In UA, EC classification showed that only 3% of the samples were categorized as good, 50% as permissible, 23% as doubtful, and 23% as unsuitable. The EC fuzzy membership maps (Figure S2) showed that only small patches located in the north and southwest were suitable for irrigation use. The TDS results showed that out of the 30 samples, only 9 samples were freshwater and 21 were brackish. The major and minor ions in irrigated water serve as macro- and micronutrients for the growth of plants [86]. Higher concentrations of ions such as Na^+ and Cl^- affect the salinity and soil permeability, reduce crop yields, and hinder plant growth [87–89]. The Na^+ concentration is observed to be higher in brackish/saline waters and treated wastewater [90]. Ozturk [75] recommends the limitation of the usage of RO water with higher Na^+. Higher EC, indicative of high salinity, results in a drought scenario, where plants fail to compete with different ions in the soil [27]. The soil structure and crop growth are critically impacted by EC when irrigation water contains high salinity [91]. The higher EC concentration may be due to the process of weathering of rocks, leaching, and salt dissolution during rainfall. An increase in the salinity in water leads to a higher ionic concentration around the roots and the impact of osmotic processes in this zone, resulting in changes in biochemical and metabolic processes [92,93]. The irrigation of brackish groundwater with higher salinity increases the rate of chlorophyll degradation, prevents the synthesis of proteins, and inhibits the activity of enzymes and rates of photosynthesis [94].

The cationic ratio between the alkaline and alkaline-earth elements, where Na^+ is considered as the main alkali, is referred to as Kelly's ratio (KR). The KR determines the ISGW based on the concentration of Na^+ and that of Ca^{2+} and Mg^{2+} ions [53,95]; K^+ is often neglected due to lesser concentrations. This ratio determines the excess sodium in the cations, such as Na%, indicating the probability of salt formation, thereby affecting the soil properties and crop yield. Na^+ ions adsorb onto clay particles at high concentrations, expelling Mg^{2+} and Ca^{2+} ions. Substituting Na^+ for Ca^{2+} and Mg^{2+} causes the soil to have a weak internal channel and reduces water and air movement in wet conditions [96]. Drought conditions cause the soil to harden into unworkable clods that destroy crops. According to the KR, only 44.7% of samples in the RA region were considered safe and fit for irrigation use. The fuzzy membership map of the KR in RA indicated that higher values were represented in the central and southern parts of the region (Figure S1). The northern part of the region had lower values, indicating that it was suitable for irrigation use. Similarly, in UA, 43% of samples were considered suitable and safe for irrigation based on the KR. It was observed from the KR fuzzy maps (Figure S2) that low membership values were only present in the NW and SW regions.

A negative impact on crop cultivation is reported due to the volume of water held between clay and excess Mg^{2+} [92], which increases the alkalinity, reducing the percolation rate by damaging the soil structure [97]. The crop yield can also be affected by the increase in soil alkalinity developed by excess Mg^{2+} in water, and it is referred to as the magnesium hazard [98]. Stomatal changes, leaf burn, and the varying uptake of Ca and Mg are associated with higher concentrations of these ions in brackish or RO brines, also modifying the osmotic variations to regulate the plant uptake of water [99]. Calcium and magnesium ions are usually balanced in groundwater [100], but they act differently in soil. Especially in brackish to saline water, the exchange of Mg^{2+} with Na^+ in irrigated soils disperses soil assemblages and damages the soil structure [101], thus reducing crop yields [102,103]. In the RA field, all samples in the region were placed in the suitable category based on the MAR classification [104]. The MAR fuzzy membership map showed that the groundwater in the region was suitable and safe for irrigation (Figure S1). Similarly, the MAR values of groundwater samples in the UA field indicated that it was suitable and excellent for irrigation. The fuzzy membership maps of the MAR (Figure S2) revealed that the membership values were low throughout UA.

Similarly, the parameters associated with high Na^+ in groundwater, such as SAR and Na%, can also reduce the permeability of the soil and the water-absorbing capacity of crops [91]. Soil properties are generally affected by sodium, especially by SAR [105]. The relative percentage of Na^+ to that of other cations in irrigated water affects crops and the soil structure. The permeability, soil aeration, and physical structure of soil are affected by higher Na% in irrigated water [106]. Higher Na^+ in water may be due to its brackish nature or derived from the pressure of ion exchange, the dissolution leaching of anthropogenic salts, or even agrochemicals or the intrusion of seawater in coastal regions. The sodicity issue is linked to permeability in the Food and Agriculture Organization (FAO) recommendations [58]. Regions with excess Na^+ in their groundwater produce crops with inadequate water due to low infiltration rates resulting from the poor hydraulic conductivity of the soil. Montmorillonitic clays are more sensitive to Na^+ than in kaolinite regions. Because sodium ions (Na^+) are adsorbed onto soil exchange sites, they cause nonhomogeneous aggregates and thus decrease the soil permeability [107]. The potential effect of Na^+ may be slightly amplified in Mg^{2+}-dominated water, especially when Mg/Ca is >1 [80]. The Na% indicated that only 2% of the RA groundwater was excellent for irrigation use, whereas 28.9% fell within the good category, 21% in permissible, with 23.6% in the doubtful and unsuitable categories [85]. The fuzzy map of Na% showed that low membership values were seen in the northern part of the RA region, along with some small patches in the central part of the area (Figure S1). Thus, the samples from the central part of the RA field were considered permissible, whereas those along the eastern and southeastern parts were considered unsuitable for irrigation purposes.

In UA, according to the Na%, samples were categorized as excellent (2%), good (26%), permissible (30%), and doubtful (36%). From the fuzzy maps of Na% (Figure S2), it was observed that good groundwater was detected in the southwest, with moderate membership values in the central part of the UA field. Moreover, the eastern part of the region showed unsuitable groundwater quality for irrigation use.

Seedling growth is affected by decreasing soil aeration [86], which may also result from the irrigation of the soil with water with higher ionic concentrations [108]. The index of permeability is typically compared to the total concentration of ions, and if the index value exceeds 75%, the sample is suitable and fit for irrigation, while less than 25% is considered unsuitable [109]. Soil permeability is reduced, and hardening occurs with higher salt concentrations in the surface and on the pore spaces [110]. As stated earlier, the inability of plants to absorb water and nutrients via osmotic processes and metabolic responses is another route by which high salinity can harm plant growth [101]. Soil permeability and water quality for irrigation are influenced by the long-term presence of sodium, calcium, magnesium, and bicarbonate ions [111]. Most of the RA samples (55%) fell under the good category, with 42% being classified as suitable and only 2.6% in the unsuitable class. Furthermore, the fuzzy map of PI showed that RA was mostly characterized by low membership values, and high membership values were only observed in the southeastern part (Figure S1). However, among the UA samples, only 10% were found to be suitable and 83% of the samples represented the good category, whereas only 6% of the samples were unsuitable. The fuzzy membership map of PI (Figure S2) showed that the entire region was represented by low membership values, reflecting the suitability for irrigation in the UA area.

The potential salinity parameter Is specifically determined by the groundwater chloride ions containing half of the sulfate ions. Since chloride is a strongly electronegative ion, it facilitates the conductivity of water and then associates with other ions in arid environments to form salts. In RA, the PS values indicated that only 5% of the samples were suitable, 31.5% were moderate, and 63% were considered unsuitable for irrigation [56]. In addition, the PS fuzzy map showed low to moderate membership values along the northern part of the RA area, with minor patches in the central and southern parts (Figure S1). In the UA region, however, only 3% of the samples were suitable, 50% moderate, and 46% were considered unsuitable for irrigation based on the PS values [56]. In the fuzzy map of PS (Figure S2), most of the membership values were high, except for a small portion along the northern and southwestern parts of UA that had low membership values.

The surplus of carbonates or bicarbonates in irrigated water may result in the formation of salts, which precipitate in the soil pore spaces, thereby affecting its permeability, as reflected by the SI values of the carbonate minerals. The excess HCO_3 relative to Ca^{2+} and Mg^{2+} tends to precipitate as $NaHCO_3$ in the irrigated soil, affecting the soil fertility and thus the crop yields [112,113]. Water classification for irrigation according to the residual sodium carbonate [52] in the RA region showed that only 7% of the groundwater samples were in the medium category, and the rest (92%) was observed to represent the good category. The RSC spatial map showed that higher fuzzy values were represented in the entire RA region (Figure S1). Moreover, in the UA region, the RSC values and fuzzy membership map (Figure S2) revealed that all the groundwater samples were suitable for irrigation use.

The sodium absorption ratio can be used to evaluate the sodium in excess compared to magnesium and calcium, which decreases the permeability of the soil, resulting in water limitations for plants. An increase in SAR can reduce the water absorption capacity of plants [92], in addition to the impact on the soil structure and fertility. The irrigated water based on SAR was classified as unsuitable, doubtful, good, and excellent, based on the values of >26, 18–26, 10–18, and <10, respectively. The irrigation of plants with sodium-enriched water results in ion exchange reactions with Ca^{2+} and Mg^{2+} [78]. However, sodium that is bound to the adsorption sites on clay particles is generally released during irrigation with Ca-enriched water. According to the SAR classification [52], most of the RA samples (63%) were in the excellent category, with 21%, 13%, and 2% in the good, fair, and poor categories, respectively. The spatial map of SAR showed that the RA region was predominantly characterized by low fuzzy values, with a minor representation of high membership values in the east (Figure S1). Similarly, most of the UA samples (83%) were excellent for irrigation, whereas 6% were good, 6% were fair, and only 3% were in the unsuitable category. The SAR spatial distribution map (Figure S2) showed that most of the UA area had low fuzzy values, with a minor representation of high membership values along the eastern part of the area.

According to the classification of SSP [57], 2.6% of the RA samples were considered excellent, 28.9% good, 21% permissible, 23.6% doubtful, and 23.6% unsuitable. The fuzzy membership map of SSP in the RA region (Figure S1) showed that lower values were only noticed in the north, and they gradually increased towards the southeast. Similarly, the SSP values in UA showed that 10% of samples had excellent suitability, whereas 26.6% were good, 26.6% were permissible, and 36.6% were doubtful for irrigation use. Furthermore, the fuzzy membership map of UA (Figure S2) showed low values along the southwestern part of the area, with a gradual increase in value towards the eastern part of the area.

3.6. Final Fuzzy Overlay Map

In order to identify contamination levels, contaminated regions, and sources, and to map the extent of the contaminated areas, groundwater quality evaluation and spatial mapping using GIS approaches are required [114,115]. For this purpose, researchers across the world rely on inverse distance weighting (IDW) interpolation [30,37,116]. The fuzzy membership maps for the considered parameters were overlaid to create a final groundwater suitability map for irrigation purposes (Figure 7). Based on the values of fuzzy membership, the groundwater suitability for irrigation purposes was categorized into four classes: excellent, good, moderate, and poor.

In the final overlay map of the RA region (Figure 7A), the excellent to good ISGW was dispersed from the northern to the southern parts, whereas medium to poor groundwater quality was predominantly observed along the eastern portion of the wellfield. However, in the UA region (Figure 7B), suitable zones of excellent to good ISGW were observed in the northern and eastern portions of the wellfield and the suitability gradually changed from moderate to poor towards the east and south of the area.

The final fuzzy overlay map of the RA field revealed that the region was categorized as excellent (8.36%), good (56.64%), medium (32.17%), and poor (2.81%). Meanwhile, the UA field was categorized as excellent (5.90%), good (46.50%), medium (29.70%), and poor (17.70%) (Table 3). Hence, the final overlay map of these two wellfields indicated that samples from the RA wellfield were relatively more suitable for irrigation purposes.

Table 3. Area coverage of the Rawdhatain and Umm Al Aish regions and their water suitability categories.

Category	Rawdhatain		Umm Al Aish	
	Area (sq km)	Percentage	Area (sq km)	Percentage
Excellent	4.44	8.40%	2.64	5.90%
Good	30.1	56.6%	20.76	46.5%
Medium	17.1	32.2%	13.25	29.7%
Poor	1.50	2.81%	7.93	17.7%

Figure 7. Final fuzzy overlay map of (**A**) Rawdhatain and (**B**) Umm Al Aish regions, depicting the spatial distribution of different categories of groundwater for irrigation purposes (excellent, good, medium, and poor).

3.7. Livestock Suitability

The health of biota can be deteriorated by the prolonged usage of drinking water with high salinity, and it can even lead to mortality [117]. Due to the high salt content of groundwater, desalination units are used by local poultry farms. The suitability of groundwater for animal rearing based on the salinity in the research area [58] was studied (Figure 8). The results indicated that among the 38 samples from RA, only 36 samples were rated as excellent (28.9%) and very satisfactory (52.6%) for livestock and poultry, whereas 13.1% were satisfactory for livestock and unfit for poultry. Similarly, out of the 31 samples from UA, 27 samples were rated as excellent (26.6%) and very satisfactory (46.6%) for all classes of poultry and livestock, whereas 13.3% were considered satisfactory for livestock and unsuitable for poultry, 10% could be considered for limited use for livestock and unsuitable for poultry, and 3.3% were not recommended for livestock and poultry consumption.

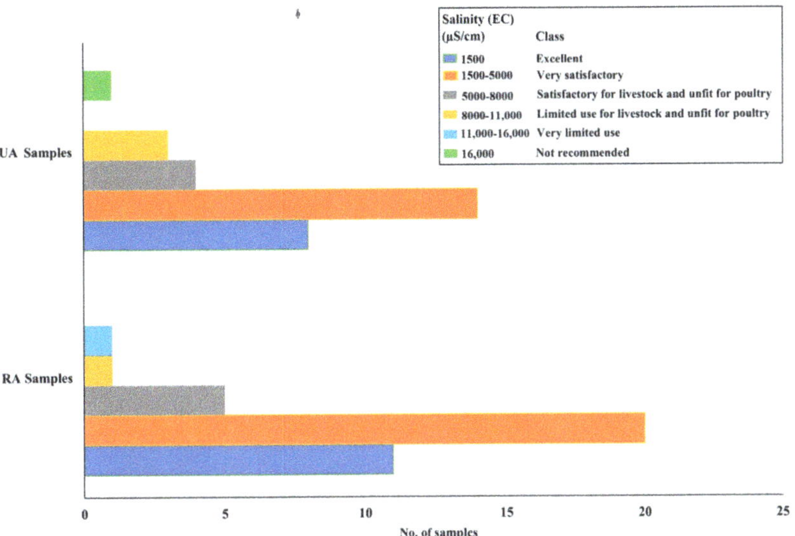

Figure 8. Groundwater samples in UA and RA compared to the classification of Ayer and Westcot [58] regarding groundwater quality for livestock and poultry.

The composition of the diet has a considerable impact on the water salinity tolerance for goats and sheep [118,119]. Sheep and goats have different water salinity tolerances [118,119]. Sheep of different breeds and sizes should be provided drinking water with TDS generally lower than 5000 mg/L [117] (lactating females, dry adults, and young). EC of 6000 to 10,000 µS/cm for young and 9300 to 21,800 µS/cm is suggested as optimal for goats (dry adults) [117]. However, water with EC higher than 10,000 µS/cm should be used with caution. Under these circumstances, sheep and goats could use 97% and 90% of the groundwater samples from the RA and UA fields for drinking purposes, respectively.

3.8. Management of Brackish Groundwater

Since water is scarce and unevenly distributed, globally, eighteen countries are considered to be at "serious risk" [120]. Fifteen of these countries are in the Middle East. Water resource management is becoming increasingly difficult because of national development, irrigated agriculture, and water competition between urban areas, farming, and industry [121]. Groundwater has been seriously disrupted from its normal state due to irregular rainfall, scant recharge, and uncontrolled abstraction patterns, leading to the major depletion of underground aquifers and long-term water quality issues.

The demand for food and water increases with the population [122]. The salinity of groundwater used for irrigation may also lead to seed dormancy [123]. The selection of appropriate crops can offer a solution to soil salinity issues [124]. The salinity thresholds for different crops vary and thus suitable crops can be identified based on the soil fertility and electrical conductivity of the groundwater, although certain glycophytes [125], such as coconut, are moderately salinity-tolerant. The quantity of water used and the value of the threshold limit of salinity can be helpful in the utilization of brackish groundwater for irrigation. Similarly, the different crops grown annually have different criteria with respect to salinity and demand [126,127]. One of the main types of salt-tolerant crops are halophytes [128], which could be used for several purposes. The cultivation of halophytes could be a viable solution for disposed inland RO reject brines and brackish groundwater.

Certain plants used for animal feed, with more nutrients and carbohydrates, such as forage cactus, can be cultivated with brackish groundwater individually or in integrated multiple agricultural systems, along with aquaculture or through intercropping with *Gliricidia* [129,130]. The yields of crop are increased without the formation of salts in the soil and water use can be maintained through supplemental irrigation techniques. This method also enhances the rate of photosynthesis. The hydroponic production of tilapia can be performed under salinity with minimal water requirements [131]. Integrated systems involving halophytes such as forage cactus and seedling production to supplement irrigation along with pisciculture can be considered for brackish groundwater agriculture [131].

3.9. Fit for Purpose (FFP)

Brackish and freshwater mixing can be considered for certain crops that are salt-tolerant, such as cotton, but is not appropriate for freshwater-dependent crops such as corn [132]. Management strategies can also consider the utility of the available water resources based on their quality, referred to as "fit for purpose" (FFP) [133,134]. The FFP method addresses the UN's Sustainable Developmental Goals by advocating for the use of recycling [135], the integration of different management techniques to sustain water resources [136], the implementation of water policies and modeling techniques [137], consideration of the use of the available water resources [138], and desalination and wastewater treatment [139]. Recent studies have tried to integrate these resources to address the FFP in water management and governance strategies as a step towards achieving SDG 6.

Developing a new water management system that takes into account rising temperatures, shifting precipitation patterns, along with shifts in the volume and distribution of rainfall is essential to satisfy future global water needs. To find workable and generally accepted water resource allocation solutions for different users and to speed up the development of non-conventional water resources, such as water reuse and the desalination of seawater, future studies should focus on innovative water resource management systems such as integrated water resource management (IWRM) [140]. The *insitu* monitoring of soil and water in irrigated regions is a basic requirement for planning and management strategies.

4. Conclusions and Limitations

The salination of soil is a key issue in arid regions, as it alters the physiochemical properties of the soil and reduces the crop yields. The study of the ISGW of fresh to brackish samples collected from the two adjacent water wellfields in RA and UA indicates Ca-Na water type in the central portions of the wellfields and are observed to be of the Na-Cl type along the southern and eastern portions of the study area, reflecting the increase in salinity along the regional groundwater flow. The geochemical modeling of the composition of the groundwater in both wellfields indicates that the saturation states of carbonates are greater, in contrast to sulfates, reflecting the probability of the precipitation of carbonate salts in the soil pores, affecting the permeability of the soil and thereby the crop yields. The probability salt formation on the soil surfaces and the pores is inferred to increase during high-temperature months. Further, the formation of salts is also identified to be governed by the ionic strength, pCO_2, and changes in the thermodynamic properties of the system, which lead to drastic variations in salt formation, thus affecting permeability. The integrated plot of the ISGW shows that UA samples are more suitable for irrigation than RA samples. Cation exchange between groundwater and clay minerals, and the dissolution and precipitation of secondary carbonate minerals, affect the groundwater chemistry.

The ISGW for these two regions were evaluated by considering nine parameters (EC, KR, MAR, Na%, PI, PS, SAR, RSC, and SSP) adopting the fuzzy logic technique. The fuzzy membership maps of the nine parameters were then integrated into a final fuzzy overlay map of the RA and UA fields to investigate the ISGW. Based on the values of the fuzzy membership, the groundwater suitability for irrigation purposes was categorized into four classes: excellent, good, moderate, and poor. Na played a key role in most of these parameters and hence the higher values of Na^+ ions in samples affected the irrigation quality of the water. The RA field 8.3% of samples were categorized as excellent,56.6% as good,32.1% as moderate, and 2.8% as poor. In the UA field, 5.9% of samples were excellent, 46.5% were good, 29.7% were moderate, and 17.7% were poor. It is inferred from the maps that most of the groundwater in RA can be used for irrigation, except those from the eastern side of the study area, with poor water quality. The groundwater in the northern and western parts of the UA region is suitable for agriculture, whereas that from the eastern and southern parts is unsuitable for irrigation. It is suggested that salt-tolerant crops could be prioritized in regions with poor suitability. Further, the water provided to the different parts of plants from the soil and subsurface is also affected due to the effect of salinity on the osmotic pressure of the stomatal cells. The study also indicates that most of the groundwater in the wellfield is suitable for consumption by livestock. However, this depends on the type of livestock, their age, size, breed, gender, etc., which also varies the quantity of consumption.

Due to irregular precipitation in arid regions and minimal aquifer recharge along with extensive abstraction patterns, groundwater has been substantially disturbed from its natural state, resulting in significant aquifer depletion and long-term water quality challenges. Therefore, soil and water monitoring in irrigated regions, along with innovative water resource management systems such as IWRM, are crucial for the sustainability of resources in arid regions such as Kuwait. Further, monitoring of the soil and quality of irrigated water in arid regions with brackish groundwater helps to maintain the crop yields and fertility of the soil. Moreover, in regions with inland desalination units, it is advisable to develop a monitoring system to observe the leaching of rejects and impacts on soil and to determine the maximum depth of leaching. This monitoring system would aid in conserving the fertility of the soil and existing groundwater reserves.

The hydrogeochemical interpretation of the saturation states regarding the precipitation of salts was not validated with XRD data, and this could be considered a major limitation of the study. The geochemical processes inferred during infiltration though vadose zone could be confirmed with litho-logs and core samples in future studies. Such interpretations could also help in assessing the geochemical environment of the topsoil region and the capillary zone. The water level data of all the wells were not obtained in the study, but these could have clearly illustrated the variations in the geochemistry along the flow direction. Furthermore, the isotopic evaluation of groundwater would assist in the interpretation of evaporation, recharge, and contaminated sources. Although the overall process of the region and its suitability was established in the current study, the abovementioned additional data and their integrated analysis would yield more concrete solutions for the future groundwater resource management of the region.

Supplementary Materials: The following supporting information can be downloaded at: https://www.mdpi.com/article/10.3390/w15142674/s1, Figure S1: Rawdhatain fuzzy membership maps. Figure S2: Umm Al Aish membership maps.

Author Contributions: A.A.-R.—Conceptualization, Writing—Original Draft, Review and Editing, Methodology, Software; C.S.—Conceptualization, Project Administration, Supervision, Writing—Original Draft, Review and Editing; D.R.S.—Data Curation, Methodology, Formal analysis, Writing—Original Draft; B.A.—Writing—Review and Editing, Software, Visualization; T.R.—Sampling, Methodology, Writing. All authors have read and agreed to the published version of the manuscript.

Funding: This research was funded by the International Atomic Energy Agency through TC project KUW7010 and the Kuwait Institute for Scientific Research through project WM084C.

Data Availability Statement: Data will be provided on request.

Acknowledgments: The authors would like to extend their appreciation to the International Atomic Energy Agency (IAEA) for their in-kind contribution to the study through IAEA-KUW7010. The support of the Kuwait Institute for Scientific Research's (KISR's) management (WM084C) was pivotal in carrying out the various tasks of the project.

Conflicts of Interest: The authors declare no conflict of interest.

References

1. Mahdian, M.; Hosseinzadeh, M.; Siadatmousavi, S.M.; Chalipa, Z.; Delavar, M.; Guo, M.; Abolfathi, S.; Noori, R. Modelling Impacts of Climate Change and Anthropogenic Activities on Inflows and Sediment Loads of Wetlands: Case Study of the Anzali Wetland. *Sci. Rep.* **2023**, *13*, 5399. [CrossRef] [PubMed]
2. Malekmohammadi, B.; Uvo, C.B.; Moghadam, N.T.; Noori, R.; Abolfathi, S. Environmental Risk Assessment of Wetland Ecosystems Using Bayesian Belief Networks. *Hydrology* **2023**, *10*, 16. [CrossRef]
3. Khosravi, K.; Rezaie, F.; Cooper, J.; Kalantari, Z.; Abolfathi, S.; Hatamiafkoueieh, J. Soil Water Erosion Susceptibility Assessment Using Deep Learning Algorithms. *J. Hydrol.* **2023**, *618*, 129229. [CrossRef]
4. Bouwer, H. Integrated Water Management: Emerging Issues and Challenges. *Agric. Water Manag.* **2000**, *45*, 217–228. [CrossRef]
5. Rajendiran, T.; Sabarathinam, C.; Panda, B.; Elumalai, V. Influence of Dissolved Oxygen, Water Level and Temperature on Dissolved Organic Carbon in Coastal Groundwater. *Hydrology* **2023**, *10*, 85. [CrossRef]
6. Kolahchi, Z.; Jalali, M. Effect of Water Quality on the Leaching of Potassium from Sandy Soil. *J. Arid Environ.* **2007**, *68*, 624–639. [CrossRef]
7. Tewari, R. Management of Declining Groundwater Resources and the Role of Policy Planning in Semi-Arid Economies: The Case of Texas High Plains. In *Emerging Issues in Groundwater Resources. Advances in Water Security*; Springer: Cham, Switzerland, 2016; pp. 365–381. [CrossRef]
8. Freeze, A.R.; Cherry, J.A. *Groundwater*; Prentice Hall, Inc.: Englewood Cliffs, NJ, USA, 1979.
9. Tanninen, J.; Kamppinen, L.; Nystrom, M. *Pre-Treatment and Hybrid Processes: Nanofiltration-Principles and Application*; Elsevier: Oxford, UK, 2005; pp. 253–254.
10. Pimentel, D.; Bailey, O.; Kim, P.; Mullaney, E.; Calabrese, J.; Walman, L.; Nelson, F.; Yao, X. Will Limits of the Earth's Resources Control Human Numbers? *Environ. Dev. Sustain.* **1999**, *1*, 19–39. [CrossRef]
11. Qadir, M.; Ghafoor, A.; Murtaza, G. Use of Saline–Sodic Waters through Phytoremediation of Calcareous Saline–Sodic Soils. *Agric. Water Manag.* **2001**, *50*, 197–210. [CrossRef]
12. Ayars, J.E.; Tanji, K.K. Effects of Drainage on Water Quality in Arid and Semiarid Irrigated Lands. *Agric. Drain.* **1999**, *38*, 831–867. [CrossRef]
13. Steward, D.R.; Bruss, P.J.; Yang, X.; Staggenborg, S.A.; Welch, S.M.; Apley, M.D. Tapping Unsustainable Groundwater Stores for Agricultural Production in the High Plains Aquifer of Kansas, Projections to 2110. *Proc. Natl. Acad. Sci. USA* **2013**, *110*, E3477–E3486. [CrossRef]
14. Marston, L.; Konar, M.; Cai, X.; Troy, T.J. Virtual Groundwater Transfers from Overexploited Aquifers in the United States. *Proc. Natl. Acad. Sci. USA* **2015**, *112*, 8561–8566. [CrossRef] [PubMed]
15. Ferreira, J.F.; Liu, X.; Suddarth, S.R.P.; Nguyen, C.; Sandhu, D. NaCl Accumulation, Shoot Biomass, Antioxidant Capacity, and Gene Expression of *Passiflora edulis* f. Flavicarpa Deg. in Response to Irrigation Waters of Moderate to High Salinity. *Agriculture* **2022**, *12*, 1856. [CrossRef]
16. Azimi, R.; Borzelabad, M.J.; Feizi, H.; Azimi, A. Interaction of SiO Nanoparticles with Seed Prechilling on Germination and Early Seedling Growth of Tall Wheatgrass (*Agropyron elongatum* L.). *Pol. J. Chem. Technol.* **2014**, *16*, 25–29. [CrossRef]
17. Flores, A.; Schutte, B.; Shukla, M.; Picchioni, G.; Ulery, A. Time-Integrated Measurements of Seed Germination for Salttolerant Plant Species. *Seed Sci. Technol.* **2015**, *43*, 541–547. [CrossRef]
18. Robinson, B.W.; Al Ruwaih, F. The Stable-Isotopic Composition of Water and Sulfate from the Raudhatain and Umm Al Aish Freshwater Fields, Kuwait. *Chem. Geol. Isot. Geosci. Sect.* **1985**, *58*, 129–136. [CrossRef]
19. Kwarteng, A.; Viswanathan, M.; Al-Senafy, M.; Rashid, T. Formation of Fresh Ground-Water Lenses in Northern Kuwait. *J. Arid Environ.* **2000**, *46*, 137–155. [CrossRef]
20. Akber, A.; Viswanathan, M.; Al Senafy, M.; Rashed, T. *Assessment of Long-Term Pollution for the Groundwater of Raudhatain and Umm Al-Aish Areas (Phase II)*; Kuwait Institute for Scientific Research: Kuwait City, Kuwait, 2002.
21. Akber, A.; Al-Murad, M.; Mukhopadhyay, A.; Rashid, A.; Al-Qallaf, H.; Al-Haddad, A.; Bhandary, H.; Marzouk, F.; Al-Salman, B. *Long-Term Monitoring and Remediation Strategy for Hydrocarbon Pollutants in the Groundwater of Raudhatain and Umm Al-Aish Fields*; Report No. KISR9902; Kuwait Institute for Scientific Research: Kuwait City, Kuwait, 2009.
22. Mukhopadhyay, A.; Akber, A.; Rashed, T.; Kotwicki, V.; Uddin, S.; Bushehri, A. Establishing a Baseline to Evaluate Future Impacts to Groundwater Resources in North Kuwait. *Environ. Earth Sci.* **2016**, *75*, 324. [CrossRef]
23. Al-Rashed, M.; Mukhopadhyay, A.; AlSenafy, M.; Ghoneim, H. *Effect of Seawater Discharge on Groundwater Quality in the Northern Kuwait*; Kuwait Institute for Scientific Research: Kuwait City, Kuwait, 2002.

24. Mukhopadhyay, A.; Quinn, M.; Al-Haddad, A.; Al-Khalid, A.; Al-Qallaf, H.; Rashed, T.; Bhandary, H.; Al-Salman, B.; Bushehri, A.; Boota, A. Pollution of Fresh Groundwater from Damaged Oil Wells, North Kuwait. *Environ. Earth Sci.* **2017**, *76*, 145. [CrossRef]
25. Tjandra, F.L.; Kondhoh, A.; Mohammed, M. A Conceptual Database Design for Hydrology Using GIS. In Proceedings of the Asia Pacific Association of Hydrology and Water Resources, Kyoto, Japan, 13–15 March 2003; pp. 13–15.
26. Selvam, S.; Dar, F.A.; Magesh, N.; Singaraja, C.; Venkatramanan, S.; Chung, S. Application of Remote Sensing and GIS for Delineating Groundwater Recharge Potential Zones of Kovilpatti Municipality, Tamil Nadu Using IF Technique. *Earth Sci. Inform.* **2016**, *9*, 137–150. [CrossRef]
27. Naseem, S.; Hamza, S.; Bashir, E. Groundwater Geochemistry of Winder Agricultural Farms, Balochistan, Pakistan and Assessment for Irrigation Water Quality. *Eur. Water* **2010**, *31*, 21–32.
28. Kumarasamy, P.; Dahms, H.-U.; Jeon, H.-J.; Rajendran, A.; Arthur James, R. Irrigation Water Quality Assessment—An Example from the Tamiraparani River, Southern India. *Arab. J. Geosci.* **2014**, *7*, 5209–5220. [CrossRef]
29. Vadiati, M.; Nalley, D.; Adamowski, J.; Nakhaei, M.; Asghari-Moghaddam, A. A Comparative Study of Fuzzy Logic-Based Models for Groundwater Quality Evaluation Based on Irrigation Indices. *J. Water Land Dev.* **2019**, *43*, 158–170. [CrossRef]
30. Chidambaram, S.; Prasanna, M.; Venkatramanan, S.; Nepolian, M.; Pradeep, K.; Panda, B.; Thivya, C.; Thilagavathi, R. Groundwater Quality Assessment for Irrigation by Adopting New Suitability Plot and Spatial Analysis Based on Fuzzy Logic Technique. *Environ. Res.* **2022**, *204*, 111729. [CrossRef]
31. Purnamasari, R.A.; Noguchi, R.; Ahamed, T. Land Suitability Assessments for Yield Prediction of Cassava Using Geospatial Fuzzy Expert Systems and Remote Sensing. *Comput. Electron. Agric.* **2019**, *166*, 105018. [CrossRef]
32. Ostovari, Y.; Beigi-Harchegani, H.; Asgari, K. A Fuzzy Logic Approach for Assessment and Mapping of Groundwater Irrigation Quality: A Case Study of Marvdasht Aquifer, Iran. *Arch. Agron. Soil Sci.* **2015**, *61*, 711–723. [CrossRef]
33. Tafreshi, A.M.; Tafreshi, G.M. A New Approach in Qualitative Zoning of Groundwater to Assessment Suitable Irrigation Water Using Fuzzy Logic Spatial Modeling via GIS. *Res. Sq.* **2022**. preprint. [CrossRef]
34. Jha, M.K.; Shekhar, A.; Jenifer, M.A. Assessing Groundwater Quality for Drinking Water Supply Using Hybrid Fuzzy-GIS-Based Water Quality Index. *Water Res.* **2020**, *179*, 115867. [CrossRef]
35. Van Eck, N.; Waltman, L. Software Survey: VOSviewer, a Computer Program for Bibliometric Mapping. *Scientometrics* **2010**, *84*, 523–538. [CrossRef]
36. Dixon, B. Groundwater Vulnerability Mapping: A GIS and Fuzzy Rule Based Integrated Tool. *Appl. Geogr.* **2005**, *25*, 327–347. [CrossRef]
37. Brahim, F.B.; Boughariou, E.; Bouri, S. Multicriteria-Analysis of Deep Groundwater Quality Using WQI and Fuzzy Logic Tool in GIS: A Case Study of Kebilli Region, SW Tunisia. *J. Afr. Earth Sci.* **2021**, *180*, 104224. [CrossRef]
38. Dhaoui, O.; Agoubi, B.; Antunes, I.M.; Tlig, L.; Kharroubi, A. Groundwater Quality for Irrigation in an Arid Region—Application of Fuzzy Logic Techniques. *Environ. Sci. Pollut. Res.* **2023**, *30*, 29773–29789. [CrossRef]
39. Misak, R.; Hussain, W. Groundwater in Kuwait. In *The Geology of Kuwait*; Springer: Cham, Switzerland, 2022; pp. 199–214. [CrossRef]
40. Umar, A.; Umar, R.; Ahmad, M. Hydrogeological and Hydrochemical Framework of Regional Aquifer System in Kali-Ganga Sub-Basin, India. *Environ. Geol.* **2001**, *40*, 602–611. [CrossRef]
41. Samayamanthula, D.R.; Sabarathinam, C.; Alayyadhi, N.A. Trace Elements and Their Variation with PH in Rain Water in Arid Environment. *Arch. Environ. Contam. Toxicol.* **2021**, *80*, 331–349. [CrossRef]
42. Radha, S.D.; Sabarathinam, C.; Al Otaibi, F.; Al-Sabti, B.T. Variation of Centennial Precipitation Patterns in Kuwait and Their Relation to Climate Change. *Environ. Monit. Assess.* **2023**, *195*, 20. [CrossRef] [PubMed]
43. Owen, R.; Nasr, S.N. Stratigraphy of the Kuwait-Basra Area: Middle East. In *Habitat of Oil*; American Association of Petroleum Geologists: Tulsa, OK, USA, 1958.
44. Edgell, H. Aquifers of Saudi Arabia and Their Geological Framework. *Arab. J. Sci. Eng.* **1997**, *22*, 3–31.
45. Un-Escwa, B.; United Nations Economic and Social Commission for Western Asia. *Bundesanstalt Für Geowissenschaften Und Rohstoffe*; Inventory of Shared Water Resources in Western Asia: Beirut, Lebanon, 2013.
46. Parson Corporation. *Groundwater Resources of Kuwait, I–III*; Ministry of Electricity and Water: Kuwait City, Kuwait, 1963.
47. Bouwer, H.; Rice, R. A Slug Test for Determining Hydraulic Conductivity of Unconfined Aquifers with Completely or Partially Penetrating Wells. *Water Resour. Res.* **1976**, *12*, 423–428. [CrossRef]
48. Al-Ruwaih, F.M.; Shehata, M. Hydrochemical Processes and Environmental Isotopic Study of Groundwater in Kuwait. *Water Int.* **2004**, *29*, 158–166. [CrossRef]
49. Sabarathinam, C.; Bhandary, H.; Hadi, K. CHIDAM-A Software for Chemical Interpretation of the Dissolved Ions in Aqueous Media. *Groundw. Sustain. Dev.* **2021**, *13*, 100496. [CrossRef]
50. Parkhurst, D.L.; Thorstenson, D.C.; Plummer, L.N. *PHREEQE: A Computer Program for Geochemical Calculations*; US Geological Survey; Water Resources Division: Reston, VA, USA, 1982; Volume 80.
51. Eaton, F.M. Significance of Carbonates in Irrigation Waters. *Soil Sci.* **1950**, *69*, 123–134. [CrossRef]
52. Richard, L. Diagnosis and Improvement of Saline and Alkali Soils. In *Agriculture Handbook*; U.S. Government Printing Office: Washington, DC, USA, 1954; Volume 60, pp. 83–100.
53. Kelly, W.P. Permissible Composition and Concentration of Irrigated Waters. *Proc. Am. Soc. Civil Eng.* **1940**, *66*, 607–613. [CrossRef]

54. Szabolcs, I.; Darab, K. Radio-Active Technique for Examining the Improving Effect of $CaCO_3$ on Alkali (Szik) Soils. *Acta Agron. Hung* **1964**, *13*, 93–101.

55. Doneen, L. Water for Field and Truck Crops. *Calif. Agric.* **1948**, *2*, 9–16.

56. Palacios, V.O.; Aceves, N.E. *Instructions for Sampling, Data Recording and Interpretation of Water Quality for Agricultural Irrigation*; Colegio de Postgraduados: Chapingo, Mexico, 1970.

57. Todd, D.K.; Mays, L.W. *Groundwater Hydrology*; John Wiley & Sons: Hoboken, NJ, USA, 2004; ISBN 0-471-05937-4.

58. Ayers, R.S.; Westcot, D.W. *Water Quality for Agriculture*; Food and Agriculture Organization of the United Nations: Rome, Italy, 1985; Volume 29. Available online: https://www.fao.org/3/T0234E/T0234E00.htm (accessed on 28 May 2023).

59. Zadeh, L.A. Outline of a New Approach to the Analysis of Complex Systems and Decision Processes. *IEEE Trans. Syst. Man Cybern.* **1973**, *SMC-3*, 28–44. [CrossRef]

60. Bonham-Carter, G. *Geographic Information Systems for Geoscientists: Modelling with GIS*; Elsevier: Amsterdam, The Netherlands, 1994; ISBN 0-08-042420-1.

61. Tafreshi, A.M.; Tafreshi, G.M. A Novel GIS-Based Approach to Assessment Suitable Irrigation Water Using a Fuzzy-Multi Indices Method in Astaneh-Kuchesfahan Plain, Iran. *Res. Sq.* **2021**, preprint. [CrossRef]

62. Al-Alati, H.N.; Gad, M. Study The Seasonal Fluctuations of Groundwater Characteristics in Al-Raudhatain And Umm Al-Aish Depressions, North Kuwait. *J. Eng. Res. Appl.* **2018**, *8*, 43–56.

63. Al-Homidy, A.A.; Dahim, M.H.; Abd El Aal, A.K. Improvement of Geotechnical Properties of Sabkha Soil Utilizing Cement Kiln Dust. *J. Rock Mech. Geotech. Eng.* **2017**, *9*, 749–760. [CrossRef]

64. Al-Katheeri, E.; Howari, F.; Murad, A. Hydrogeochemistry and Pollution Assessment of Quaternary–Tertiary Aquifer in the Liwa Area, United Arab Emirates. *Environ. Earth Sci.* **2009**, *59*, 581–592. [CrossRef]

65. Estévez, M.H.; Sanjulián, J.J.C.; Mendizábal, A.S. Evolución Geoquímica de Las Aguas Subterráneas En Una Cuenca Sedimentaria Semiárida: Acuífero de Baza-Caniles, Granada, España. *Tierra Tecnol. Rev. Inf. Geol.* **1995**, *10*, 39–48.

66. Hidalgo, M.C.; Cruz-Sanjulián, J. Groundwater Composition, Hydrochemical Evolution and Mass Transfer in a Regional Detrital Aquifer (Baza Basin, Southern Spain). *Appl. Geochem.* **2001**, *16*, 745–758. [CrossRef]

67. Tijani, M.N. Evolution of Saline Waters and Brines in the Benue-Trough, Nigeria. *Appl. Geochem.* **2004**, *19*, 1355–1365. [CrossRef]

68. Sheikholeslami, R. Mixed Salts—Scaling Limits and Propensity. *Desalination* **2003**, *154*, 117–127. [CrossRef]

69. Sheikholeslami, R. Nucleation and Kinetics of Mixed Salts in Scaling. *AIChE J.* **2003**, *49*, 194–202. [CrossRef]

70. Nancollas, G.H.; Zieba, A. Constant Composition Kinetics Studies of the Simultaneous Crystal Growth of Some Alkaline Earth Carbonates and Phosphates. In *Mineral Scale Formation and Inhibition*; Springer: Boston, MA, USA, 1995; pp. 1–9. [CrossRef]

71. Chidambaram, S.; Bhandary, H.; Al-Khalid, A. Modeling of Temperature Governed Saturation States and Metal Speciation in the Marine Waters of Kuwait Bay–Concern to the Desalination Process. *Desalin Water Treat* **2020**, *176*, 234–242. [CrossRef]

72. Díaz, F.; Benes, S.; Grattan, S. Field Performance of Halophytic Species under Irrigation with Saline Drainage Water in the San Joaquin Valley of California. *Agric. Water Manag.* **2013**, *118*, 59–69. [CrossRef]

73. Allen, R.G.; Pereira, L.S.; Raes, D.; Smith, M. *Crop Evapotranspiration-Guidelines for Computing Crop Water Requirements*; FAO Irrigation and Drainage Paper 56; FAO: Rome, Italy, 1998; Volume 300, p. D05109.

74. Adhikari, P.; Shukla, M.; Mexal, J. Spatial Variability of Hydraulic Conductivity and Sodium Content of Desert Soils: Implications for Management of Irrigation Using Treated Wastewater. *Trans. ASABE* **2012**, *55*, 1711–1721. [CrossRef]

75. Ozturk, O.F. Irrigation Using Brackish Groundwater and RO Concentrate: Effects on Germination, Emergence, and Growth of Halophytes. Ph.D. Thesis, New Mexico State University, Las Cruces, NM, USA, 2016.

76. Wigley, T.; Plummer, L.; Pearson, F., Jr. Mass Transfer and Carbon Isotope Evolution in Natural Water Systems. *Geochim. Cosmochim. Acta* **1978**, *42*, 1117–1139. [CrossRef]

77. Njitchoua, R.; Dever, L.; Fontes, J.C.; Naah, E. Geochemistry, Origin and Recharge Mechanisms of Groundwaters from the Garoua Sandstone Aquifer, Northen Cameroon. *J. Hydrol.* **1997**, *190*, 123–140. [CrossRef]

78. Stigter, T.; Van Ooijen, S.; Post, V.; Appelo, C.; Dill, A.C. A Hydrogeological and Hydrochemical Explanation of the Groundwater Composition under Irrigated Land in a Mediterranean Environment, Algarve, Portugal. *J. Hydrol.* **1998**, *208*, 262–279. [CrossRef]

79. Appelo, C.; Postma, D. *Geochemistry, Groundwater and Pollution*; CRC Press: Boca Raton, FL, USA, 1999.

80. Jalali, M. Salinization of Groundwater in Arid and Semi-Arid Zones: An Example from Tajarak, Western Iran. *Environ. Geol.* **2007**, *52*, 1133–1149. [CrossRef]

81. Bohn, H.; Brain, L.; George, A.; O'Connor, G. *Soil Chemistry*; John Wiley & Sons: New York, NY, USA, 1979; Volume 129, p. 389.

82. Mukhopadhyay, A.; Al-Sulaimi, J.; Al-Awadi, E.; Al-Ruwaih, F. An Overview of the Tertiary Geology and Hydrogeology of the Northern Part of the Arabian Gulf Region with Special Reference to Kuwait. *Earth-Sci. Rev.* **1996**, *40*, 259–295. [CrossRef]

83. Mokeddem, I.; Belhachemi, M.; Merzougui, T.; Nabbou, N.; Lachache, S. Hydrochemical Assessment and Groundwater Pollution Parameters in Arid Zone: Case of the Turonian Aquifer in Béchar Region, Southwestern Algeria. *J. Water Land Dev.* **2018**, *39*, 109–117. [CrossRef]

84. Sabarathinam, C.; Rashed, T.; Dashti, F.; Bhandary, H. Equilibrium States of Groundwater Chemistry in Coastal Region of Kuwait. *Desalination Water Treat.* **2022**, *263*, 248–256. [CrossRef]

85. Wilcox, L. *Classification and Use of Irrigation Waters*; US Department of Agriculture: Washington, DC, USA, 1955.

86. Khalid, S. An Assessment of Groundwater Quality for Irrigation and Drinking Purposes around Brick Kilns in Three Districts of Balochistan Province, Pakistan, through Water Quality Index and Multivariate Statistical Approaches. *J. Geochem. Explor.* **2019**, *197*, 14–26. [CrossRef]

87. Jang, C.-S.; Chen, J.-S. Probabilistic Assessment of Groundwater Mixing with Surface Water for Agricultural Utilization. *J. Hydrol.* **2009**, *376*, 188–199. [CrossRef]

88. Wu, J.; Li, P.; Qian, H.; Fang, Y. Assessment of Soil Salinization Based on a Low-Cost Method and Its Influencing Factors in a Semi-Arid Agricultural Area, Northwest China. *Environ. Earth Sci.* **2014**, *71*, 3465–3475. [CrossRef]

89. Li, P.; Wu, J.; Qian, H. Hydrochemical Appraisal of Groundwater Quality for Drinking and Irrigation Purposes and the Major Influencing Factors: A Case Study in and around Hua County, China. *Arab. J. Geosci.* **2016**, *9*, 15. [CrossRef]

90. Adhikari, P.; Shukla, M.K.; Mexal, J.G. Spatial Variability of Soil Properties in an Arid Ecosystem Irrigated with Treated Municipal and Industrial Wastewater. *Soil Sci.* **2012**, *177*, 458–469. [CrossRef]

91. Nematollahi, M.J.; Ebrahimi, P.; Ebrahimi, M. Evaluating Hydrogeochemical Processes Regulating Groundwater Quality in an Unconfined Aquifer. *Environ. Process.* **2016**, *3*, 1021–1043. [CrossRef]

92. Tahmasebi, P.; Mahmudy-Gharaie, M.H.; Ghassemzadeh, F.; Karimi Karouyeh, A. Assessment of Groundwater Suitability for Irrigation in a Gold Mine Surrounding Area, NE Iran. *Environ. Earth Sci.* **2018**, *77*, 766. [CrossRef]

93. Chen, J.; Huang, Q.; Lin, Y.; Fang, Y.; Qian, H.; Liu, R.; Ma, H. Hydrogeochemical Characteristics and Quality Assessment of Groundwater in an Irrigated Region, Northwest China. *Water* **2019**, *11*, 96. [CrossRef]

94. Torres Mendonça, A.J.; Silva, A.A.R.d.; Lima, G.S.d.; Soares, L.A.d.A.; Nunes Oliveira, V.K.; Gheyi, H.R.; Lacerda, C.F.d.; Azevedo, C.A.V.d.; Lima, V.L.A.d.; Fernandes, P.D. Salicylic Acid Modulates Okra Tolerance to Salt Stress in Hydroponic System. *Agriculture* **2022**, *12*, 1687. [CrossRef]

95. Paliwal, K.; Singh, S. Effect of Gypsum Application on the Quality of Irrigation Waters. *Madras Agric. J.* **1967**, *59*, 646–647.

96. Subba Rao, N. Seasonal Variation of Groundwater Quality in a Part of Guntur District, Andhra Pradesh, India. *Environ. Geol.* **2006**, *49*, 413–429. [CrossRef]

97. Hussain, Y.; Ullah, S.F.; Akhter, G.; Aslam, A.Q. Groundwater Quality Evaluation by Electrical Resistivity Method for Optimized Tubewell Site Selection in an Ago-Stressed Thal Doab Aquifer in Pakistan. *Model. Earth Syst. Environ.* **2017**, *3*, 15. [CrossRef]

98. Ravikumar, P.; Somashekar, R.; Angami, M. Hydrochemistry and Evaluation of Groundwater Suitability for Irrigation and Drinking Purposes in the Markandeya River Basin, Belgaum District, Karnataka State, India. *Environ. Monit. Assess.* **2011**, *173*, 459–487. [CrossRef] [PubMed]

99. Flores, A.M.; Shukla, M.K.; Schutte, B.J.; Picchioni, G.; Daniel, D. Physiologic Response of Six Plant Species Grown in Two Contrasting Soils and Irrigated with Brackish Groundwater and RO Concentrate. *Arid Land Res. Manag.* **2017**, *31*, 182–203. [CrossRef]

100. Hem, J.D. *Study and Interpretation of the Chemical Characteristics of Natural Water*; US Geological Survey; Department of the Interior: Reston, VA, USA, 1985; Volume 2254.

101. Rao, N.S.; Rao, P.S.; Reddy, G.V.; Nagamani, M.; Vidyasagar, G.; Satyanarayana, N. Chemical Characteristics of Groundwater and Assessment of Groundwater Quality in Varaha River Basin, Visakhapatnam District, Andhra Pradesh, India. *Environ. Monit. Assess.* **2012**, *184*, 5189–5214. [CrossRef] [PubMed]

102. Narsimha, A.; Sudarshan, V. Assessment of Fluoride Contamination in Groundwater from Basara, Adilabad District, Telangana State, India. *Appl. Water Sci.* **2017**, *7*, 2717–2725. [CrossRef]

103. Tiwari, A.K.; Singh, A.K.; Singh, A.K.; Singh, M. Hydrogeochemical Analysis and Evaluation of Surface Water Quality of Pratapgarh District, Uttar Pradesh, India. *Appl. Water Sci.* **2017**, *7*, 1609–1623. [CrossRef]

104. Lloyd, J.W.; Heathcote, J. *Natural Inorganic Hydrochemistry in Relation to Ground Water*; U.S. Department of Energy Office of Scientific and Technical Information: Oak Ridge, TN, USA, 1985.

105. McNeal, B.; Layfield, D.; Norvell, W.; Rhoades, J. Factors Influencing Hydraulic Conductivity of Soils in the Presence of Mixed-salt Solutions. *Soil Sci. Soc. Am. J.* **1968**, *32*, 187–190. [CrossRef]

106. Singaraja, C.; Chidambaram, S.; Anandhan, P.; Prasanna, M.; Thivya, C.; Thilagavathi, R.; Sarathidasan, J. Hydrochemistry of Groundwater in a Coastal Region and Its Repercussion on Quality, a Case Study—Thoothukudi District, Tamil Nadu, India. *Arab. J. Geosci.* **2014**, *7*, 939–950. [CrossRef]

107. Tijani, M.N. Hydrogeochemical Assessment of Groundwater in Moro Area, Kwara State, Nigeria. *Environ. Geol.* **1994**, *24*, 194–202. [CrossRef]

108. Singh, A.K.; Mondal, G.; Kumar, S.; Singh, T.; Tewary, B.; Sinha, A. Major Ion Chemistry, Weathering Processes and Water Quality Assessment in Upper Catchment of Damodar River Basin, India. *Environ. Geol.* **2008**, *54*, 745–758. [CrossRef]

109. Doneen, L. *Notes on Water Quality in Agriculture Published as a Water Science and Engineering*; Department of Water Sciences and Engineering, Paper 4001; University of California: Berkeley, CA, USA, 1964.

110. Islam, S.D.-U.; Majumder, R.K.; Uddin, M.J.; Khalil, M.I.; Ferdous Alam, M. Hydrochemical Characteristics and Quality Assessment of Groundwater in Patuakhali District, Southern Coastal Region of Bangladesh. *Expo. Health* **2017**, *9*, 43–60. [CrossRef]

111. Davraz, A.; Özdemir, A. Groundwater Quality Assessment and Its Suitability in Çeltikçi Plain (Burdur/Turkey). *Environ. Earth Sci.* **2014**, *72*, 1167–1190. [CrossRef]

112. Joshi, D.M.; Kumar, A.; Agrawal, N. Assessment of the Irrigation Water Quality of River Ganga in Haridwar District. *Rasayan J. Chem.* **2009**, *2*, 285–292. [CrossRef]
113. Selvakumar, S.; Ramkumar, K.; Chandrasekar, N.; Magesh, N.; Kaliraj, S. Groundwater Quality and Its Suitability for Drinking and Irrigational Use in the Southern Tiruchirappalli District, Tamil Nadu, India. *Appl. Water Sci.* **2017**, *7*, 411–420. [CrossRef]
114. Manoj, S.; Thirumurugan, M.; Elango, L. An Integrated Approach for Assessment of Groundwater Quality in and around Uranium Mineralized Zone, Gogi Region, Karnataka, India. *Arab. J. Geosci.* **2017**, *10*, 557. [CrossRef]
115. Elumalai, V.; Brindha, K.; Sithole, B.; Lakshmanan, E. Spatial Interpolation Methods and Geostatistics for Mapping Groundwater Contamination in a Coastal Area. *Environ. Sci. Pollut. Res.* **2017**, *24*, 11601–11617. [CrossRef]
116. Mirzaei, R.; Sakizadeh, M. Comparison of Interpolation Methods for the Estimation of Groundwater Contamination in Andimeshk-Shush Plain, Southwest of Iran. *Environ. Sci. Pollut. Res.* **2016**, *23*, 2758–2769. [CrossRef]
117. Rajmohan, N.; Niazi, B.A.; Masoud, M.H. Evaluation of a Brackish Groundwater Resource in the Wadi Al-Lusub Basin, Western Saudi Arabia. *Environ. Earth Sci.* **2019**, *78*, 451. [CrossRef]
118. *Drought Feeding and Management of Sheep A Guide for Farmers and Land Managers*, 2015th ed.; Victorian Government Department of Primary Industries: Melbourne, Australia, 1997.
119. Scarlett, T.; Carberry, P. Drought Feeding of Goats. New South Wales Agricultrue: Sydney. Available online: http//www.agric. nsw.gov.au/reader/5337 (accessed on 22 February 2002).
120. Maplecroft Climate Change and Environmental Risk. Available online: http://maplecroft.com (accessed on 16 July 2023).
121. World Health Organization. *Guidelines for Drinking-Water Quality: Second Addendum. Vol. 1, Recommendations*; World Health Organization: Geneva, Switzerland, 2008; ISBN 92-4-154760-X.
122. Ladeiro, B. Saline Agriculture in the 21st Century: Using Salt Contaminated Resources to Cope Food Requirements. *J. Bot.* **2012**, *2012*, 310705. [CrossRef]
123. Dalling, J.W.; Davis, A.S.; Schutte, B.J.; Elizabeth Arnold, A. Seed Survival in Soil: Interacting Effects of Predation, Dormancy and the Soil Microbial Community. *J. Ecol.* **2011**, *99*, 89–95. [CrossRef]
124. Keiffer, C.H.; Ungar, I.A. Germination and Establishment of Halophytes on Brine-affected Soils. *J. Appl. Ecol.* **2002**, *39*, 402–415. [CrossRef]
125. Santos, M.M.S.; Lacerda, C.F.; Neves, A.L.R.; de Sousa, C.H.C.; de Albuquerque Ribeiro, A.; Bezerra, M.A.; da Silva Araújo, I.C.; Gheyi, H.R. Ecophysiology of the Tall Coconut Growing under Different Coastal Areas of Northeastern Brazil. *Agric. Water Manag.* **2020**, *232*, 106047. [CrossRef]
126. Ribeiro, A.D.A.; Lacerda, C.F.; Rocha Neves, A.L.; Sousa, C.H.C.D.; Braz, R.D.S.; Oliveira, A.C.D.; Pereira, J.M.G.; Ferreira, J.F.D.S. Uses and Losses of Nitrogen by Maize and Cotton Plants under Salt Stress. *Arch. Agron. Soil Sci.* **2021**, *67*, 1119–1133. [CrossRef]
127. Dourado, P.R.M.; de Souza, E.R.; Santos, M.A.d.; Lins, C.M.T.; Monteiro, D.R.; Paulino, M.K.S.S.; Schaffer, B. Stomatal Regulation and Osmotic Adjustment in Sorghum in Response to Salinity. *Agriculture* **2022**, *12*, 658. [CrossRef]
128. Panta, S.; Flowers, T.; Lane, P.; Doyle, R.; Haros, G.; Shabala, S. Halophyte Agriculture: Success Stories. *Environ. Exp. Bot.* **2014**, *107*, 71–83. [CrossRef]
129. Miranda, K.R.D.; Dubeux, J.C.B.; Mello, A.C.L.D.; Silva, M.D.C.; Santos, M.V.F.D.; Santos, D.C.D. Forage Production and Mineral Composition of Cactus Intercropped with Legumes and Fertilized with Different Sources of Manure. *Ciênc. Rural* **2019**, *49*. [CrossRef]
130. Da Silva Brito, G.S.M.; Santos, E.M.; de Araújo, G.G.L.; de Oliveira, J.S.; Zanine, A.D.M.; Perazzo, A.F.; Campos, F.S.; de Oliveira Lima, A.G.V.; Cavalcanti, H.S. Mixed Silages of Cactus Pear and Gliricidia: Chemical Composition, Fermentation Characteristics, Microbial Population and Aerobic Stability. *Sci. Rep.* **2020**, *10*, 6834. [CrossRef]
131. Lessa, C.I.N.; de Lacerda, C.F.; Cajazeiras, C.C.d.A.; Neves, A.L.R.; Lopes, F.B.; Silva, A.O.d.; Sousa, H.C.; Gheyi, H.R.; Nogueira, R.d.S.; Lima, S.C.R.V.; et al. Potential of Brackish Groundwater for Different Biosaline Agriculture Systems in the Brazilian Semi-Arid Region. *Agriculture* **2023**, *13*, 550. [CrossRef]
132. Karim, A.; Gonzalez Cruz, M.; Hernandez, E.A.; Uddameri, V. A GIS-Based Fit for the Purpose Assessment of Brackish Groundwater Formations as an Alternative to Freshwater Aquifers. *Water* **2020**, *12*, 2299. [CrossRef]
133. Wagner, J.; Williams, S.; Webster, C. Biomarkers and Surrogate End Points for Fit-for-purpose Development and Regulatory Evaluation of New Drugs. *Clin. Pharmacol. Ther.* **2007**, *81*, 104–107. [CrossRef] [PubMed]
134. Metcalf, L.; Benn, S. The Corporation Is Ailing Social Technology: Creating a 'Fit for Purpose' Design for Sustainability. *J. Bus. Ethics* **2012**, *111*, 195–210. [CrossRef]
135. Hurlimann, A.; McKay, J.M. What Attributes of Recycled Water Make It Fit for Residential Purposes? The Mawson Lakes Experience. *Desalination* **2006**, *187*, 167–177. [CrossRef]
136. Muller, M. Fit for Purpose: Taking Integrated Water Resource Management Back to Basics. *Irrig. Drain. Syst.* **2010**, *24*, 161–175. [CrossRef]
137. Thrysøe, C.; Arnbjerg-Nielsen, K.; Borup, M. Identifying Fit-for-Purpose Lumped Surrogate Models for Large Urban Drainage Systems Using GLUE. *J. Hydrol.* **2019**, *568*, 517–533. [CrossRef]
138. Ries, M.; Trotz, M.; Vairavamoorthy, K. 'Fit-for-Purpose' Sustainability Index: A Simplified Approach for US Water Utility Sustainability Assessment. *Water Pract. Technol.* **2016**, *11*, 35–47. [CrossRef]

139. Cecconet, D.; Callegari, A.; Hlavínek, P.; Capodaglio, A.G. Membrane Bioreactors for Sustainable, Fit-for-Purpose Greywater Treatment: A Critical Review. *Clean Technol. Environ. Policy* **2019**, *21*, 745–762. [CrossRef]
140. Hötzl, H. Water resources management in the Middle East under aspects of climatic changes. In *Climatic Changes and Water Resources in the Middle East and North Africa*; Springer: Berlin/Heidelberg, Germany, 2008; pp. 77–92. [CrossRef]

Article

A Study on Water Rights Allocation in Transboundary Rivers Based on the Transfer and Inequality Index of Virtual Water

Xia Xu [1], Jing Yuan [1,*] and Qianwen Yu [2,*]

[1] Industrial Engineering Department, Architectural Engineering School, Tongling University, Tongling 244000, China; 027627@tlu.edu.cn
[2] Logistics Department, Business School, Suzhou University of Science and Technology, Suzhou 215009, China
* Correspondence: 154427@tlu.edu.cn (J.Y.); yuqianwen1992@163.com (Q.Y.)

Abstract: Virtual water exerts an essential effect on water resources, yet such effect is rarely considered in current studies on water rights allocation in transboundary rivers. Hence, this paper ran a case study on Taihu Lake Basin, collecting data from 2017 to make clear the physical water rights of four regions—Jiangsu Province, Zhejiang Province, Anhui Province, and Shanghai City—in the Basin. After that, the multiregional input–output (MRIO) approach was utilized to measure the trade in value-added (TiVA) transfer and virtual water transfer (VWT) and construct an inequality index of VWT (VWI). Next, water efficiency coefficient was employed to convert the VWT into riparian level. Finally, VWT and VWI were incorporated into the water rights allocation model to form up a water rights allocation scheme for Taihu Lake Basin. Results showed: (1) Jiangsu enjoys the most allocated physical water rights, followed by Zhejiang, and Anhui ranks the lowest; (2) Anhui and Jiangsu are net virtual water exporters (2.259 billion m^3 and 1.78 billion m^3, respectively), while Zhejiang and Shanghai are net importers (2.344 billion m^3 and 1.695 billion m^3, respectively); (3) Anhui suffers the most inequality—0.4401—followed by 0.5076 of Jiangsu, while Zhejiang has the most equal environment—0.7012; (4) after the inclusion of virtual water, the quantity of water rights allocation changes, whereas Anhui experiences the largest growth—144 million m^3—due to the dual effects from the highest VWT and inequality. In conclusion, the effect of virtual water is indispensable, so VWT and VWI should both be considered in the physical water rights allocation of transboundary rivers.

Keywords: virtual water; inequality index of virtual water transfer; water rights allocation

Citation: Xu, X.; Yuan, J.; Yu, Q. A Study on Water Rights Allocation in Transboundary Rivers Based on the Transfer and Inequality Index of Virtual Water. *Water* 2023, 15, 2379. https://doi.org/10.3390/w15132379

Academic Editor: Adriana Bruggeman

Received: 24 May 2023
Revised: 21 June 2023
Accepted: 24 June 2023
Published: 28 June 2023

1. Introduction

Water shortage leads to disputes over water resources between upper and lower parts of a drainage basin or even regional intense competition for water rights, such as the Maipo River in Nile and the Tennessee Valley in USA [1–4]. Hence, with the water rights fairly determined and allocated, such conflicts could be greatly relieved, beneficial to establish an equitable water trading market and transregional ecological compensation. In light of overseas attempts, the key to initial water rights allocation is to adhere to multiple objectives of social equity, economic benefit, ecological preservation, and risk control [5–7]. Scholars have discovered a variety of allocation methods such as linear programming, dynamic programming, multi-model methodology, the multi-objective method, collaboration, and game, and gained experience in exploring rules for initial water rights allocation in drainage basins [8–24].

Virtual water refers to the water hidden in trading products and varies according to regional trade cooperation [25–39]. Frequent trading leads to an increase in virtual water transfer (VWT) between upper and lower parts of a basin, which exerts an indirect impact on the actual amount of water resources [25–39]. Therefore, it is necessary to incorporate virtual water into initial water rights allocation [25,40–42]. However, most current studies are centered around physical water resources, evading the invisible effect of virtual water [8–24]. In this case, this paper included virtual water into the exploration of water rights allocation in transboundary river basins.

To include virtual water in physical water rights allocation, an underlying issue—unfairness between economic benefit and environmental cost—must be considered. So, this paper is grounded in the inequality index of VWT (VWI). Methodologies such as the Gini coefficient, Theil index, Lorenz curve, and coefficient of variation were utilized to measure inequality [43–48], and focused on the holistic perspective of trade–environmental inequality. Some scholars have probed into the equivalence between the implicit resource cost and economic value added of trade [49–52]. On such basis, this paper employed the environmentally extended multiregional input–output (MRIO) model to measure VWT and trade in value-added (TiVA) transfer to construct VWI, to analyze the imbalance between them and integrate it into an allocation model.

The principles of consensus among scholars regarding physical water rights allocation are domestic water first, food security, respect for the history and status quo, and sustainable development [5–7]. In terms of methodology, this paper, drawing from previous studies, thus constructed an index system grounded in principles of status quo, equity, efficiency, sustainability, and macro regulation. Second, as for VWT methodology, many scholars found that the multiregional input–output approach can clarify the interdependence among sections of an entire supply chain in an economy [53–60]. It provides a clearer quantification of the amount of water deployed in trade and makes the virtual water calculation more intuitive and accurate [53–60]. Therefore, this paper drew an input–output tablet for modelling to calibrate water resources allocation in trade [35–45]. Third, VWT is measured among provinces, while water rights allocation is measured in the riparian areas of a transboundary drainage basin. So, VWT needs to be converted. However, there are only a few studies to refer to. Some scholars have adopted water efficiency coefficient to convert from a qualitative perspective [41,61,62]. Considering this, we also measured and converted VWT using the proportion of the water efficiency coefficient of basin area to that of the corresponding province.

In general, many studies have proven the VWT included in trade and its influence, yet still mainly taken physical water into consideration while formulating water rights allocation schemes. Therefore, to fill this gap, this paper first makes clear the physical water rights allocation in the transboundary river, then calculates provincial VWT and constructed VWI and convert provincial VWT into riparian VWT. Last, the converted VWT and VWI are integrated into the water rights allocation model.

So, based on the preceding analysis, the research framework underpinning this paper is meticulously assembled (Figure 1). First, according to the allocation rules, this paper makes clear the physical water rights allocation in the transboundary river. Next, the MRIO approach is employed to measure the VWT and TiVA transfer, and VWI is constructed accordingly. Third, considering that the provincial VWT was predicted at the provincial level, while the physical water allocation at the basin level within the provincial jurisdiction, a conversion of provincial VWT into riparian VWT with water efficiency coefficient is necessitated to reduce errors. Last, the converted VWT and VWI are integrated into the water rights allocation model.

This paper makes the following contributions: (1) integrating virtual water into physical water rights allocation to enrich the theories of initial water rights allocation of transboundary rivers; (2) constructing VWI to make allocation schemes more equal and reasonable.

The rest of this paper is arranged as follows: Section 2 introduces the methodology and data collection; Section 3 summarizes the main results; Section 4 is discussion; Section 5 presents the conclusion.

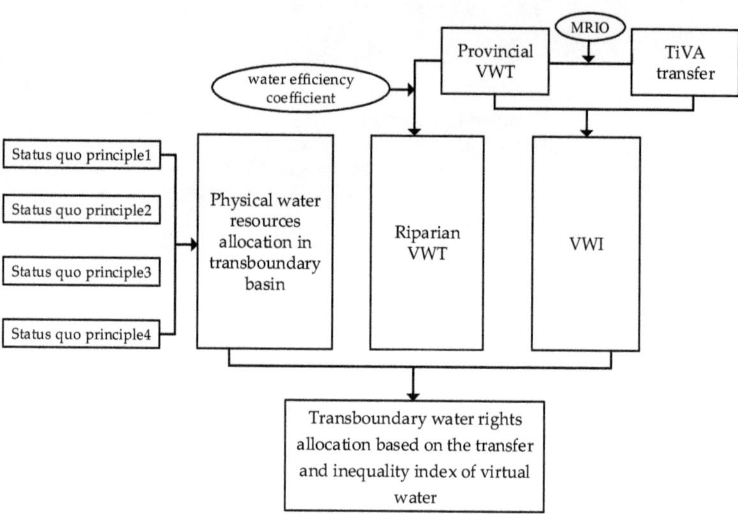

Figure 1. Research framework.

2. Methodology and Data Collection

2.1. Methodologies

2.1.1. Modelling of Physical Water Rights Allocation

Constructing an Index System of Physical Water Rights Allocation

Drawing from previous research [5–8], this paper builds an index system of physical water rights allocation (Table 1).

Table 1. Index system of physical water rights allocation.

Scheme	Principle	Indicator	Units	Symbol	Attribute
Physical water rights allocation scheme A	Status quo B_1	Current water use	Billion m^3	P_{11}	Benefit-based
		Water use per capita	m^3/person	P_{12}	Benefit-based
		Water use per farmland unit	m^3/mu	P_{13}	Benefit-based
		Current water supply scale	10,000 m^3	P_{14}	Benefit-based
	Equity B_2	Annual average runoff volume	Billion m^3	P_{21}	Benefit-based
		Population	10,000 people	P_{22}	Benefit-based
		Effective Irrigated area	m^3	P_{23}	Benefit-based
	Efficiency B_3	GDP per capita	10,000 yuan	P_{31}	Benefit-based
		Industrial output per capita	10,000 yuan	P_{32}	Benefit-based
		Agricultural output per capita	10,000 yuan	P_{33}	Benefit-based
		Water consumption per 10,000 yuan GDP	m^3	P_{34}	Cost-based
		Water consumption per 10,000 yuan agricultural output	m^3	P_{35}	Cost-based
		Water consumption per 10,000 yuan industrial output	m^3	P_{36}	Cost-based
	Sustainability B_4	Economic growth rate	%	P_{41}	Benefit-based
		Greening rate	%	P_{42}	Benefit-based
		Population growth rate	%	P_{43}	Benefit-based
		Proportion of waste water meeting discharge standards	%	P_{44}	Benefit-based
	Macro regulation B_5	Priority of regional development	points	P_{51}	Benefit-based
		Protection of vulnerable groups	points	P_{52}	Benefit-based

In Table 1, the lower grades of cost-based coefficients mean that more initial water rights should be allocated, while the higher grades of benefit-based coefficients mean that more initial water rights should be allocated.

Allocating Physical Water Rights

(1)   Decision making matrix

Suppose n regions in a drainage basin are involved in the initial water rights allocation, each having a total of $m(m = 1, 2, \cdots, 19)$ indicators, so m indicators of n regions constitute a decision making matrix below:

$$
X = \begin{bmatrix} x_{11} & x_{12} & \cdots & x_{1n} \\ x_{21} & x_{22} & \cdots & x_{2n} \\ \vdots & \vdots & \cdots & \vdots \\ x_{m1} & x_{m2} & \cdots & x_{mn} \end{bmatrix} = (x_{ij})_{m \times n}
\tag{1}
$$

(2)   Data normalization

The matrix X needs to be normalized to unitize the indicator standards.

For cost-based indicators, Formula (2) is used for standardization to eliminate the influence of inconsistent dimensions:

$$
y_{ij} = \frac{\max\limits_{i} x_{ij} - x_{ij}}{\max\limits_{i} x_{ij} - \min\limits_{i} x_{ij}}, i = 1, 2, \cdots, m; j = 1, 2, \cdots, n
\tag{2}
$$

For benefit-based indicators, Formula (3) is used for standardization to eliminate the influence of inconsistent dimensions:

$$
y_{ij} = \frac{x_{ij} - \min\limits_{i} x_{ij}}{\max\limits_{i} x_{ij} - \min\limits_{i} x_{ij}}, i = 1, 2, \cdots, m; j = 1, 2, \cdots, n
\tag{3}
$$

where $\max\limits_{i} x_{ij}$ represents the maximum of indicator i in region n, and $\min\limits_{i} x_{ij}$ is the minimum.

(3)   Project indicator function

Matrixes of both positive-ideal and negative-ideal solutions are below:

$$
\begin{aligned}
Y_j^+ &= \max\limits_{j}\{y_{1j}, y_{2j}, \cdots, y_{mj}\} \\
Y_j^- &= \min\limits_{j}\{y_{1j}, y_{2j}, \cdots, y_{mj}\}
\end{aligned}
\tag{4}
$$

Distance from every region to the positive-ideal solution is:

$$
d_j^+ = \left[\sum_{i=1}^{m} \omega_i (y_{ij} - y_{ij}^+)^2\right]^{0.5}
\tag{5}
$$

where ω_i stands for the weight of indicator i, and the smaller d_j^+ is, the closer region $j, j = 1, 2, \cdots, n$ is to the positive-ideal point, and the more water rights should be allocated to the region.

Similarly, the distance from every region to the negative-ideal solution is:

$$
d_j^- = \left[\sum_{i=1}^{m} \omega_i (y_{ij} - y_{ij}^-)^2\right]^{0.5}
\tag{6}
$$

where the bigger d_j^- is, the further region j is from the negative-ideal point, and the more water rights should be allocated to the region.

Relative proximity is used according to TOPSIS:

$$D_j = \frac{d_j^-}{d_j^- + d_j^+} \tag{7}$$

In terms of ω_j, qualitative indicators (p_{51}, p_{52}) and quantitative indicators are included, so an analytic hierarchy process (AHP) combining both indicators is employed to calculate the weights of indicators.

(4) Water rights allocation scheme

Based on Equations (5)–(7), the quantity of water rights allocated for every region is:

$$\varphi_j = \frac{D_j}{\sum\limits_{j=1}^{n} D_j} \tag{8}$$

Quantity of initial water rights is:

$$C_r = \varphi_r \cdot C_0 \tag{9}$$

where C_0 stands for the available water resources, and $(C_1, C_2 \cdots C_n)$ is the initial water rights allocation schemes.

2.1.2. Model for Calculating VWT and VWI
2.1.2.1. Provincial VWT Calculation

Provincial VWT is calculated based on the MRIO model. By constructing input–output matrices between regions, it can determine the virtual water input–output relationship generated in the trade of intermediate and final products between regions and reflect the distribution of virtual water in different sectors of different regions. Additionally, according to the CEADs' MRIO database, the VWT of each region is calculated. The specific steps are as follows:

A transboundary basin contains n provinces –province $1, 2, \ldots, n$. Other provinces are treated as province $(n + 1)$. In addition, this paper modified the MRIO table, as shown in Table A1 in Appendix A.

Based on the modified MRIO table, we can obtain

$$Z^{rs} = \sum\limits_{p=1}^{n+1} W^r L^{rp} F^{ps} \tag{10}$$

To be specific, r and s refer to basin provinces, p stands for the provinces trading with province r and s, and $n + 1$ is the number of provinces trading with province r and s. When $p = r$, VWT is included in the direct trade from province r to province s. When $p \neq r$, province r exports semifinished products to province p, which then further process into final products and export to province s, transferring water resources from province r to province s. It is, in other words, indirect VWT from province r to province s.

Z^{rs} represents VWT from province r to province s, W the direct water coefficient matrix, and W^r the direct water coefficient matrix of province r, which is directly obtained from the *China Statistic Yearbook*. L stands for the well-known Leontief inverse, representing the gross output generated throughout the production process of one unit of consumption; L^{rp} refers to the submatrices of Leontief inverse matrix for province r to province p, which is calculated by MRIO; F is the final demand matrix, and F^{ps} represents the submatrices of the final demand matrix for province p to province s, which is directly obtained by MRIO.

In the same way, VWT from province s to province r can be attained:

$$Z^{sr} = \sum_{p=1}^{n+1} W^s L^{sp} F^{pr} \tag{11}$$

where W^s stands for the direct water coefficient matrix of province s, L^{sp} the submatrices of the Leontief inverse matrix for province s to province p, and F^{pr} the submatrices of the final demand matrix for province p to province r.

In conclusion, the net VWT from province r to other provinces via trade is represented as follows:

$$kz^r = \sum_{s=1}^{n} (Z^{rs} - Z^{sr}) \tag{12}$$

If $kz^r > 0$, province r is a net exporter of virtual water; if $kz^r < 0$, it is a net importer.

VWI

Trade brings economic benefits to all parties involved, and the most direct benefits is domestic economic growth, namely, an increase in domestic value added. However, the value added of each country exists in two forms: imports and exports. Imports represent an increase in value added in other countries, while exports represent an increase in value added in one's own country. The net transfer of value added can be expressed as the difference between the value added of exports and imports. A positive value indicates that the country is in generally economic benefit, while a negative value results in economic losses.

From an environmental perspective, trade brings virtual water to different countries in either imported or exported forms. Countries with a net import of virtual water are the beneficiaries, while those with a net export are the losers. However, when two trading partners divide up the work differently in the global supply chain, the "hidden" water and added value in their traded goods also differ greatly. This means that the water and environmental costs that the trading countries pay can be much more than the economic benefits they receive, which is known as the unfairness of virtual water transfer, also known as the unfairness of trade value-added and virtual water transfer.

Therefore, this paper constructed the VWI based on the input–output of virtual water and trade value-added, and measured the gains (losses) of regions.

(1)   Model of provincial TiVA transfer

By using a MRIO model, we can calculate the impact of each region's trade on the value added of other regions, that is, TiVA transfer. The TiVA transfer of each region is decomposed to track the impact of final demand from other regions on the implied TiVA of the region. Meanwhile, if a region has a net output of value-added in its trade with other regions, it indicates that other regions are driving the economic growth; conversely, if a region is driving the economic growth of other regions, it indicates that the region is a net importer of value-added.

In the way mentioned in Section 2.1.2.1, TiVA transfer matrix (from province r to province s) is obtained:

$$V^{rs} = \sum_{p=1}^{n+1} (\mathrm{TiVA})^r L^{rp} F^{ps} \tag{13}$$

where $(\mathrm{TiVA})^r$ stands for the trade value added matrix of province r, which is obtained from MRIO directly. V^{rs} means TiVA transfer matrix.

The TiVA transfer matrix from province s to province r is

$$V^{sr} = \sum_{p=1}^{n+1} (\mathrm{TiVA})^s L^{sp} F^{pr} \tag{14}$$

Net TiVA transfer from province r to other provinces via trade is

$$kv^r = \sum_{s=1}^{n}(V^{rs} - V^{sr}) \tag{15}$$

If $kv^r > 0$, province r is a net exporter of TiVA; if $kv^r < 0$, province r is a net importer.

(2) Model of Inequality Index

TiVA estimates the value added in the production of goods and services for trade. Net TiVA transfer equals export minus import—if the result is positive then the province profits; if negative, the province loses. The relations between VWT and TiVA transfer, as shown in Figure 2, mainly fall into two categories: same direction and opposite direction. On such basis, suppose province r has positive net VWT and either positive or negative net TiVA transfer, then three relations can be formed—AA', BB' and CC'—as shown in Figure 3.

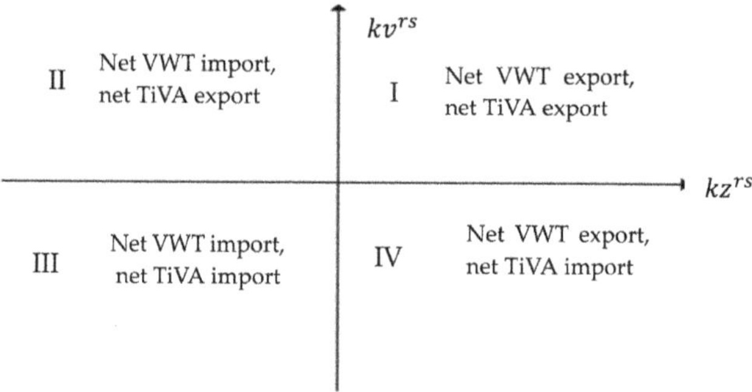

Figure 2. Relations between VWT and net TiVA transfer.

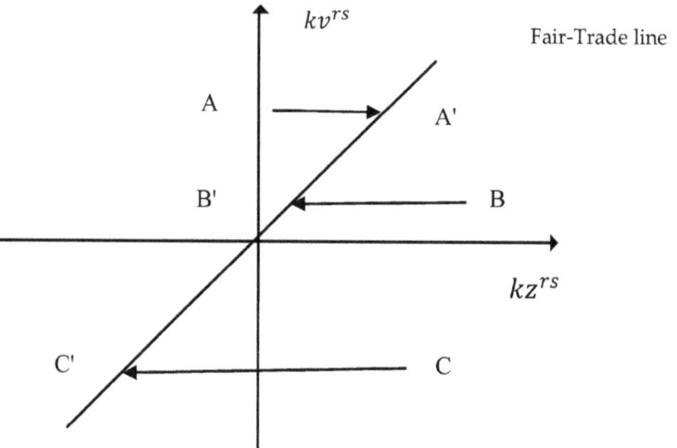

Figure 3. VWI.

In general, suppose the straight line cutting through the origin in Figure 3 is the fair-trade line, and the dots on it stand for the national mean of the relation between VWT and TiVA transfer. Namely, the net TiVA export (import) from unit net virtual water export (import) is up to the national mean. So, TiVA transfer caused by unit net VWT—slope of the fair trade line, is:

$$\beta = \left. \frac{\sum_{r=1}^{m}\sum_{s=1,r\neq s}^{m}|kv^{rs}|}{} \middle/ \sum_{r=1}^{m}\sum_{s=1,r\neq s}^{m}|kz^{rs}| \right. \tag{16}$$

where β represents the slope of the fair trade line for TiVA-VWT and $\beta \geq 0$; $|kv^{rs}|$ represents the absolute value of TiVA transfer from province r to province s; $|kz^{rs}|$ represents the absolute value of VWT from province r to province s; $\sum_{r=1}^{m} \sum_{s=1,r\neq s}^{m} |kv^{rs}|/2$ stands for the sum of net TiVA transfer of all provinces/cities in China; and $\sum_{r=1}^{m} \sum_{s=1,r\neq s}^{m} |kz^{rs}|/2$ stands for the sum of net VWT of all provinces/cities in China.

This paper hypothesized $kz^{r} > 0$ so that the dots only show up to the right of the Y-axis in Figure 3, namely Quadrant I. The dots A, B, and C away from the fair trade line are counterparts of A′, B′, and C′ on the line, and segments AA′, BB′, and CC′ refer to the net VWT to be added or subtracted for the dots to reach the line. Furthermore, DP, deflection distance, represents the deviation degree from the fair trade line:

$$DP = \begin{cases} \dfrac{\frac{kv^{rs}}{\beta} - kz^{rs}}{\frac{kv^{rs}}{\beta}} = 1 - \dfrac{\beta kz^{rs}}{kv^{rs}}, \ \dfrac{kv^{rs}}{\beta} > kz^{rs} > 0, 0 \leq DP < 1 \\[3mm] \dfrac{kz^{rs} - \frac{kv^{rs}}{\beta}}{kz^{rs}} = 1 - \dfrac{kv^{rs}}{\beta kz^{rs}}, \ 0 < \dfrac{kv^{rs}}{\beta} < kz^{rs}, 0 \leq DP < 1 \\[3mm] \dfrac{kz^{rs} - \frac{kv^{rs}}{\beta}}{kz^{rs}} = 1 - \dfrac{kv^{rs}}{\beta kz^{rs}}, \ \dfrac{kv^{rs}}{\beta} < 0, \ DP > 1 \end{cases} \tag{17}$$

where if the TiVA transfer and VWT are in Quadrant I, DP is $[0, 1)$; if in Quadrant IV, DP is $[1, \infty)$; the higher DP is, the further away the dot is from the fair trade line; if DP = 0, the dot is on the line, standing for equality.

On such basis, this paper constructed the exponential function $y = e^{-x}$ to build the VWI within $(0, 1]$ as follows:

$$\text{VWI} = e^{-DP} = \begin{cases} e^{-(1 - \frac{\beta kz^{rs}}{kv^{rs}})}, \ \frac{kv^{rs}}{\beta} \geq kz^{rs}, \text{ Dot above fair}-\text{trade line} \\[3mm] e^{-(1 - \frac{kv^{rs}}{\beta kz^{rs}})}, \ \frac{kv^{rs}}{\beta} \leq kz^{rs}, \text{ Dot below fair}-\text{trade line} \end{cases} \tag{18}$$

The closer VWI gets to 1, the more equal it is; the closer VWI gets to 0, the more unequal it is. In conclusion, the inequality between VWT and TiVA transfer is obtained and integrated into the initial water rights allocation model to make the model more equal and reasonable.

2.1.3. Coupling of Physical and Virtual Water
Regional VWT

Based on water use coefficient, the provincial VWT was then converted into the riparian level—VWT of the riparian area of each province:

$$KZ_{b}^{r} = \tau KZ^{r} \tag{19}$$

where τ is the conversion percentage and KZ_{b}^{r} is the net VWT of basin r.

Model Integrated with Virtual Water

Higher VWI means higher influence from VWT. As a result, once the physical water transfer, VWT, and VWI are made clear, a model of water rights allocation in a transboundary river included with virtual water is constructed below:

$$C_{r}^{\prime} = C_{r} + kz_{b}^{r} \times \left| \frac{1}{n} - \frac{VWI_{r}}{\sum\limits_{r=1}^{n} VWI_{r}} \right| \tag{20}$$

where $C_{r}^{\prime}$ is the amount of water allocation for region r with virtual water included.

2.2. Data Collection

2.2.1. Study Area

Taihu Lake Basin covers Jiangus Province, Zhejiang Province, Anhui Province, and Shanghai City. Additionally, the Taihu Lake Basin lies in the central area of the Yangtze River Delta region, bordered by the Yangtze River to the north, the East China Sea to the east, the Qiantang River to the south, and Tianmu and Maoshan Mountain to the west. It covers a water surface area of 2338 square kilometers, with an average water depth of 1.89 m and a maximum of 2.60 m. Its annual average water level and annual average water storage capacity are 3.21 m and 4.956 billion m^3 [63]. Water areas and administrative boundaries within the Taihu Lake Basin are shown in Figure 4.

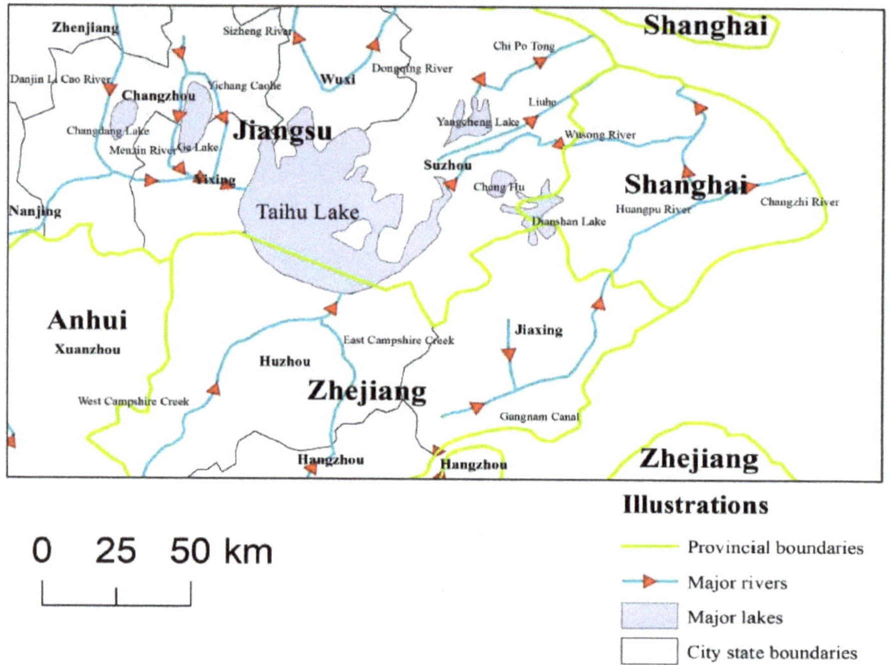

Figure 4. Map of Water Areas and Administrative Boundaries in the Taihu Lake Basin.

Additionally, Taihu Lake is situated in the subtropical zone with a mild and humid monsoon climate and interacts with more than 50 major inflowing and outflowing rivers. Figure 5 depicts the main water inflow and outflow in the Taihu Lake in 2021 [63].

In addition, in order to better manage the Taihu Lake Basin, China established the Taihu Lake Basin Authority, which oversees multiple administrative regions. It monitors the Taihu Lake Basin areas across three provinces and one city. Since the surface water is main water source for Taihu Lake, with an average annual supply 33.78 billion m^3 (groundwater sources and other sources supply 0.023 billion m^3 and 0.526 billion m^3, respectively) [63]. Therefore, the Authority mainly considers the surface water of these regions in its assessment indicators to ensure that the total amount of surface water used is within acceptable limits.

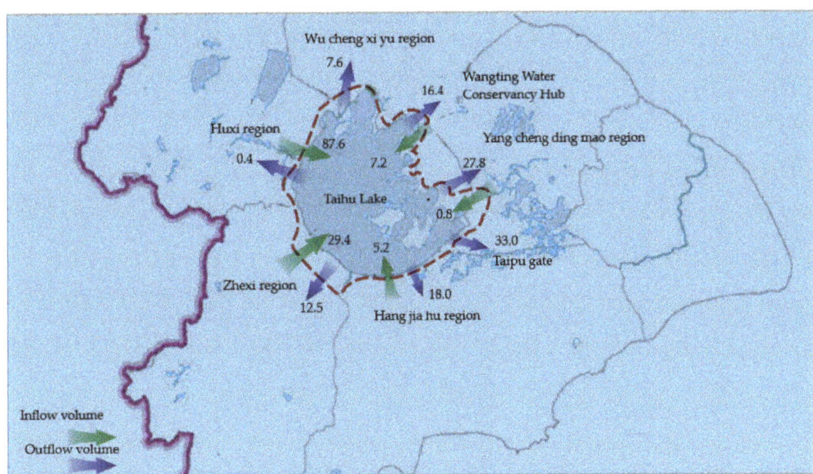

Figure 5. Water inflow and outflow quantity in the Taihu Lake in 2021 (100 million m^3).

In 2021, the total population of the Taihu Lake Basin was 68.11 million, accounting for 4.8% of China's total population. The regional gross domestic product (GDP) was 11.2736 trillion yuan, accounting for 9.9% of the national GDP. The per capita GDP was 165,000 yuan, double the national average [63]. As of 2021, the annual water resource consumption along the Basin reached 34.23 billion m^3, while it only supplied 28.99 billion m^3 of water, leading to a huge supply–demand gap [63]. The specific water consumption of the three provinces and one city is shown in Table 2.

Table 2. The specific water consumption of the three provinces and one city (billion m^3).

Region	Water Use	Domestic Water	Industrial Water	Total Water Consumption
Anhui		0.002	0.016	0.018
Jiangsu		1.59	18.27	20.02
Zhejiang		0.77	3.43	4.32
Shanghai		1.37	8.42	9.87

We know that the Basin is under tremendous pressure of worsening water scarcity, water pollution, and carrying capacity. Therefore, to facilitate cross-regional ecological compensation and water rights trading in the Taihu Lake Basin, it is imperative to establish an initial allocation of water rights based on the allocation plans from various provinces and regions. A well-designed water rights allocation plan is essential for promoting the development of the water rights market and ensuring its sustainability [8].

Meanwhile, the four regions constitute integral parts of the Regional Integrated Development Plan for Yangtze River Delta and contribute to many VWT and transactions, which produces an invisible effect on the actual virtual water amount. Therefore, this paper incorporated VWI and VWT into the initial water rights allocation scheme of the Taihu Lake Basin.

2.2.2. Data Sources

This paper chose data on virtual water in 2017—the latest public data on virtual water from Carbon Emission Accounts & Datasets (CEADs) [64]. The CEADs' input–output intermediate use table covers the input and output scenarios of various industries, and these industries can be classified as manufacturing, agriculture, and services. Therefore, this paper mainly combined specific industries of agriculture, manufacturing, and services for VWT calculation.

Data on physical water of Taihu Lake Basin from the communiques on water of the four regions and the Taihu Basin Authority in 2017 as well as the *China Statistic Yearbook* (2017) [63,65–68].

3. Result

3.1. Physical Water Allocation

3.1.1. Characteristic Value of Indicators

The characteristic values of indicators in Table 1 were attained based on relevant datasets, as detailed in Table 3.

Table 3. Indicator characteristic value of physical water allocation.

Indicator	Anhui N_1	Jiangsu N_2	Zhejiang N_3	Shanghai N_4
P_{11}	0.24	195	47.1	98.2
P_{12}	503	802	370	418
P_{13}	331	446	381	524
P_{14}	1.3	94.6	82.6	28.4
P_{21}	0.12	10.68	6.66	2.84
P_{22}	36.35	3186.51	1987.02	848.12
P_{23}	11.75	1030.43	642.55	274.26
P_{31}	2.8	7.02	13.6	18.6
P_{32}	1.91	4.81	7.97	9.82
P_{33}	0.37	0.45	0.24	0.06
P_{34}	214	52	35	33
P_{35}	1.85	0.011	0.01	0.0052
P_{36}	72	82	21	75
P_{41}	8.5	7.2	7.8	6.9
P_{42}	0.082	0.028	0.06	0.028
P_{43}	58.03	39.84	37.8	30
P_{44}	58.3	60.4	57.2	70.53
P_{51}	6	8	8	7
P_{52}	7	6	6	6

3.1.2. Normalization

Next, the characteristic values above were normalized using Equations (2) and (3) to formulate the following matrix:

$$X = \begin{bmatrix}
0.00 & 1.00 & 0.24 & 0.50 \\
0.31 & 1.00 & 0.00 & 0.11 \\
0.00 & 0.60 & 0.26 & 1.00 \\
0.00 & 1.00 & 0.87 & 0.29 \\
0.00 & 1.00 & 0.62 & 0.26 \\
0.00 & 1.00 & 0.62 & 0.26 \\
0.00 & 1.00 & 0.62 & 0.26 \\
0.00 & 0.28 & 0.68 & 1.00 \\
0.00 & 0.37 & 0.77 & 1.00 \\
0.79 & 1.00 & 0.46 & 0.00 \\
0.00 & 0.90 & 1.0 & 1.00 \\
0.00 & 0.10 & 0.10 & 1.00 \\
0.16 & 0.00 & 1.00 & 0.11 \\
1.00 & 0.19 & 0.56 & 0.00 \\
1.00 & 0.00 & 0.593 & 0.00 \\
1.00 & 0.35 & 0.28 & 0.00 \\
0.08 & 0.24 & 0.00 & 1.00 \\
0.00 & 1.00 & 1.00 & 0.50 \\
1.00 & 0.00 & 0.00 & 0.00
\end{bmatrix}$$

3.1.3. Indicator Weight

Then, AHP was utilized to make clear of the indicator weights, as detailed in Table 4.

Table 4. Indicator weight.

	B_i	P_i	ω_i
	B_1	P_{11}	0.1472
		P_{12}	0.0503
		P_{13}	0.0503
		P_{14}	0.0172
	B_2	P_{21}	0.0983
		P_{22}	0.1967
		P_{23}	0.1967
	B_3	P_{31}	0.0272
A		P_{32}	0.0272
		P_{33}	0.0272
		P_{34}	0.0136
		P_{35}	0.0272
		P_{36}	0.0136
	B_4	P_{41}	0.0363
		P_{42}	0.0140
		P_{43}	0.0140
		P_{44}	0.0054
	B_5	P_{51}	0.0251
		P_{52}	0.0125

3.1.4. Physical Water Allocation

The quantity and ratio of physical water allocation are calculated using Equations (6)–(10). Specifically, Anhui accounts for the least quantity (1.683 billion m^3) and ratio (5.51%) in the Taihu Lake Basin, whereas Jiangsu ranks the first, taking up 14.491 billion m^3 of physical water, followed by Zhejiang of 9.018 billion m^3 and then Shanghai of 6.318 billion m^3 (17.43%). Details are shown in Table 5.

Table 5. Result of physical water allocation in Taihu Lake Basin.

Region	Distance to Positive-Ideal Solution d_k^+	Distance to Negative-Ideal Solution d_k^-	Relative Proximity D_k	Ratio φ_k	Quantity (Billion m^3)
Anhui	0.8659	0.0992	0.1028	5.51%	1.683
Jiangsu	0.1070	0.8233	0.8850	47.50%	14.491
Zhejiang	0.2810	0.3444	0.5507	29.56%	9.018
Shanghai	0.4767	0.2293	0.3248	17.43%	6.318

3.2. VWT among Regions in Taihu Lake Basin

3.2.1. Provincial VWT Calculation

The import, export, and net import of virtual water among regions in Taihu Lake Basin are calculated using Equations (10)–(12), as detailed in Figures 6 and 7.

In accordance with Figures 4 and 5, Anhui exports the most virtual water, 4.285 billion m^3; Anhui and Jiangsu are the top net exporters −2.259 billion m^3 and −1.78 billion m^3, respectively—and they share a similar mutual VWT. Zhejiang, on the other hand, imports the most virtual water, 3.424 billion m^3, while Anhui the least, 2.505 billion m^3. Zhejiang and Shanghai are net importers 2.344 billion m^3 and 1.695 billion m^3, respectively.

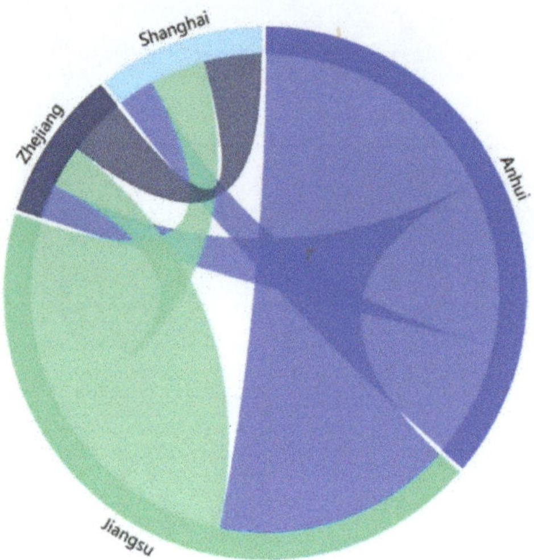

Figure 6. VWT of Taihu Lake Basin (100 million m^3).

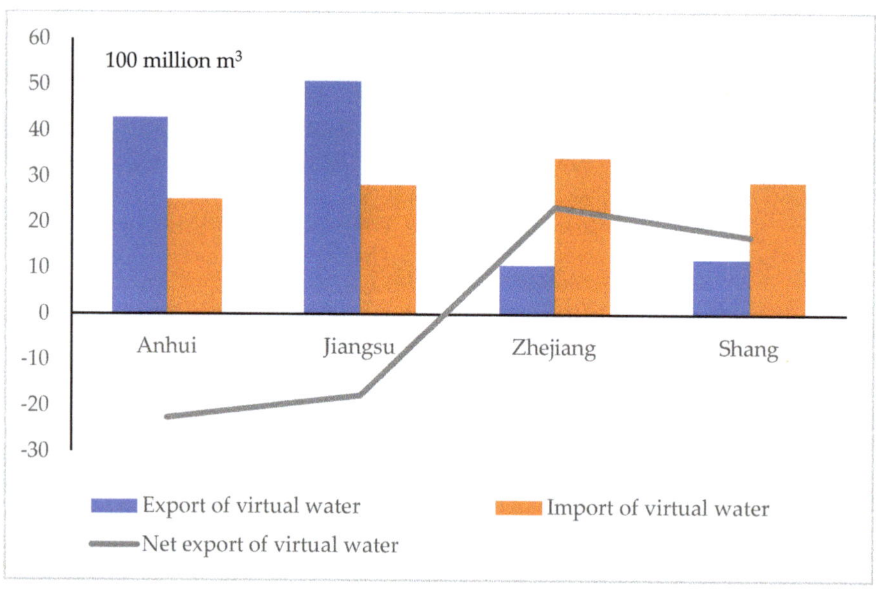

Figure 7. VWT within Taihu Lake Basin (100 million m^3).

3.2.2. VWI

VWI of the four regions was measured using their TiVA and VWT. Anhui has the lowest VWI 0.4401, indicating that it suffers the highest inequality between TiVA and VWT, while Zhejiang has the highest VWI 0.7012, indicating that VWT and TiVA in Zhejiang are in the most equal relation (Table 6).

Table 6. VWI of the four regions.

Region	VWI	$\dfrac{VWI}{\sum\limits_{r=1}^{n} VWI}$	$\dfrac{1}{n} - \dfrac{VWI}{\sum\limits_{r=1}^{n} VWI}$
Anhui	0.4401	0.1957	0.0543
Jiangsu	0.5076	0.2257	0.0243
Zhejiang	0.7012	0.3118	−0.0618
Shanghai	0.5999	0.2668	−0.0168

3.3. The Water Allocation Coupling Physical and Virtual Water Resources

Equations (19) and (20) lead to the result of water rights allocation considering both VWT and VWI, as detailed in Table 7.

Table 7. Water rights allocation of Taihu Lake Basin with virtual water included.

Region	Net VWT (KZ^{r}) (Billion m^3)	τ	Ratio φ_k	Quantity (Billion m^3)
Anhui	2.259	1.17	5.73%	1.827
Jiangsu	1.780	0.96	45.58%	14.533
Zhejiang	−2.344	1.03	28.15%	8.869
Shanghai	−1.695	1.34	19.93%	6.280

In detail, in terms of absolute quantity, Jiangsu still enjoys the most allocated water—14.533 billion m^3, accounting for 45.58%—followed by Zhejiang—8.869 billion m^3 and 28.15%. Anhui suffers the highest inequality and exports the most virtual water, so it witnesses the largest growth in water allocation, increasing by 1.683 billion m^3 to 1.827 billion m^3, whereas Shanghai drops to 6.28 billion m^3. Moreover, Zhejiang, even though boasting the most equal VWT, declines by 149 million m^3 of allocated water, even more than that of Shanghai, because it exports the most virtual water.

4. Discussion

4.1. Physical Water Allocation

To sum up, the allocation results in this paper are fundamentally in line with the findings of other scholars, except Anhui. Anhui Province was not considered in previous studies for its little-to-no share of the basin area. As for the allocation quantity of other regions, Jiangsu has the most allocated water, followed by Zhejiang and Shanghai; as for allocation ratio, Jiangsu declines by 6.44% while Zhejiang and Shanghai grow by 5.30% and 6.65%, respectively, in comparison with previous studies. When it comes to total ratio, however, Jiangsu still takes the largest proportion, 47.50%, while Shanghai takes the least, 17.43%, as shown in Table 8. The results above are basically in line with the actual conditions—Jiangsu takes the largest share of cities in the Basin and needs the most water; Anhui, on the other hand, only takes 0.6% of the basin area so it needs the least water, which also accords with the principle of status quo.

Table 8. Physical water rights allocation of Taihu Lake Basin.

Allocation Amount　　　　　Region	Anhui	Jiangsu	Zhejiang	Shanghai	Units
Amount in this paper	16.83	144.91	90.18	53.18	100 million m^3
Amount in previous studies	0	147.33	125.07	86.39	100 million m^3
Proportion in this paper	5.52	47.50	29.56	17.43	%
Proportion in previous studies	0	41.06	34.86	24.08	%

Comparing with Table 8, it is found that the absolute amount of water allocation and the allocation ratio in this paper is in line with that of previous studies. This can explain that the results of the physical water allocation in this paper are applicable and consistent with reality (Jiangsu has the largest water area, followed by Zhejiang, Shanghai, and Anhui). The difference is that previous studies ignored the Anhui. This is unfair to Anhui province, especially in the context of China's strict water resources management and control over water withdrawals in various regions. Therefore, we included the Anhui region in the allocation process. Additionally, this is one main contribution of this paper in terms of physical water allocation.

4.2. VWT

As for VWT of the four regions, Anhui exports the most virtual water (see Table 7 and Figure 7); as for VWI, Anhui trades a relatively large number of VWT for little economic benefit, indicating that Anhui accommodates many water-consuming yet few high-tech and water-saving industries, and thus is in greater need of economic structure reshaping. Meanwhile, Jiangsu, in comparison with Zhejiang and Shanghai, is also a net exporter, with its inequality ranking second (see Table 6 and Figure 7), mainly because Jiangsu also houses many water-consuming industries, especially in northern Jiangsu. As a result, Anhui and Jiangsu need to be allocated more water rights. On the other hand, Zhejiang and Shanghai are net importers—2.344 billion m^3 and 1.695 billion m^3, respectively. Hence, the two regions need to relinquish some water rights to Jiangsu and Anhui.

Additionally, the increased (decreased) amount in water rights allocation of the four regions experience certain changes after the inclusion of virtual water and VWI, as detailed in Figure 8. It illustrates the significant impact of virtual water on physical water and proves the necessity of including virtual water in physical water rights allocation. At the same time, after including the effect of virtual water, Anhui suffers the most loss in trade with the other regions and thus has the largest compensation, as Figure 8 suggests. It conforms to China's Regional Integrated Development Plan for Yangtze River Delta.

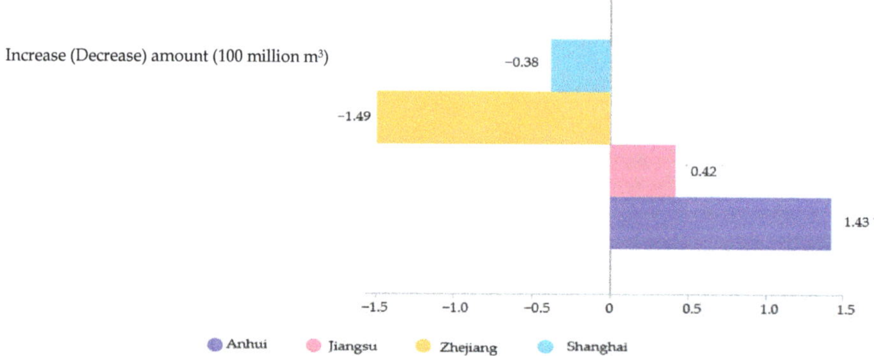

Figure 8. Increased (decreased) amount in water rights allocation of Taihu Lake Basin before and after embedded virtual water.

According to Tables 5 and 7, and Figure 8, we can see that the change quantity of allocation of the three provinces and one city is not significant before and after virtual water transfer. Although the Anhui, Jiangsu, Zhejiang, and Shanghai changes in volume are close to 100 million m^3, especially Anhui and Zhejiang, the changes exceeded 100 million m^3. In the case of abundant water, 100 million m^3 is a small proportion of Taihu Lake. However, in the case of water scarcity, it becomes extremely important. For example, in 2022, a severe drought occurred in the Yangtze River Basin, and water resource was scarce in each region; 100 million m^3 of water could provide water for one million people for one year. Therefore, 100 million m^3 of water becomes extremely precious.

Additionally, as strict water resource management and total water consumption control policies are implemented in China, the water consumption in Anhui, Zhejiang, and Shanghai have been included in the assessment. Failure to pass the assessment results in punishment. Therefore, even slight fluctuations in water consumption alert local governments.

Thirdly, the data in this paper are based on 2018. With the implementation of the *Yangtze River Integration Strategy*, the trade between these four regions will become closer. The VWT may increase, and the impact of virtual water would also become greater. This is also a topic that will be studied in the future to dynamically track the impact of virtual water.

Finally, the data in this paper are based on riparian VWT. If only the impact of provincial VWT was considered, the impact would be greater than it is now.

So, the transfer of interprovincial virtual water has an invisible impact on the absolute amount of water resources in the regions surrounding Taihu Lake. Therefore, virtual water should be incorporated into the allocation of physical water resources.

To conclude, virtual water yields a considerable invisible effect on physical water. If a region produces higher net export (net import) of virtual water with higher inequality, it suffers a higher imbalance between economy and environment, and it needs more compensation. Thus, virtual water must be included in ecological compensation as well as water rights allocation of transboundary rivers.

5. Conclusions

Increased transactions lead to higher interregional VWT, which further yields a higher impact on regional water resources. However, the role of virtual water is not considered in current studies on water rights allocation of transboundary rivers. On such basis, this paper ran a case study on Taihu Lake Basin to make clear the physical water rights of the four regions—Jiangsu, Zhejiang, Anhui, and Shanghai—along the Basin, then built a model on VWI. After that, the water use coefficient was employed to convert the provincial VWT into riparian VWT. Last, VWT and VWI were included in the model of physical water rights allocation, to form up a comprehensive allocation scheme for the Basin.

(1) Jiangsu enjoys the most allocated water, followed by Zhejiang, Shanghai, and then Anhui.
(2) Anhui and Jiangsu are net exporters of virtual water (Anhui > Jiangsu), whereas Zhejiang and Shanghai are net importers (Zhejiang > Shanghai).
(3) Anhui suffers the highest inequality, while Zhejiang boasts the most equal environment where economic benefit and environment are most matched.
(4) VWT and VWI exert an impact on the water rights allocation of the four regions. Anhui in particular experiences the largest growth in allocated water rights due to the dual effects from VWT and VWI.
(5) Anhui and Jiangsu are net exporters of virtual water, indicating that the two regions need economic structure reshaping more urgently.

Author Contributions: X.X.: Writing; J.Y.: Providing case and idea; Q.Y.: Providing revised advice. Additionally, the contributions of J.Y. and Q.Y. are the same, so both are recognized as corresponding authors. All authors have read and agreed to the published version of the manuscript.

Funding: This paper was supported by the National Natural Science Foundation of China, Yuan J, grant number 42271301; Anhui University Excellent Research and Innovation Project, Yuan J, grant number 2022AH010094; Anhui Provincial Education Department Humanities Key Fund, Xu X, grant number SK2021A0652; and Tongling College Talent Fund, Xu X, grant number 2021tlxyr15. Besides, the APC was funded by Wanjiang scholarship, Yuan J.

Data Availability Statement: The relevant data can be found in this article.

Acknowledgments: The Ministry of Education of Humanities and Social Science Project (no. 21YJCZH206).

Appendix A

Table A1. Multi regional input–output table.

Input \ Output			Intermediate Use — Basin Province 1, Industry 1	Intermediate Use — Basin Province 1, Industry m	Intermediate Use — Basin Province (n), Industry 1	Intermediate Use — Basin Province (n), Industry m	Intermediate Use — Basin Province (n + 1), Industry 1	Intermediate Use — Basin Province (n + 1), Industry m	Final Demand — Basin Province 1	Final Demand — Basin Province n	Final Demand — Other Province (n + 1)	Total Output
Intermediate use	Basin province 1	Industry 1	y_{11}^{11}	y_{1m}^{11}	y_{11}^{1n}	y_{1m}^{1n}	$y_{11}^{1(n+1)}$	$y_{1m}^{1(n+1)}$	f_1^{11}	f_1^{1n}	$f_1^{1(n+1)}$	y_1^1
		Industry m	y_{m1}^{11}	y_{mm}^{11}	y_{m1}^{1n}	y_{mm}^{1n}	$y_{m1}^{1(n+1)}$	$y_{mm}^{1(n+1)}$	f_m^{11}	f_m^{1n}	$f_m^{1(n+1)}$	y_m^1
	Basin province n	Industry 1	y_{11}^{n1}	y_{1m}^{n1}	y_{11}^{nn}	y_{1m}^{nn}	$y_{11}^{n(n+1)}$	$y_{1m}^{n(n+1)}$	f_1^{n1}	f_1^{nn}	$f_1^{1(n+1)}$	y_1^n
		Industry m	y_{m1}^{n1}	y_{mm}^{n1}	y_{m1}^{nn}	y_{mm}^{nn}	$y_{m1}^{n(n+1)}$	$y_{mm}^{n(n+1)}$	f_m^{n1}	f_m^{nn}	$f_m^{1(n+1)}$	y_m^n
	Other province (n + 1)	Industry 1	$y_{11}^{(n+1)1}$	$y_{1m}^{(n+1)1}$			$y_{11}^{n(n+1)}$	$y_{1m}^{(n+1)(n+1)}$	$f_1^{(n+1)1}$	$f_1^{n(n+1)}$	$f_1^{(n+1)(n+1)}$	$y_1^{(n+1)}$
		Industry m	$y_{m1}^{(n+1)1}$	$y_{mm}^{(n+1)1}$			$y_{m1}^{(n+1)(n+1)}$	$y_{mm}^{(n+1)(n+1)}$	$f_m^{(n+1)1}$	$f_m^{(n+1)n}$	$f_m^{(n+1)(n+1)}$	y_m^{n+1}
Added value			v_1^1	v_m^1			v_1^{n+1}	v_m^{n+1}				
Total input			y_1^1	y_m^1			y_1^{n+1}	y_m^{n+1}				
Direct water input						w_n^r						

References

1. Zuo, Q.T.; Zhang, Z.Z.; Ma, J.X. The relationship between the utilization level of water resources in the Yellow River Basin and economic and social development China Population. *Resour. Environ.* **2021**, *31*, 10.
2. Straffin, P.D.; Heaney, J.P. Game theory and the tennessee valley authority. *Int. J. Game Theory* **1981**, *10*, 35–43. [CrossRef]
3. Connell, D.; Grafton, R.Q. Water reform in the Murray-Darling Basin. *Water Resour. Res.* **2011**, *47*, 1–9. [CrossRef]
4. Genova, P.; Wei, Y. A socio-hydrological model for assessing water resource allocation and water environmental regulations in the Maipo River basin. *J. Hydrol.* **2023**, *617*, 129159. [CrossRef]
5. Hu, J.L.; Ge, Y.X. The allocation mode and coordination mechanism of water resources in the Yellow River—Also on the construction and management of the Yellow River water rights market. *Manag. World* **2004**, *8*, 43–60.
6. Feng, J. Optimal allocation of regional water resources based on multi-objective dynamic equilibrium strategy. *Appl. Math. Model.* **2021**, *90*, 1183–1203. [CrossRef]
7. Wu, D.; Liu, M.G. Research progress on initial water rights allocation methods in river basins. *People's Pearl River* **2022**, *43*, 7.
8. Ge, M.; Wu, F.P.; You, M. Initial provincial water rights dynamic projection pursuit allocation based on the most stringent water resources management: A case study of Taihu Basin, China. *Water* **2017**, *9*, 35. [CrossRef]
9. Wu, D.; Wu, F.P.; Chen, Y.P. A Double Layer Optimization Model for the Composite System of Initial Water Rights Allocation in a Basin. *Syst. Eng. Theory Pract.* **2012**, *32*, 196–202.
10. Van Campenhout, B.; D'Exelle, B.; Lecoutere, E. Equity–efficiency optimizing resource allocation: The role of time preferences in a repeated irrigation game. *Oxf. Bull. Econ. Stat.* **2015**, *77*, 234–253. [CrossRef]
11. Wang, L.; Fang, L.; Hipel, K.W. Basin-wide cooperative water resources allocation. *Eur. J. Oper. Res.* **2008**, *190*, 798–817. [CrossRef]
12. Zhang, W.; Wang, Y.; Peng, H.; Li, Y.; Tang, J.; Wu, K.B. A coupled water quantity–quality model for water allocation analysis. *Water Resour. Manag.* **2010**, *24*, 485–511. [CrossRef]
13. Read, L.; Madani, K.; Inanloo, B. Optimality versus stability in water resource allocation. *J. Environ. Manag.* **2014**, *133*, 343–354. [CrossRef]
14. Zeng, X.T.; Li, Y.P.; Huang, G.H.; Liu, J. Modeling of water resources allocation and water quality management for supporting regional sustainability under uncertainty in an arid region. *Water Resour. Manag.* **2017**, *31*, 3699–3721. [CrossRef]
15. Wu, D.; Wang, Y.H. A Multilayer Hierarchical Decision Model for Initial Water Rights Allocation in Watersheds under Dual Control Action. *China Popul. Resour. Environ.* **2017**, *27*, 10.
16. Milly, P.C.D.; Betancourt, J.; Falkenmark, M.; Hirsch, R.M.; Kundzewicz, Z.W.; Lettenmaier, D.P.; Stouffer, R.J. Stationarity is dead: Whither water management? *Science* **2008**, *319*, 573–574. [CrossRef]
17. Null, S.E.; Prudencio, L. Climate change effects on water allocations with season dependent water rights. *Sci. Total Environ.* **2016**, *571*, 943–954. [CrossRef]
18. Dau, Q.V.; Momblanch, A.; Adeloye, A.J. Adaptation by himalayan water resource system under a sustainable socioeconomic pathway in a high-emission context. *J. Hydrol. Eng.* **2021**, *26*, 50–55. [CrossRef]

19. Zhang, L.N.; Wu, F.P. A study on the coupling configuration model of quantity and quality of initial water rights in provinces and regions based on GSR theory. *Resour. Sci.* **2017**, *39*, 12.

20. Wang, H.M.; Yu, R.; Niu, W.J. Design of Water Quantity Coordination Scheme for Cross boundary Water Resources Conflict in the Zhanghe River Basin Based on Strong Reciprocity. *Theory Syst. Eng. Theory Pract.* **2014**, *34*, 2170–2178.

21. Lou, S.; Wang, H.M.; Niu, W.J.; Xu, Y. Research on group decision-making of water resource allocation based on interval intuitionistic fuzzy sets. *Resour. Environ. Yangtze River Basin* **2014**, *23*, 9.

22. Xevi, E.; Khan, S. A multi-objective optimisation approach to water management. *J. Environ. Manag.* **2005**, *77*, 269–277. [CrossRef] [PubMed]

23. Hong, S.; Xia, J.; Chen, J.; Wan, L.; Ning, L.; Shi, W. Multi-object approach and its application to adaptive water management under climate change. *J. Geogr. Sci.* **2017**, *27*, 259–274. [CrossRef]

24. Dadmand, F.; Naji-Azimi, Z.; Farimani, N.M.; Davary, K. Sustainable allocation of water resources in water-scarcity conditions using robust fuzzy stochastic programming. *J. Clean. Prod.* **2020**, *276*, 123812. [CrossRef]

25. Allan, J.A. Virtual water: A strategic resource global solutions to regional deficits. *Ground Water* **1998**, *36*, 545–546. [CrossRef]

26. Ye, Q.; Li, Y.; Zhuo, L.; Zhang, W.; Xiong, W.; Wang, C.; Wang, P. Optimal allocation of physical water resources integrated with virtual water trade in water scarce regions: A case study for Beijing, China. *Water Res.* **2018**, *129*, 264–276. [CrossRef]

27. Duarte, R.; Pinilla, V.; Serrano, A. The water footprint of the Spanish agricultural sector: 1860–2010. *Ecol. Econ.* **2014**, *108*, 200–207. [CrossRef]

28. Brindha, K. International virtual water flows from agricultural and livestock products of India. *J. Clean. Prod.* **2017**, *161*, 922–930. [CrossRef]

29. Hoekstra, A.Y.; Mekonnen, M.M. Imported water risk: The case of the UK. *Environ. Res. Lett.* **2016**, *11*, 055002. [CrossRef]

30. Wang, R.; Zimmerman, J. Hybrid analysis of blue water consumption and water scarcity implications at the global, national, and basin levels in an increasingly globalized world. *Environ. Sci. Technol.* **2016**, *50*, 5143–5153. [CrossRef]

31. Weinzettel, J.; Pfister, S. International trade of global scarce water use in agriculture: Modeling on watershed level with monthly resolution. *Ecol. Econ.* **2019**, *159*, 301–311. [CrossRef]

32. Zhao, X.; Liu, J.; Liu, Q.; Tillotson, M.R.; Guan, D.; Hubacek, K. Physical and virtual water transfers for regional water stress alleviation in China. *Proc. Natl. Acad. Sci. USA* **2015**, *112*, 1031–1035. [CrossRef] [PubMed]

33. Zhao, Y.; He, G.; Wang, J.; Gao, X.; Li, H.; Zhu, Y.; Jiang, S. Water stress assessment integrated with virtual water trade and physical transfer water: A case study of Beijing, China. *Sci. Total Environ.* **2020**, *708*, 431–478. [CrossRef] [PubMed]

34. Zhao, H.; Qu, S.; Guo, S.; Zhao, H.; Liang, S.; Xu, M. Virtual water scarcity risk to global trade under climate change. *J. Clean. Prod.* **2019**, *230*, 1013–1026. [CrossRef]

35. Xu, C.X.; Ma, C.; Tian, G.L.; Xie, W. Research on the Mechanism and Contribution Share of Virtual Water Trade on Regional Economy. *China Soft Sci.* **2011**, *12*, 110–119.

36. Islam, M.S.; Oki, T.; Kanae, S.; Hanasaki, N.; Agata, Y.; Yoshimura, K. A grid-based assessment of global water scarcity including virtual water trading. *Water Resour. Manag.* **2007**, *21*, 19–37. [CrossRef]

37. Brindha, K. National water saving through import of agriculture and livestock products: A case study from India. *Sustain. Prod. Consum.* **2019**, *18*, 63–71. [CrossRef]

38. Yu, C.; Yue, D.M.; Gong, M.D.; Wang, L.; Cheng, B. Log water footprint accounting and spatial flow pattern of countries along the "the Belt and Road". *For. Econ.* **2021**, *42*, 16–27.

39. Angelis, E.; Metulini, R.; Bove, V.; Riccaboni, M. Virtual water trade and bilateral conflicts. *Adv. Water Resour.* **2017**, *110*, 549–561. [CrossRef]

40. Tian, G.L.; Wang, X.W. A study on the virtual water flow relationship between China and countries along the Mekong River driven by agricultural trade. *J. North China Univ. Water Resour. Hydropower* **2018**, *39*, 16–23.

41. Wu, P.; Gao, X.R.; Zhao, X.N.; Wang, Y.B.; Sun, S.K. Basic framework of "two-dimensional and three-dimensional" coupled flow theory of solid water virtual water. *J. Agric. Eng.* **2016**, *32*, 1–10.

42. Heil, M.T.; Wodon, Q.T. Inequality in CO_2 emissions between poor and rich countries. *J. Environ. Dev.* **1997**, *6*, 426–452. [CrossRef]

43. Yang, T.; Liu, W. Inequality of household carbon emissions and its influencing factors: Case study of urban China. *Habitat Int.* **2017**, *70*, 61–71. [CrossRef]

44. Hedenus, F.; Azar, C. Estimates of trends in global income and resource inequalities. *Ecol. Econ.* **2006**, *55*, 351–364. [CrossRef]

45. Padilla, E.; Serrano, A. Inequality in CO_2 emissions across countries and its relationship with income inequality: A distributive approach-science direct. *Energy Policy* **2006**, *34*, 1762–1772. [CrossRef]

46. Wang, D.; Nie, R.; Wang, S.Z. Measurement and Decomposition of Regional Inequality in Carbon Dioxide Emissions in China: From the Perspective of Interpersonal Equity. *Sci. Res.* **2012**, *30*, 1662–1670.

47. Shang, H.Y. Comparison of methods for measuring the unfairness of virtual water consumption. *Rural. Water Resour. Hydropower China* **2011**, *5*, 1–5.

48. Wang, H.H.; Liu, H.C.; He, X.J.; Zeng, W. Research on Carbon Emission Fairness from the Perspective of Historical Accumulated Carbon Emissions. *China Popul. Resour. Environ.* **2016**, *26*, 22–25.

49. Sun, H.; Liu, Y.Y. Analysis of Interregional Carbon Emission Differences and Profit and Loss Deviation in China. *Manag. Rev.* **2016**, *28*, 89–96.

50. Yan, M.; Sun, H.A. Study on the Implied Carbon Transfer and Profit and Loss Deviation in Domestic Trade: Based on the Perspective of Economic Internal Circulation. *Technol. Econ. Manag. Res.* **2021**, *12*, 12–16.
51. Zhang, W.; Liu, Y.; Feng, K.; Hubacek, K.; Wang, J.; Liu, M.-M.; Jiang, L.; Jiang, H.; Liu, N.; Zhang, P.; et al. Revealing environmental inequality hidden in China's inter-regional trade. *Environ. Sci. Technol.* **2018**, *52*, 7171–7181. [CrossRef]
52. Leontief, W. Environmental repercussions and the economic structure: An input-output approach. *Rev. Econ. Stat.* **1970**, *52*, 262–271. [CrossRef]
53. Chapagain, A.K.; Hoekstra, A.Y. *Virtual Water Flows between Nations in Relation to Trade in Livestock and Livestock Products*; UNESCO-IHE: Delft, The Netherlands, 2003.
54. Antonelli, M.; Roson, R.; Sartori, M. Systemic input-output computation of green and blue virtual water 'flows' with an illustration for the Mediterranean region. *Water Resour. Manag.* **2012**, *26*, 4133–4146. [CrossRef]
55. Deng, G.; Lu, F.; Wu, L.; Xu, C. Social network analysis of virtual water trade among major countries in the world. *Sci. Total Environ.* **2021**, *753*, 142043. [CrossRef]
56. Zhu, Q.R.; Yang, L.; Liu, X. Research on water footprint of China's export trade and trade structure optimization. *Quant. Econ. Technol. Econ. Res.* **2016**, *12*, 42–60.
57. Dietzenbacher, E.; Velazquez, E. Analysing Andalusian virtual water trade in an Input-Output Framework. *Reg. Stud.* **2007**, *41*, 185–196. [CrossRef]
58. Lutter, S.; Pfister, S.; Giljum, S.; Wieland, H.; Mutel, C. Spatially explicit assessment of water embodied in European trade: A product-level multi-regional input-output analysis. *Glob. Environ. Chang.* **2016**, *38*, 171–182. [CrossRef]
59. Chen, Z.M.; Chen, G.Q. Virtual water accounting for the globalized world economy: National water footprint and international virtual water trade. *Ecol. Indic.* **2013**, *28*, 142–149. [CrossRef]
60. Xu, X.; Wu, F.; Yu, Q.; Chen, X.; Zhao, Y. Invisible Effect of Virtual Water Transfer on Water Quantity Conflict in Transboundary Rivers—Taking Ili River as a Case. *Int. J. Environ. Res. Public Health* **2022**, *19*, 8917. [CrossRef]
61. Xu, X.; Wu, F.; Yu, Q.; Chen, X.; Zhao, Y. Analysis on Management Policies on Water Quantity Conflict in Transboundary Rivers Embedded with Virtual Water—Using Ili River as the Case. *Sustainability* **2022**, *14*, 9406. [CrossRef]
62. Bureau of Taihu Lake Basin. Available online: http://www.tba.gov.cn/slbthlyglj/sj/sj.html (accessed on 10 April 2023).
63. Carbon Emission Accounts & Datasets. Available online: https://www.ceads.net/user/login.php?lang=cn (accessed on 10 March 2023).
64. China Statistical Yearbook. Available online: http://www.stats.gov.cn/sj/ndsj/2017/indexch.htm (accessed on 10 April 2023).
65. Anhui Provincial Department of Water Resources. Available online: http://slt.ah.gov.cn/tsdw/swj/szyshjjcypj/119406621.html (accessed on 10 March 2023).
66. Jiangsu Provincial Department of Water Resources. Available online: http://jssslt.jiangsu.gov.cn/art/2018/7/31/art_84437_10430538.html (accessed on 10 March 2023).
67. Zhejiang Provincial Department of Water Resources. Available online: http://slt.zj.gov.cn/col/col1229243017/index.html (accessed on 10 March 2023).
68. Shanghai Municipal Water Affairs Bureau. Available online: http://swj.sh.gov.cn/szy/20200910/138eca2189d24dfda9e6f2b40a24797a.html (accessed on 10 March 2023).

water

Article

Effects of Groundwater Depth on Vegetation Coverage in the Ulan Buh Desert in a Recent 20-Year Period

Ting Lu [1,2,3], Jing Wu [1,3,4], Yangchun Lu [1,2,3], Weibo Zhou [1,3] and Yudong Lu [1,2,3,*]

1 Key Laboratory of Subsurface Hydrology and Ecological Effects in Arid Region of the Ministry of Education, Chang'an University, Xi'an 710054, China; 2019029004@chd.edu.cn (T.L.); endless5wj@163.com (J.W.); 18142399823@163.com (Y.L.); zwbzyz823@163.com (W.Z.)
2 Key Laboratory of Mine Geological Hazards Mechanism and Control, Ministry of Natural Resources, Xi'an 710054, China
3 School of Water and Environment, Chang'an University, Xi'an 710054, China
4 Qinghai Institute of Geo-Environment Monitoring, Xining 810008, China
* Correspondence: luyudong@chd.edu.cn

Abstract: As a typical desert in the Inner Mongolia Autonomous Region, the Ulan Buh Desert has a dry climate and scarce precipitation all year round. Groundwater has become the main factor limiting the growth of vegetation in this region. It is of great significance to study the influence of groundwater depth on the spatial distribution pattern of vegetation in this region. Based on the PIE-Engine platform and using long-term time-series Landsat data, this paper analyzed the spatial–temporal distribution characteristics and trends in vegetation coverage in the Ulan Buh Desert in the last 20 years using a pixel dichotomy model and the image difference method. The Kriging interpolation method was used to interpolate the groundwater depth data from 106 monitoring wells in the Ulan Buh Desert over the past 20 years, and the spatial distribution characteristics of groundwater depth in the Ulan Buh Desert were analyzed. Finally, the correlation coefficient between changes in vegetation coverage and changes in groundwater depth was calculated. The results showed the following: (1) The vegetation coverage in the Ulan Buh Desert was higher in the periphery and lower in the center of the desert. The overall vegetation level showed an increasing trend year by year; the growth rate was 4.73%/10 years, and the overall vegetation cover showed an improving trend. (2) The overall groundwater depth in the Ulan Buh Desert was deep in the southwest and shallow in the northeast. In the past 20 years, the groundwater depth in the Ulan Buh area has become shallower, and the ecological condition has gradually improved. (3) On the whole, the vegetation coverage varied with the groundwater depth, and the shallower the groundwater depth, the greater the vegetation coverage. When the groundwater depth increased to more than 4 m, the change in the groundwater depth had a significant effect on the vegetation coverage. However, when the groundwater depth was greater than 6 m, the change in the groundwater depth had no significant effect on the change in vegetation coverage.

Keywords: Ulan Buh Desert; vegetation coverage; pixel binary model; groundwater depth; correlation coefficient

Citation: Lu, T.; Wu, J.; Lu, Y.; Zhou, W.; Lu, Y. Effects of Groundwater Depth on Vegetation Coverage in the Ulan Buh Desert in a Recent 20-Year Period. *Water* **2023**, *15*, 3000. https://doi.org/10.3390/w15163000

Academic Editors: Peiyue Li and Jianhua Wu

Received: 12 July 2023
Revised: 9 August 2023
Accepted: 11 August 2023
Published: 20 August 2023

1. Introduction

Vegetation has always been the material basis for human survival and development [1]. It is also an important part of the terrestrial ecosystem and plays an irreplaceable role in the sustainable development of global and regional ecosystems [2–4]. In recent years, climate change [5,6] and human activities [7–9] have significantly altered the dynamics of terrestrial plants.

There has been a large amount of research on surface water in various regions of the world [10–13], but relatively little research on groundwater. Inner Mongolia is in a transition zone from a humid area to an arid and semi-arid area in the north of China, with an uneven distribution of water resources and great variations in runoff [14,15], both of which are very sensitive to changes in the ecological environment, making it one of the ideal regions to study changes in regional vegetation [16,17]. As a typical desert in Inner Mongolia, the Ulan Buh Desert has attracted the attention of many scholars for a long time. Due to its proximity to the border of the East Asian summer monsoon region, this region is more sensitive to the fluctuation in monsoon intensity and is one of the most seriously desertified areas in China. The climate in this region is arid all year long with minimal precipitation [18,19], and the growth and development of vegetation are highly dependent on the groundwater burial depth [20–22]. Exploring the relationship between vegetation and groundwater burial depth in this region can provide a certain reference value for the study of desert vegetations in northern China.

Groundwater has potential ecological consequences in the Ulan Buh Desert [23–25]. When the groundwater depth is deep, the soil moisture content becomes low, and vegetation growth is limited. When the groundwater depth is shallow, the soil moisture content increases, and the vegetation biomass changes dramatically under the action of groundwater capillarity. However, when the burial depth is too shallow, the salinization of shallow soil will inhibit plant growth to a certain extent [26,27]. Therefore, it is of great significance for vegetation protection in the Ulan Buh area to conduct the quantitative analysis of vegetation coverage in relation to different buried depths and explore the influence of groundwater depth on the vegetation spatial distribution pattern.

At present, there are two main methods used to study the correlation between groundwater depth and vegetation. One is to study the appropriate ecological water level of different vegetation populations after vegetation population division in the study area [28,29]. Cheng Yan et al. [30] adopted a vegetation quadrat survey to obtain vegetation feature information in the study area and used a Gaussian model to conduct the statistical analysis of vegetation features and groundwater depth. Zhang et al. [31] statistically analyzed the critical water level of ecological vegetation succession after groundwater development using the vegetation structure map analysis method. Although these methods are simple, intuitive, and highly applicable, field investigations require a lot of manpower and material resources, and the accuracy of the samples directly determines the reliability of the results [32]. The second method is to use the vegetation index as a regional ecological evaluation factor to study the response relationship between it and the groundwater depth [33–35]. Jin et al. [34] analyzed the correlation between the *NDVI* (normalized difference vegetation index) and groundwater depth in the Yinchuan Plain and found that the groundwater level depth had a significant control effect on vegetation growth. Song et al. [36] explored the correlation between vegetation and various influencing factors in a desert grassland area in Inner Mongolia using multiple sources of remote sensing satellite data and groundwater data. Within a certain threshold range, there is a clear linear relationship between the *NDVI* and groundwater depth. As a measure of the surface vegetation cover condition [37,38], vegetation coverage can reveal things about the regional ecological environment and evaluate the regional ecological quality [39,40]. At present, a large number of studies have shown that the *NDVI* is the most commonly used variable and is a highly useful index to calculate vegetation coverage [41–43].

This study used long-term Landsat data to extract vegetation coverage information using the PIE (Pixel Information Expert) Engine platform in order to study the changes in the vegetation cover in the Ulan Buh Desert over the past twenty years. The Kriging interpolation method was used to process the measured groundwater data in the Ulan Buh Desert, and the correlation coefficient between vegetation coverage and the groundwater depth was calculated to clarify the spatial response relationship between vegetation coverage and groundwater depth in the Ulan Buh area in order to provide a scientific basis for vegetation restoration and groundwater resource management in the Ulan Buh area.

2. Materials and Methods

2.1. Study Area

The Ulan Buh Desert (Figure 1) is one of the major deserts in China. It is part of the northwest desert region of our country, located in the Alashan League and Bayannur City in the west of the Inner Mongolia Autonomous Region, bordering the Yellow River in the east, reaching the northern foothills of the Helan Mountain in the south, and extending to the Wolfshan-Bayannur Mountain Range in the west [44]. The elevation of the Ulan Buh Desert area ranges from 971 to 1353 m, with a relative elevation difference of 340 m, and the relative height of local dunes can reach 50–60 m [18]. The landform is mainly dominated by denudation hills, accumulated platforms, accumulated basins, the Yellow River valley, and alluvial plains. The Ulan Buh Desert is located in the temperate, semi-arid to arid climate transition zone. The climate is characterized by sufficient sunshine, little rain, hot summers, cold winters, large daily temperature differences, strong evaporation, strong winds, and a short frost-free period. The average annual temperature reaches 8.6 °C. The highest temperature in July is 38.7 °C, the lowest temperature in January is −32.8 °C, the average annual precipitation reaches 116–162 mm, and the average annual evaporation is 2560–3200 mm [45,46]. The main vegetation is sea buckthorn, sand holly, white thorn, overlord, red yarn, and reed.

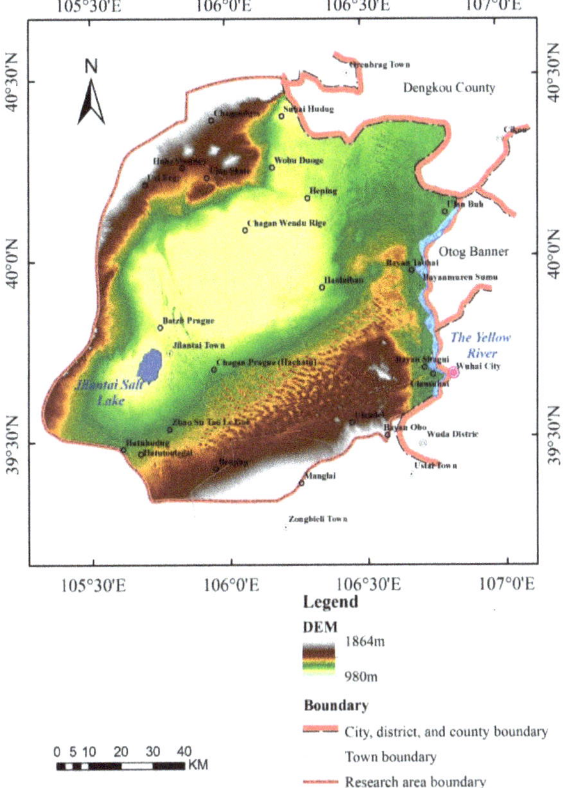

Figure 1. Overview of the study area.

2.2. Data Source and Processing

The Ladsat-5 multispectral data and Ladsat-8 multispectral data used in this paper are from the Landsat program of the National Aeronautics and Space Administration (NASA), SRTM (Shuttle Radar Topography Mission, which was established with support from the

German Aerospace Center (DLR)), and the Italian Space Agency (ASI). An international project was formed by NASA and the National Geospatial-Intelligence Agency (NGA) to acquire ordnance survey data through space shuttle-mounted radar [47]. The meteorological data are from the National Meteorological Information Center (http://data.cma.cn/ accessed on 3 April 2023), using average annual temperatures and precipitation from 2000 to 2020.

The groundwater monitoring data used in this study consist of the measured groundwater depth levels at 107 monitoring wells in Jilantai, Chaganwendugrige, Maiwulajia, Bayanaobao, Heping, and other regions in 2000, 2005, 2010, 2015, and 2020 [48].

The PIE-Engine platform is a professional PaaS/SaaS cloud computing service platform built on container cloud technology and independently developed by Aerospace Hongtu, which is similar to the GEE (Google Earth Engine) and has powerful data storage and high-performance analysis and calculation capabilities. Cheng Wei demonstrated the effectiveness of the spatiotemporal remote sensing cloud computing platform of the PIE-Engine Studio in their paper "Research and Application" [49]. Remote sensing data used in this study were processed online with the PIE-Engine platform, including Landsat-5 multispectral data, Landsat-8 multispectral data, and SRTM digital elevation data. The optical composite images were created from data obtained during the months of August and September in 2000, 2005, 2010, 2015, and 2020. This period was chosen because it provides the most cloud-free data and is within the growth period of vegetation, which can retain more vegetation information. To mitigate the effects of cloud pollution, the percentage of cloud cover was limited (<20%) when synthesizing cloudless images. Then, the Landsat cloud mask algorithm was used to calculate the image in the specified time and space range, and the median synthesis method was used to reconstruct the minimum cloud coverage composite image. Benefiting from the PIE-Engine platform's data operation and management mechanism, all remote sensing data used in this study were sampled to 30 m, and the PIE-Engine ensured geometric registration accuracy between different data sources by using a unified coordinate system based on the embedded algorithm. Using the PIE-Engine platform, the near-infrared band (NIR, 0.76~0.96 μm) and visible *RED* band (*RED*, 0.62~0.69 μm) were calculated to obtain the *NDVI* of each pixel in the study area. Finally, the *NDVI* remote sensing images with outliers removed were obtained with the calculation formula of outliers.

2.3. Research Method

2.3.1. Calculation of *NDVI*

The *NDVI* is used as a long-term monitoring tool to evaluate the growth status of plant coverage and is the most common standardized method to measure vegetation cover [50]. It is sensitive to vegetation growth and change and is the most commonly used index by analysts at present [51]. Formula (1) for calculating the *NDVI* is as follows:

$$NDVI = \frac{NIR - RED}{NIR + RED} \tag{1}$$

NIR is the near-infrared wave band, and *RED* refers to infrared wave band. This research adopts the Landsat TM/ETM images' *RED*-corresponding wave band. The Landsat OLI images' *NIR*-corresponding band is Band5 and the *RED*-corresponding band is Band4.

For outliers, with *NDVI* values greater than 1 and less than −1, a mask calculation on the PIE-Engine platform was used to remove them.

2.3.2. Estimation of Vegetation Coverage

The pixel binary model is a simple and practical remote sensing estimation model. It is often used to calculate vegetation cover because it can reduce the influence of atmosphere and water on remote sensing images [52]. The principle of the binary pixel model is to divide the spectral information of a remote sensing image into two parts, namely vegetation cover and no vegetation cover. The specific calculation formula is as follows:

$$VFC = \frac{S - S_{soil}}{S_{veg} - S\_soil} \tag{2}$$

Here, VFC is vegetation coverage, S is mixed pixels information, S_{soil} means no vegetation information, and S_{veg} indicates vegetation-like meta information. When the binary pixel model is used to analyze vegetation information, the $NDVI$ is usually used as an estimate. Replacing pixel information with the $NDVI$ can reduce the error caused by radiation. The formula is as follows:

$$VFC = \frac{NDVI - NDVI_{soil}}{NDVI_{veg} - NDVI_{soil}} \tag{3}$$

$NDVI_{soil}$ means no vegetation information and $NDVI_{veg}$ indicates vegetation-like meta information. Usually, many scholars intercept the upper and lower thresholds of the $NDVI$ within a certain confidence interval according to the gray distribution of the $NDVI$ in the whole image to approximately represent $NDVIv$ and $NDVIs$ [53,54]. In this paper, according to the frequency statistical chart of $NDVI$ data, the $NDVI$ value with a cumulative frequency of 5% is taken as $NDVIs$, and the $NDVI$ value with a cumulative frequency of 95% is taken as $NDVIv$.

Five periods of remote sensing images from 2000 to 2020 were processed based on the PIE-Engine platform, and five sets of vegetation coverage data were obtained in 2000, 2005, 2010, 2015, and 2020. The vegetation coverage data were processed and the change trend of the vegetation coverage in the Ulan Buh Desert was analyzed.

According to the vegetation characteristics of the Ulan Buh Desert and the technical regulations of the Land Use Status Investigation, Technical Specifications of Chinese Desert Cataloging and National Ecological and Environmental Standards of the People's Republic of China [55], the vegetation coverage of the Ulan Buh Desert was divided into four levels: very low coverage ($0 \leq VFC < 0.2$), low coverage ($0.2 \leq VFC < 0.3$), medium coverage ($0.3 \leq VFC < 0.6$), and high coverage ($0.6 \leq VFC < 1$).

2.3.3. Difference Comparative Analysis

The image difference method involves subtracting or dividing the remote sensing images of two time phases. The principle is that the unchanged part of the image generally has an equal or similar gray value in the remote sensing image of the two phases, and when the two images change, the gray value of the corresponding position will be greatly different. It can be conducted using grayscale values or feature values to obtain the difference image [56]. The spatiotemporal changes in vegetation in the study area could be obtained through the image difference method, and the positive and negative values of the difference could reflect the increase or decrease in vegetation [57]. A positive difference indicates an increase in vegetation in the study area, a negative difference indicates a decrease in vegetation in the study area, and a difference of 0 indicates no change in vegetation cover status. The specific formula is as follows (4):

$$\Delta VFC = VFC_{year2} - VFC_{year1} \tag{4}$$

In the formula, ΔVFC represents the change in vegetation coverage, VFC_{year2} represents the vegetation coverage in the following year, and VFC_{year1} represents the vegetation coverage in the previous year. By comparing the vegetation cover map of the Ulan Buh Desert in 2000 with the vegetation cover map in 2020, the spatial changes in the vegetation cover of the Ulan Buh Desert were obtained. According to the method of standard deviation, the spatial change in vegetation cover in the Ulan Buh Desert was divided into 7 levels, including extreme improvement ($0.67 \leq \Delta VFC < 1$), moderate improvement ($0.33 \leq \Delta VFC < 0.67$), slight improvement ($0 \leq \Delta VFC < 0.33$), unchanged ($\Delta\Delta VFC = 0$), slight decline ($-0.33 \leq \Delta VFC < 0$), moderate decline ($-0.67 \leq \Delta VFC < -0.33$), and extreme decline ($-1 \leq \Delta VFC < -0.67$).

2.3.4. Data Gridding

Data gridding refers to the method of converting point positioning data into surface data through spatial topology analysis [58]. The purpose of gridding is to make each data point more standardized for statistical purposes. This article used the fishing net tool and a square grid as the grid shape for the data grid. A total of 1703 regular square grids with an area of 6.25 km^2 were established within the study area.

2.3.5. Kriging Interpolation

Kriging interpolation is based on the concept of spatial autocorrelation, that is, the closer the points are, the stronger the correlation between them. It is a method of assigning weight to each sample according to its spatial distribution position and the degree of correlation between the samples, and it estimates the average value of the samples on unknown sample points in a weighted-average manner [59]. The ordinary Kriging interpolation method is the most basic and widely used interpolation method among all Kriging interpolation methods. It first considers the variation distribution of spatial attributes in the spatial position, determines the distance range that affects the value of a point to be interpolated, and then estimates the attribute value of the point to be interpolated with the sampling points in this range. Its basic principle is to estimate data from other unobserved positions in space through regularly distributed sample data. Therefore, it is necessary to fit an empirical semi-variogram model to reflect the relevant characteristics of the spatial data, and then obtain weights for prediction [60]. The calculation formula for the most basic semi variogram is as follows (5) [61]:

$$\gamma(h) = \frac{1}{2N(h)} \sum_{i=1}^{N(h)} [z(x_i + h) - z(x_i)]^2 \tag{5}$$

In the formula, h is the sample spacing, N *(h)* is the logarithm of sample points separated by a distance h in space, and $z(x_i)$ and $z(x_i + h)$ are the variable values at points x_i and x_i+h, respectively. For $\gamma(h)$, as the sample spacing h increases, the half square difference of all lag distance pairs reaches a relatively stable constant value from a non-zero value.

This study used Kriging interpolation to analyze the spatial distribution characteristics of groundwater depth data from 106 monitoring wells in the Ulan Buh Desert region from 2000 to 2020 and summarized the spatiotemporal changes in groundwater depth in the region over a five-year period. This study aimed to analyze the correlation between groundwater depth and vegetation coverage.

2.3.6. Correlation between Groundwater Depth and Vegetation Coverage

Correlation analysis refers to the analysis of two or more correlated variable elements to measure the degree of correlation between the two factors [62]. In order to study the impact of groundwater depth on vegetation coverage, this study used pixels as the calculation unit to calculate the correlation coefficients of vegetation coverage changes and groundwater depth changes between 2000 and 2020. The calculation Formula (6) is as follows [63–66]:

$$R = \frac{\Sigma(x_i - \bar{x})(y_i - \bar{y})}{\Sigma(x_i - \bar{x})^2 \Sigma(y_i - \bar{y})^2} \tag{6}$$

In the formula, x_i represents the vegetation coverage in 2000 and 2020, y_i represents the groundwater depth in 2000 and 2020, $\bar{x}$ is the average vegetation coverage over the past 20 years, and $\bar{y}$ is the average groundwater depth over the past twenty years.

3. Results and Discussion

3.1. Temporal and Spatial Variation in Vegetation Coverage

3.1.1. Temporal Distribution Characteristics of the Vegetation Cover

The trend of vegetation cover change in the Ulan Buh Desert over the past 20 years is shown in Figure 2. During the period from 2000 to 2020, the annual average vegetation cover of the Ulan Buh Desert showed an overall increasing trend year by year, ranging from 0.30 to 0.46, with an increase rate of 4.73% per decade. The highest vegetation cover occurred in 2020, with a vegetation coverage of 0.4560. The lowest vegetation coverage occurred in 2000, with a coverage of 0.3377. The vegetation coverage showed a slight downward trend in 2010. Except for 2010, the vegetation coverage in other years showed an upward trend compared to the previous period, and the growth rate of vegetation coverage was the fastest from 2000 to 2020, with a growth rate of 11.88% per decade.

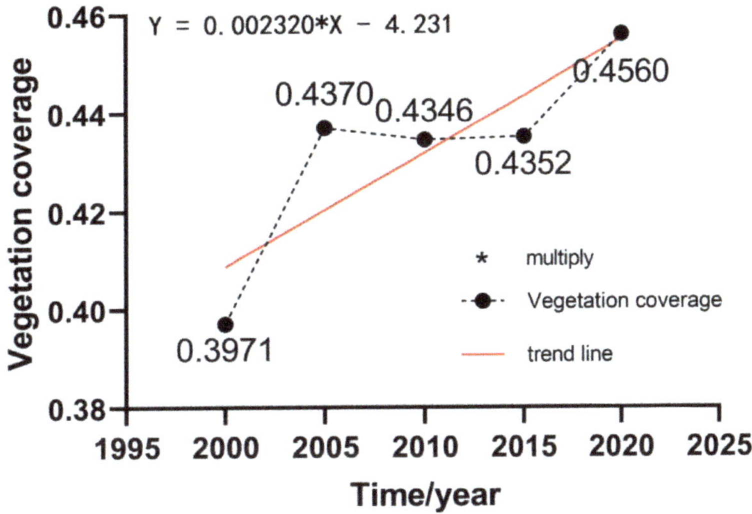

Figure 2. Interannual Changes in Vegetation Coverage.

3.1.2. Spatial Dynamic Change Characteristics of the Vegetation Cover

Through the analysis of the vegetation coverage classification map (Figure 3) of the Ulan Buh Desert in 2000, 2005, 2010, 2015, and 2020, it was found that the degree distribution of vegetation coverage in each year in the study area showed a certain regularity, that is, the vegetation coverage of the Ulan Buh Desert was generally distributed in a large area, with high coverage around the outer sections and low coverage in the middle, and the high vegetation coverage was mainly concentrated in Aolunbulag Town, Jilantai Salt Lake, and the surrounding areas of the Yellow River basin. The vegetation coverage in the central region was low, especially in the towns of Qulantai and Ustai, where the middle and low vegetation coverage areas surrounded the growth of low vegetation coverage areas. The vegetation coverage level gradually increased as it diverged outward from the low vegetation coverage area.

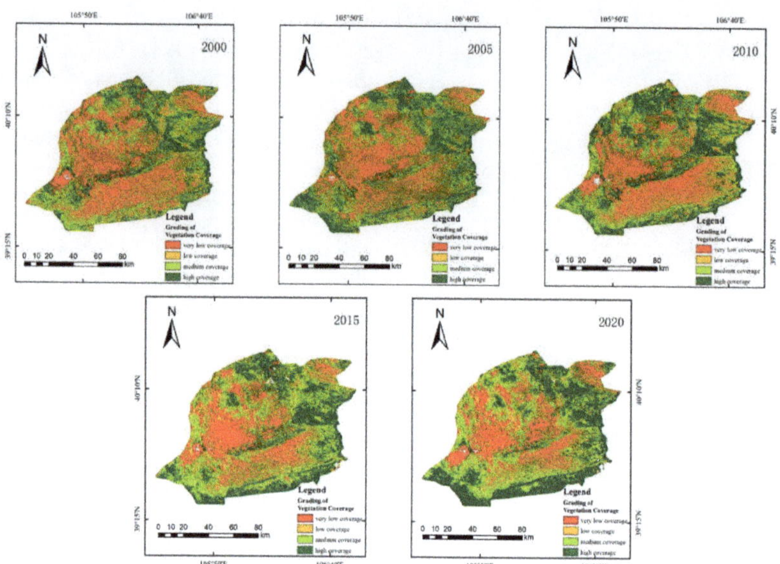

Figure 3. Vegetation Coverage Grading Maps of the Ulan Buh Desert.

3.1.3. Analysis of the Vegetation Coverage Change Trend

The absolute spatial distribution map of vegetation coverage in the Ulan Buh Desert is shown in Figure 4. The overall vegetation coverage in the Ulan Buh Desert showed an improvement trend, with 67.76% of the areas showing improvement and 2.47% of the areas showing extreme improvement. It was mainly distributed in patches near the Yellow River basin and many urban residential areas. The proportion of areas with moderate improvement was 16.79%, mainly distributed around extreme improvement areas, with distribution at both ends and boundaries. The areas with slight improvement accounted for 48.50%, spread over the entire Ulan Buh Desert, most widely distributed in the central and southern regions. The proportion of areas with declining vegetation coverage was 32.23%, with areas of extreme decline accounting for 0.29%. They were mainly distributed at the upper boundary of the Ulan Buh Desert and presented a local patchy pattern. The proportion of areas with moderate recession was 4.03%, mainly distributed in the upper boundary area; however, there was also a scattered distribution in the lower area. The proportion of areas with a slight recession was 27.91%, distributed above the central region, presenting a large-scale blocky distribution. It can be seen that areas showing improvement were greater than those showing a decline, and the overall vegetation coverage of the Ulan Buh Desert showed an improvement trend.

3.2. Changes in Groundwater Depth

In this study, five time nodes (2000, 2005, 2010, 2015, and 2020) were selected to obtain groundwater depth data from groundwater level monitoring wells in the Ulan Buh Desert, and spatial interpolation was carried out to obtain the spatial variation characteristic maps of the groundwater level in the Ulan Buh Desert from 2000 to 2020, as shown in Figure 5.

From a spatial perspective, the groundwater depth in the Ulan Buh Desert was deep in the southwest and shallow in the northeast. The underground water depths in the central and northeast regions of the Ulan Buh area were shallow, with depths of 2–4 m in most regions and 0–2 m in some regions. The depth of groundwater in the southwest was relatively deep; the depth of groundwater in most areas was more than 6 m, and the depth in the farthest southern region was more than 10 m.

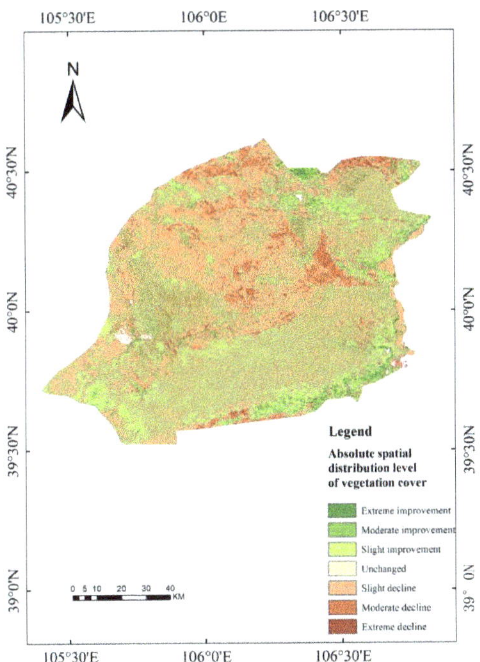

Figure 4. Absolute spatial distribution of vegetation coverage in the Ulan Buh Desert.

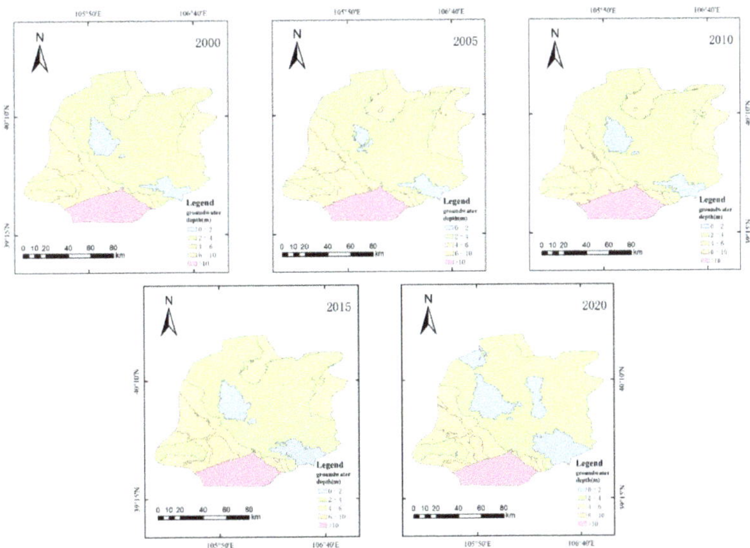

Figure 5. Groundwater Depth from 2000 to 2020.

From the perspective of time, from 2000 to 2020, the area with groundwater depth less than 4 m increased by 1280 m^2, accounting for about 11% of the total area of the Ulan Buh Desert, which indicates that the overall groundwater depth of the Ulan Buh Desert showed a trend of becoming shallower. However, from 2000 to 2005, the groundwater depth became deeper. The area of 0–2 m groundwater depths decreased significantly, from

726 m^2 in 2000 to 464 m^2 in 2005, a reduction of about 36%, and the area of 2–4 m buried depths also decreased by 674 m^2. The areas with underground water depths of 4–6 m and 6–10 m increased to a certain extent, and the underground water depth of the Ulan Buh Desert showed a trend of deepening from 2000 to 2005. This may be due to the continuous decrease in average precipitation between 2001 and 2003, as shown in Figure 5.

From the changes in groundwater depth from 2000 to 2020, we can see that, except for the years 2000–2005, the groundwater depth of the Ulan Buh Desert showed an overall trend of becoming shallower. Considering the changes in vegetation coverage in the above section, it is not difficult to see that the ecological situation of the Ulan Buh Desert showed a gradual trend of improvement. Moreover, the shallower the buried area, the more sensitive it was to the change in climate and precipitation, being more likely to change under the influence of climate and precipitation and other factors.

3.3. Correlation between Changes in Groundwater Depth and Vegetation Coverage

In order to study the relationship between the groundwater depth change and vegetation coverage change, this paper first obtained the difference in groundwater depth and vegetation coverage of each pixel in the two time nodes of the study area in 2020 and 2000 through difference calculation, then normalized the two difference values to calculate the correlation coefficient between them. The distribution of correlation coefficients is shown in Figure 6. It can be seen that the proportion of areas with positive correlation between vegetation coverage changes and groundwater depth changes is 53%, while the proportion of areas with negative correlation is 47%. It can be found that when the groundwater depth rose to more than 4 m, the change in groundwater depth had a significant effect on the vegetation coverage. When the groundwater depth was greater than 6 m, the change in groundwater depth had no obvious effect on the change in vegetation coverage.

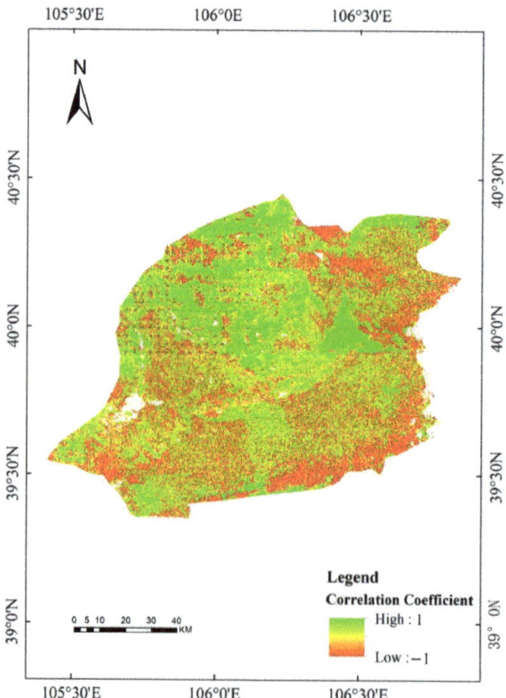

Figure 6. Distribution of correlation coefficients between changes in groundwater depth and vegetation coverage in the Ulan Buh Desert.

Using 2020 groundwater depth data and 2020 vegetation coverage data, taking vegetation coverage as the horizontal coordinate and groundwater depth data obtained from monitoring wells as the vertical coordinate, a rectangular coordinate system was established to obtain the scatter plot in Figure 7. This scatter plot reflects the influence of groundwater depth on the vegetation index. It can be seen that under the same groundwater depth, there were great differences in vegetation coverage, which can be caused by climate, soil type, vegetation type, and other factors. However, overall, there was a certain regularity in the variation in vegetation coverage with the depth of groundwater. The shallower the groundwater depth, the greater the vegetation coverage. Through analysis, it can be found that when the groundwater depth increased to over 4 m, the change in groundwater depth had a significant improvement effect on vegetation coverage. When the groundwater depth was greater than 6 m, there was no significant impact of changes in groundwater depth on vegetation coverage.

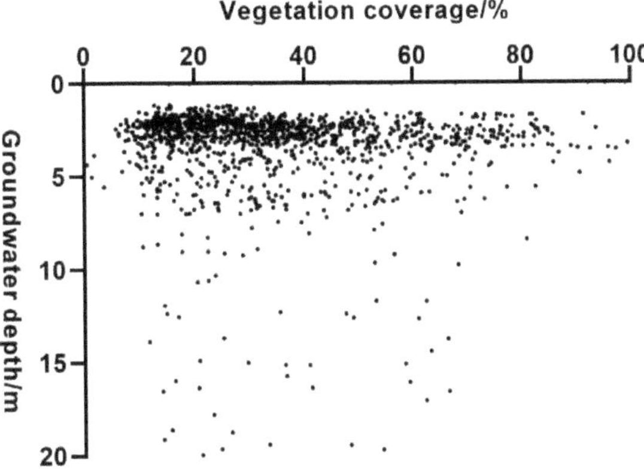

Figure 7. Characteristics of vegetation coverage changes with groundwater in the Ulan Buh Desert research area.

3.4. Discussion

3.4.1. Influence of Temperature

Li Yan [67] explored the relationship between soil temperature and soil moisture content in the Ulan Buh area and utilized an evaluation index system for soil moisture content, groundwater depth, capillary rise height of sandy soil, annual variation in groundwater depth, groundwater mineralization, and soil salt content. The ecological suitability of vegetation was analyzed, and the suitable area was divided into more suitable areas, less suitable areas, and unsuitable areas. Several studies have proven that the growth suitability of vegetation is related to temperature and groundwater depth. Reza Amiri [68] and others retrieved the surface temperature of the Tabriz metropolitan area in Iran using Landsat satellite data and established the temperature vegetation index (TVX). The results showed that, over time, the TVX had migrated from low temperature dense vegetation to high temperature sparse vegetation, indicating that the heat island effect generated by urban development had to some extent affected the growth of vegetation. Based on the above article, we investigated the response of the vegetation in the Ulan Buh area to temperature and groundwater.

As shown in Figure 8, From 2000 to 2017, the average annual temperature in the Ulan Buh area showed an overall upward trend, with some fluctuations over the years. Due to the cold resistance, heat resistance, dryness preference, and strong ecological adaptability of vegetation in desert areas, a slight increase in temperature can promote the growth of vegetation in desert areas to some extent.

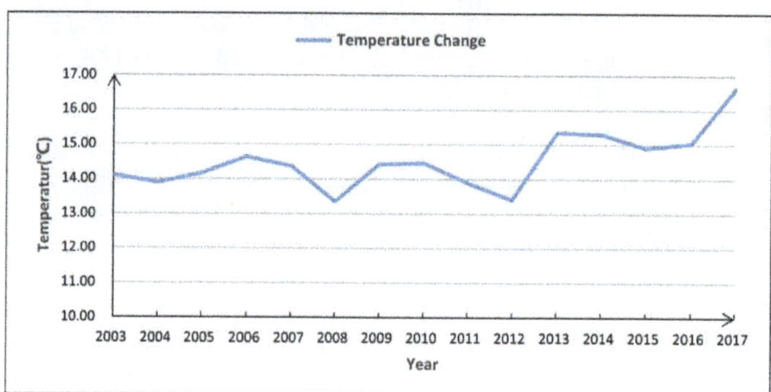

Figure 8. Trend of temperature changes in the Ulan Buh Desert.

3.4.2. Influence of Precipitation

Figure 9 shows a line chart of average precipitation for the years 2000–2020. It can be seen that the precipitation decreased significantly from 2001 to 2003. At the same time, compared with 2000, the vegetation coverage decreased in 2005, and the groundwater depth also showed the same trend. The area with the most suitable groundwater depth decreased by 674.3 km^2. From 2005 to 2008, the precipitation was at a relatively high level, and the vegetation growth trend was good. The area with the most suitable groundwater burial depth increased by 498.8 km^2. After 2011, the annual precipitation stabilized, vegetation coverage also steadily increased, and the most suitable areas for groundwater burial slowly increased. It is not difficult to see that an appropriate increase in precipitation contributes to the growth of vegetation. It is reasonable to infer that precipitation affects the depth of groundwater and thus affects the growth of vegetation.

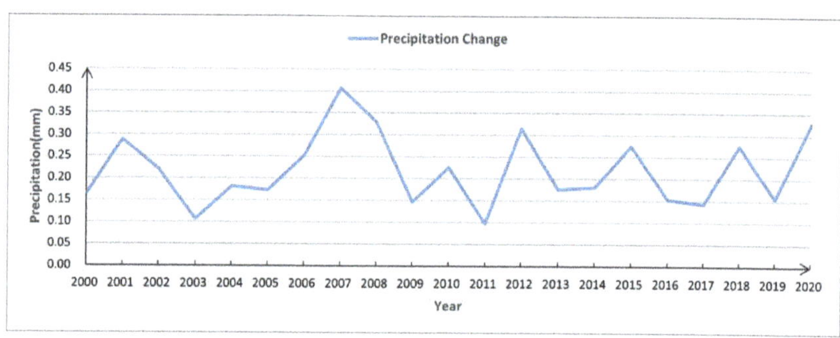

Figure 9. Trend of precipitation changes in the Ulan Buh Desert.

3.4.3. Other Influencing Factors and Analysis

The eastern and northern parts of the Ulan Buh Desert are close to the Yellow River; therefore, the local government has been carrying out wind prevention, sand fixation, and afforestation projects around the edge of the Ulan Buh Desert for many years. This has resulted in high vegetation coverage in the periphery of the Ulan Buh Desert and low vegetation coverage in the center of the desert.

In this paper, the Ulan Buh Desert, a typical desert in northern China, was selected as a research area. By revealing the correlation between the groundwater depth change and vegetation coverage change in the Ulan Buh Desert, the influence mechanism of groundwater depth on vegetation in desert areas can be obtained. Through vegetation as a key ecological environment element, we can also obtain the general change trend of the local ecological environment, which can provide a certain reference for the research and analysis of the ecological environment in similar desert areas.

3.4.4. Limitations

There are some limitations in the Landsat data used in this paper. Future studies can use data with higher accuracy, such as Sentinel data, to reduce the errors that may be introduced in the process of data acquisition and processing. When using the Kriging interpolation method, it is necessary to consider whether the attribute is stationary in space and whether the sample points are evenly distributed to deal with the uncertainty. In order to increase the reliability and accuracy of the research results, subsequent research can use multi-source data combined with different interpolation methods. Future studies can add more influencing factors for analysis, such as human factors, soil factors, etc., to further reveal the influencing mechanism behind the change in the desert ecological environment.

4. Conclusions

This study used the PIE-Engine software for spatial interpretation, quantitatively analyzed the spatiotemporal changes in vegetation coverage and water depth in the Ulan Buh Desert over a 20-year period, and explored the mechanism of the response of groundwater depth to vegetation coverage.

(1)   In terms of time, the vegetation coverage of the Ulan Buh Desert has shown an overall trend of increasing year by year over the past 20 years, with an increase rate of 4.73%/10 years. The highest vegetation coverage appeared in 2020 and represented a 35% increase compared to the year with the lowest vegetation coverage (2000). The vegetation coverage in 2010 showed a slight downward trend. Except for 2010, the vegetation coverage in other years showed an upward trend compared to the previous period, and the growth rate of vegetation coverage was the fastest from 2000 to 2020, with a growth rate of 12% per decade. The downward trend of vegetation coverage in 2010 may have been influenced by precipitation and temperature.

(2)   In space, the degree distribution of vegetation coverage in the Ulan Buh Desert in each year showed a certain regularity. The vegetation coverage in the drainage basin was high around the periphery and low across a large area in the middle. The vegetation coverage of the Ulan Buh Desert showed an overall improvement trend, with 68% of the areas showing improvement in vegetation coverage, including 3% showing extreme improvement, 16.79% showing moderate improvement, and 48.50% showing slight improvement. The proportion of areas with a declining vegetation cover was 32%, with areas of extreme decline accounting for less than 1%, areas of moderate decline accounting for 4%, and areas of slight decline accounting for 28%. The overall vegetation coverage in the Ulan Buh Desert was relatively stable. The vegetation coverage in most areas of the Ulan Buh Desert basin showed a moderate-to-strong variation, with only a small number of areas experiencing a weak variation. Strong-variation areas were mainly distributed in the central region, while moderate variation areas almost covered the Yellow River, which flows through the Ulan Buh Desert region and the Salt Lake region. Weak variation areas were mainly scattered at the eastern and southwestern boundaries.

Based on the analysis of groundwater depth, there was a certain regularity in the variation in vegetation coverage with groundwater depth. The shallower the groundwater depth, the greater the vegetation coverage. The proportion of areas with a positive correlation between vegetation coverage changes and groundwater depth changes was 53%, while the proportion of areas with a negative correlation was 47%. When the groundwater depth increased to more than 4 m, the change in groundwater depth had a significant improvement effect on vegetation coverage. When the groundwater depth was greater than 6 m, there was no significant impact of changes in groundwater depth on vegetation coverage.

Author Contributions: Conceptualization, J.W. and T.L.; methodology, W.Z. and Y.L. (Yangchun Lu); investigation, T.L., J.W. and Y.L. (Yudong Lu); resources, W.Z.; data curation, J.W. and Y.L. (Yangchun Lu); writing—original draft preparation, T.L. and J.W.; writing—review and editing, Y.L. (Yudong Lu) and T.L.; project administration, Y.L. (Yudong Lu); funding acquisition, Y.L. (Yudong Lu). All authors have read and agreed to the published version of the manuscript.

Funding: This work was funded by the National Natural Science Foundation of China (Grant No. U2243204).

Data Availability Statement: Not applicable.

Conflicts of Interest: The authors declare no conflict of interest.

References

1. Jia, L.; Li, Z.; Xu, G.; Ren, Z.; Li, P.; Cheng, Y.; Zhang, Y.; Wang, B.; Zhang, J.; Yu, S. Dynamic change of vegetation and its response to climate and topographic factors in the Xijiang River basin, China. *Environ. Sci. Pollut. Res.* **2020**, *27*, 11637–111648. [CrossRef]
2. Liu, H.; Jiao, F.; Yin, J.; Li, T.; Gong, H.; Wang, Z.; Lin, Z. Nonlinear relationship of vegetation greening with nature and human factors and its forecast—A case study of Southwest China. *Ecol. Indic.* **2019**, *111*, 106009. [CrossRef]
3. Duan, H.; Xue, X.; Wang, T.; Kang, W.; Liao, J.; Liu, S. Spatial and Temporal Differences in Alpine Meadow, Alpine Steppe and All Vegetation of the Qinghai-Tibetan Plateau and Their Responses to Climate Change. *Remote Sens.* **2021**, *13*, 669. [CrossRef]
4. Forkel, M.; Carvalhais, N.; Rödenbeck, C.; Keeling, R.; Heimann, M.; Thonicke, K.; Zaehle, S.; Reichsrtein, M. Enhanced seasonal CO_2 exchange caused by amplified plant productivity in northern Ecosystems. *Science* **2016**, *351*, 696–699. [CrossRef]
5. Zhu, Z.; Piao, S.; Myneni, R.; Huang, M.; Zeng, Z.; Canadell, J.; Ciais, P.; Sitch, S.; Friedlingstein, P.; Arneth, A.; et al. Greening of the Earth and its drivers. *Nat. Clim. Change* **2016**, *6*, 791–795. [CrossRef]
6. Yuan, W.; Zheng, Y.; Piao, S.; Ciais, P.; Lombardozzi, D.; Wang, Y.; Ryu, Y.; Chen, G.; Dong, W.; Hu, Z.; et al. Increased atmospheric vapor pressure deficit reduces global vegetation growth. *Sci. Adv.* **2019**, *5*, eaax1396. [CrossRef]
7. Qu, S.; Wang, L.; Lin, A.; Zhu, H.; Yuan, M. What drives the vegetation restoration in Yangtze River basin, China: Climate change or anthropogenic factors? *Ecol. Indic.* **2018**, *90*, 438–450. [CrossRef]
8. Li, S.; Yan, J.; Liu, X.; Wan, J. Response of vegetation restoration to climate change and human activities in Shaanxi-Gansu-Ningxia Region. *J. Geogr. Sci.* **2013**, *23*, 98–112. [CrossRef]
9. Chen, C.; Park, T.; Wang, X.; Piao, S.; Xu, B.; Chaturvedi, R.; Fuchs, R.; Brovkin, V.; Ciais, P.; Fensholt, R.; et al. China and India lead in greening of the world through land-use management. *Nat. Sustain.* **2019**, *2*, 122–129. [CrossRef]
10. Kurwadkar, S.; Sethi, S.S.; Mishra, P.; Ambade, B. Unregulated discharge of wastewater in the Mahanadi River Basin: Risk evaluation due to occurrence of polycyclic aromatic hydrocarbon in surface water and sediments. *Mar. Pollut. Bull.* **2022**, *179*, 113686. [CrossRef]
11. Kurwadkar, S.; Dane, J.; Kanel, S.; Nadagouda, M.; Cawdrey, R.; Ambade, B.; Struckhoff, G.; Wilkin, R. Per- and polyfluoroalkyl substances in water and wastewater: A critical review of their global occurrence and distribution. *Sci. Total Environ.* **2022**, *809*, 151003. [CrossRef] [PubMed]
12. Ambade, B.; Sethi, S.; Kumar, A.; Sankar, T. Solvent Extraction Coupled with Gas Chromatography for the Analysis of Polycyclic Aromatic Hydrocarbons in Riverine Sediment and Surface Water of Subarnarekha River and Its Tributary, India. In *Miniaturized Analytical Devices: Materials and Technology*; John Wiley & Sons Ltd.: Hoboken, NJ, USA, 2021. [CrossRef]
13. Ambade, B.; Sethi, S.; Kumar, A.; Kurwadkar, S. Health Risk Assessment, Composition, and Distribution of Polycyclic Aromatic Hydrocarbons (PAHs) in Drinking Water of Southern Jharkhand, East India. *Arch. Environ. Contam. Toxicol.* **2021**, *80*, 120–133. [CrossRef]
14. Miao, C.; Gou, J.; Fu, B.; Tang, Q.; Duan, Q. High-quality reconstruction of China's natural streamflow. *Sci. Bull.* **2022**, *67*, 547–556. [CrossRef] [PubMed]
15. Gou, J.; Miao, C.; Duan, Q.; Tang, Q.; Di, Z. Sensitivity Analysis-Based Automatic Parameter Calibration of the VIC Model for Streamflow Simulations over China. *Water Resour. Res.* **2020**, *56*, e2019WR025968. [CrossRef]
16. Ma, Z.; Sun, P.; Yao, R. Temporal and Spatial Variation of Drought and Its Impact on Vegetation in Inner Mongolia. *J. Soil Water Conserv.* **2022**, *36*, 231–240. [CrossRef]
17. Mu, S.; Li, J.; Chen, Y. Spatial Differences of Variations of Vegetation Coverage in Inner Mongolia during 2001–2010. *Acta Geogr. Sin.* **2012**, *67*, 1255–1268. [CrossRef]
18. Wang, N.; Chun, X. Research progress on the quaternary environmental evolution in the Ulan Buh Desert. *J. Desert Res.* **2022**, *42*, 175–183.
19. Chun, X.; Chen, F.; Fan, Y. Formation of Ulan Buh Desert and Its Environmental Evolution. *J. Desert Res.* **2007**, *27*, 927–931. [CrossRef]
20. Cooper, D.; Sanderson, J.; Stannard, D. Effects of long-term water table drawdown on evapotranspiration and vegetation in an arid region phreatophyte community. *J. Hydrol.* **2006**, *325*, 21–34. [CrossRef]
21. He, J.; Zhao, T.; Chen, Y.; Shang, X.; Liu, S. Effect of Subsurface Water on Spatial Pattern of Vegetation Coverage in Mu Us Sandy Area. *J. Soil Water Conserv.* **2023**, *37*, 90–99. [CrossRef]

22. Eamus, D.; Froend, R.; Loomes, R. A functional methodology for determining the groundwater regime needed to maintain the health of groundwater-dependent vegetation. *Aust. J. Bot.* **2006**, *54*, 97–114. [CrossRef]

23. Ambade, B.; Sethi, S.; Kurwadkar, S.; Kumar, A.; Sankar, T. Toxicity and health risk assessment of polycyclic aromatic hydrocarbons in surface water, sediments and groundwater vulnerability in Damodar River Basin. *Groundw. Sustain. Dev.* **2021**, *13*, 100553. [CrossRef]

24. Shaibur, M.R.; Ahmmed, I.; Sarwar, S.; Karim, R.; Hossain, M.M.; Islam, M.S.; Shah, M.S.; Khan, A.S.; Akhtar, F.; Uddin, M.G. Groundwater Quality of Some Parts of Coastal Bhola District, Bangladesh: Exceptional Evidence. *Urban Sci.* **2023**, *7*, 71. [CrossRef]

25. Kodate, J.; Dhurvey, V.; Dhawas, S.; Urkude, M.; Ragini, M. Assessment of groundwater quality with special emphasison fluoride contamination in some villages of Chandrapur district of Maharashtra, India. *IOSR J. Environ. Sci. Toxicol. Food Technol.* **2016**, *10*, 2319–2399.

26. Yin, X.; Feng, Q.; Li, Y.; Deo, R.; Liu, W. An interplay of soil salinization and groundwater degradation threatening coexistence of oasis-desert ecosystems. *Sci. Total Environ.* **2022**, *806*, 150599. [CrossRef]

27. Cui, Y.; Shao, J. The Role of Ground Water in Arid/Semiarid Ecosystems, Northwest China. *Ground Water* **2005**, *43*, 471–477. [CrossRef]

28. Li, F.; Qin, X.; Xie, Y.; Chen, X.; Hu, J. Physiological mechanisms for plant distribution pattern: Responses to flooding and drought in three wetland plants from Dongting Lake, China. *Limnology* **2013**, *14*, 71–76. [CrossRef]

29. Chen, Y.; Zilliacus, H.; Li, W. Ground-water level affects plant species diversity along the lower reaches of the Tarim river, Western China. *J. Arid. Environ.* **2006**, *66*, 231–266. [CrossRef]

30. Cheng, Y.; Chen, L.; Yin, J. Depth Interval Study of Vegetation Ecological Groundwater in the Water Source Area at Manaz River Valley. *Environ. Sci. Technol.* **2018**, *41*, 26–33.

31. Zhang, E.; Tao, Z.; Wang, X. A study of vegetation response to groundwater on regional scale in northern Ordos Basin based on structure chart method. *Geol. China* **2012**, *39*, 811–817. [CrossRef]

32. Zhai, J.; Dong, Y.; Qi, S. Advances in Ecological Groundwater Level Threshold in Arid Oasis Regions. *J. China Hydrol.* **2021**, *41*, 7–14. [CrossRef]

33. Lv, J.; Wang, X.; Zhou, Y.; Qian, K.; Wan, L. Groundwater-dependent distribution of vegetation in Hailiutu River catchment, a semi-arid region in China. *Ecohydrology* **2013**, *6*, 142–149. [CrossRef]

34. Jin, X.; Schaepman, M.; Clevers, J. Groundwater Depth and Vegetation in the Ejina Area, China. *Arid Land Res. Manag.* **2011**, *25*, 194–199. [CrossRef]

35. Hu, M.; Mao, F.; Sun, H. Study of normalized difference vegetation index variation and its correlation with climate factors in the three-river-source region. *Int. J. Appl. Earth Obs. Geoinf.* **2011**, *13*, 24–33. [CrossRef]

36. Song, Y.; Guo, Z.; Lu, Y.; Yan, D.; Liao, Z. Pixel-Level Spatiotemporal Analyses of Vegetation Fractional Coverage Variation and Its Influential Factors in a Desert Steppe: A Case Study in Inner Mongolia, China. *Water* **2017**, *9*, 478. [CrossRef]

37. Cao, R.; Chen, J.; Shen, M. An improved logistic method for detecting spring vegetation phenology in grasslands from MODIS EVI time-series data. *Agric. For. Meteorol.* **2015**, *200*, 9–20. [CrossRef]

38. Liu, X.; Zhang, J.; Zhu, X.; Pan, Y.; Liu, Y. Spatiotemporal changes in vegetation coverage and its driving factors in the Three-River Headwaters Region during 2000–2011. *J. Geogr. Sci.* **2014**, *24*, 288–302. [CrossRef]

39. Liu, Y.; Zeng, P.; Sun, F. Vegetation coverage change of the demonstration area in the Yangtze River Delta, China on ecologically friendly development based on GEE and BRT during 1984–2019. *J. Appl. Ecol.* **2020**, *32*, 1033–1043. [CrossRef]

40. Ding, M.; Guo, Y.; Chen, S. Evaluation of Vegetation Coverage Land Reclaimation Based on RS. *Remote Sens. Technol. Appl.* **2010**, *25*, 102–106. [CrossRef]

41. Hou, W.; Hou, X. Spatial–temporal changes in vegetation coverage in the global coastal zone based on GIMMS NDVI3g data. *Int. J. Remote Sens.* **2019**, *41*, 1118–1138. [CrossRef]

42. Tu, Y.; Jia, K.; Wei, X.; Yao, Y.; Xia, M. A Time-Efficient Fractional Vegetation Cover Estimation Method Using the Dynamic Vegetation Growth Information from Time Series GLASS FVC Product. *IEEE Geosci. Remote Sens. Lett.* **2020**, *17*, 1672–1676. [CrossRef]

43. Wang, X.; Jia, K.; Liang, S.; Li, Q.; Wei, X. Estimating Fractional Vegetation Cover From Landsat-7 ETM+ Reflectance Data Based on a Coupled Radiative Transfer and Crop Growth Model. *IEEE Trans. Geosci. Remote Sens.* **2017**, *55*, 5539–5546. [CrossRef]

44. Li, X.; Ma, Y.; Li, X. Plant community heterogeneity and its influencing factors in the Ulan Buh Desert. *J. Desert Res.* **2022**, *42*, 187–194.

45. Dong, X.; Xin, Z.; Duan, R. Species diversity of typical shrubs and niches of dominant shrub species in the UlanBuh Desert. *Arid Zone Res.* **2020**, *37*, 1009–1017. [CrossRef]

46. Niu, Y.; Ren, G.; Lin, G.; Di Biase, L.; Fattorini, S. Fine-Scale Vegetation Characteristics Drive Insect Ensemble Structures in a Desert Ecosystem: The Tenebrionid Beetles (Coleoptera: Tenebrionidae) Inhabiting the Ulan Buh Desert (Inner Mongolia, China). *Insects* **2023**, *11*, 6–19. [CrossRef]

47. Rabus, B.; Eineder, M.; Roth, A. The shuttle radar topography mission—A new class of digital elevation models acquired by spaceborne radar. *ISPRS-J. Photogramm. Remote Sens.* **2003**, *57*, 241–262. [CrossRef]

48. Gou, J.; Miao, C.; Samaniego, L. CNRD v1.0: A High-Quality Natural Runoff Dataset for Hydrological and Climate Studies in China. *Bull. Am. Meteorol. Soc.* **2021**, *102*, E929. [CrossRef]

49. Cheng, W.; Qian, X.; Li, S.; Ma, H.; Liu, D.; Liu, F.; Liang, J.; Hu, J. Research and application of PIE Engine Studio, a spatiotemporal remote sensing cloud computing platform. *Can. J. Remote Sens.* **2022**, *26*, 335–347. [CrossRef]

50. Schnell, J. Monitoring the vernal advancement and retrogradation (greenwave effect) of natural vegetation. In *Nasa/Gsfct Type Final Report*; Remote Sensing Center Texas A&M University: College Station, TX, USA, 1974.

51. Lu, Q.; Zhao, D.; Wu, S.; Dai, E.; Gao, J. Using the NDVI to analyze trends and stability of grassland vegetation cover in Inner Mongolia. *Theor. Appl. Climatol.* **2019**, *135*, 1629–1640. [CrossRef]

52. Li, F.; Chen, W.; Zeng, Y.; Zhao, Q.; Wu, B. Improving Estimates of Grassland Fractional Vegetation Cover Based on a Pixel Dichotomy Model: A Case Study in Inner Mongolia, China. *Remote Sens.* **2014**, *6*, 4705–4722. [CrossRef]

53. Miao, Z.; Liu, Z.; Wang, Z.; Song, K.; Ren, C. Dynamic Monitoring of Vegetation Fraction Change in Jilin Province Based on MODIS NDVI. *Remote Sens. Technol. Appl.* **2010**, *25*, 387–393.

54. Zhang, C.; Lou, Y.; Li, Y. Change of Vegetation Coverage in Funiu Mountain Regions Based on the Dimidiate Pixel Model. *Res. Soil Water Conserv.* **2020**, *27*, 301–307. [CrossRef]

55. HJ 1170-2021; Technical Specification for Investigation and Assessment of National Ecological Status—Field Observation of Desert Ecosystem. Ministry of Ecology and Environment: Beijing, China, 2021.

56. Wen, H.; Song, C.; Xiang, X. Optical remote sensing change detection method for the identification of landslide clusters induced by heavy rainfall. *Sci. Surv. Mapp.* **2022**, *47*, 193–202. [CrossRef]

57. Murashko, F.; Ryzhkova, E.; Vlasenko, O. Search for an Object in an Image by Image Difference Method to Find Contours of a Natural Leather Blank in Pattern Cutting Process. *Fibre Chem.* **2018**, *50*, 38–41. [CrossRef]

58. Wu, M.; Wang, J.; Wang, X. Spatio-temporal analysis of urban built-up area extension in Yantai City based on multi-source remote sensing images. In Proceedings of the 2011 19th International Conference on Geoinformatics, Shanghai, China, 24–26 June 2011. [CrossRef]

59. Yu, Y. Application of ArcGIS Kriging Interpolation Method in Spatial Characteristics Analysis of Precipitation Evolution in Linghe River Basin. *Heilongjiang Hydraul. Sci. Technol.* **2022**, *50*, 160–163. [CrossRef]

60. Arfaoui, M.; Hédi, M. Advantages of using the kriging interpolator to estimate the gravity surface, comparison and spatial variability of gravity data in the El Kef-Ouargha region (northern Tunisia). *Arab. J. Geosci.* **2013**, *6*, 3139–3147. [CrossRef]

61. Zhang, Y. Common mode error filtering method for GPS coordinate time series based on Kriging interpolation. *Geomat. Technol. Equip.* **2022**, *24*, 28–31.

62. Bao, G.; Gao, Q. Correlation Analysis between the Emotion and Aesthetics for Chinese Classical Garden Design Based on Deep Transfer Learning. *J. Environ. Public Health* **2022**, *2022*, 1828782. [CrossRef]

63. Liu, Y.; Shen, X.; Zhang, J.; Wang, Y.; Wu, L.; Ma, R.; Lu, X.; Jiang, M. Temporal and Spatial Variation in Vegetation Coverage and Its Response to Climatic Change in Marshes of Sanjiang Plain, China. *Atmosphere* **2022**, *13*, 2077. [CrossRef]

64. Xu, F.; Li, P.; Chen, W.; He, S.; Li, F.; Mu, D.; Elumalai, V. Impacts of land use/land cover patterns on groundwater quality in the Guanzhong Basin of northwest China. *Geocarto Int.* **2022**, *37*, 16769–16785. [CrossRef]

65. Wang, D.; Li, P.; He, X.; He, S. Exploring the response of shallow groundwater to precipitation in the northern piedmont of the Qinling Mountains, China. *Urban Clim.* **2023**, *47*, 101379. [CrossRef]

66. Xu, F.; Li, P.; Du, Q.; Yang, Y.; Yue, B. Seasonal hydrochemical characteristics, geochemical evolution, and pollution sources of Lake Sha in an arid and semiarid region of northwest China. *Expo. Health* **2023**, *15*, 231–244. [CrossRef]

67. Li, Y. *Water Vapor Heat Coupling Migration Law and Ecological Significance of Vegetation in Ulanbuhe Desert*; Chang'an University: Xi'an, China, 2013.

68. Amiri, R.; Weng, Q.; Alimohammadi, A.; Alavipanah, S.K. Spatial–temporal dynamics of land surface temperature in relation to fractional vegetation cover and land use/cover in the Tabriz urban area, Iran. *Remote Sens. Envion.* **2009**, *113*, 2606–2617. [CrossRef]

water

MDPI

Article

Evolution Characteristics of Rainfall and Runoff in the Upper Reaches of Zhang River Basin

Lijuan Du [1,2,*], Guangyao Wang [3] and Bo Lei [1,2]

[1] State Key Laboratory of Simulation and Regulation of Water Cycle in River Basin, A-1 Fuxing Road, Haidian District, Beijing 100038, China; bolei1228@hotmail.com

[2] China Institute of Water Resources and Hydropower Research (IWHR), A-1 Fuxing Road, Haidian District, Beijing 100038, China

[3] College of Water Conservancy and Environment, University of Jinan, 336 Nanxinzhuang West Road, Shizhong District, Jinan 250022, China; wgy19991125@foxmail.com

* Correspondence: iwhr_dulj@foxmail.com; Tel.: +86-13611369282

Abstract: It is of great significance to study and analyze the surface water resources and their change trend in the groundwater overexploitation area of the North China Plain, which is of great significance to solve the shortage of water resources in the groundwater overexploitation area of the North China Plain, promote the exploitation of groundwater, and realize the sustainable development of water resources. This paper takes Minyou Irrigation District of Handan City, a typical overexploitation area in the North China Plain, as an example. Based on the measured rainfall and runoff data from 1957 to 2020, the Mann–Kendall trend test, cumulative anomaly method, double cumulative curve method, and Morlet wavelet transform were used to analyze and predict the trend of water resources in the irrigation area and the individual contribution of climate change and human activities to runoff change. The results show that the annual rainfall and annual runoff in the irrigation area have a significant downward trend and significant cyclical changes throughout the study period. In 1977, the annual runoff showed a sudden change, and the average contribution rates of climate change and human activities to its change were 40.55% and 59.46%, respectively. In the future (2020–2035), runoff will remain stable and rainfall will show an increasing trend. The research results can provide scientific reference for the development, utilization, and rational allocation of surface water resources in the groundwater overexploitation area of the North China Plain.

Keywords: rainfall/runoff variation; surface water resources; evolution characteristics; attribution analysis; trend prediction; North China Plain over-exploitation area

check for
updates

Citation: Du, L.; Wang, G.; Lei, B. Evolution Characteristics of Rainfall and Runoff in the Upper Reaches of Zhang River Basin. *Water* **2023**, *15*, 2521. https://doi.org/10.3390/w15142521

Academic Editors: Peiyue Li and Jianhua Wu

Received: 29 May 2023
Revised: 28 June 2023
Accepted: 4 July 2023
Published: 10 July 2023

1. Introduction

The North China Plain is an important wheat and corn production base in China, which plays an important role in ensuring China's food security. However, due to the insufficient amount of water resources in the region, the unbalance of precipitation and water resource distribution within the region under the influence of climate change and human activities has intensified, resulting in an increasingly serious agricultural drought and water deficiency. The exploitation of groundwater for irrigation for a long time and on a large scale has become one of the important reasons for groundwater overexploitation in the North China Plain. Although, with the promotion of water-saving irrigation technology and the implementation of comprehensive management measures for water-saving compressive exploitation since 2014, agricultural water use and groundwater exploitation in North China have shown a downward trend, the dominant position of groundwater in farmland irrigation has not yet changed. Therefore, further governance actions have been carried out by the State, such as the Ministry of Water Resources, the Ministry of Finance, the National Development and Reform Commission, and the Ministry of Agriculture and

Rural Affairs jointly issued the "Action Plan for Comprehensive Management of Groundwater over exploitation in North China" and the government of Hebei Province issued the "Five-year Implementation Plan for Comprehensive Management of Groundwater over exploitation in Hebei Province (2018–2022)," clearly requiring to strictly restrict the groundwater exploitation. In this regard, it is particularly urgent to study the evolution rule of surface water resources in North China and predict the variation trend of surface water resources in the future in a situation where the contradiction between the supply and demand of water resources is intensifying.

As an important constituent part and transformation form of water resources, runoff is not only the key link within the surface water cycle but also the main source of available surface water resources in irrigation districts. Human activities such as building large-scale water conservancy projects to change the regional runoff yield conditions have an impact on the regional hydrological process, and the combined effects of different climate change scenarios and various human activities make the runoff yield mechanism of the basin more complicated [1]. Under the current situation of increasing shortages of water resources caused by global climate change and human activities, runoff variation and attribution prediction have become one of the hot issues in the field of modern water science [2] and also an important basic work for the comprehensive management of surface water resources in irrigation districts [3]. In recent years, scholars have continued to explore the new methods from multiple perspectives, including the tendency [4–6], mutability [7–9], wet and dry abundance [10–12], periodicity [13–15], and attribution analysis [16–20] for rainfall-runoff evolution. For example, Panditharathne et al. [21] used the Mann–Kendall test to study the relationship among the variation trend of rainfall-runoff, the change points and runoff variation, and the rainfall-runoff in the Nirvara River Basin in southern Sri Lanka. Li et al. [22] used multiple hydrological statistical methods such as the Mann–Kendall test, cumulative departure, and rainfall-runoff double mass curve to study the evolution characteristics of hydrological elements in the Dawen River Basin. Ji et al. [23] used the cumulative departure method to identify the years of runoff mutation and the multiple linear regression method to evaluate the contribution rate of climate change and human activities to vegetation change in the upper reaches of the Yellow River. Banda et al. [24] used the Mann–Kendall test and double mass curve technology to explore the impact of human activities and climate variability on runoff variations in Rietspruit sub-basin of South Africa. Based on the double mass curve method, Wang et al. [25] performed attribution analysis on the runoff variation in Nanxiaohegou watershed, and quantitatively revealed the contribution rate of climate change and human activities to runoff variation. Zhang et al. [26] used Mann–Kendall method and the Morlet wavelet analysis method to study the annual variation characteristics of precipitation in Zhengzhou, and the results showed that the mutation and periodicity of precipitation series in Zhengzhou were significant. The Mann–Kendall trend test method can test the variation trend and mutation points of the sequence without being disturbed by a few outliers and has wide applicability, but it can only make qualitative judgments on the sequence. The cumulative departure method can quantitatively count and analyze the trend of the sequence and visually present the mutations. The double mass curve method can analyze the correlation of rainfall-runoff and distinguish the influence of climate and human activities on natural runoff throughout the observation period. Morlet wavelet analysis can show the variation of the wet and dry cycles of the sequence on multiple time scales, and on this basis, it can predict the trend variations in the future.

Based on the above research methods, this paper intends to study the variation trend characteristics and driving factors of rainfall runoff in the surface water source of Handan public irrigation area (Zhanghe River Basin above Yuecheng Reservoir), a typical groundwater overexploitation area in the North China Plain, to explore the main driving factors causing runoff changes, and on this basis, to analyze and predict the periodic changes of rainfall runoff. The research results can provide scientific reference for the development and utilization of surface water resources and rational allocation in North China.

2. Overview of the Research Area

As the main tributary of the South Canal water system in the Hai River Basin, the Zhang River has many branches in a fan-shaped distribution along its upper reaches, including the Qingzhang River and the Zhuozhang River. Flowing through the Yuecheng Reservoir built at its outlet, the Zhang River leaves the mountain and enters the plain eastward. The section of the Zhang River down Yuecheng Reservoir flows eastward through the irrigation district of Handan City, Hebei Province, with a length of 117.4 km. The total basin area of the Zhang River is 19,220 km^2, of which the catchment area of the Yuecheng Reservoir is 18,100 km^2, accounting for 94.2% of the total basin area and being the main water source of the Minyou irrigation district downstream. Located in Handan City, Hebei Province, and in the middle of the North China Plain, the total control area of Minyou irrigation district is 3480.6 km^2, and the effective irrigation area is 1340 km^2, with an average annual rainfall of 580 mm and an average annual evaporation of 1120 mm. With the topography being higher in the southwest and lower in the northeast, the main surface water source of the irrigation district comes from the Yuecheng Reservoir in the upper reaches of the Zhang River. However, since the surface water resources cannot meet the irrigation needs of the irrigation district, the groundwater is exploited for irrigation all year round in the middle and lower reaches of the irrigation district, which leads to the continuous decline of groundwater level and even the formation of groundwater funnel areas in some areas. The location of the irrigation district and its upstream meteorological and hydrological stations are shown in Figure 1.

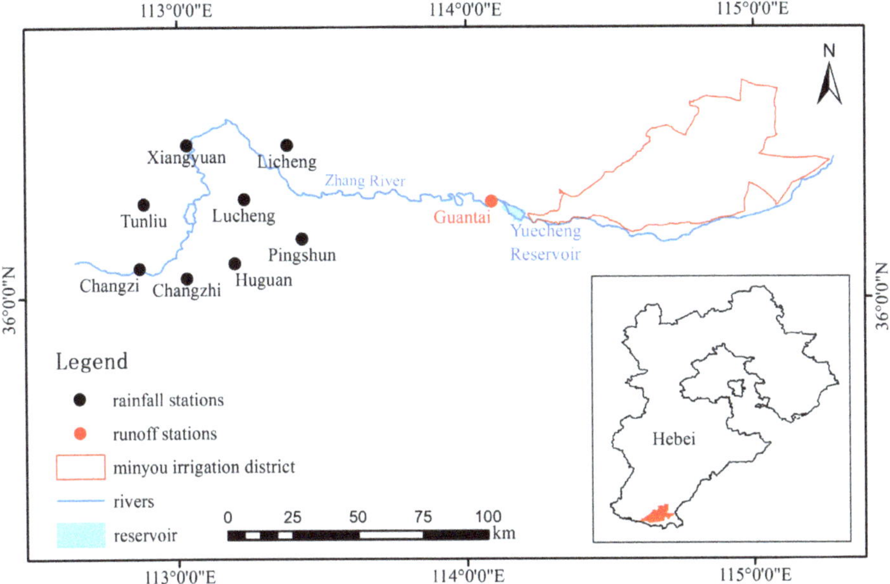

Figure 1. Locations of meteorological stations in the Zhang River Basin and its upstream.

3. Research Methods and Data Sources

3.1. Research Methods

In this paper, the Mann–Kendall (MK) trend test method [27–29], cumulative departure method [30–32], double mass curve method [33,34], and Morlet wavelet transform method [35,36] are used to perform the trend mutation analysis, attribution analysis, and periodic analysis prediction of runoff data and rainfall of meteorological stations in the research area. The details of each research topic are as follows:

(1) MK Trend Test

The MK trend test is a nonparametric test method recommended by the World Meteorological Organization. Under the precondition that the sequences are independent of each other with the same continuous distribution, the calculation formula is as follows:

For hydrological sequence x, an order sequence is constructed:

$$S_k = \sum_{i=1}^{k} r_i (k = 2, 3, 4, \cdots, n) \tag{1}$$

$$r_i = \begin{cases} +1 & x_i > x_j \\ +0 & x_i < x_j \end{cases} (j = 1, 2, \cdots, n) \tag{2}$$

where the order column S_k is the cumulative count of the number of values when the ith greater time value is greater than that at time j. Under the assumption of random independence of time series, the statistical magnitude is constructed:

$$UF_k = \frac{S_k - E(S_k)}{\sqrt{var(S_k)}} (k = 1, 2, \cdots, n) \tag{3}$$

where $UF_1 = 0$; $E(S_k)$, $var(S_k)$ is the mean and variance of the cumulative count S_k. When $x_1, x_2, \cdots, x_n$ are independent of each other and have the same distribution, it can be calculated by the following formula:

$$E(S_k) = \frac{k(k+1)}{4} \tag{4}$$

$$var(S_k) = \frac{k(k-1)(2k+5)}{72} \tag{5}$$

$$S = \sum_{i=2}^{n} \sum_{j=1}^{i-1} \text{sign}(X_i - X_j) \tag{6}$$

$$\begin{cases} Z = \dfrac{S-1}{\sqrt{n(n-1)(2n+5)/18}} & S > 0 \\ Z = 0 & S = 0 \\ Z = \dfrac{S-1}{\sqrt{n(n-1)(2n+5)/18}} & S < 0 \end{cases} \tag{7}$$

where sign ( ) is a sign function, S is a normal distribution, UF is a standard normal distribution, and $UB = -UF$. Given the significance level $\alpha = 0.05$, if $|UF|$ or $|UB|$ is greater than the critical value of the test statistic 1.96, it indicates that there is a significant trend change in the sequence; given the significance level $\alpha = 0.01$, if $|UF|$ or $|UB|$ is greater than the critical value of the test statistic 2.38, it indicates that the trend change of the sequence is extremely significant, and the part exceeding the critical line is the time of mutation [37]. Z is a positive value indicating an increasing trend and a negative value indicating a decreasing trend. For a given significance level α, if $|Z|$ is greater than or equal to 1.96 and 2.38, it indicates that it has passed the significance test with confidence levels of 95% and 99%, respectively.

(2) Cumulative departure method

As a commonly used method to intuitively judge the variation trend by the curve, the cumulative departure method can reflect the evolution trend of the elements [38,39]. When the cumulative departure curve shows an upward trend, it indicates that the departure value increases, and vice versa [40]. This index can be used to judge the variation trend of long-term rainfall and runoff in the Zhang River Basin, and the approximate time of runoff mutation can also be judged from the Figure. For sequence x, the cumulative departure of time t can be calculated by the following formula:

$$X = \sum_{i=1}^{t}\left(x - \sum_{i=1}^{t} x_i\right)(t = 1, 2, \cdots, n) \tag{8}$$

where X is the cumulative departure value from year 1 to year t (t is the time series, $t \leq n$).

(3)　Double mass curve

The double mass curve method is the simplest, most intuitive, and most widely used method for analyzing the consistency or long-term evolution trend of hydrometeorological elements. In the double mass curve of rainfall and runoff, the change in cumulative rainfall in a limited period of time is a natural change, and the influence of human activities is relatively small. Cumulative runoff is affected by both human activities and rainfall. Therefore, using cumulative rainfall as a reference variable, the influence of human activities on natural runoff can be distinguished by the double mass curve. The formula is as follows:

$$S_{ri} = \sum_{i=1}^{n} r_i \tag{9}$$

$$S_{pi} = \sum_{i=1}^{n} p_i \tag{10}$$

where S_{ri} is the cumulative runoff depth of the first i years during n years in the basin (mm), S_{pi} is the cumulative rainfall of the first i year during n years in the basin (mm), r_i is the runoff depth of the ith year (mm), p_i is the rainfall of the ith year (mm).

(4)　Morlet wavelet transform

Wavelet transform analysis is a multi-resolution analysis method that can perform the analysis of the time domain and frequency domain simultaneously and enables the hydrological time series being studied to reveal the law of system variation at different levels [41]. The calculation formula is:

$$\int_{-\infty}^{+\infty} \psi(t)dt = 0 \tag{11}$$

$$\psi_{a,b}(t) = |a|^{-1/2}\psi\left(\frac{(t-b)}{a}\right) \quad b \in R, a \in R, a \neq 0 \tag{12}$$

$$\omega_f(a,b) = |a|^{-1/2}\int_{-\infty}^{+\infty} f(t)\psi\left(\frac{(t-b)}{a}\right)dt \tag{13}$$

$$Var(a) = \int_{-\infty}^{+\infty} \left|\omega_f(a,b)\right|^2 db \tag{14}$$

where $\psi(t)$ is the basis wavelet function, $\psi_{a,b}(t)$ is a sub-wavelet, a is the scale factor, b is the time factor, $f(t)$ is the given original signal, $\omega_f(a,b)$ is the wavelet transform coefficient, $Var(a)$ is the wavelet variance, reflecting that the energy of the signal fluctuation is changing with the scale, based on which the periodic variation of the time series at different scales can be determined [42], and then the runoff variation of the basin in the next stage can be predicted as per the current period.

3.2. Data Sources

In this paper, the rainfall sequence data from 1957 to 2020 are selected from 8 meteorological stations (Pingshun station, Lucheng station, Huguan station, Changzhi station, Changzi station, Tunliu station, Xiangyuan station, and Licheng station) in the upper reaches of the Zhang River, together with the runoff data from 1962 to 2020 of Yuecheng Reservoir from the hydrologic station for inflow monitoring (Guantai Station). The site information is shown in Table 1. The rainfall data are derived from the China Meteorological Science Data Sharing Service Network (http://data.cma.cn, accessed on 13 April 2023), and the runoff data are from the Deliverables of Hydrological Data Compiled by Handan Zhangfu River Irrigation District.

Table 1. Overview of meteorological and hydrological stations in the irrigation district.

Type	Station Name	Station Number	Longitude (E)	Latitude (N)	Time Period of Measured Data
rainfall station	Pingshun station	53,888	113.26	36.12	1957–2020
	Lucheng station	53,880	113.14	36.2	1976–2020
	Huguan station	53,885	113.12	36.07	1961–2020
	Changzhi station	53,882	113.02	36.04	1973–2020
	Changzi station	53,873	112.52	36.06	1965–2020
	Tunliu station	53,879	112.53	36.19	1971–2020
	Xiangyuan station	53,884	113.02	36.31	1957–2020
	Licheng station	53,878	113.23	36.31	1958–2020
hydrologic station	Guantai station	36,255	114.08	36.32	1962–2020

4. Results and Discussion

4.1. Analysis of Trend Mutability

4.1.1. MK Trend Test

(1) Rainfall trend analysis

The rainfall series from 1962 to 2020 were selected for Mann-Kendall trend testing and analysis from 8 meteorological stations (Pingshun station, Lucheng station, Huguan station, Changzhi station, Zhangzi station, Tunliu station, Xiangyuan station, and Licheng station) in the upper reaches of the Zhang River. The results are shown in Figure 2a–h. On the whole, the rainfall at meteorological stations in the past 60 years has had a significant decreasing trend. The summary results are shown in Table 2, and the analysis of each station is as follows:

Table 2. The mutation test results of the rainfall-runoff trend in the irrigation district.

Type	Station Name	Mann−Kendall Trend Test				Accumulated Variance
		Trend Test Statistic	Year of Mutation Point	Mutation Point Statistic	Trends	Year of Mutation Point
rainfall station	Pingshun station	−0.49	1977 *, 1988 *, 2000 *	+2.18, −2.06, −2.02	decreased	1976
	Lucheng station	1.63	1977 *, 1979 *	+2.08, −2.04	raised	1992
	Huguan station	−0.39	1977 *	+1.98	decreased	1976
	Changzhi station	0.18	\	\	raised	1976
	Changzi station	−0.53	\	\	decreased	1976
	Tunliu station	−0.66	1981 **, 2013 *	−2.57, −1.98	decreased	1976
	Xiangyuan station	−0.71	1987 *	−2.05	decreased	1976
	Licheng station	−1.06	\	\	decreased	1976
hydrologic station	Guantai station	−3.92 **	1969 *, 1977 **	−2.29, −2.39	decreased	1977

Notes: * Statistically significant trends at the 5% significance level. ** Statistically significant trends at the 1% significance level. The statistical value is positive, indicating an increasing trend, and vice versa.

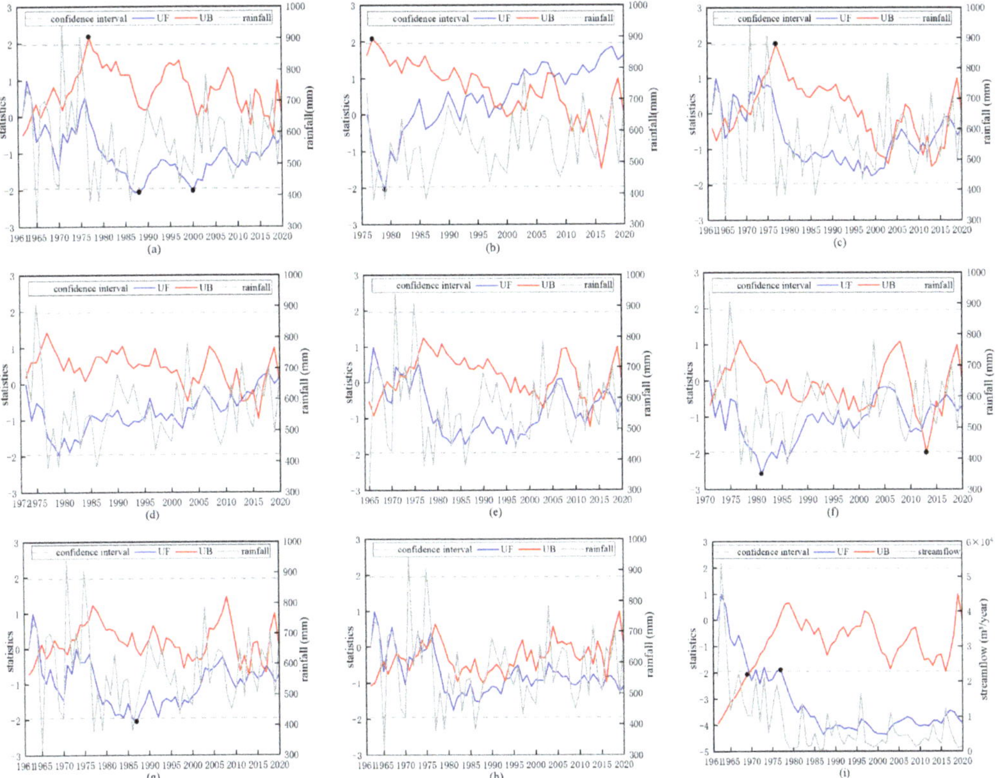

Figure 2. The results of the MK trend analysis of rainfall-runoff in the Zhang River Basin. (**a**) Pingshun station; (**b**) Lucheng station; (**c**) Huguan station; (**d**) Changzhi station; (**e**) Zhangzi station; (**f**) Tunliu station; (**g**) Xiangyuan station; (**h**) Licheng station; (**i**) Guantai station.

The rainfall of 7 stations at Pingshun, Huguan, Tunliu, Xiangyuan, Changzhi, Changzi, and Licheng showed a decreasing trend ($UF < 0$). All the rainfall exceeded the critical line at the significance level of 0.05, respectively, at Pingshun station in 1977, 1988, and 2000, at Huguan station in 1977, at Tunliu station in 1981 and 2013, and at Xiangyuan station in 1987, indicating that the decreasing trend of rainfall was very significant, of which the statistical value of rainfall at Tunliu station in 1981 even exceeded the significant level of 0.01, showing the extremely significant decreasing trend of rainfall. It can be seen from the UF curve and UB curve that the mutation points occurred at the four stations of Pingshun, Huguan, Tunliu, and Xiangyuan, respectively, in 1977, 1988, 2000, 1977, 1981, 2013, and 1987, while no mutation point occurred at Changzhi, Zhangzi, and Licheng.

The rainfall at Lucheng Station showed a decreasing trend ($UF < 0$) from 1976 to 1985 and an increasing trend after 1985 ($UF > 0$). According to the UF curve and UB curve, the mutation points of the rainfall sequence at Lucheng station were in 1977 and 1979.

(2)   Runoff trend analysis

The Mann–Kendall (MK) trend test was used to perform the trend analysis of the runoff series from 1962 to 2020 at Guantai hydrological station. As shown in Figure 2i, the runoff at Guantai hydrological station generally showed a decreasing trend ($UF < 0$) with an obvious decreasing trend from 1969 to 1977, which exceeded the critical line at the significance level of 0.01 after 1977, indicating an extremely significant decreasing trend of runoff. According to the UF curve and UB curve, the mutation points of the runoff sequence at Guantai station were in 1969 and 1977. Overall, from 1962 to 1964, the runoff was increasing ($UF > 0$), while since 1964, it has maintained a decreasing trend ($UF < 0$). The variation trend of runoff at Guantai hydrological station and the variation trend of rainfall in the upper reaches of the Zhang River showed a downward trend. Compared with the variation of rainfall, the variation of runoff's decreasing trend was more intense.

4.1.2. Cumulative Departure Method

(1) Analysis of rainfall

The Mann–Kendall trend test only made a qualitative judgment on the variation trend of rainfall-runoff in the Zhang River Basin. In order to further reveal the mutation characteristics of rainfall-runoff in the basin, the cumulative departure method was used to further analyze the relationship between the amount of water resources in the irrigation district and the rainfall in the upper reaches of the Zhang River. Figure 3 shows the cumulative departure variation process of rainfall at 8 stations in the upper reaches of the Zhang River. In general, it is divided into two stages in the late 1970s: the period of increasing rainfall before the 1970s and that of decreasing rainfall after the 1970s.

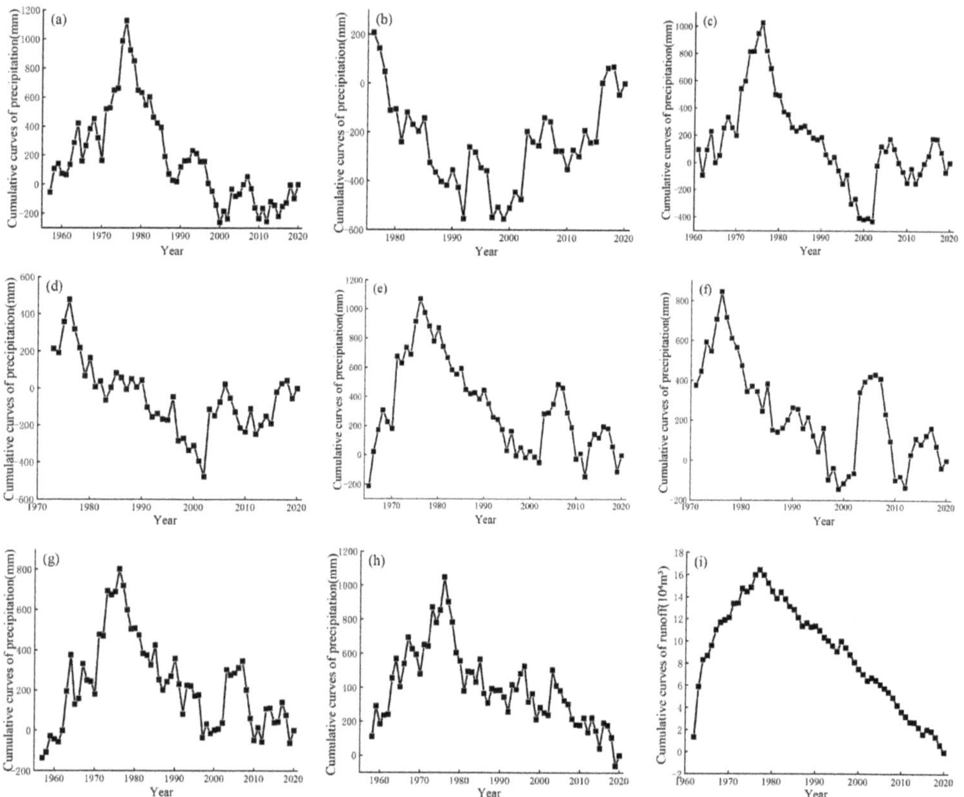

Figure 3. The results of MK trend analysis of rainfall-runoff in the Zhang River Basin. (**a**) Pingshun station; (**b**) Lucheng station; (**c**) Huguan station; (**d**) Changzhi station; (**e**) Zhangzi station; (**f**) Tunliu station; (**g**) Xiangyuan station; (**h**) Licheng station; (**i**) Guantai station.

(2) Analysis of runoff

The cumulative anomaly change process of runoff at the observation station is shown in Figure 3i. The changes in runoff and rainfall are similar. Taking the middle and late

1970s as the boundary, there are obvious increasing and decreasing periods of runoff. In summary, since 1957, the runoff variation process at Guantai station has been closely related to the rainfall in the upper reaches of the Zhang River. Their overall pattern is basically the same, and the amount of water resources increases first and then decreases during the research period.

4.1.3. Results of Trend Mutation

The analysis results of the MK trend test method and the cumulative departure method are shown in Table 2. A comprehensive analysis of the mutation points can determine the mutation years. The results show that there are some differences in the mutation points obtained by the two methods. It is determined through comprehensive analysis that there was one runoff mutation point in the Zhang River Basin from 1962 to 2020, and it was in 1977. Finally, it is determined that the years from 1962 to 1977 were the runoff base period of the Zhang River Basin, and those from 1977 to 2020 were the runoff variation period of the Zhang River Basin.

4.2. Results of Attribution Analysis

The variation in runoff series is the result of the combined effect of climate change and human activities, of which climate change will cause the variations in "natural runoff", while the difference between "measured runoff" and "natural runoff" is dominated by human activities. The two key steps of the runoff variation attribution method are the diagnosis of the runoff mutation point and the reduction of natural runoff, for which improving the accuracy of natural runoff simulation has been an important direction of hydrological research in recent years. According to the principle of attribution analysis, human activities will change the rainfall-runoff relationship in the region, showing the slope variation on the rainfall-runoff double accumulation curve. The rainfall-runoff regression analysis conducted before and after the mutation can quickly achieve quantitative attribution.

According to the results from the previous trend analysis, the regression analysis was performed for the base period (1962–1977) and the variation period (1977–2020), and according to Formula (2), the double mass curve of runoff at Guantai station from 1962 to 2020 and the annual rainfall at Pingshun station in the upper reaches of the Zhang River were plotted (Figure 4).

Because Lucheng station was built in 1976, compared with other stations, the data time series are quite different and not representative. Therefore, this paper only makes attribution analysis at the remaining 7 stations. From the blue and red in Figure 4, which represent the changes of the base period and the change period respectively, it can be seen that the slope of the curve is obviously inconsistent, and there is an inflection point around 1977. The slope of the curve in the base period is steeper, and the slope of the curve in the change period is obviously alleviated. It shows that under the same rainfall condition, the runoff in the base period is greater than that in the change period; that is, the underlying surface conditions of the basin before and after the inflection point are quite different, and human activities are more obvious. The contribution rate of rainfall change and human activities to the runoff change in the study area is shown in Table 3. Combined with Table 3, the attribution analysis of the runoff change at hydrological stations in the study area is as follows: The factors contributing to the annual runoff reduction in the Zhang River Basin during the change period are mainly human activities, accounting for more than 50%. Pingshun station, Tunliu station, and Xiangyuan station accounted for more than 60%, and Licheng station accounted for more than 70%.

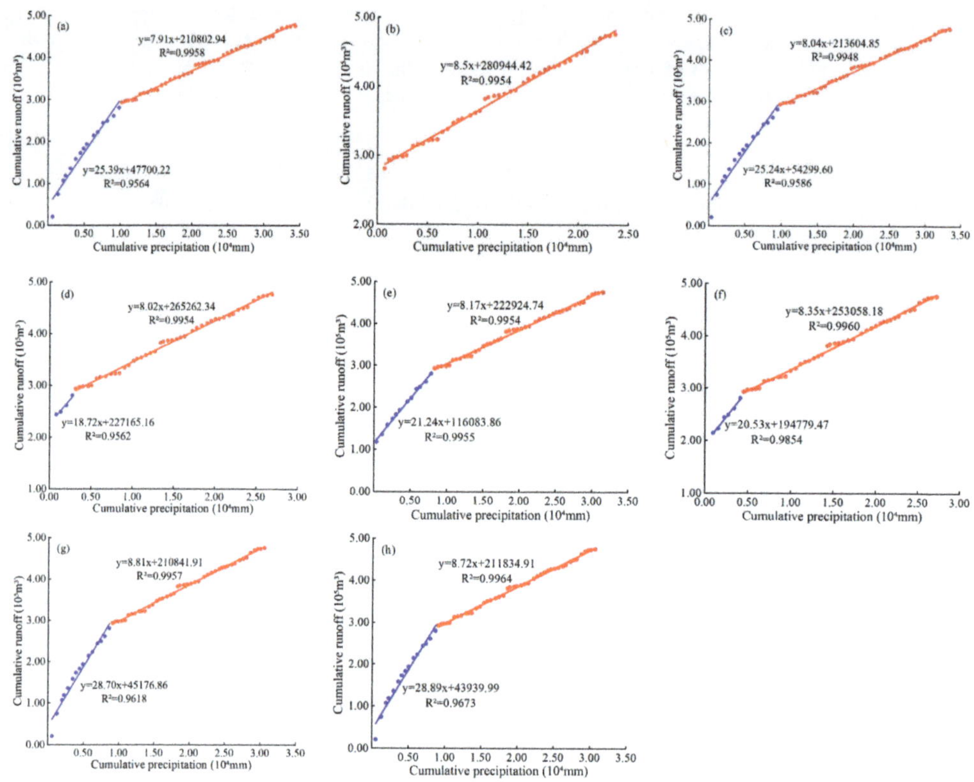

Figure 4. The double mass curve of annual runoff at Guantai station and annual rainfall in the upper reaches of the Zhang River. (**a**) Pingshun station; (**b**) Lucheng station; (**c**) Huguan station; (**d**) Changzhi station; (**e**) Zhangzi station; (**f**) Tunliu station; (**g**) Xiangyuan station; (**h**) Licheng station.

Table 3. The contribution rate of rainfall variation and human activities to runoff variation in the research area.

Rainfall Station-Hydrological Station	Period	Observed Value (m³/s)	Simulated Value (m³/s)	Runoff Variation Value (m³/s)	Rainfall		Human Activities	
					Variation (m³/s)	Contribution Rate (%)	Variation (m³/s)	Contribution Rate (%)
Pingshun Station-Guantai Station	1962–1977	11,034.60	12,049.70	215.85	83.73	38.79	132.12	61.21
	1977–2020	10,818.75	10,950.87					
Huguan Station-Guantai Station	1962–1977	11,034.60	11,748.48	215.85	105.12	48.70	110.73	51.30
	1977–2020	10,818.75	10,929.48					
Changzhi Station-Guantai Station	1973–1977	11,034.60	10,835.78	215.85	103.69	48.04	112.16	51.96
	1977–2020	10,818.75	10,930.91					
Zhangzi Station-Guantai Station	1965–1977	11,034.60	11,105.96	215.85	91.94	42.60	123.91	57.40
	1977–2020	10,818.75	10,910.70					
Tunliu Station-Guantai Station	1971–1977	11,034.60	10,899.02	215.85	82.37	38.16	133.48	61.84
	1977–2020	10,818.75	10,901.13					
Xiangyuan Station-Guantai Station	1962–1977	11,034.60	11,770.28	215.85	82.80	38.36	133.05	61.64
	1977–2020	10,818.75	10,901.56					
Licheng Station-Guantai Station	1962–1977	11,034.60	11,905.22	215.85	63.13	29.25	152.72	70.75
	1977–2020	10,818.75	10,881.89					
Average				215.85	87.54	40.55	128.31	59.46

Human activities have played a role in reducing the runoff from the Zhang River Basin. The change in rainfall runoff in the study area coincides with the three-year "Great Leap Forward" from 1957 to 1960. A large number of large and medium-sized reservoirs and irrigation areas have been built. Small-scale farmland water conservancy projects are blooming everywhere, and the time series of meteorological stations began in the whole period. Due to the impact of the "Cultural Revolution" in China from 1966 to 1979, the construction of farmland water conservancy was once stalled, coupled with climate change, rainfall and runoff increased year by year. Until the end of the "Cultural Revolution" in 1976, the existing engineering facilities and farmland water conservancy projects were restored. By the end of 1979, a large number of small and medium-sized farmland water conservancy projects had been renovated, and the runoff had been reduced year by year. It is necessary to study the variation law of rainfall runoff in the Zhang River Basin, give full play to the regulating effect of water conservancy projects on runoff, and improve the water resources management system so as to rationally allocate water resources in time and space.

4.3. Results of Periodicity

In order to understand the overall changing rule of a hydrological sequence, in addition to trend analysis, periodicity is also one of the important analysis contents. Because water resources and rainfall are not only affected by atmospheric circulation and temperature variations but also by human activities in the evolution process, there should be certain rules or changes in periodicity among these influencing factors, so it is necessary to study the periodicity of rainfall and total water resources. In order to predict the periodic variation of rainfall-runoff in the Zhang River Basin in the future, the Morlet wavelet transform is performed on the runoff data of the irrigation district and the rainfall data of the upper reaches of the Zhang River in Matlab to obtain the wavelet coefficient real part contour map (Figure 5, Table S1) and the Morlet wavelet coefficient variance diagram (Figure 6, Table S2).

Figure 5 is the wavelet coefficient real-part contour map. The abscissa is the year, the ordinate is the time scale, and the equivalent curve in the figure is the real part value of the wavelet coefficient. When the real part value of the wavelet coefficient is positive, it represents the wet period of runoff, plotted in blue in the figure when it is negative, it indicates the dry period of runoff, plotted in red in the figure.

There are significant periodic changes in the interannual scale of rainfall changes in the upper reaches of the river. As shown in Figure 5a–h, there are three types of periodic changes of 16~30 years, 10~16 years, and 6~9 years in the evolution of rainfall in the basin. On the three types of scales, there are multiple oscillations of dry-wet alternation, and the oscillation period is about 5 years. At the same time, the periodic changes of the above three scales are very stable and global throughout the whole analysis period. The periodic change of 16~30 years is relatively stable. The negative phase of the 16–30-year scale has approached closure by 2020, and the new oscillation period begins, indicating that the annual rainfall will show a fluctuating upward trend from 2020 to 2035.

The change in runoff in the irrigation area also has significant periodic changes on an interannual scale. As shown in Figure 5i, there are two types of periodic changes of 9~32 years and 4~10 years in the evolution of runoff in the basin. There are multiple oscillations of dry-wet alternation on both scales, and the oscillation period is about 2–5 years. However, since 2000, due to the influence of coal mining, the periodic variation trend of runoff has been weakening. The runoff from the observation station tends to be stable, and the dry-wet change is not obvious. Especially after 2020, the short fluctuation period has disappeared, indicating that the runoff remains stable from 2020 to 2035.

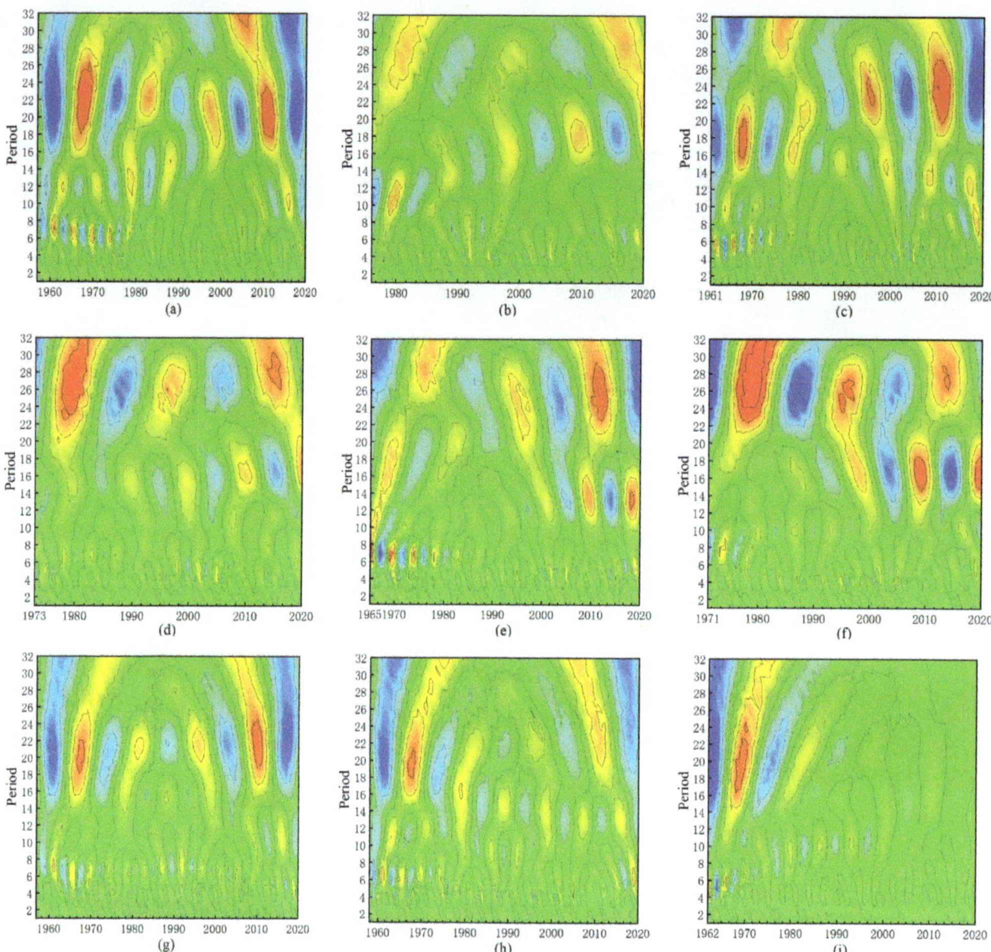

Figure 5. The contour map of wavelet coefficients of the rainfall stations and hydrological stations in the upper reaches of the Zhang River. (**a**) Pingshun station; (**b**) Lucheng station; (**c**) Huguan station; (**d**) Changzhi station; (**e**) Zhangzi station; (**f**) Tunliu station; (**g**) Xiangyuan station; (**h**) Licheng station; (**i**) Guantai station.

Figure 6 is the wavelet coefficient variance diagram; the abscissa is the number of years of the cycle, and the ordinate is the relative size of the wavelet variance. From the diagram, the main cycle of water resource change in the irrigation area can be seen. Figure 6 shows that there are three peaks in the wavelet variance of annual runoff. The first peak corresponds to a time scale of 20 years, which is the first main cycle of annual runoff change in the basin, indicating that the annual runoff in this area has the strongest periodic oscillation of about 20 years. The second peak is 11 years, which is the second main cycle of annual runoff change. The third peak is 5 years, which is the third main cycle of annual runoff change. This indicates that the fluctuation of the above three cycles controls the variation characteristics of runoff in the whole time series.

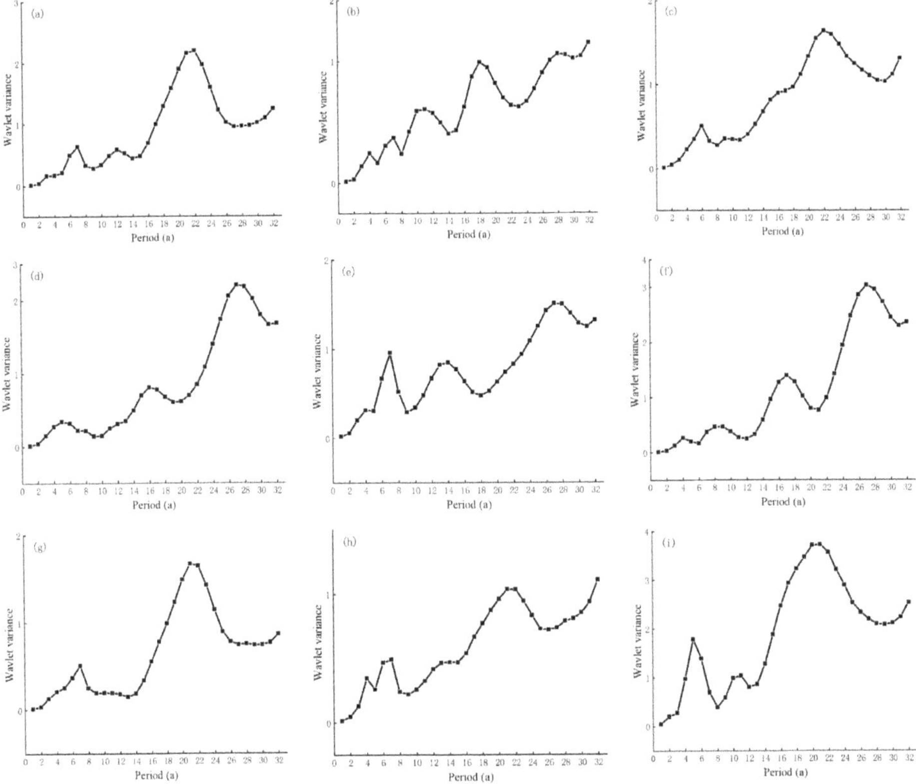

Figure 6. The map of wavelet variance of the rainfall stations and hydrological stations in the upper reaches of the Zhang River. (**a**) Pingshun station; (**b**) Lucheng station; (**c**) Huguan station; (**d**) Changzhi station; (**e**) Zhangzi station; (**f**) Tunliu station; (**g**) Xiangyuan station; (**h**) Licheng station; (**i**) Guantai station.

5. Conclusions

In this paper, trend mutation analysis, attribution analysis, periodic analysis, and prediction of rainfall and runoff in the research area are carried out based on the runoff data from 1962 to 2020 at Guantai hydrological station in Zhang River Basin and the rainfall data from 1957 to 2020 of its upstream meteorological stations, combined with the methods of Mann–Kendall trend test, cumulative departure method, double mass curve method, and Morlet wavelet transform, to obtain more scientific and reasonable conclusions with the main contents as follows:

In the past 60 years, the variation process of water resources in the irrigation district has been closely related to the rainfall in the upper reaches of the Zhang River, with the overall pattern being basically the same, that is, a decreasing trend as a whole in terms of rainfall and runoff in the irrigation district. Except for the late-built Lucheng station, the measured rainfall-runoff from the other seven stations has changed significantly, and the mutation year of rainfall-runoff was around 1977. The runoff variation can be divided into two stages: the basis period (1962~1977) and the variation period (1977~2020).

The results from an attribution analysis of runoff variation in the research area show that the factors contributing to annual runoff reduction in the Zhang River Basin are mainly due to human activities, with the contribution rate accounting for more than 50%. The rainfall-runoff data was collected during the "Great Leap Forward" period in the late 1950s. The rainfall runoff increased year by year during the "Cultural Revolution" period from the

1960s to the 1970s and decreased year by year until the end of the "Cultural Revolution" in the late 1970s. The hydrological effects caused by human activities such as a large number of farmland water conservancy projects, groundwater exploitation, and various water storage projects built in the late 1970s are the main reasons for the decrease in runoff.

There is a long fluctuation period of about 20 or 15 years and a short fluctuation period of 5 years on the interannual scale of rainfall and runoff variation in the upper reaches of the Zhang River. During the research period, there are many oscillations of dry-wet alternation, with the oscillation period being about 5 years. The long fluctuation period of rainfall was approaching closure by 2020, while the new oscillation period initiated, indicating that the annual rainfall may represent a fluctuating upward trend from 2020 to 2035. Since 2000, due to the influence of coal mining, the periodic variation trend of runoff in the irrigation district has been weakening. The runoff at Guantai station tends to be stable, and the dry-wet variation is not obvious, especially since the short fluctuation period almost disappears, indicating that the runoff has remained stable since 2000.

Supplementary Materials: The following supporting information can be downloaded at: https://www.mdpi.com/article/10.3390/w15142521/s1, Table S1: Figure 5 Pingshun station. Table S2: Figure 6's wavelet variance.

Author Contributions: Conceptualization, L.D. and G.W.; methodology, L.D.; software, G.W.; validation, L.D., G.W. and B.L.; formal analysis, G.W.; investigation, B.L.; resources, G.W.; data curation, G.W.; writing—original draft preparation, G.W.; writing—review and editing, L.D.; visualization, G.W.; supervision, B.L.; project administration, B.L.; funding acquisition, L.D. All authors have read and agreed to the published version of the manuscript.

Funding: This research was funded by the Water Resources Demonstration Project of Zhangfu River Irrigation District in Handan City, Hebei Province, funder: Lijuan Du, funding number ID120202A0022022, and The APC was funded by Lijuan Du.

Data Availability Statement: The rainfall data are derived from the China Meteorological Science Data Sharing Service Network (http://data.cma.cn), and the runoff data are from the Deliverables of Hydrological Data Compiled by Handan Zhangfu River Irrigation District.

Acknowledgments: This research has been supported by funds from the Water Resources Demonstration Project of Zhangfu River Irrigation District in Handan City, Hebei Province (ID120202A0022022).

Conflicts of Interest: The authors declare no conflict of interest.

References

1. Wang, G.Q.; Zhang, J.Y.; Guan, X.X.; Bao, Z.X.; Liu, Y.L.; He, R.M.; Jin, J.L.; Liu, C.S.; Chen, X. Quantifying attribution of runoff change for major rivers in China. *Adv. Water Sci.* **2020**, *31*, 313–323.
2. Liu, N.; Harper, R.J.; Smettem, K.R.J.; Dell, B.; Liu, S. Responses of streamflow to vegetation and climate change in southwestern Australia. *J. Hydrol.* **2019**, *572*, 761–770. [CrossRef]
3. Liu, Y.; Han, X.X.; Li, L.J. Characteristics and attribution analysis of runoff changes in the middle and upper reaches of Cao'e River from 1980 to 2018. *J. Water Resour. Water Eng.* **2022**, *33*, 53–61.
4. Nyikadzino, B.; Chitakira, M.; Muchuru, S. Rainfall and runoff trend analysis in the Limpopo river basin using the Mann Kendall statistic. *Phys. Chem. Earth* **2020**, *117*, 102870. [CrossRef]
5. Dan'azumi, S.; Ibrahim, U.A. Trend analysis of observed precipitation, temperature, and streamflow for Hadejia-Nguru wetlands catchment, Nigeria. *Theor. Appl. Climatol.* **2023**, *151*, 195–207. [CrossRef]
6. Clavera, G.R.; Quintana, S.P.; Palazón, L.; Zabaleta, A.; Cenobio, O.; Barella, O.A.; Beguería, S. Streamflow trends of the Pyrenees using observations and multi-model approach (1980–2013). *J. Hydrol. Reg. Stud.* **2023**, *46*, 101322.
7. Liu, J.P.; Wang, W.; Zhou, J.J.; Zou, X.B. The Research of Runoff Variation Characteristics of Wei River Basin. In Proceedings of the 3rd International Conference on Advances in Energy and Environmental Science, Zhuhai, China, 25–26 July 2015.
8. Shen, Z.T.; Zhao, Q.; Li, X.F. Analysis of Rainfall-Runoff Variation Characteristics and Influencing Factors in the Dawen River Basin in the Past 50 Years. *IOP Conf. Ser. Earth Environ. Sci.* **2020**, *619*, 012091. [CrossRef]
9. Huang, Z.Q.; Shen, K.Q.; Zhao, C.; Shan, M. Trend Analysis of Annual Runoff in Yong'an Creek Basin of Zhejiang. *IOP Conf. Ser. Earth Environ. Sci.* **2021**, *811*, 012003.
10. Lu, H.X.; Ma, G.W.; Wang, X.D.; Ping, F.; Li, F.W. Effects of Abundant-and-Low-Flow Property of Annual Rainfall on Multi-Source Water Intake System Allocation. *Appl. Mech. Mater.* **2013**, *2545*, 353–356. [CrossRef]

11. Li, Z.Y.; Sun, Z.L.; Sun, L.X.; Liu, J.; Xiong, W.H.; Dong, H.Y.; Zheng, H.L. Hydrological variation and hydro-sediment interrelation of the Luozha River in the Lancang River Basin. *Water Supply* **2022**, *22*, 4839–4851. [CrossRef]
12. Lyu, J.Q.; Yin, S.S.; Sun, Y.T.; Wang, K.X.; Luo, P.P.; Meng, X.L. Flood Runoff Simulation under Changing Environment, Based on Multiple Satellite Data in the Jinghe River Basin of the Loess Plateau, China. *Remote Sens.* **2023**, *15*, 550. [CrossRef]
13. He, B.; Miao, C.Y.; Shi, W. Trend, abrupt change, and periodicity of streamflow in the mainstream of Yellow River. *Environ. Monit. Assess.* **2013**, *185*, 6187–6199. [CrossRef] [PubMed]
14. Wu, L.L.; Zhang, W.; Wu, T. Analysis of the Periodicity of Annual Extreme Runoff at Datong Station, Yangtze River, China. *Appl. Mech. Mater.* **2012**, *256–259*, 2606–2610. [CrossRef]
15. Lamborn, C.C.; Smith, W.J. Human perceptions of, and adaptations to, shifting runoff cycles: A case-study of the Yellowstone River (Montana, USA). *Fisheries Research* **2019**, *216*, 96–108. [CrossRef]
16. Ji, G.X.; Wu, L.Y.; Wang, L.D.; Yan, D.; Lai, Z.Z. Attribution Analysis of Seasonal Runoff in the Source Region of the Yellow River Using Seasonal Budyko Hypothesis. *Land* **2021**, *10*, 542. [CrossRef]
17. Wang, L.; Cao, H.; Li, Y.R.; Feng, B.F.; Qiu, H.; Zhang, H.R. Attribution Analysis of Runoff in the Upper Reaches of Jinsha River, China. *Water* **2022**, *14*, 2768. [CrossRef]
18. Zhang, C.; Wu, C.S.; Peng, Z.D.; Kuai, S.Y.; Zhang, S.H. Synergistic Effects of Changes in Climate and Vegetation on Basin Runoff. *Water Resour. Manag.* **2022**, *36*, 3265–3281. [CrossRef]
19. Mo, C.X.; Lai, S.F.; Yang, Q.; Huang, K.K.; Lei, X.B.; Yang, L.F.; Yan, Z.W.; Jiang, C.H. A comprehensive assessment of runoff dynamics in response to climate change and human activities in a typical karst watershed, southwest China. *J. Environ. Manag.* **2023**, *332*, 117380. [CrossRef]
20. Wang, H.X.; Ma, Y.C.; Yang, H.; Hong, F.T.; Guo, W.X. Quantitative evaluation of the impact of climate change and human activities on Jialing river runoff changes in the past 60 years, China. *J. Water Clim. Change* **2023**, *14*, 590–609. [CrossRef]
21. Panditharathne, R.; Gunathilake, M.B.; Chathuranika, I.M.; Rathnayake, U.; Babel, M.S.; Jha, M.K. Trends and Variabilities in Rainfall and Streamflow: A Case Study of the Nilwala River Basin in Sri Lanka. *Hydrology* **2022**, *10*, 8. [CrossRef]
22. Li, Y.; Zhao, L.; Zhang, Z.; Li, J.X.; Hou, L.; Liu, J.Q.; Wang, Y.B. Research on the Hydrological Variation Law of the Dawen River, a Tributary of the Lower Yellow River. *Agronomy* **2022**, *12*, 1719. [CrossRef]
23. Ji, G.X.; Gao, H.S.; Huang, J.C.; Yang, X.; Zhang, Y.L. Research on Runoff Variation Characteristics and Attribution Analysis in the Upper Yellow River Basin. *J. Henan Norm. Univ. (Nat. Sci. Ed.)* **2023**, *01*, 12–19.
24. Banda, V.D.; Dzwairo, R.B.; Singh, S.K.; Kanyerere, T. Separating anthropogenic and climate contributions to streamflow variations in Rietspruit sub-basin, South Africa. *Phys. Chem. Earth* **2022**, *127*, 103200. [CrossRef]
25. Wang, L.; Song, X.Y.; Li, L.J.; Zhang, L.; Liu, Y.; Li, H.Y.; Li, Y.L. Variation trend and attribution analysis of runoff in typical small watershed in gully region of the loess. *Res. Soil Water Conserv.* **2021**, *28*, 48–53+69.
26. Zhang, H.Z.; Zhao, W.J. Quantitative Analysis of Annual Precipitation in Extreme Precipitation Areas. *Environ. Resour. Ecol. J.* **2022**, *6*, 90–94.
27. Das, S.; Scaringi, G. River flooding in a changing climate: Rainfall-discharge trends, controlling factors, and susceptibility mapping for the Mahi catchment, Western India. *Nat. Hazards* **2021**, *109*, 2439–2459. [CrossRef]
28. Talukdar, S.; Pal, S.; Shahfahad; Naikoo, M.W.; Parvez, A.; Rahman, A. Trend analysis and forecasting of streamflow using random forest in the Punarbhaba River basin. *Environ. Monit. Assess.* **2022**, *195*, 153. [CrossRef] [PubMed]
29. Zhang, L.R.; He, Y.H.; Tang, Y.P.; Wang, G.Q. Research on key technologies of Haihe River Basin runoff change trend and change attribution. In Proceedings of the Future-Oriented Water Security and Sustainable Development—Proceedings of the 14th China Water Forum, Changchun, China, 25 August 2016; pp. 117–125.
30. Liu, W.L.; Liu, L.N.; Wu, B. Analysis of dry runoff change characteristics and influencing factors in the lower reaches of the main stream of the Ganjiang River. *Hydrology* **2022**, *42*, 89–96.
31. Pan, B.; Han, M.; Wei, F.; Tian, L.X.; Liu, Y.T.; Li, Y.L.; Wang, M. Analysis of the Variation Characteristics of Runoff and Sediment in the Yellow River within 70 Years. *Water Resour.* **2021**, *48*, 676–689. [CrossRef]
32. Zhang, G.H.; Ding, W.F.; Liu, H.Y.; Liang, Y.; Xu, L.; Zhang, O.Y. Quantifying Climatic and Anthropogenic Influences on Water Discharge and Sediment Load in Xiangxi River Basin of the Three Gorges Reservoir Area. *Water Resour.* **2021**, *48*, 204–218. [CrossRef]
33. Jin, S.F.; Zheng, Z.Y.; Ning, L.K. Separating variance in the runoff in Beijing's river system under climate change and human activities. *Phys. Chem. Earth* **2021**, *123*, 103044. [CrossRef]
34. Shao, Y.T.; He, Y.; Mu, X.M.; Zhao, G.J.; Gao, P.; Sun, W.Y. Contributions of climate change and human activities to runoff and sediment discharge reductions in the Jialing River, a main tributary of the upper Yangtze River, China. *Theor. Appl. Climatol.* **2021**, *145*, 1437–1450. [CrossRef]
35. Qi, L.L.; Guo, Z.L.; Qi, Z.X.; Guo, J.J. Prospects of Precipitation Based on Reconstruction over the Last 2000 Years in the Qilian Mountains. *Sustainability* **2022**, *14*, 10615. [CrossRef]
36. Guo, W.X.; He, N.; Ban, X.; Wang, H.X. Multi-scale variability of hydrothermal regime based on wavelet analysis—The middle reaches of the Yangtze River, China. *Sci. Total Environ.* **2022**, *841*, 156598. [CrossRef]
37. Sun, N. Consistency Test of Natural Runoff Series. *Shanxi Water Conserv. Sci. Technol.* **2015**, *04*, 68–70.
38. Yang, D.; Liu, H.M.; Guo, P.P.; Zheng, F.J.; Liu, Q. The change of rainfall in Liaoning Province in recent 53 years from 1956 to 2008. *Resour. Environ. Arid. Reg.* **2011**, *25*, 002.

39. Lozowski, E.P.; Charlton, R.B.; Nguyen, C.D.; Wilson, J.D. The Use of Cumulative Monthly Mean Temperature Anomalies in the Analysis of Local Interannual Climate Variability. *J. Clim.* **1989**, *2*, 1059–1068. [CrossRef]
40. Shu, M.Z.; Liu, L.H. Temporal and spatial variation characteristics of rainfall in the Haihe River Basin in recent 51 years. *South–North Water Divers. Water Conserv. Sci. Technol.* **2015**, *13*, 009.
41. Hu, D.L.; Yan, D.H.; Song, X.S.; Zhang, M.Z.; Yu, X.; Yang, S.Y. Analysis on the changing trend of water resources in the Yangtze River Basin above Yibin. *South–North Water Divers. Water Conserv. Sci. Technol.* **2008**, *02*, 21.
42. Li, L.; Wu, X.J.; Zhou, X.D.; Dong, Y.; Zhang, F.P.; Zhang, Y.N.; Liu, J. Analysis of river runoff evolution characteristics in coal mining area based on Morlet wavelet theory. *Hydropower Energy Sci.* **2022**, *40*, 37–40.

water

MDPI

Article

Assessing Feasibility of Water Resource Protection Practice at Catchment Level: A Case of the Blesbokspruit River Catchment, South Africa

Koleka Makanda [1,2,*], Stanley Nzama [3] and Thokozani Kanyerere [2]

1 Water Resource Classification, Department of Water and Sanitation, Pretoria 0001, South Africa
2 Department of Earth Sciences, University of the Western Cape, Cape Town 7535, South Africa; tkanyerere@uwc.ac.za
3 Reserve Determination, Department of Water and Sanitation, Pretoria 0001, South Africa; nzamas@dwa.gov.za
* Correspondence: cornie.makanda@gmail.com; Tel.: +27-123-368-406

Abstract: The operationalization of water resource protection initiatives for surface water resource quality and equitable water quality allocation is critical for sustainable socio-economic development. This paper assessed Blesbokspruit River Catchment's water quality status, using the South African Water Quality standards and Water Quality Index (WQI). Protection levels for quality, and waste discharge for point sources were set and evaluated using the total maximum daily loads (TMDLs) and chemical mass balance (CMB) techniques, respectively. The study found that the water quality results for the analysed physico-chemical parameters (Na^+, Ca^{2+}, Mg^{2+}, Cl^-, F^-, pH, EC, SO_4^{2-}) of the data collected from 2015 to 2022 were within the limits of the water quality standards, except for NO_3^- and PO_4^{2-}. The water quality from the study area was categorized as acceptable for drinking purposes with the WQI of 54.80. The application of the TMDL approach resulted in the 77.96 mS/m for electrical conductivity (EC), 9.92 mg/L for phosphate (PO_4^{2-}), and 15.16 mg/L for nitrate NO_3^- being set as the protection levels for the catchment. The CMB was found to be a useful tool for the evaluation of point source discharges into water resources. The study recommends the application of TMDL and CMB techniques in water resource protection practice.

Keywords: chemical mass balance; protection limits; total maximum daily loads; water; quality allocation; water quality index; water resource protection

Citation: Makanda, K.; Nzama, S.; Kanyerere, T. Assessing Feasibility of Water Resource Protection Practice at Catchment Level: A Case of the Blesbokspruit River Catchment, South Africa. *Water* **2023**, *15*, 2394. https://doi.org/10.3390/w15132394

Academic Editors: Peiyue Li and Jianhua Wu

Received: 24 May 2023
Revised: 25 June 2023
Accepted: 26 June 2023
Published: 28 June 2023

1. Introduction

Water resources are crucial for sustaining a sufficient food supply, a fruitful habitat for biodiversity, and a healthy environment for all living things. Fundamentally, water is one of the most essential needs for life [1,2]. Globally, surface water quality has become a complex issue, one that remains very important for long-term economic development, welfare, and environmental viability [3,4]. Surface water quality, which is impacted by natural processes and anthropogenic activities [5–8], requires efficient protection for pollution prevention, especially in areas where freshwater is limited [9–13].

In South Africa, freshwater resources are the most essential resources for human existence and growth [14]. In the country, water quality challenges related to freshwater resources are well reported in the literature [15–17]. Such challenges linked to water quality deterioration were confirmed by [3], who reported that several water management areas in the country are experiencing water shortages and quality deterioration while natural systems are put under enormous pressure. Changes in river flow patterns have also been identified as one of the factors that negatively influence water quality, in addition to influences from anthropogenic activities [18]. Ref. [19] argued that the adoption of water resources management strategies that aim to strike a balance between water resource conservation and their sustainable utilization is essential considering that surface water resources are susceptible to contamination and overexploitation due to their accessibility and fragility.

The South African water quality standards for drinking water [20] and other South African Water Quality Guidelines (SAWQG) for various water uses are available and extensively used in the county to ensure that their requirements in terms of water quality are met. Water resource assessment using water quality guidelines provides information on whether water from a source is suitable for meeting fundamental human needs, such as being fit for drinking, or any other water uses, and in cases of unsuitability treatments, processes can be recommended before consumption [21–23]. However, numerical limits prescribed in the water quality guidelines are the same nationally, and they are a requirement for a specific water use such as domestic applications, industrial applications, and in aquatic ecosystem. The numerical limits prescribed in the guidelines are user-specific, and do not necessarily reflect the spatial or temporal variability of a catchment. Therefore, in order to improve water resource protection at the catchment level, numerical limits for water quality formulated on the bases of the prevailing conditions of a particular catchment must be determined (catchment-specific protection levels). In the country, the protection of water resources is ensured by undertaking studies on resource- directed measures such as resource quality objectives (RQOs). The numerical limits for RQOs are set by taking into consideration the background conditions and spatial variability of individual catchments; hence, they differ from one catchment to another. Therefore, while water quality guidelines set numerical limits for various water uses, resource-directed measures (RDMs) such as RQOs set limits for a particular resource to protect its current conditions or improve them. In this regard, several studies on RDMs have been concluded in the country [24]. Results from such studies and projects involving resource-directed actions are reported in the official publications from the government. The published indicators and numerical limits for water quality are prescribed for implementation at the catchment level [25,26].

According to [17], one of the main pressing issues regarding water resource protection practices at the catchment level is a clear link between the water use license conditions set for users and numerical limits for water quality set as protection levels for a catchment. The understanding of how water resource protection limits for surface water quality and water use conditions are set in a catchment can significantly improve water resource protection practices at catchment level. Previously, researchers have used programming models (simulations) for the purpose of waste allocation taking into consideration established water resource protection requirements [27–33].

However, in the absence of more advanced modelling software to model scenarios of a system's water quality parameters in a catchment, the application of water resource protection initiatives at the basin level becomes a challenging task for water resource managers. At the Blesbokspruit River Catchment, there is limited knowledge of how to operationalize water resource protection initiatives set for the catchment, a situation which requires addressing. Therefore, the aim of this investigation was to showcase how water resource protection initiatives can be practically applied at a water resource level using the techniques of water quality index (WQI), total maximum daily loads (TMDLs), and chemical mass balance (CMB) through a case study of the Blesbokspruit River Catchment. For the desired outcome of the investigation to be achieved, the following specific objectives were established: (i) to assess and evaluate the surface water quality status of the Blesbokspruit River Catchment using WQI; (ii) to estimate the TMDL allowable for use as protection limits for surface water resources in the catchment; and (iii) to estimate the waste load allocation for surface water quality using the CMB model.

2. Materials and Methods

The methodological approach applied in the study is shown in Figure 1 and comprises the following: (1) data collection and analysis; (2) water quality index calculation; (3) total maximum daily load estimation; and (4) waste load allocation.

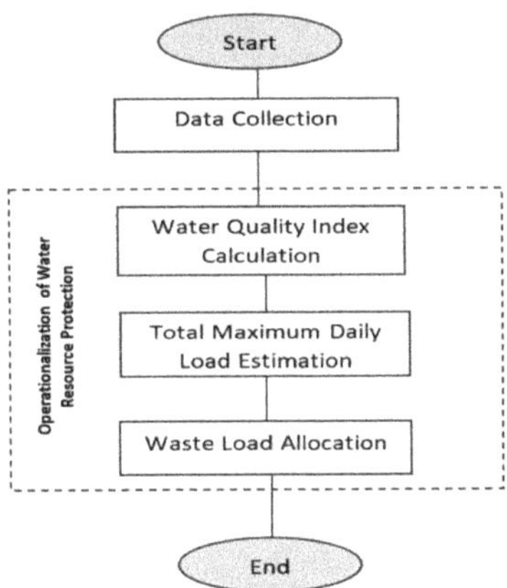

Figure 1. Flowchart for water resource protection practices in the Blesbokspruit River Catchment.

2.1. Study Area Description

The study area which is the Blesbokspruit River Catchment (BRC) is completely located to the south of the equator between latitudes 26°1.91′ S–26°36.97′ S and longitudes 28°23.41′ E–28°19.97′ E (Figure 2). The catchment area is within the Upper Vaal Water Management Area (UV-WMA) of the broader Vaal River System with the sub-catchments C21D, C21E, and C21F. The BRC, which is found 40 km to the south-east of Johannesburg, holds a delicate position within the Gauteng Province due toits location within the East Rand Region. The catchment is surrounded by 5 towns, namely: Springs, Benoni, Boksburg, Brakpan, and Nigel, which form part of South Africa's densely urbanized and industrialised hub [34]. The catchment drains into the Vaal River system which is the main source of water supply to the residents of Gauteng Province [3]. A considerable region of formal and informal urban development surrounds the basin. About 45% of the watershed is urbanized, with the remainder comprising mining, industrial, and agricultural activity. It has been reported that the quality of the BRC has dramatically deteriorated due to the discharge of mining effluent and sewerage, as well as other pollution linked with urbanization, industrialization, and agricultural growth in recent decades. [3,34]. The catchment was identified as Resource Unit 62 (RU62) within the delineated integrated unit of analysis UI of the UV-WMA [24]. The catchment was prioritised for water resource protection, and numerical limits for nitrate, phosphate, and electrical conductivity were prescribed and legalized through the government gazette [24]. This makes the BRC an ideal case study area for testing the feasibility of RDM operationalization.

2.2. Data Collection

This study relied on the secondary data sourced from the national water quality database called the Water Management System (WMS) of the South African Department of Water and Sanitation (DWS). This study utilised data from 2015 to 2022 from 9 in-stream assessment sites and examined 34 groundwater sites within the BRC; the sampling locations are shown in Figure 2a. These datasets were generated through the field data collection under the National Chemical Monitoring Programme (NCMP) of the DWS. The NCMP uses standard methods as outlined in [35,36] to collect data directly from the aquatic environment for surface water and from the field for groundwater. Briefly, water samples for physico-chemical analysis were collected from the sites within the main perennial Blesbokspruit tributary using the grab sampling technique [35]. Groundwater samples were obtained using a bailer from groundwater locations (boreholes) after boreholes had been purged and the water quality field characteristics (temperature, pH, electrical conductivity (EC)) had stabilized [21,36]. Water samples were subsequently put into the entire capacity of sterile 250 mL polyethylene sampling vials for reducing the headspace volume and were labelled accordingly. To protect the samples from microbial activity that could result in changes in chemical constituents and concentrations within a water sample, one ampoule of mercury chloride (HgCl) was added to each sample bottle. Samples were packed in a cooler box with ice packs (to keep the temperatures low) and transported to the Resource Quality Information Services (RQIS) national laboratory for analysis. Samples were stored in a dark cooler room at a temperature below 4 °C until analysis was performed by the laboratory. The chemical analyses were undertaken using Aquakem 250 Photometric Analyzer, Flow Injection Analyzer (FIA), and Flammable Atomic Absorption Spectrophotometry (FAAS). Following that, the chemical analysis findings were recorded in the WMS database for future use.

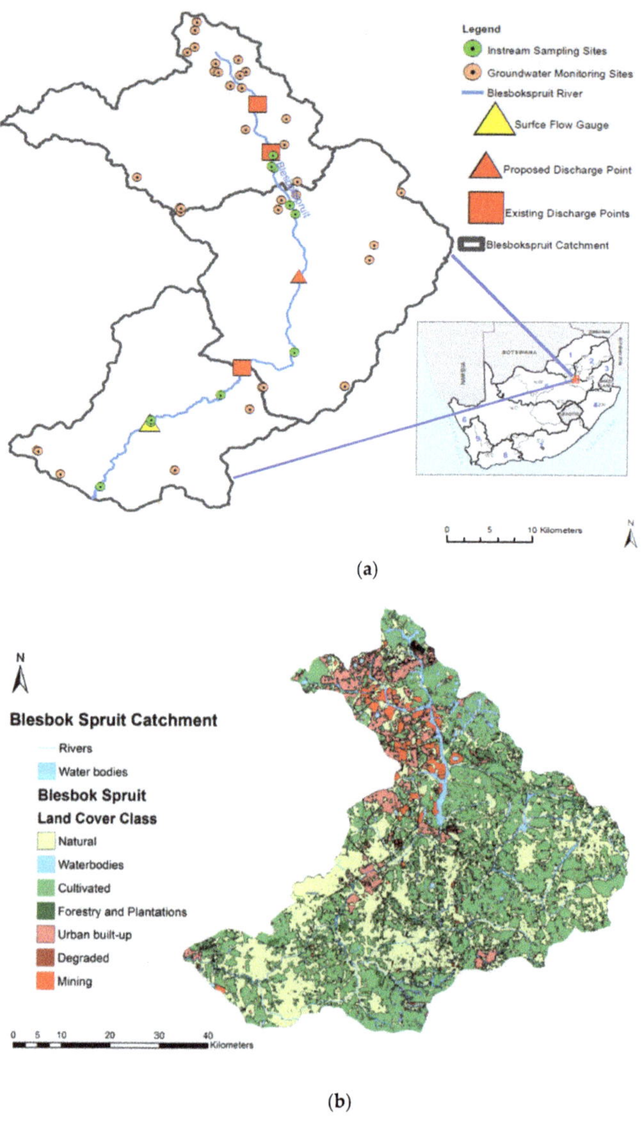

Figure 2. (**a**) Locality map of the Blesbokspruit River Catchment; and (**b**) Study area showing land cover activities, adopted from [3].

2.3. Data Analysis

Water quality data from the WMS database were transferred to the Microsoft Excel 2016 spreadsheet for descriptive statistical analysis. For each parameter, the annual averages and standard deviations were calculated. Table 1 displays the descriptive statistics (maximum, minimum, median, and standard deviation) for all 10 water quality parameters that were analysed. The water quality metrics evaluated for investigation include common main cations (Na^+, Ca^{2+}, Mg^{2+}), anions ($NO_3{}^-$, Cl^-, F^-, $SO_4{}^{2-}$ $PO_4{}^-$) and physical water quality parameters such as electrical conductivity (EC) and pH. The specified water quality parameters are critical for assessing drinking water quality in accordance with the national standards [19,20], and for aquatic ecosystem assessment [37].

Table 1. Water quality standards for domestic use and aquatic ecosystem [20,37].

WQ Variables	Measurement Units	Class 0	Class I	Class II	Class III
pH	pH units	6–9	5–6 and 9–9.5	4–5 and >9.5–10	<4 and >10
EC	mS/m	<70	70–150	150–370	>370
Ca	mg/L	<80	80–150	150–300	>300
Cl	mg/L	<100	100–200	200–600	>600
F	mg/L	<0.7	0.7–1.0	1.0–1.5	>1.5
Mg	mg/L	<70	70–100	100–200	>200
NO_3	mg/L	<6	6–10	10–20	>20
Na	mg/L	<100	100–200	200–400	>400
SO_4	mg/L	<200	200–400	400–600	>600
PO_4	mg/L	<0.005	0.005–0.025	0.025–0.25	>0.25

2.4. Water Quality Index Calculation

In assessing the surface water quality status in the research area, the limits prescribed in the water quality standards (Table 1) were utilized as a measure of unprocessed water quality. Water quality variables are described as class 0, class I, class II, and class III. The water quality classes describe raw water that is considered to be ideal water for domestic use, acceptable water for domestic use, tolerable water for domestic use, and unacceptable water for domestic use, respectively. Therefore, the average values of the concentrations for the 10 water quality parameters were compared to the limits prescribed in the guidelines.

Analysed data were also used to establish monthly changes in the concentrations of water quality parameters during the period from 2015 to 2022. This was achieved by doing trend analysis, and monthly changes were presented in the graphs of monthly averages for each chemical or physical parameter. Trend analysis was performed to establish changes in the concentrations of water quality constituents during the 7-year period. In order to obtain a comprehensive picture of the overall surface water quality within the catchment, the water quality index (WQI) was used to establish the overall water quality status of the catchment. WQI is used to simplify and convey scientific water quality information by combining the influence of various water quality parameters into a single-digit score that describes the overall water quality in a watershed [38]. The index delivers meaningful and understandable water quality information to policymakers and the general public on river quality status with a scientific basis [21,39–41]. WQI models have four stages: (1) selecting the water quality parameters of interest; (2) generating sub-indices for each parameter; (3) calculating the parameter weighting values; and (4) aggregating sub-indices to compute the overall water quality index. All four stages are demonstrated in Figure 3. The WQI was calculated for the effective nine selected parameters of the surface water quality (pH, EC, Ca, Cl, F, Mg, NO_3, Na, SO_4) by following a five-step procedure as per [21,42].

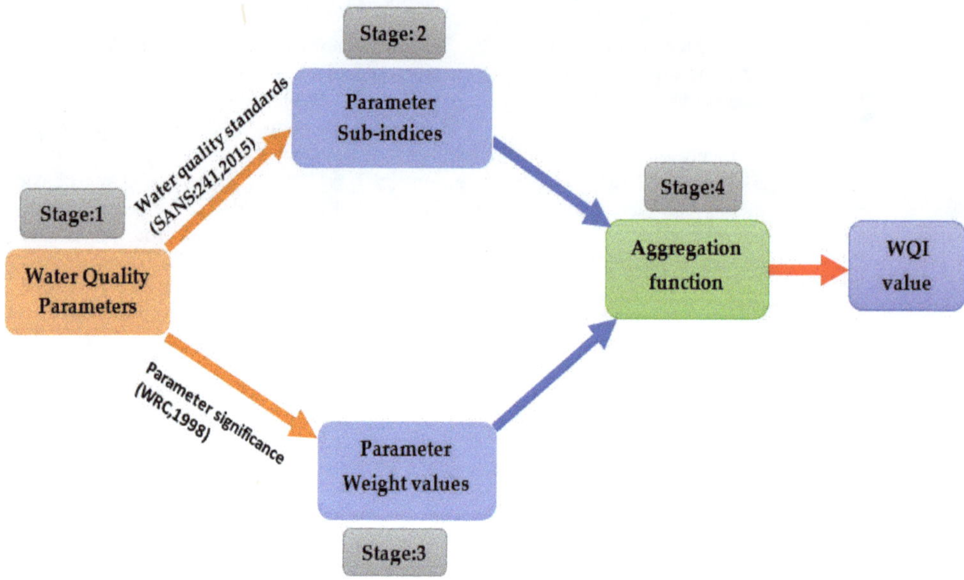

Figure 3. A general structure of the WQI model, adapted and modified from [38].

The initial step was to allocate weight (*wi*) to the chosen water quality parameters. According to [43,44] a weight value ranging from 1 to 5 was assigned to water quality parameters for estimating the water quality index, where nitrate is usually assigned with the maximum weight of 5. A weight of 4 was assigned to Cl^-, F^-, and SO_4^{2-}, with 3 assigned to the Na^+, Ca^{2+} and Mg^{2+}, while pH and EC were assigned a weight of 2 according to [21]. In the second stage, Equation (1) was used to determine a relative weight (*Wi*) for each of the selected water quality parameters.

$$Wi = wi / \sum_{i=1}^{n} wi \tag{1}$$

The third stage was to calculate and assign a quality rating scale (*qi*) for each parameter by dividing the concentration of each water quality parameter (*Ci*) by the South African water quality standard for domestic use (*Si*). When the [20] guideline was applied, numerical limits for class I were considered, whereas when the [19] standard was utilized, numerical limits as specified in the standard were considered. Equation (2) was used to convert the results into percentages.

$$qi = \left(\frac{Ci}{Si}\right) \times 100 \tag{2}$$

The fourth stage involved calculating the sub-index (*SI$_i$*) for each water quality metric using the algorithm in Equation (3).

$$SI_i = W_i \cdot q_i \tag{3}$$

The WQI for the entire study area was determined using Equation (4) in the fifth stage.

$$WQI = \sum SI_i \tag{4}$$

As shown in Table 2, the estimated WQI values for the research area were subsequently categorised into five types of water quality ratings [14,21,41] related to water quality classes [20].

Table 2. Water quality categorisation using water quality index [21] and South African water quality standards [19,20].

WQI Ratings	Definition	Water Quality Class	Definition
<50	Excellent	Class 0	Ideal water for domestic use
50–100	Good	Class I	Acceptable water for domestic use
100–200	Poor	Class II	Tolerable water for domestic use
200–300	Very poor	Class III	Unacceptable water for domestic use

2.5. Total Maximum Daily Load Estimation for Water Resource Protection

The protection limits for surface water quality in the study area were established by a simple total maximum daily loads (TMDL) approach based on the model by [45]. The model was created as a tool to assist in the restoration and protection of waterbodies where beneficial uses for aquatic life, recreation, public drinking water, or human health are impeded or threatened. Conceptually, the model is based on the fact that the pollutant loading of a water body originates from different pollutant sources such as point and non-point sources [46–48]. In the development process of TMDLs, the determination of a margin of safety (MOS) is necessary to account for any uncertainty regarding the relationship between pollution loads and receiving waterbody quality. [49]. In this current study, Equation (5) was used to establish surface water quality protection limits for EC, NO_3, and PO_4 in the study catchment.

$$TMDL = \Sigma WLA + \Sigma LA + MOS \tag{5}$$

where:

TMDL = Total maximum daily loads set as protection levels for the catchment (mg/s).
ΣWLA = Sum of waste load allocation to existing permitted point sources (mg/s).
ΣLA = Sum of load allocation for non-point sources (groundwater signature was used) (mg/s).
MOS = Margin of safety (set at 10% in this study) accounts for any uncertainty associated with attaining the protection levels for water quality.
Load = Discharge (L/s) × concentration (mg/L).
Load = mg/s.

The TMDL estimation was performed on the main river stem of the Blesbokspruit River Catchment. The estimation was based on the analysis of data derived from the instream sampling sites, discharge points, groundwater monitoring sites, and flow gauging station (B1H032) as shown in Figure 2. Data from the discharge points were used to estimate ΣWLA, while data from the instream sampling sites were used to obtain natural background levels. Data from the groundwater monitoring sites were used to estimate the levels of non-point sources. Altogether, data from the groundwater monitoring sites, instream sampling sites, and flow data were used to calculate ΣLA.

2.6. Waste Load Allocation

To demonstrate the effects of the newly set TMDLs on the evaluation of a new point discharge within the BRC, the chemical mass balance (CMB) approach was applied in a hypothetical case example. The existing point discharges survey was carried within the study area resulting in a presentation of a system diagram. This process was deemed to be necessary in this analysis to enable the identification and location of all point discharges and monitoring points necessary for new water use licences application evaluation for water quality allocation in the investigation area. The CMB approach is an indirect approach that provides a viable alternative to other conventional techniques, and it accounts for the benefit of using upstream/downstream river water quality data to estimate the load on the river and identify variations in the water quality characteristics within the river system [50]. An additional advantage of this approach is the substantial reduction in the cost involved in the analysis of a large no of water and effluent samples. The CMB method, as described by [51–53], was applied in this study. The mass load of a receiving water body is calculated based on Equation (6).

$$QdCd - QuCu = \sum_{i=1}^{n} Li - \sum losses - \sum in\ situ\ generation \tag{6}$$

where Qd and Qu represent downstream and upstream flows, Cd and Cu represent downstream and upstream concentrations in river water, and sum $\sum_{i=1}^{n} Li$ represents the sum of all individual loading into the river if losses and/or generation within the water body are insignificant. In metric units, the concentrations and flows are often expressed in mg/L and m^3/s, respectively [51]. Therefore, to determine the resultant impact of a new discharge point in terms of the concentration downstream of a water body, Equation (7) can be applied. In this study, Equation (7) was used to calculate the resultant concentrations of EC, NO_3, and PO_4 downstream as a result of a new point source discharge. The approach was necessary to assess the impact of a new discharge into the Blesbokspruit River in terms of compliance with the set protection limits for water quality.

$$C_d = \frac{(QuCu + \sum_{i=1}^{n} Li)}{Qd} \tag{7}$$

3. Results and Discussion

The water quality index, total maximum daily loads, and waste load allocation for the Blesbokspruit River Catchment are presented and discussed.

3.1. Water Quality Index for the Blesbokspruit River Catchment

Table 3 displays the findings of the investigation between 8 and 494 surface water samples for physicochemical characteristics in the research area in comparison to [19,20,37] standards.

Table 3. Summary statistics of the physico-chemical parameters determined from the catchment. Mean concentrations (mg/L), pH (standard units), and electrical conductivity (mS/m).

WQ Variables	Max	Min	Med	Std. Dev	SANS 241: 2015; SAWQG, 1996; WRC, 1998
pH	8.88	6.58	7.77	0.38	5–9.7
EC	808.00	12.42	86.90	47.20	170.00
Ca	136.82	23.99	35.96	37.18	150.00
Cl	169.40	4.90	66.19	25.12	300.00
F	14.00	0.03	0.27	0.79	1.50
Mg	66.42	5.75	30.06	12.75	100.00
NO$_3$	15.18	0.05	1.10	1.57	1.00
Na	169.08	0.05	89.40	31.46	200.00
SO$_4$	3873.00	5.80	219.03	390.12	250.00
PO$_4$	4.20	0.03	0.43	0.52	0.025

The pH of the surface water in the research area ranged from 6.58 to 8.88 with a median value of 7.77. pH in the research area falls within the limit of the water quality standards. According to [22], the pH is considered one of the main parameters used to establish the alkalinity (pH > 7), acidity (pH < 7), or neutrality (pH = 7) of an environment. Therefore, the findings of this study suggest that surface water in the study area portrays neutral conditions. The EC ranged from 12.42 to 808 mS/m with an average of 86.90 mS/m, and it complies with the limit of the water quality standards. EC is an indicator of salinity [52]; therefore, there is no evidence of salinity problems in the catchment based on the findings of this study.

The concentrations of Ca^{2+} and Mg^{2+} ranged from 23.99 to 136.82 mg/L with an average of 35.96 mg/L for Ca^{2+}, and from 5.75 to 66.42 mg/L with an average of 30.06 mg/L for Mg^{2+}. The concentrations of the two cations are well within the standards' specified limits.

A median of 89.40 mgL^{-1} was attained for Na$^+$ with a range between 0.05 and 169.08 mg/L, and the concertation falls with the requirement of the standards. The concentrations of F$^-$ and Cl$^-$ range from 0.03 to 14.00 mg/L with an average of 0.27 mg/L for F$^-$ and from 4.90 to 169.40 mg/L with an average of 66.19 mg/L for Cl$^-$. Both anions comply with the standards limits. The water quality results indicate that NO$_3^-$ concentrations averaged 1.10 mg/L with the lowest value 0.05, and the highest value of 15.18 mg/L. The median value for NO$_3^-$ is above the limits specified in the guidelines. The average concentrations of SO$_4^{2-}$ and PO$_4^{2-}$ ranged from 5.80 to 3873 mg/L with an average of 219.03 mg/L for SO$_4^{2-}$. According to [52], SO$_4^{2-}$ is an indicator of acid mine drainage and general mining impacts. Although the average falls below the stipulated limits in the standards, the highest value obtained during the period is alarming. Results for PO$_4^{2-}$ varied from 0.03 to 4.20 mg/L with an average of 0.43 mg/L which is above the required levels prescribed in the water quality standards: PO$_4^{2-}$ is used as an indicator of nutrient enrichment, and potential for eutrophication [53]. This suggests that the land use activities associated with agriculture have a negative impact on the water quality in the study area, thus requiring intervention.

Table 4 provides results of the WQI calculated to establish the overall water quality status on the entire catchment. The index was calculated to be 54.80 indicating, that water in the catchment is good according to WQI ratings (Table 2). According to the South African water quality criteria [20], the overall water quality class of the study area falls in class I, which translates to water suitable for domestic use.

Table 4. Water quality index calculated for the research area.

WQ Variables	Parameter Concentration (Ci)	Standard Limit (Si)	Weight (wi)	Relative Weight (Wi)	Quality Rating Scale (Qi)	Sub-Index (SIi)	WQI
pH	7.77	7.35	2	0.061	105.71	6.45	
EC	86.90	170.00	2	0.061	51.12	3.12	
Ca	35.96	150.00	3	0.100	23.97	2.40	
Cl	66.19	300.00	4	0.133	22.06	2.94	
F	0.27	1.50	4	0.133	18.00	2.39	54.80
Mg	30.06	100.00	3	0.100	30.06	3.01	
NO$_3$	1.10	1.00	5	0.167	110.00	18.37	
Na	89.40	200.00	3	0.100	44.70	4.47	
SO$_4$	219.03	250.00	4	0.133	87.61	11.65	

Figures 4–6 show the findings of the historical trend analysis of the physicochemical parameters. The historical trends of Ca^{2+} are depicted in Figure 4a, indicating a steady concentration increase from 2015 to 2020. The concentration of Mg^{2+} (Figure 4b) fluctuated significantly and increased from 2015 to 2022, with the lowest concentration of 23 mg/L and the highest concentration during the period was 48 mg/L. Although the Na$^+$ concentration (Figure 4c) during the same period fluctuated significantly, with the highest levels recorded in 2019 and the lowest recorded in 2021, the trend has not significantly increased. A similar trend is evident for Cl$^-$ (Figure 5a), with the highest concentration recorded in the year 2017. The concentration of NO$_3^-$ (Figure 5b) was the highest in 2015 with significant spikes between 2015 and 2021; however, there has recently been a decline in the concentration. Notably, a significant spike in the concentration of SO$_4^{2-}$ (Figure 5c) was observed between 2017 and 2019. Concentrations of F$^-$ (Figure 6a) and EC (Figure 6c) spiked very strongly in the same year (2018). According to [52], EC and SO$_4^{2-}$ reached undesirable management target limits in the catchment after the Eastern Basin Chemical Acid Mine Drainage treatment plant came into operation. In terms of PO$_4^{2-}$ concentration (Figure 6b), the highest levels were recorded in 2016, 2017, and 2019 and the trend has been slightly declining. The concentration for pH (Figure 6d) was the lowest in 2017 and the highest in 2021, and the trend indicates no significant increase or decline in suggesting that the catchment experiences neutral conditions.

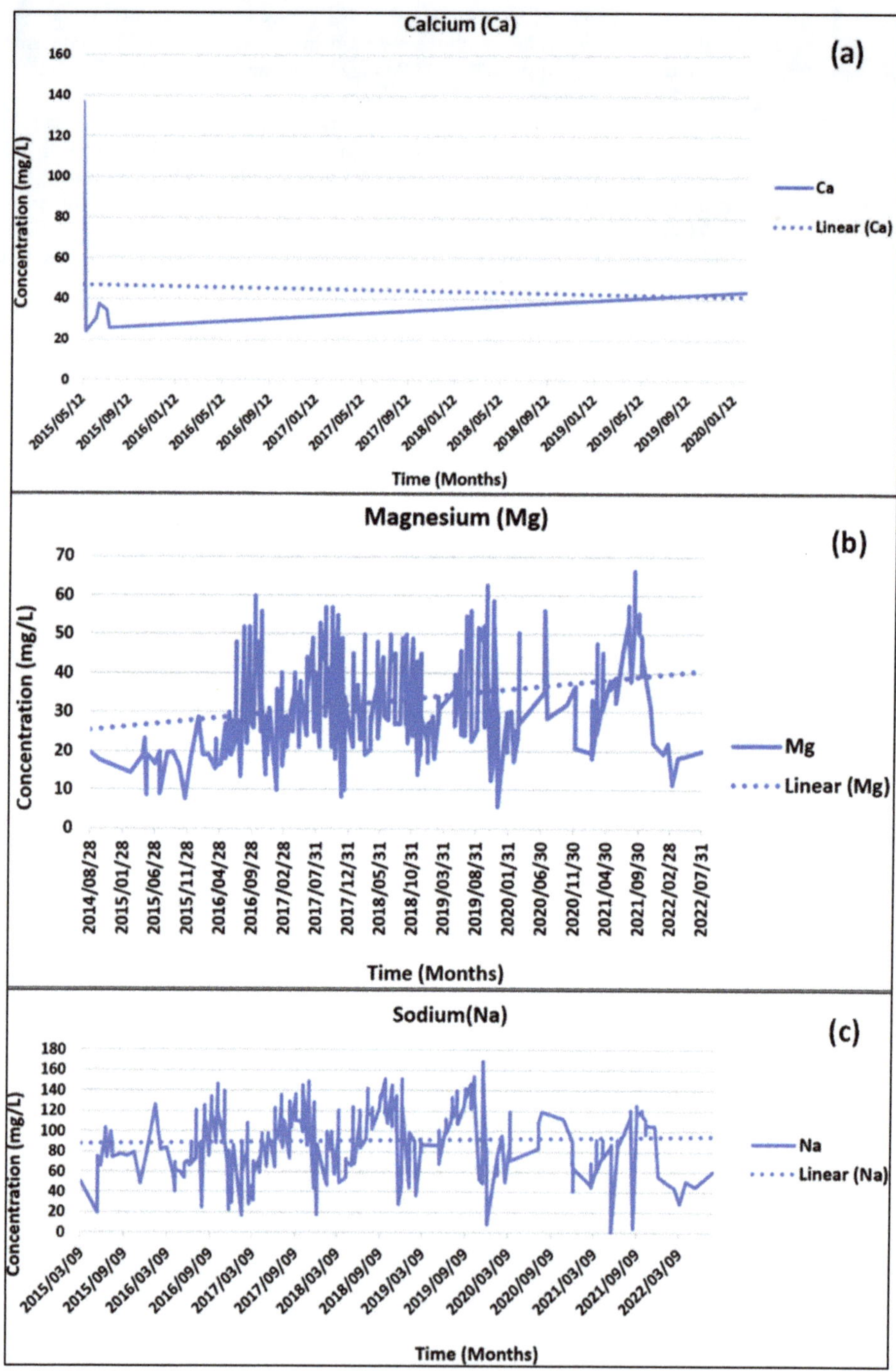

Figure 4. Trend analysis of calcium (**a**), magnesium (**b**), and sodium (**c**) from 2015 to 2022 across the study area.

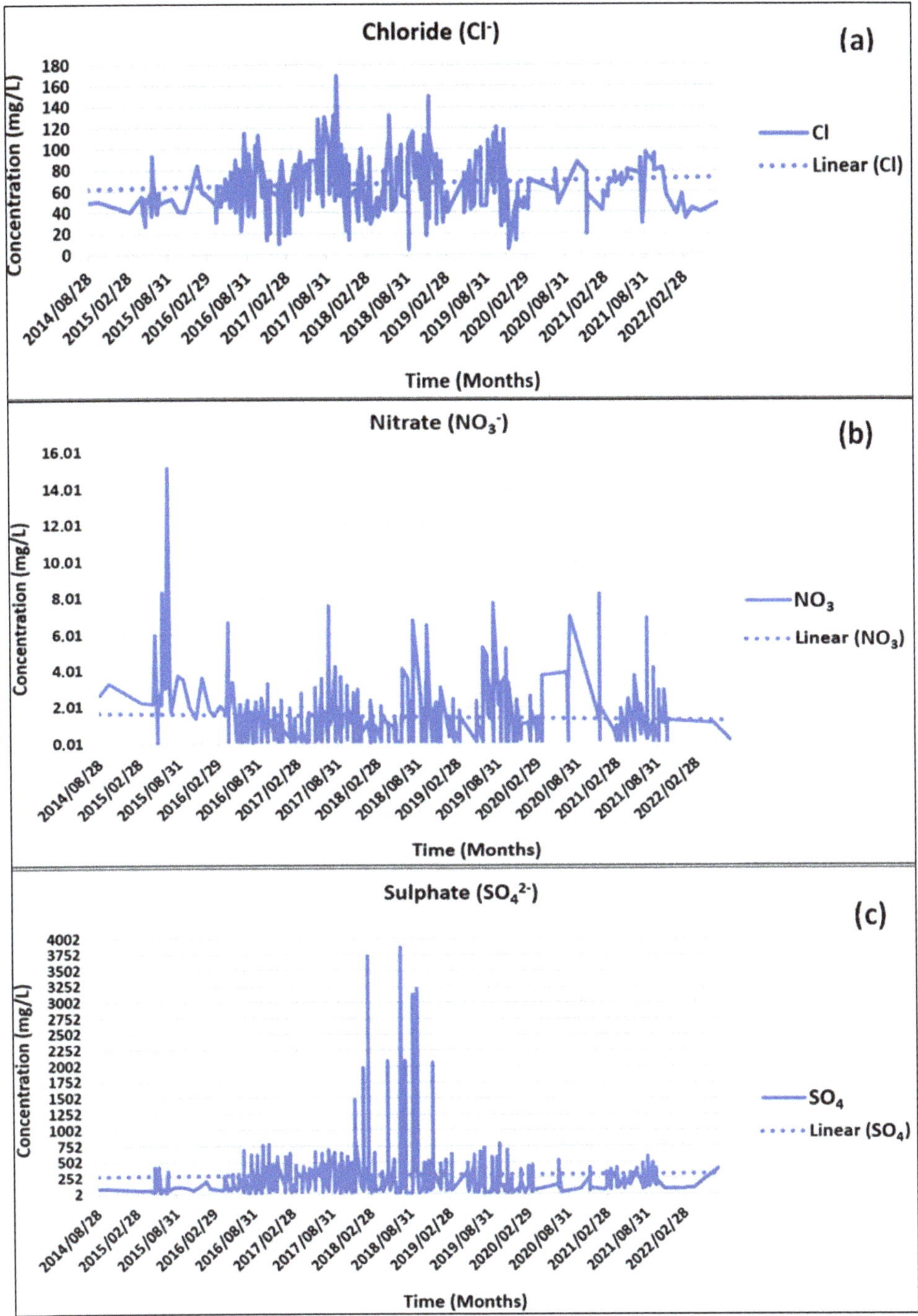

Figure 5. Trend analysis of chloride (**a**), nitrate (**b**), and sulphate (**c**) from 2015 to 2022 across the study area.

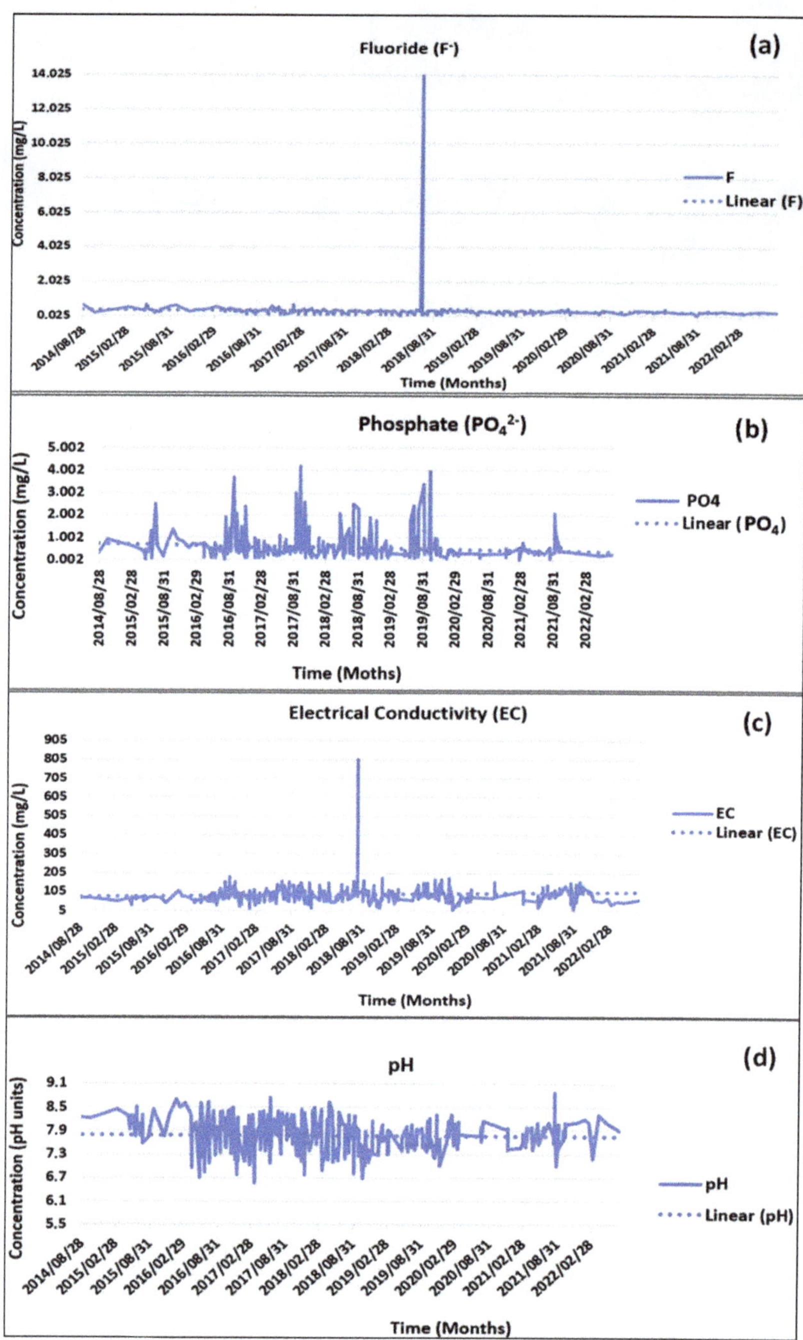

Figure 6. Trend analysis of fluoride (**a**), phosphate (**b**), electrical conductivity (**c**), and pH (**d**) from 2015 to 2022 across the study area.

Although results from water quality status assessment using the water quality index indicated that the water quality in the catchment did not adversely deteriorate during the period from 2015 to 2022. However, the results also gave evidence of NO_3^- and PO_4^- not complying with the standard limits for drinking water, suggesting that the catchment is impacted by agricultural land use activities and wastewater treatment works. NO_3^- and PO_4^{2-} are indicators of sewage and agricultural contamination [53]. The results of the trend analysis for pH, EC, and SO_4^{2-} gave an indication of the mining impact on the catchment which requires intervention. The results concur with earlier findings by [54] which reported that mine dumps in the border of the catchment provide additional source of SO_4^{2-}. The result of this study suggests that there is an urgent need for treatment of mine waste before discharge into the stream. Strong compliance monitoring and enforcement with the established water use licences conditions for the user is critical to control pollution from the source. Furthermore, the rehabilitation of the catchment may be considered in the case of exceedance of the protection limits set for the catchment.

3.2. Total Maximum Daily Loads for the Blesbokspruit River Catchment

The statistical analysis of water quality data for surface water resource protection in the Blesbokspruit River Catchment is provided in Table 5 for discussion.

Table 5. TMDL estimated for the catchment.

Water Quality Parameter	ΣWLA (mg/s)	ΣLA (mg/s)	MOS (mg/s)	TMDL (mg/s)	Concentration	Gazetted RQOs *
EC	118,410.00	4635.40	12,304.54	135,349.94	77.96	180.89
PO$_4$	15,653.20	9.24	1566.24	17,228.69	9.92	3.90
NO$_3$	23,796.00	129.36	2392.54	26,317.90	15.16	10.23

Note(s): * All measurements are in mg/L, except for the EC, which is in mS/m [25].

The water quality parameters that we considered for estimation of TMDLs in the study area were EC, PO_4^{2-}, and NO_3^- (Table 5). Some of the existing point source load (ΣWLA) was estimated at 118,410.00 mS/m/s, 15,653.20 gm/s, and 23,796.00 mg/s for EC, PO_4^-, and NO_3^-, respectively. The sum of the load allocation for non-point sources (ΣLA) for EC, $PO4^{2-}$, and NO_3^- was estimated at 4635.40 mS/m/s, 9.24 mg/s, and 129.36 mg/s, respectively. The margin of safety (MOS) was estimated at 12,304.54 mS/m/s for EC, 1566.24 mg/s for PO_4^{2-}, and 2392.54 mg/s for NO_3^-. The TMDLs for EC, PO_4^{2-}, and NO_3^- were estimated at 135,349.94 mS/m/s, 17,228.69 mg/s for PO_4^{2-}, and 26,317.90 mg/s for NO_3^-. The concentration limits derived from the TMDLs were 77.96 mS/m for EC, 9.92 mg/L for PO_4^{2-}, and 15.16 mg/L for NO_3^-.

When the results of the study were compared with the findings of the earlier study by [25], undertaken in the same catchment, it gave comparable outcomes of NO_3^- only varying from one another by a magnitude of approximately 5 mg/L. The results of the PO_4^- concentration varied by a magnitude of approximately 6 mg/L. Both these results recorded higher concentration limits from the current study compared to the [25]. However, the concentration of EC derived from the TMDL obtained in the current study was lower that the EC level that had been set by the earlier study, suggesting that the TMDL technique applied in the present study was conservative in terms of EC protection levels for the catchment. The results of this study underline the need for the consideration of existing point and non-point sources when determining protection levels for surface water quality in a catchment which is central to the TMDL approach. The consideration of both point and non-point sources of pollution are critical in the management of land use impact into the resource. Therefore, it is suggested that the total maximum daily loads approach should be employed to establish the protection levels for surface water quality of a resource in a catchment.

3.3. Waste Load Allocation for Point Source Discharge

Table 6 provides data that were considered to assess the impacts of a new proposed discharge point to the already established TMDLs. The hypothetical values derived from the catchment for the Upstream Discharge Point 1 were 70.00 mS/m for EC, 3.00 mg/L for

NO_3^-, and 0.40 mg/L for PO_4^{2-} with a discharge of 130.00 L/s. The upstream discharge point 2 had hypothetical values of 70.00 mS/m for EC, 15.00 mg/L for NO_3^-, 10.00 mg/L for PO_4^{2-}, and a discharge of 1560.00 L/s. For upstream discharge point 3, the values were 55.00 mS/m, 3.00 mg/L, 0.60 mg/L, and 2.00 L/s for EC, NO_3^-, PO_4^{2-}, and the discharge. The TMDLs in terms of concentration for EC, NO_3^-, and PO_4^{2-} estimated for the catchment were initially recorded in Table 5 as 70 mS/m, 14.4 mg/L, and 9.24 mg/L, respectively, with a median discharge of 1736.00 L/s recorded at the flow gauging station (B1H032). The new proposed discharge point was intended to discharge 3.00 L/s of waste with concentrations of 85.00 mS/m of EC, 2.30 mg/L of NO_3^-, and 6.00 mg/L of PO_4^{2-}. The effects of the proposed discharge point to the downstream monitoring point for the catchment resulted in the new concentrations of 68.36 mS/m for EC, 13.71 mg/L for NO_3^-, and 9.03 mg/L for PO_4^{2-} estimated using Equation (7). The results indicate that the proposed discharge point resulted in the decrease in the concentrations of EC, NO_3^-, PO_4^{2-} in the catchment with new concentration levels being lower than the TMDLs that were set for the catchment as protection limits. This suggests that the proposed discharge point will have beneficial effects on the catchment. Therefore, the proposed discharge point can be allowed to discharge waste with a concentration of 85.00 mS/m of EC, 2.30 mg/L of NO_3^-, and 6.00 mg/L of PO_4^{2-} at a discharge rate of 3.00 L/s.

Table 6. Existing data from the catchment considered for waste load allocation in the catchment.

Existing and New Point Discharges	Water Quality			Discharge (L/s)
	EC (mS/s)	NO_3 (mg/L)	PO_4 (mg/L)	
Upstream discharge point 1	70.00	3.00	0.40	130.00
Upstream discharge point 2	70.00	15.00	10.00	1560.00
Upstream discharge point 3	55.00	3.00	0.60	2.00
Proposed discharge point	*85.00 **	*2.30 **	*6.00 **	*3.00 **
[a] TMDL	70	14.4	9.24	1736.00
Effects of the proposed discharge point	*68.36*	*13.71*	*9.03*	

Note(s): * Hypothetical values for a new discharge point. [a] Values obtained from Table 5.

The evaluation of the impact induced by a new discharge point in terms of the resultant flows in the river indicated that the total discharge was 1.695 m^3/s. When Equation (7) was applied to calculate the resultant concentrations of EC, NO_3^-, and PO_4^{2-} at a downstream monitoring point because of a new point discharging into the river, the results were as follows. The calculated/resultant concentrations were found to be 70 mg/L for EC, 14.4 mg/L for NO_3^-, and 9.24 mg/L PO_4^{2-} (Table 6). The calculated/resultant concentration of 70 mg/L for EC is below the 180.89 mS/m set as the protection limits for the catchment by 61%. Therefore, the new discharge point can be allowed to discharge the proposed 85.00 mS/m concentration of EC into the river. However, the calculated/resultant concentration of 14.4 mg/L for NO_3^- is above the 10.23 mg/L set as the protection limits for the catchment by 41%. In this case, the new discharge point cannot be allowed to discharge the proposed 2.30 mg/L concentration of NO_3^- as this will results into the protection levels set for the catchment being exceeded. In terms of PO_4^{2-}, the calculated/resultant concentration of 9.24 mg/L far exceeds (137%) the protection limits of 3.9 mg/L set for the catchment. Therefore, the new discharge point cannot be allowed to discharge the proposed 6.00 mg/L of PO_4^{2-}; instead, the waste will have to be treated to lower concertation levels before discharge, otherwise the protection levels set for the catchment would be compromised.

4. Conclusions

The operationalization of the water resource protection initiatives within the Blesbokspruit River Catchment was assessed. The study discovered that the overall water quality status for the physico-chemical parameters (Na^+, Ca^{2+}, Mg^{2+}, Cl^-, F^-, pH, EC, SO_4^{2-}, PO_4^-) is within the limits of the water quality standards, except for NO_3- and PO_4-. The concentration levels of parameters including EC, SO_4^{2-}, NO_3^-, and PO_4^{2-} vary significantly as a result of the mining activities, and waste discharge from wastewater treatment works. The application of the TMDL and CMB approaches facilitates water resource protection practices at the catchment level. Continuous water resource monitoring for surface water quality is critical for compliance monitoring against established protection levels in a catchment. The findings of this study provide critical evidence on the feasibility of resource-directed measures' implementation at the catchment level for water resources' protection in South Africa. To improve water resource protection practices at the catchment level, this study recommends the application of TMDL and CMB techniques, and that active adaptive management actions should form part of any water resource management plans in a catchment. The study also recommends further research into the application of adaptive management tools such as treatability index techniques for improved water resource protection practices.

Author Contributions: K.M. was responsible for data collection, data analysis, and drafting the manuscript. S.N. and T.K. were responsible for the review and editing of the manuscript. S.N. is the academic co-supervisor of the corresponding author and T.K. is the academic supervisor of the corresponding author. All authors have read and agreed to the published version of the manuscript.

Funding: This research was funded by the Department of Water and Sanitation (South Africa).

Data Availability Statement: Not applicable.

Acknowledgments: This paper is part of a Ph.D. project by the corresponding author. The author gratefully acknowledges the Department of Water and Sanitation (DWS), Water Ecosystem Management and Resource Quality Information Services (RQIS) for availing the historical data.

Conflicts of Interest: The authors declare no conflict of interest.

References

1. Al Kafy, A. Importance of Surface Water Bodies for Sustainable Cities: A Case Study of Rajshahi City Corporation. In *Souvenir of World Town Planning Day*; Bangladesh Institute of Planners (BIP): Dhaka, Bangladesh, 2018; pp. 113–119.
2. Kılıç, Z. The importance of water and conscious use of water. *Int. J. Hydrol.* **2020**, *4*, 239–241. [CrossRef]
3. Du Plessis, A.; Harmse, T.; Ahmed, F. Quantifying and Predicting the Water Quality Associated with Land Cover Change: A Case Study of the Blesbok Spruit Catchment, South Africa. *Water* **2014**, *6*, 2946–2968. [CrossRef]
4. Abed, S.A.; Ewaid, S.H.; Al-Ansari, N. Evaluation of Water quality in the Tigris River within Baghdad, Iraq using Multivariate Statistical Techniques. *J. Phys.* **2019**, *1294*, 072025. [CrossRef]
5. Noori, R.; Berndtsson, R.; Hosseinzadeg, M. A critical review on the application of the National Sanitation Foundation Water Quality Index. *Environ. Pollut.* **2019**, *244*, 275–587. [CrossRef] [PubMed]
6. Akhtar, N.; Ishak, M.I.S.; Bhawani, S.A.; Umar, K. Various Natural and Anthropogenic Factors Responsible for Water Quality Degradation: A Review. *Water* **2021**, *13*, 2660. [CrossRef]
7. Khan, U.; Janjuhah, H.T.; Kontakiotis, G.; Rehman, A.; Zarkogiannis, S.D. Natural Processes and Anthropogenic Activity in the Indus River Sedimentary Environment in Pakistan: A Critical Review. *J. Mar. Sci. Eng.* **2021**, *9*, 1109. [CrossRef]
8. Mnyango, S.S.; Thwala, M.; Oberholster, P.J.; Truter, C.J. Using Multiple Indices for the Water Resource Management of a Monomictic Man-Made Dam in Southern Africa. *Water* **2022**, *14*, 3366. [CrossRef]
9. Gossweiler, B.; Wesstrom, I.; Messing, I.; Romero, A.M.; Joel, A. Spatial and Temporal Variations in Water Quality and Land Use in a Semi-Arid Catchment in Bolivia. *Water* **2019**, *11*, 2227. [CrossRef]
10. Abdelal, Q. Floating PV; an assessment of water quality and evaporation reduction in semi-arid regions. *Int. J. Low-Carbon Technol.* **2020**, *16*, 732–739. [CrossRef]
11. Khadija, D.; Hicham, A.; Rida, A.; Hicham, E.; Nordine, N. Surface water quality assessment in the semi-arid area by a combination of heavy metal pollution indices and statistical approaches for sustainable management. *Environ. Chall.* **2021**, *5*, 100230. [CrossRef]
12. Perrin, J.L.; Rais, N.; Chahinian, N.; Moulin, P.; Ijjaali, M. Water quality assessment of highly polluted rivers in a semi-arid Mediterranean zone Oued Fez and Sebou River (Morocco). *J. Hydrol.* **2014**, *510*, 26–34. [CrossRef]
13. Santana, C.S.; Olivares, D.M.M.; Silva, V.H.C.; Luzardo, F.H.L.; Velasco, F.G.; de Jesus, R.M. Assessment of water resources pollution associated with mining activity in a semi-arid region. *J. Environ. Manag.* **2020**, *273*, 111148. [CrossRef] [PubMed]
14. Adom, R.K.; Simatele, M.D. Analysis of public policies and programmes towards water security in post-apartheid South Africa. *J. Water Policy* **2021**, *23*, 503–520. [CrossRef]
15. Van Rensburg, S.J.; Barnard, S.; Krüger, M. Challenges in the potable water industry due to changes in source water quality: Case study of Mid -Vaal Water Company, South Africa. *J. Water S. Afr.* **2016**, *42*, 633–640. [CrossRef]
16. Dlamini, S.; Gyedu-Ababio, T.K.; Slaughter, A. The loading capacity of the Elands River: A case study of the Waterval Boven wastewater treatment works, Mpumalanga Province, South Africa. *J. Water Resour. Prot.* **2019**, *11*, 1049–1063. [CrossRef]

17. Odume, O.N.; Slaughter, A.; Griffin, N.; Chili, A. *Case Study for Linking Water Quality License Conditions with Resource Quality Objectives for the Leeutaaiboschspruit Industrial Complex Situated within the Vaal Barrage Catchment: Volume 1*; Water Research Commission: Gezina, South Africa, 2021.

18. Verlicchi, P.; Grillini, V. Surface Water and Groundwater Quality in South Africa and Mozambique: Analysis of the Most Critical Pollutants for Drinking Purposes and Challenges in Water Treatment Selection. *Water* **2020**, *12*, 305. [CrossRef]

19. Makanda, K.; Nzama, S.; Kanyerere, T. Assessing the Role of Water Resources Protection Practice for SustainableWater Resources Management: A Review. *Water* **2022**, *14*, 3153. [CrossRef]

20. *SANS 241-1:2015*; Edition 2 Drinking Water. Part 1: Microbiological, Physical, Aesthetic and Chemical Determinants. South African Bauru of Standards (SABS): Pretoria, South Africa, 2015.

21. Water Research Commission (WRC). *Quality of Domestic Water Supplies—Volume 1: Assessment Guide*, 2nd ed.; Water Research Commission Report No: TT 101/98; Water Research Commission: Gezina, South Africa, 1998.

22. Nzama, S.M.; Kanyerere, T.O.B.; Mapoma, H.W.T. Using groundwater quality index and concentration duration curves for classification and protection of groundwater resources: Relevance of groundwater quality of reserve determination, South Africa. *Water Recourse. Manag.* **2021**, *7*, 31. [CrossRef]

23. Lalumbe, L.; Oberholster, P.J.; Kanyerere, T. Feasibility Assessment of the Application of Groundwater Remediation Techniques in Rural Areas: A Case Study of Rural Areas in the Soutpansberg Region, Limpopo Province, South Africa. *Water* **2022**, *14*, 2365. [CrossRef]

24. Makanda, K.; Nzama, S.; Kanyerere, T. Policy Implementation for Water Resources Protection: Assessing Spatio-Temporal Trends of Results from Process-Based Outcomes of Resource-Directed Measures Projects in South Africa. *Water* **2022**, *14*, 3322. [CrossRef]

25. Department of Water and Sanitation (DWS). *Determination of Resource Quality Objectives in the Upper Vaal Water Management Area (WMA8): Sub-Component Prioritisation and Indicator Selection Report*; Report No.: RDM/WMA08/00/CON/RQO/0114; Department of Water and Sanitation: Pretoria, South Africa, 2014.

26. Department of Water and Sanitation (DWS). *Classes and Resource Quality Objectives of Water Resources for Catchments of the Upper Vaal*; Department of Water and Sanitation (DWS): Pretoria, South Africa, 2016.

27. Kwonta, O.I.; Dzwairo, B.; Otieno, F.A.O.; Adeyemo, J.A. A review on water resources yield model. *S. Afr. J. Chem. Eng.* **2017**, *23*, 107–115.

28. Mohamed, M.M.; El-Shorbagy, W.; Kizhisseri, M.I.; Chowdhury, R.; McDonald, A. Evaluation of policy scenarios for water resources planning and management in an arid region. *J. Hydrol.* **2020**, *32*, 100758. [CrossRef]

29. Huang, D.; Liu, J.; Han, G.; Huner-Lee, A. Water-energy nexus analysis in an urban water supply system based on a water evaluation and planning model. *J. Clean. Prod.* **2023**, *403*, 136750. [CrossRef]

30. Keupers, I.; Willems, P. Development and Testing of a Fast Conceptual River Water Quality Model. *J. Water Res.* **2017**, *113*, 62–71. [CrossRef] [PubMed]

31. Martin, J.L.; McCutcheon, S.C. *Hydrodynamics and Transport for Water Quality Modeling*; CRC Press, Technology and Engineering: Boca Raton, FL, USA, 2018.

32. Odume, O.N.; Griffin, N.; Mensah, P.K. *Literature Review and Terms of Reference for Case Study for Linking the Setting of Water Quality License Conditions with Resource Quality Objectives and/or Site-Specific Conditions in the Vaal Barrage Area and Associated Rivers within the Lower Sections of the Upper Vaal River Catchment*; Water Research Commission: Gezina, South Africa, 2018.

33. Karaoui, I.; Arioua, A.; Elhamdouni, D.; Nouaim, W.; Ouhamchich, K.A.; Hssaisoune, M. Assessing Water Quality Status Using a Mathematical Simulation Model of El Abid River (Morocco). *J. Water Manag. Model.* **2022**, *30*, 491. [CrossRef]

34. Ambani, A.E.; Annegan, H. A reduction in mining and industrial effluents in the Blesbokspruit Ramsar wetland, South Africa: Has the quality of the surface water in the wetland improved? *Water SA* **2015**, *41*, 5. [CrossRef]

35. Department of Water Affairs and Forestry (DWAF). *Water Quality Sampling Manual for the Aquatic Environment*; Institute for Water Quality Studies (IWQS): Pretoria, South Africa, 1997.

36. Weaver, J.M.C.; Cavé, L.; Talma, A.S. *Groundwater Sampling. A Comprehensive Guide for Sampling Methods*, 2nd ed.; Prepared for the Water Research Commission by Groundwater Sciences, Council for Scientific Industrial Research, South Africa; WRC Report No TT 303/07; Water Research Commission: Gezina, South Africa, 2007.

37. Department of Water Affairs and Forestry (DWAF). *South African Water Quality Guidelines (SAWQG). Volume 7: Aquatic Ecosystems*, 1st ed.; Department of Water Affairs and Forestry (DWAF): Pretoria, South Africa, 1996.

38. Uddin, M.G.; Nash, S.; Olbert, A.I. A review of water quality index models and their use for assessing surface water quality. *J. Ecol. Indic.* **2021**, *122*, 107218. [CrossRef]

39. Banda, T.D.; Kumarasamy, M.V. Development of water quality indices (WQIs): A review. *Pol. J. Environ. Stud.* **2019**, *29*, 2011–2021. [CrossRef]

40. Wong, Y.J.; Shimizu, Y.; He, K. Comparison among different ASEAN water quality indices for the assessment of the spatial variation of surface water quality in the Selangor River basin, Malaysia. *J. Environ. Monit. Assess.* **2020**, *192*, 644. [CrossRef]

41. Wong, Y.J.; Shimizu, Y.; Kamiya, A. Application of artificial intelligence methods for monsoonal river classification in Selangor River basin, Malaysia. *J. Environ. Monit. Assess.* **2021**, *193*, 438. [CrossRef]

42. Atta, H.S.; Omar, M.A.-S.; Tawfik, A.M. Water quality index for assessment of drinking groundwater purpose case study: Area surrounding Ismailia Canal, Egypt. *J. Eng. Appl. Sci.* **2022**, *69*, 83. [CrossRef]

43. Oseke, F.I.; Anornu, G.K.; Adjei, K.A.; Eduvie, M.O. Assessment of water quality using GIS techniques and water quality index in reservoirs affected by water diversion. *Water-Energy Nexus* **2021**, *4*, 25–34. [CrossRef]
44. Ramakrishnaiah, C.; Sadashivaiah, C.; Ranganna, G. Assessment of water quality index for the groundwater in Tumkur Taluk, Karnataka State, India. *J. Chem.* **2009**, *6*, 523–530. [CrossRef]
45. United States Environmental Protection Agency (USEPA). *Guidelines for Reviewing TMDLs under Existing Regulations Issued in 1992*; U.S. EPA Watershed Branch Office of Wetlands: Washington, DC, USA, 2002. Available online: https://www.epa.gov/tmdl/guidelines-reviewing-tmdls-under-existing-regulations-issued-1992 (accessed on 17 March 2023).
46. Hou, G.; Zheng, J.; Cui, X.; He, F.; Zhang, Y.; Wang, Y.; Li, X.; Fan, C.; Tan, B. Suitable coverage and slope guided by soil and water conservation can prevent non-point source pollution diffusion: A case study of grassland. *Ecotoxicol. Environ. Saf.* **2022**, *241*, 113804. [CrossRef] [PubMed]
47. Zhang, X.; Chen, P.; Han, Y.; Dai, S. Quantitative analysis of self-purification capacity of non-point source pollutants in watersheds based on SWAT model. *Ecol. Indic.* **2022**, *143*, 109425. [CrossRef]
48. Hayward, E.E.; Gillis, P.L.; Bennet, C.J.; Prosser, R.S.; Salerno, J.; Liang, T.; Robertson, S.; Metcalfe, C.D. Freshwater mussels in an impacted watershed: Influences of pollution from point and non-point sources. *Chemosphere* **2022**, *307*, 135966. [CrossRef]
49. Dors, K.M.; Tsatsaros, J. Determining Margin of Safety for TMDLS. *Proc. Water Environ. Fed.* **2002**, *2*, 1892–1901. Available online: https://www.researchgate.net/publication/233592431_DETERMINING_MARGIN_OF_SAFETY_FOR_TMDLS (accessed on 17 March 2023). [CrossRef]
50. Jain, C.K. Application of chemical mass balance approach to determine nutrient loading. *J. Hydrol. Sci.* **2000**, *45*, 577–588. [CrossRef]
51. Purandara, B.K.; Varadarajan, N.; Kumar, C.P. Application of Chemical Mass Balance to Water Quality Data of Malaprabha River. *J. Hydrol.* **2004**, *4*, 1–24.
52. Lourenco, M.; Curtis, C. The influence of a high-density sludge acid mine drainage (AMD) chemical treatment plant on water quality along the Blesbokspruit Wetland, South Africa. *Water SA* **2021**, *47*, 35–44. [CrossRef]
53. Retief, H.; Pollard, S. Overview of Water Quality and Quantity in Olifants River Catchment. Association for Water and Rural Development (AWARD). 2020. Available online: https://award.org.za/wp/wp-content/uploads/2020/05/AWARD-BROCHURE-Overview-of-water-quality-and-quantity-in-Olifants-River-catchment-2020-v2.pdf (accessed on 30 March 2023).
54. McKay, T.J.M. The Blesbokspruit Wetland, South Africa: A High Altitude Wetland under Threat. Researchgate. 2018. Available online: https://www.researchgate.net/publication/326441349 (accessed on 19 June 2023).

MDPI

St. Alban-Anlage 66

4052 Basel

Switzerland

www.mdpi.com

Water Editorial Office

E-mail: water@mdpi.com

www.mdpi.com/journal/water